URBAN STUDIES: BORDER AND MOBILITY

PROCEEDINGS OF THE 4TH INTERNATIONAL CONFERENCE ON URBAN STUDIES (ICUS 2017), UNIVERSITAS AIRLANGGA, SURABAYA, INDONESIA, 8–9 DECEMBER 2017

Urban Studies: Border and Mobility

Edited by

Thor Kerr
Curtin University, Australia

Bekisizwe Ndimande
The University of Texas at San Antonio, USA

Jan Van der Putten
Universität of Hamburg, Germany

Daniel F. Johnson-Mardones
Universidad Alberto Hurtado, Chile

Diah Ariani Arimbi
Universitas Airlangga, Indonesia

Yuni Sari Amalia
Universitas Airlangga, Indonesia

Routledge
Taylor & Francis Group
LONDON AND NEW YORK

First published 2019 by CRC Press/Balkema

2 Park Square, Milton Park, Abingdon, Oxon, OX14 4RN
605 Third Avenue, New York, NY 10017

Routledge is an imprint of the Taylor & Francis Group, an informa business

First issued in paperback 2020

Typeset by V Publishing Solutions Pvt Ltd., Chennai, India

ISBN: 978-1-138-58034-3 (hbk)
ISBN: 978-0-367-73231-8 (pbk)

Urban Studies: Border and Mobility – Kerr et al. (Eds)
© 2019 Taylor & Francis Group, London, ISBN 978-1-138-58034-3

Table of contents

Urban Studies: Border and Mobility – Kerr et al. (Eds)
© 2019 Taylor & Francis Group, London, ISBN 978-1-138-58034-3

Preface/Foreword

Assalamualaikum Wr. Wb.

Welcome to the 4th International Conference on Urban Studies held by the Universitas Airlangga, Surabaya. The theme of this year's conference is "Borders and Mobility." We delve into and challenge our preconceptions of order and borders. We think about the ways borders create place, space, identity, and discourse in areas as diverse as the academia, advocacy, politics, socio-economic and security locally, nationally or internationally. This conference aims to assess how the notions of boundaries continue to shape and contextualize our approaches to gender roles, race, ethnicity, policies and security.

We aim for our conference to be a celebration of knowledge and idea sharing between researchers, educators, students, NGOs and representatives from the Indonesian Government. We hope the discussions at the conference will further extend dialogues to better understand issues related to borders and mobility at both local and global levels.

Finally, as you attend a number of programs and meet people in this conference, keep in mind that other people can also serve as doorways to new worlds. Listening and understanding someone else's background and experiences can often bring fascinating discoveries that can educate us and may affect us profoundly. Thus, please take advantage of this precious gathering and a series of dialogues to meet and talk with one another and learn from each other.

I welcome you once again to our 4th International Conference on Urban Studies. Have a productive and resourceful conference and enjoy your stay in Surabaya.

Wassalamualaikum Wr. Wb.

Diah Ariani Arimbi
Dean of Faculty of Humanities, Universitas Airlangga, Indonesia

Urban Studies: Border and Mobility – Kerr et al. (Eds)
© 2019 Taylor & Francis Group, London, ISBN 978-1-138-58034-3

Committees

ORGANIZING COMMITTEE

Committee Chair

- Yuni Sari Amalia

Members

- Viqi Ardaniah
- Rizki Andini
- Dahlia

SCIENTIFIC COMMITTEE

Moch. Amin Alamsjah (*Universitas Airlangga, Indonesia*)
Brenda Yeoh (*National University of Singapore, Singapore*)
I Made Andi Arsana (*Universitas Gadjah Mada, Indonesia*)
Michele Ford (*University of Sydney, Australia*)
Denny Arnos Kwary (*Universitas Airlangga, Indonesia*)
Thor Kerr (*Curtin University, Australia*)
Melani Budianta (*Universitas Indonesia, Indonesia*)
Bekisizwe Ndimande (*University of Texas at San Antonio, USA*)
Jan Van Der Putten (*University of Hamburg, Germany*)
Diah Ariani Arimbi (*Universitas Airlangga, Indonesia*)
Daniel Johnson-Mardones (*Universidad de Chile, Chile*)
Leonardus Sudibyo (*University of Illinois at Urbana-Champaign, USA*)

Urban Studies: Border and Mobility – Kerr et al. (Eds)
© 2019 Taylor & Francis Group, London, ISBN 978-1-138-58034-3

About the editors

Dr. Thor Kerr researches representation in public controversies around new island and waterfront reclamation projects. His books include *To the Beach: Community Conservation and its Role in 'Sustainable Development'* (2015) and *Setting up the Nyoongar Tent Embassy: A Report on Perth Media* (2013). Thor coordinates a master course at Curtin University's School of Media, Culture and Creative Arts, where he is also a member of the Australia-Asia-Pacific Institute (AAPI). He recently co-edited a volume for AAPI titled: *Indian Ocean Futures: Communities, Sustainability and Security* (2016). Thor developed and managed architectural media and information services in Southeast Asia in the late 1990s and early 2000s. His relationship with Indonesian media goes back to 1996, when he began working at *The Jakarta Post* newspaper.

Bekisizwe S. Ndimande earned his PhD. in Curriculum & Instruction from the University of Wisconsin-Madison, USA. He is currently Associate Professor of Curriculum and Instruction in the Department of Interdisciplinary Learning and Teaching at the University of Texas at San Antonio. His research interests intersect issues of equity, diversity, decolonizing methodology, and social justice education. He conducts this research within the fields of curriculum studies, education policy, multicultural education, children's rights, and immigrant education. He has published several journal articles and book chapters, including Pedagogy of the Township, in Sonia Nieto (Ed.), *Dear Paulo: Letters from Those who Dare Teach*; Race and Resources, *Race Ethnicity and Education*; The Role of Indigenous Languages and Focus Groups in Qualitative Inquiry: Experiences from the Global South, in Norm Denzin & Michael Giardina (Eds.). *Global dimensions of qualitative inquiry*; and Lutas Docentes nas Escolas Públicas para negros na.

África do Sul pós-apartheid, *Cadernos de Educação*, a Brazilian journal. Dr. Ndimande's article, From Bantu Education to the Fight for Socially Just Education, which appeared in *Equity & Excellence in Education* journal, was among the 18 articles selected by Taylor & Francis Press as representative of the interdisciplinary nature of social justice studies. His new book, *Privatization and the education of marginalized children: Policies, impacts ad global lessons*, (co-edited with Dr. Chris Lubienski) was published by Routledge in 2017.

Jan van der Putten is Professor Austronesistik in the Department of Southeast Asia (Asien-Afrika-Institut) at the University of Hamburg where he teaches on Southeast Asian literatures and cultures. Traditional Malay writings is one of his main research projects affiliated with the Centre for the Study of Manuscript Cultures (CSMC) in Hamburg. He also ventures in other types and periods of Malay traditions and literary expressions. In general, his research explores the meaning of traditional and popular Malay texts and their distribution among peoples and exchange between cultures. Some recent publications are:

- A Nation in a State of Suspended Disbelief: The Construction and Unravelling of the Indonesian Massacres of 1965–66 as a Site of Memory, in *Traditions Redirecting Contemporary Indonesian Cultural Productions.* (co-edited with Monika Arnez, Arndt Graf and Edwin Wieringa). Newcastle upon Tyne: Cambridge Scholars Publishing, 2017, pp. 99–126.
- 'On the edge of a tradition. Some prolegomena to paratexts in Malay rental manuscripts', *Indonesia and the Malay World*, 45, no. 132 (2017): 1–20.
- *Translation in Asia. Theories, Practices, Histories* (co-edited with Ronit Ricci, 2011), Manchester UK: St Jerome.

Daniel Johnson-Mardones is Professor in the Department of Education FASCO, University of Chile. He earned his PhD in Curriculum & Instruction from the University of Illinois, Urbana-Champaign, USA, his Master's Degree in the field of Education, Curriculum y Comunidad Educativa, in the University of Chile (2010) and his degree in Historia y Geografia at the University of Concepción (1998). His PhD. scholarship was from Becario Fullbright (2011–2015). He is also a scholarship recipient of Becario Tinker (2012) for research in Latin Amercia and Becario CONICYT (2008–2009) and for the Master's Program in Education, Curriculum, and Communicative Community. Whilst studying for his doctorate, he was an Instructor in Social Sciences focusing on Problems and Diversity in the Teaching of Social Sciences (UIUC 2014–2016). He has also been an editor of the Qualitative Research Journal and a reviewer on Cultural Studies and Critical Methodologies, Educational Thought, Themes of Education. Some recent publications include:

- Fernández, B.; Johnson, D. (2015). Investigación-acción en formación de profesores: Desarrollo histórico, supuestos epistemológicos y diversidad metodológica. *Psicoperspectivas: Individuo y Sociedad* 14 (3) 93–105. doi:10.5027/Psicoperspectivas-Vol14-Issue3-Fulltext-626
- Johnson-Mardones, D. (2015). Understating Curriculum as Phenomenon, Field and Design: A Multidimensional conceptualization. *Journal of International Dialogues in Education* 2 (2) 1–9.
- Johnson-Mardones, D. (2015). Formar ciudadanos interculturales en un mundo global: Algunas notas desde los estudios curriculares. *Diálogo Andino*. 47 (2) 7–14.
- Johnson-Mardones, D. (2015). Freire and the U.S Reconceptualization: Remembering Curriculum as International Conversation. *Transnational Curriculum Inquiry*. 12(1) 3–12.
- Johnson-Mardones, D. (2015). Understanding Critical-Pedagogical-Performative Autoethnography. *Cultural Studies and Critical Methodologies*, 15 (3) 190–191. doi: 10.1177/1532708614562884

Diah Ariani Arimbi lectures in English literature at the Airlangga University in Surabaya, Indonesia. She received her PhD from UNSW, Sydney, Australia in 2006. Her current researches include images of women and the conception of beauty in teenage magazines, and the ways women are portrayed in Orientalist discourses. One of her published books entitled "Reading Contemporary Indonesian Muslim Women Writers: Representation, Identity and Religion of Muslim Women in Indonesian Fiction" has been cited numerously. She is currently officiated as the Dean of Faculty of Humanities, Universitas Airlangga.

Dr. Yuni Sari Amalia is lecturer in the Faculty of Humanities at the Universitas Airlangga, Indonesia. She is also the head of the Center for Innovation in Learning and Certification [PIPS UNAIR]. The center oversees the university's digital learning program along with the development of Airlangga University's E-Learning Application (AULA).

She earned her Master's and Doctoral degrees in the Department of Curriculum and Instruction at the University of Illinois at Urbana-Champaign, USA. She is a Fulbright fellow and has received several international awards, including the William Bagley Award, James Machin Award, and the Indonesian Ambassador's Award for Excellence. Her research interests include multicultural education, teaching methodologies, learning innovation, and evaluation in higher education. She has presented her research internationally, including at the Oxford University and Harvard University. Collaboratively, she published a book chapter in the international handbook of interpretation in educational research methods. She also presented and published a paper entitled *coming to know theory* at the American Educational Research Association (AERA) conference.

- Dialogue in narrative inquiry. In P. Smeyers, D. Bridges, N. Burbules, & M. Griffiths, (Eds). *International handbook of interpretation in educational research methods.* Thousand Oaks, CA: SAGE Publications. 2015.

Borders and mobility in arts, history, and well-being

Urban Studies: Border and Mobility – Kerr et al. (Eds)

Comparison of curcumin content and antioxidant activity of turmeric samples collected from Indonesia and Thailand: Considerations for the future sharing of the natural resource

A. Dechakhamphu, J. Junlatat, M. Agil, B. Prajogo & N. Pursariwati
Ubon Ratchathani Rajabhat University, Thailand

ABSTRACT: Turmeric has been used as a spice and a medicine in traditional Thai and Indonesian medicine since ancient times. It contains antioxidant, anti-inflammatory, anti-cancer and anti-bacterial properties. As both Thailand and Indonesian are producers of turmeric in the world market, this makes the sharing of this resource and the creation of trade bargain possible. However, the geographical area may have an impact on the medicinal properties of this plant. Therefore, this article aims to study curcumin content and some biological properties of turmeric, including its antioxidant and anti-bacterial properties, by using *in vitro* experiments. Turmeric samples were collected from Madura Island, Indonesia and from Ubonratchathani Province, Thailand during May–June 2015. Curcumin content was measured by using the UV spectrum. Antioxidant activity was measured by using DPPH assay. The experimental data showed that turmeric samples collected from Indonesia and Thailand contain similar curcumin content and biological properties. From the data from this study, it can be concluded that sharing of the turmeric resource between the studied areas is possible, not only for material use but also for scientific knowledge.

Keywords: biological property, sharing resource, traditional Thai medicine, traditional Indonesian medicine, turmeric

1 INTRODUCTION

Turmeric (*Curcuma longa* L.) is a perennial herb which is distributed throughout the tropical and subtropical regions of India, South-East Asia and China (Sahne, F. 2016). The underground rhizome not only imparts a distinctive flavor to the food but also provides a deep, indelible orange color. Turmeric is often sold to customers in developed countries in the form of a fine, dried, yellow powder. It is used in a wide variety of South Asian cuisines, but locally it is also used as an antiseptic for skin abrasion (Robbins, P., 1995). In Indonesia and Thailand, it is used as a food ingredient and natural remedy. Turmeric has also been proven to have powerful anti-diabetic, anti-asthmatic, anti-peptic ulcer, and wound healing effects, as well as the ability to improve brain conditions such as Alzheimer's disease (Lim, G.P., 2001).

India is the leading producer, consumer, and exporter of turmeric in the world. It is followed by other Asian producers such as Bangladesh, Pakistan, Sri Lanka, Taiwan, China, Myanmar, and Indonesia (Weiss, E.A., 2002). The major importers are the Middle East and North African countries. The United States imports turmeric from India to the amount of 97%, and the rest is supplied by the islands of the Pacific and Thailand (ASTA, 2002). The increasing demand for natural products as food additives makes turmeric an ideal candidate as a food colorant. In addition, recent medical research has demonstrated the anti-cancer and anti-viral activities of turmeric, thereby increasing its demand in Western countries (Weiss, E.A. 2002). Although India is the largest producer of turmeric in the world (846,700 tons), it exports only 6% of the total production. During 2006–2007, it exported 51,500 tons of turmeric. It also exported some amounts of turmeric to ASEAN countries such as Malaysia (2,263 tons) and Singapore (622 tons) (Apeda.gov, 2017). The global consumption rate of turmeric is expected to increase in the near future. The export of turmeric may increase the local income of some countries, including Indonesia and Thailand. Nowadays, Indian turmeric is preferred to that of other countries due to its high curcumin content, a major bioactive compound of turmeric which plays a crucial role in preventing and treating diseases. Therefore, the standardization of this raw material is also important. In general, high curcumin content and good biological activities are the main factors for its commercial advantage.

Several studies have reported the genetic diversity and variation found in the active compounds of turmeric collected from different parts of Thailand (Thaikert, R. and Paisooksantivatana, Y. 2009). A study conducted in India revealed that bioactive compounds depend on the region of cultivation (Sinkar, P.V., 2005). Ashraf, K. et al. (2012) reported that major bioactive compounds of turmeric were mainly found in different geographical regions. Therefore, the superior quality of turmeric with high bioactive properties is regarded as a strategy for turmeric trade. From these reports, it can be concluded that ASEAN countries such as Indonesia and Thailand are likely to gain profit from turmeric trade. Evidence for testing of major bioactive components and biological properties will be a source of competitive advantage. Therefore, this study aims to verify curcumin content and antioxidant activity of turmeric collected from Indonesia and Thailand. The results from this study provide not only scientific data but also the prospect of sharing turmeric between the two countries. In addition, this will pave the way for working together to create trade bargain at the international level.

2 METHODOLOGY

This study was conducted under the cooperation between the Faculty of Thai Traditional and Alternative Medicine, Ubon Ratchathani Rajabhat University, Thailand and the Faculty of Pharmacy, University of Airlangga, Indonesia. The rhizome of turmeric was collected in triplicates from Madura Island, Indonesia and from Ubonratchathani Province, Thailand during May–June 2015. Bioactive compounds were extracted from the samples by using a standard protocol. Curcumin content was measured by using the UV spectrum. Antioxidant activity was measured by using DPPH assay.

The comparison of turmeric samples collected from Indonesia and Thailand is presented in Table 1.

3 DISCUSSION

Turmeric is commonly used as a spice for preparing food in India and ASEAN countries such as Indonesia and Thailand. It is also used in the traditional medicine for various therapeutic purposes, including antioxidant, anti-cancer, anti-inflammation, and improved brain functions. Therefore, turmeric is considered as a natural panacea, which is supported by scientific research. Consequently, the world consumption rate of turmeric is expected to increase in the near future. Although turmeric crops may become a good income for local cultivators, high curcumin content and good biological activities are the key determining factors for the purchaser. Many reports have demonstrated that curcumin content varied depending on geographical areas. Hence, the objective of this study was to verify curcumin content and antioxidant activity of turmeric collected from Indonesia and Thailand. The results are summarized in Table 1.

Table 1. Comparison of curcumin content and antioxidant activity of turmeric samples collected from Madura Inland, Indonesia and Ubonratchathani Province, Thailand.

Sample collection areas	Curcumin content (%)	Antioxidant activity (SC_{50}, ug/ml)
Indonesia	5.9	1.69
Thailand	5.1	1.74

According to the data from Table 1, curcumin content and antioxidant activity of turmeric samples collected from Madura Island, Indonesia were similar to those of the samples collected from Ubonratchathani, Thailand. Kanjilal et al. (2002) and Chandra et al. (2005) reported the highest curcumin content of 6.8–7.3% in turmeric samples collected from the regions of Meghalaya and Lakadong, India. Kotoky et al. (1999) reported that Tamenglong turmeric had the highest curcumin content (7.3%) among the seven *Curcuma longa* cultivars grown in Manipur, India. Compared with these reports, curcumin content found in this study was slightly lower. Nevertheless, it was classified as good quality. Therefore, we anticipate the approach of sharing the turmeric resource, scientific research, and trade competition. To increase more quality and reliability, influencing factors of curcumin content and biological activity such as harvest times, climate conditions, cultivation handling, storing conditions and age of vegetable materials should be considered. In addition, biological activities of turmeric such as anti-cancer, antimicrobial and anti-inflammatory effects should be further investigated.

4 CONCLUSION

Turmeric is used both as a spice and a medicine in the world. Although India is the leading producer and exporter of turmeric, other countries such as Indonesia and Thailand are also in the trade race. This study conducted an experimental analysis to verify the quality of turmeric collected from Madura Island, Indonesia and Ubonratchathani Province, Thailand. The results indicated that curcumin content and antioxidant activity of turmeric samples collected from both studied areas were similar. Therefore, it can be concluded that sharing of the turmeric resource between the studied areas is possible, not only for material use but also for scientific knowledge.

REFERENCES

Ashraf, K., Mujeeb, M., Altaf, A., Amir, M., Mallick, M.N., Sharma, D. 2012. Validated HPTLC analysis method for quantification of variability in content of curcumin in Curcuma longa L. (turmeric) collected from different geographical region of India. Asian Pacific J Trop Biomed, S584-S588.

ASTA 2002. A concise guide to spices, herbs, seeds, and extractives. American Spice Trade Association, 48–50.

Chandra, R., Yadav, D.S., Rai N. and Sarma P. 2005. Megha Turmeric 1: a new turmeric for Meghalaya. Indian Horticult, 50(2), 18.

Kanjilal, P.B., Kotoky, R. and Singh, R.S. 2002. Morphological and chemical parameter of certain cultivars of chilli, ginger and turmeric grown in Meghalaya. Adv in Plant Sci, 15(1), 225–229.

Kotoky, R., Kanjilal, P.B., Singh, R.S., Pathak, M.G. and Mazid, E. 1999. Studies on curcumin and essential oil content of different cultivars of turmeric (Curcuma zanga L.) grown in Manipur. Indian Journal of Arecanut, Spicesand Medicinal Plants. 1, 91–92.

Lim, G.P., Chu, T., Yang, F., Beech, W., Frautschy, S.A., et al. 2001. The curry spice curcumin reduces oxidative damage and amyloid pathology in an Alzheimer transgenic mouse. J Neurosci, 21, 8370–8377.

Robbins, P. 1995. Tropical commodities and their markets (A Guide and Directory), 229–230.

Sahne, F., Mohammadi, M., Najafpour G.D. and Moghadamnia, A.A. 2016. Extraction of bioactive compound curcumin from turmeric (Curcuma longa L.) via different routes: A comparative study. Pak. J. Biotechnol, 13(3), 173–180.

Sinkar, P.V., Haldankar, P.M. Khandekar, R.G., Ranpise, S.A., Joshi, G.D., Mahale B.B. 2005. Preliminary evaluation of turmeric (Curcuma longa) varieties at Konkan region of Maharashtra. J Spices Aromatic Crops, 14, 28–33.

Thaikert, R., Paisooksantivatana, Y. 2009. Variation of total curcuminoids content, antioxidant activity and genetic diversity in turmeric (Curcuma longa L.) collections. Kasetsart J. (Nat. Sci.), 43, 507–518.

Turmeric trade. Available at http//www.agriexchange.apeda.gov.in/Market%20Profile/MOA/ Product/Turmeric.pdf. (Accessed 5 November 2017).

Weiss, E.A. 2002. Spice Crops. CAB International publishing, Oxon, UK.

Urban Studies: Border and Mobility – Kerr et al. (Eds)

Efficiency of household accounting: A case study of a model village in Thailand

N. Thongprasert & S. Mala
Ubon Ratchathani Rajabhat University, Thailand

ABSTRACT: This study aimed to investigate the relationship between the efficiency of household accounting based on the philosophy of sufficiency economy and the quality of life. For this purpose, we conducted a case study of the people in the model village with sufficient level of living in Bu Pueai sub-district, Nam Yuen district of the Ubon Ratchathani province, Thailand. The study sample comprised 364 villagers of the Bu Pueai sub-district who exercised household accounting. A systematic sampling technique was employed to obtain the samples. Frequency, percentage, arithmetic mean, standard deviation and Pearson's Correlation Coefficient test were used to analyze the data. It was found that the samples were fond of exercising household accounting based on the sufficiency economy philosophy at a high level. The investigation, as a whole, revealed that the relationship between efficiency of household accounting and the people's quality of life was highly positive. A detailed examination indicated that the relationship between the efficiency of household accounting and the aspects related to economics and learning conditions was positive at a high level, while that between the efficiency and the aspects concerning mental, social, natural resource and environment conditions was positive at a medium level. The statistical significance level of this study was set at 0.01.

Keywords: Border town, Household Accounting, Quality of Life

1 INTRODUCTION

Thailand has been severely encountering economic crisis since 1997. People have noticed its influence on the stability of their careers, incomes, expenses, household debts as well as mental and physical health. It also affected people's way of living and the society at large. To provide a solution to the increasing problems in communities, King Bhumibhol Adulayadej's philosophy of sufficiency economy is being adopted. Communities, in general, are considered an important foundation mechanism that helps in boosting the country's progress. Philosophy of sufficiency economy has been carried out through many channels in these communities. Many activities including household accounting, based on the King's initiation that aims at people's self-reliance and honest careers, have been undertaken and supported. This is done to alleviate poverty, and make sure people have enough to eat and live. The philosophy also aims at simple living (being economic or not living luxuriously), and at basing one's life on Lord Buddha's middle path principles (realizing sufficiency and satisfaction with what one has) (Sumeth Tantivejakul, 2006: 53).

Keeping record of the incomes and expenses in the household account promotes careful spending and at the same time discourages over spending as the log of each expense provides an insight into the balance, incomes and expenses of a family. Reducing unnecessary expenses after the analysis can result in savings. If spending is done within one's means and resources falling in debt traps can be eluded.

Bu Pueai, a sub-district in Nam Yuen district of the Ubon Ratchathani province, is a model village of sufficiency economy at a sufficient level of living supported by The Ministry of Interior, 2014 (The Ministry of Interior, 2016:1). The people in this area live their life based on the philosophy of sufficiency economy. Training the people on accounting fundamentals is organized by the Accounting Preparation Promotion Unit, Cooperative Auditing Department, the Ministry of Agriculture and Cooperatives. It is considered to be one of the guidelines to drive the philosophy in those specific areas. However, not many studies investigated the efficiency of exercising household accounting in terms of how this affected people's lives in the model villages. This inspired us to study the relationship between the efficiency of the household accounting based on the philosophy of sufficiency economy and people's way of life, and to examine if the obtained results of the implementation achieved the criteria set by the Ministry of Interior. The study can help set up guidelines for solving problems and improving people's quality of life.

2 METHODOLOGY

Quantitative type questionnaires of 44 questions each were used to collect the data. The questions were related to samples' demographic data, living conditions under the philosophy of sufficiency economy, and the efficiency of household accounting. Validation of the questionnaire was done by a panel of 5 experts. To obtain a more reliable data, the questionnaire was presented to 30 samples in the Kradian sub-district, Trakarn Puechpol district of the Ubon Ratchathani province. Cronbach's alpha coefficient was calculated from the obtained data. The reliability score of the questionnaire was 0.09.

2.1 *Population and samples*

The study was conducted on a sample population of 7,287 people of Bu Pueai sub-district, Nam Yuen district of the Ubon Ratchathani province. This sub-district was a model village at a sufficient living level under the philosophy of sufficiency economy scheme, The Ministry of Interior, 2014 (The Ministry of Interior, 2016:1). By using R.V. Krejcie and D.W. Morgan's table of systematic sampling technique, 364 samples were obtained.

2.2 *Variables*

Two variables were investigated. The first one was the people's way of living under the philosophy of sufficiency economy in Bu Pueai sub-district, Nam Yuen district of the Ubon Ratchathani province. Four aspects were examined: 1) mental and social conditions consisting of 7 indications, 2) economic conditions with 5 indications, 3) learning conditions with 7 indications, and 4) environmental conditions comprising four indications. The second variable was the efficiency of household accounting with 13 indications.

2.3 *Conceptual framework*

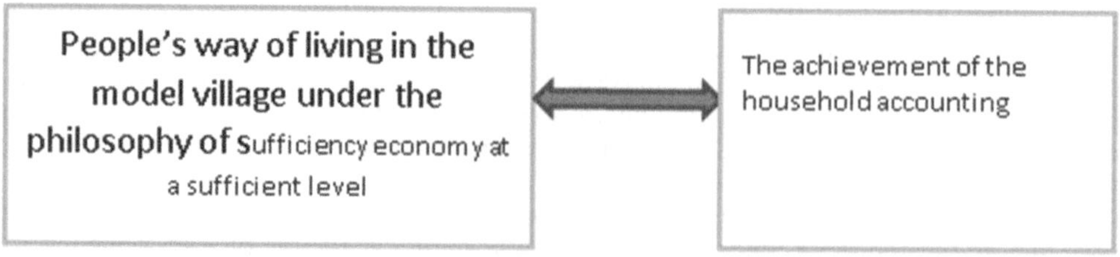

2.4 *Hypothesis*

People's quality of life under the philosophy of sufficiency economy correlated positively with the efficiency of household accounting.

2.5 *Data analysis*

After checking the returned questionnaires, the complete ones were classified according to the sample types. Statistical Package for the Social Science (SPSS) software was used to calculate and analyze the following:

1. Demographic data of the samples. Frequency distribution was made. Percentage was calculated and presented in the form of tables.
2. Four aspects of the samples' living condition under the philosophy of sufficiency economy in Bu Pueai sub-district, Nam Yuen district of the Ubon Ratchathani province. Arithmetic mean and standard deviation were used to calculate the data. Criteria for interpretation of means were set as follows (Srisa-ard, 2016:3):

Mean Range	Interpretation
4.51–5.00	Highest
3.51–4.50	High
2.51–3.50	Medium
1.51–2.50	Less
1.00–1.50	The least

3. Correlation of four aspects of the living condition under the mentioned philosophy and the efficiency of household accounting in the selected model village. Pearson's Correlation Coefficient was used for this purpose.

The set criteria of the correlation coefficient were interpreted as follows (Hinkle, William and Stephen, 1998:118):

± 0.81 –± 1.00 referred to very high relationship
± 0.61 –± 0.80 referred to high relationship
± 0.41 –± 0.60 referred to medium relationship
± 0.21 –± 0.40 referred to low relationship
± 0.00 –± 0.20 referred to very low relationship.

3 FINDINGS

Part 1: Demographic data. The study samples comprised of 364 people living in the model village of sufficiency economy in Bu Pueai sub-district, Nam Yuen district of the Ubon Ratchathani province.

Table 1 shows that the number of male informants is greater than female: 162 (51.4%): 153 (48.6). Most of them (112 or 35.6%) are 31 to 40 years old. The educational background of most of them (250 informants or 79.4%) is lower than bachelor's degree. Apparently 148 (47%) informants have income ranging from 5,001 to 10,000 baht/month. The average number of family members in 157 (49.9%) families is 3–5. Informants who are agriculturists are 119 (37.8%). Number of married are 265 (84.1%) and 288 (91.4%) have exercised household accounting for more than 1 year.

Part 2: An analysis of the relationship between the people's living condition under the philosophy of sufficiency economy and the efficiency of the household accounting. Frequency was counted. Mean and standard deviation were calculated and interpretation of agreement on each question item was made as shown in Tables 2–7.

Table 2 shows that most of the informants mentally and socially follow the philosophy of sufficiency economy at a high level ($\overline{X} = 4.24$, and S.D. = 0.26). With respect to the detailed investigation, the means of the responses to the 7 questions asked range from 4.14 to 4.40 which is at a high level. The item related to the people's observation of democracy principles shows the highest mean, followed by the ones involving observation of rules and regulations, cooperation and willingness in performing community activities, being virtuous and ethical, being free from vices, having village funds, belief in and application of the philosophy of sufficiency economy, respectively.

Table 3 shows that, economically, most informants live their life according to the philosophy of sufficiency economy at a high level. The mean of this aspect, as a whole, is 4.22

Table 1. Percentage of demographic data classified into gender, age, education, income, number of family members, occupations, marital status and household accounting experience.

Demographic data	Number (total 315)	%
Gender		
Male	162	51.4
Female	153	48.6
Age		
30 and less	58	18.4
31–40	112	35.6
41–50	61	19.4
51–60	63	20.0
61 and above	21	6.6
Education		
Lower than Bachelor's degree	250	79.4
Bachelor's degree	60	19.0
Master's degree	5	1.6
Income		
5,000 baht and less	99	31.4
5,001–10,000	148	47.0
10,001–20,000	52	16.5
20,001–30,000	11	3.5
30,001 and above	5	1.6
Family members		
1–2	76	24.1
3–5	157	49.9
More than 5	82	26.0
Occupation		
Civil servant	28	8.9
Merchant/Business owner	92	29.2
Agriculturist	119	37.8
Employee	59	18.7
House makers/Unemployed	17	5.4
Marital status		
Single	22	7.0
Married	265	84.1
Divorced	18	5.7
Separated	1	0.3
Widow	9	2.9
Household accounting experience		
Less than 1 year	27	8.6
More than 1 year	288	91.4

with standard deviation 0.35. With respect to the detailed investigation, the means of the responses to the 5 questions asked range from 4.08 to 4.25 which is at a high level. The item related to the people's activities to reduce expenses and increase the incomes secured the highest rank, followed by those related to operation of community enterprises, variety of savings, household accounting and forming groups for occupation development, respectively.

Table 4 shows that, due to the learning aspect, most informants live their life according to the philosophy of sufficiency economy at a high level. The mean of this aspect, as a whole, is 4.23 with standard deviation 0.31. Regarding the detailed investigation, the means of the response to the 7 questions asked range from 4.14 to 4.35 which is at a high level. The item related to people's search for the application of local wisdom for creating more value attained the highest rank followed by those concerning creation of development network, application of technology appropriate to the community potential, having community data, practice of self-reliance,

Table 2. Mean, standard deviation and interpretation of questionnaire items in relation to mental and social conditions of the informants living under the philosophy of sufficiency economy.

Mental and social conditions of informants living under the philosophy of sufficiency economy	$\bar{X}$	S	Level of agreement
1. People are cooperative and willingly support community activities.	4.30	00.46	much
2. Community rules and regulations are mutually set up and followed.	4.30	0.56	much
3. Welfare funds, such as Village funds, Mother of the Land funds, Million Baht funds, are provided for the members of the community.	4.19	0.44	much
4. People in the community observe democracy principles.	4.40	0.57	much
5. People in the community are virtuous and ethical.	4.21	0.55	much
6. Bu Pueai sub-district is free from all vices including gambling and quarrelling.	4.21	0.63	much
7. People in the community believe in the philosophy of sufficiency economy and apply it to their daily life.	4.14	0.59	much
Total	4.24	0.26	much

Table 3. Mean and standard deviation of questionnaire items in relation to economic condition of the informants living under the philosophy of sufficiency economy.

Economic condition of the informants living under the philosophy of sufficiency economy.	$\bar{X}$	S.D.	Level of agreement
1. The people exercise household accounting.	4.21	0.54	much
2. The people have activities to reduce expenses and increase incomes.	4.29	0.57	much
3. The people divide themselves into groups for occupation development purpose, such as basketry group and food processing group.	4.10	0.58	much
4. The people have various ways of savings such as one baht savings project.	4.22	0.61	much
5. Community enterprise or similar enterprise is operated.	4.28	0.66	much
Total	4.22	0.35	much

Table 4. Mean and standard deviation of questionnaire items in relation to learning condition of the informants living under the philosophy of sufficiency economy.

Learning condition of the informants living under the philosophy of sufficiency economy.	$\bar{X}$	S.D.	Level of agreement
1. Community information such as community history, population data and census is provided.	4.23	0.56	much
2. Community information and plan are used and benefited.	4.14	0.53	much
3. Local wisdom is searched and used to add more values.	4.35	0.63	much
4. Learning center based on the philosophy of sufficiency economy is provided in the community.	4.18	0.57	much
5. Technology appropriate to community potential is used.	4.25	0.54	much
6. An associate and a network are built for development.	4.29	0.61	much
7. Self-reliance is practiced.	4.25	0.67	much
Total	4.23	0.31	much

Table 5. Mean and standard deviation of questionnaire items in relation to natural resource and environmental conditions reflected by the informants living under the philosophy of sufficiency economy.

Natural resource and environment conditions of the informants living under the philosophy of sufficiency economy	$\overline{X}$	S.D.	Level of agreement
1. The community cooperatively conserves natural resources and the environment.	4.15	0.57	much
2. In the community, there is an organization working on conserving natural resources and the environment.	4.36	0.70	much
3. The community realizes the importance of the alternative energy and makes an utmost use of it based on the philosophy of sufficiency economy.	4.15	0.60	much
4. The community adds value to the local natural resources based on the philosophy of sufficiency economy.	4.12	0.54	much
Total	4.19	0.41	much

Table 6. Mean and standard deviation of the efficiency of the household accounting.

Efficiency of the household accounting	$\overline{X}$	S.D.	Efficiency level
1. Household accounting helps the people to know their family's income, expenses and balance.	4.14	0.64	much
2. Household accounting helps the people to know what their family's necessary expense is.	4.39	0.53	much
3. Household accounting helps in reducing the people's family debts.	4.24	0.58	much
4. Household accounting helps the people to know the causes of the income and expense imbalance.	4.36	0.63	much
5. Household accounting helps the people to reduce and to stop luxurious expenses and vices.	4.23	0.55	much
6. Household accounting helps to self-rely more, and to generate more family income.	4.34	0.53	much
7. Household accounting helps the people to save more.	4.34	0.51	much
8. Household accounting helps in generating occupations for sufficient living.	4.34	0.53	much
9. Household accounting helps the people to be able to plan for family expenditure and to generate more income.	4.20	0.50	much
10. Household accounting helps in generating community occupation enterprises such as local agricultural product processing, cooperative, weaving groups, etc.	4.29	0.53	much
11. Household accounting helps in regulating family savings resulting in having money to invest on production expansion and retail sales.	4.23	0.56	much
12. Household accounting helps in creating service groups for tourists such as agricultural tourism.	4.18	0.61	much
13. Household accounting helps the people and their family members to be able to pay debts.	4.17	0.38	much
Total	4.26	0.22	much

having center for learning the philosophy of sufficiency economy, utilization of community information and plan.

Table 5, with respect to the natural resources and environmental aspects, shows that most informants live their life in accordance with the philosophy of sufficiency economy at a high level. The mean of this aspect, as a whole, is 4.19 with standard deviation 0.41. The detailed investigation reveals that the means of the response to the four questions asked ranged from

Table 7. Summary of conditions of the people living under the philosophy of sufficiency economy and the results of household accounting.

Conditions of the people living under the philosophy of sufficiency economy	$\overline{X}$	S.D.	Level of agreement
Mental and social aspects	4.24	0.26	much
Economics aspect	4.22	0.35	much
Learning aspect	4.23	0.31	much
Natural resource and environment aspects	4.19	0.41	much
Efficiency of household accounting	4.26	0.22	much

Table 8. Relationship between aspects of people's living conditions under the philosophy of sufficiency economy and the efficiency of household accounting.

The relationship between the variables and the efficiency of household accounting	Correlation coefficient	p	Level of relationship
Mental and social aspects	0.467**	0.000	Medium
Economics aspect	0.722**	0.000	High
Learning aspect	0.666**	0.000	High
Natural resource and environmental aspects	0.539**	0.000	Medium
All aspects of living conditions under the philosophy of sufficiency economy	0.712**	0.000	High

**Statistical significance at the level of 0.01.

4.12 to 4.36 which are at a high level. The item related to community that has organization working for conserving natural resources and the environment is recognized the most, followed by people in community cooperatively conserving the resources and the environment, people's realization of importance and utilization of the alternative energy, and addition of value to the existing natural resources based on the philosophy of sufficiency economy respectively.

Table 6 shows that the efficiency of the household accounting, as a whole, is at a high level ($\overline{X}$ = 4.26, S.D. = 0.22). With respect to the detailed investigation, the means of the response to the 13 questions asked range from 4.14 to 4.39 which is at a high level. The efficiency that ranked the highest was the household accounting helped the people to know their necessary expenses, followed by the its help to see the causes of the insufficient income, to be more self-reliant, to generate more family income, to promote more savings, to generate more occupations for sufficient living, to promote community enterprises, to have more money to invest on production and sale, to have careful plan for expenditure, to initiate more service groups like agricultural tourism, and to be able to pay debts, respectively.

Table 7 summarizes the conditions of the people living under the philosophy of sufficiency economy and the efficiency of the household accounting in terms of means, standard deviation and the level of agreement and efficiency. This shows that the people's life was better when they adopted household accounting based on the philosophy of sufficiency economy.

Part 3: Analysis of relationship between aspects of people's living conditions under the philosophy of sufficiency economy and the efficiency of household accounting, using Pearson product moment correlation coefficient.

Table 8 shows that living under the philosophy of sufficiency economy positively correlates with the efficiency of the household accounting at a high level and at a statistical significance level of 0.01. The correlation coefficient (r^2) value is 0.712. A detailed examination indicates that the relationship between the efficiency of household accounting and the aspects related to economics and learning conditions is highly positive (r^2 = 0.722 and 0.666 respectively), while that between the efficiency and the aspects concerning mental, social, natural resource and environment conditions is positive at a medium level (r^2 = 0.539 and 0.467 respectively).

4 DISCUSSION

The samples of this study comprised of 315 people living in Bu Pueai sub-district, Nam Yuen district of the Ubon Ratchathani province, with the age ranging from 31 to 40, having educational background lower than bachelor's degree, earning 5,001–10,000 baht income/month, being agriculturist and having more than one-year experience in household accounting.

The analysis of the obtained data in relation to mental, social, economic, learning, natural resource and environmental aspects indicated that the people living under the philosophy of sufficiency economy was quite good and the efficiency of household accounting was quite high too. This showed that the people in Bu Pueai sub-district realized the importance and the benefit of household accounting. Moreover, the analysis of the relationship between living under the philosophy of sufficiency economy and the efficiency of household accounting showed a highly positive relationship as hypothesized, assuring that household accounting was beneficial to the people. The finding of this study was in accordance with Ubonsri and Pannun's (2012) study applying economy to lifestyles and use of resources at community levels. They found that application of the philosophy of sufficiency economy affected the economic status of the community. The people learned and participated in planning and solving the community problems. In addition, they took part in group activities, building up more immunity for themselves, their family and community. Household accounting helped the people determine necessary and unnecessary expenses, so they could cut off the unnecessary ones. This finding was in accordance with the results of the study conducted by Naipinit, Promsaka Na Sakolnakorn and Kroeksakul (2013) who investigated "Sufficiency Economy for Social and Environmental Sustainability: A Case Study of Four Villages in Rural Thailand." They found that the philosophy of sufficiency economy was a guideline that the people could apply to their daily life in terms of earning for their family by planting vegetables for self-consumption and sale, raising cattle and poultry and exercising saving in the village cooperatives. The finding also accorded with Vyas, Snow, Roe and Brereton's (2016) study on "Social Organization of Household Finance: Understanding Artful Financial Systems in the Home" in terms of the usage of instruments such as spreadsheets and diaries to exercise household accounting. They suggested that those instruments could be utilized if they were developed to meet the family values, relationship and daily life of members of each family.

However, the relationship between the efficiency of the household accounting and the living condition related to natural resource and the environment, mental and social aspects was found positive at a medium level. This finding was similar to that of Ubonsri and Pannun's (2012). They investigated application of the philosophy of sufficiency economy on lifestyles and use of resources at community levels and found that the people's good quality of life resulted from adding value to the existing natural resources to generate more family income, and taking part in planning and solving community problems related to utilizing natural resources appropriately. As for why the relationship on this matter was found positive at a medium level which was a bit less than those of the other aspects, it can be explained that even though the people lived their life under the philosophy of sufficiency economy and kept the household accounting continuously, they might not believe in it fully and they did not have enough knowledge to analyze the recording data for application on planning their expenditure as they were from the model village at sufficient living level who might have less immunity than the villages with better living conditions. Thongprasert (2017: abstract) in her study entitled "Behavior of Using Agriculturist Credit Card and Quality of Life of Agriculturists Living Under Philosophy of Sufficiency Economy" found that if the agriculturists lived their life in a sufficient way as learned from the philosophy, not following the others in overspending, they would be happy with their life and in fact have better quality of life. On the contrary, if one overspent and lacked self-discipline, they would be in debts resulting in adverse mental health and degraded quality of life. The study on "relationship of efficiency of household accounting and quality of life under the philosophy of sufficiency economy of the people in Mahasarakham province by Musik (2014) also confirmed that if the authority wanted the people to be mentally and socially happy, they should encourage the people to exercise household accounting to build up immunity of sufficiency realization, to be careful in spending their money, to know what was necessary to purchase and what was not, and not to be luxurious".

REFERENCES

Bridgman, B., Dugan, A. Mikhael, L., Osborne, M. and Villones, S. 2012. Accounting for Household Production in the National Accounts: the paper develops a satellite account that adjusts gross domestic product (GDP) for household production. Survey of Current Business, 92(5), 23–36.

Dhaval, V., Snow, S. and Brereton, M. 2016. Social Organization of Household Finance: Understanding Artful Financial Systems in the Home. 19th ACM conference. CA: Queensland University of Technology, 1777–1789.

Hinkle D.E., William W, and Stephen G.J. 1998. Applied Statistics for the Behavior Sciences. 4th ed. New York: Houghton Mifflin.

Interpretation of Rating Scale Data (In Thai). Available at www.watpon.com/boonchom/ 05.doc (Accessed 22 March 2016).

Naipinit, A., Promsaka Na Sakolnakorn, T. and Kroeksakul, P. 2013. Sufficiency Economy for Social and Environmental Sustainability: A Case Study of Four Villages in Rural Area, Thailand (In Thai). Asian Social Science, 10(2), 102–111.

Naratawan, M. 2014. Relationship of Efficiency of Household Accounting and Quality of Life under Philosophy of Sufficiency Economy of People in Mahasarakham Province (In Thai). Journal of Graduate School of Commerce Burapa Review. 9(1), 45–58.

Report of Census and Houses (In Thai). Available at www.stat.bora.dopa.go.th/stat/statnew/statTDD/ (Accessed 2 October 2016).

Tantivejakul, S. 2006. Philosophy of Sufficiency Economy Adapted for Self-application and Economics Development (In Thai). BOI Magazine, 8–12.

Thongprasert, N. 2017. Behavior of Using Agriculturist Credit Card and Quality of Life of Agriculturists Living under Philosophy of Sufficiency Economy (In Thai). Panyapiwat, 9(2), 83–94.

Ubonsri, B. and Pannun, A. 2012. A study of applying sufficiency economy to lifestyles and use of resources at community levels. The 3rd International Conference on Sustainable Future for Human Security. Kyoto: Kyoto University.

Urban Studies: Border and Mobility – Kerr et al. (Eds)
© 2019 Taylor & Francis Group, London, ISBN 978-1-138-58034-3

Forecasts for trans-border mobility: A case study of agricultural products imported from Laos to Thailand via the Chong-Mek border

N. Nanthasamroeng
Ubon Ratchathani Rajabhat University, Thailand

ABSTRACT: This study aimed to 1) examine the overall data of agricultural products imported from Laos to Thailand via the Chong-Mek border, 2) find a reliable forecasting model for imported agricultural products, and 3) forecast the value of selected agricultural products to be imported from Laos to Thailand. All data were collected from the website of Chong-Mek customs house. The results indicate that 79.75% of the total products imported from Laos to Thailand were agricultural products. From the fiscal year 2008–2016, cabbage was the best imported product with an import value of 2,681 million USD. Chipped cassava was the second best with an import value of 1,970 million USD. Several forecasting models were applied to the total imported data, including 1) simple linear regression, 2) moving average, 3) exponential smoothing, and 4) double exponential smoothing. We found that simple linear regression and exponential smoothing were the most reliable forecasting models with minimum forecast errors. The 5-year forecast results indicate that the demand for cabbage, chipped cassava, sweet potato, and green banana from Laos is expected to increase to 782.52, 904.25, 149.93, and 121.97 million USD in the fiscal year 2021.

Keywords: Agribusiness, forecasting model, imported agricultural products, Laos border

1 INTRODUCTION

An import and export business across the border of neighboring countries has an enormous impact on the socio-economic conditions of each country. Thailand has border trade with four neighboring countries, including Malaysia, Laos, Myanmar, and Cambodia. The values of import and export via borders are shown in Figure 1.

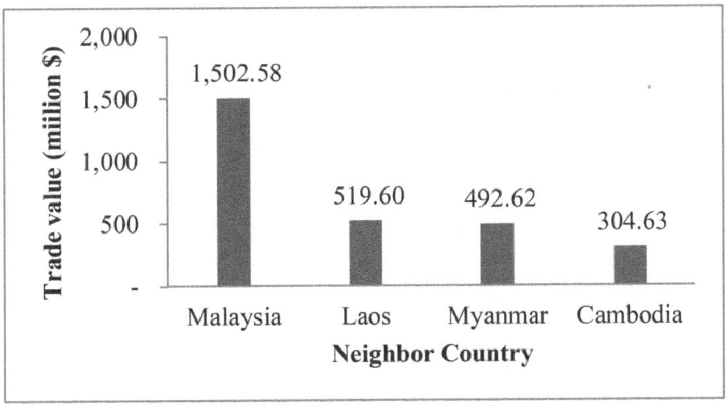

Figure 1. Values of import and export via borders during January–August 2017.

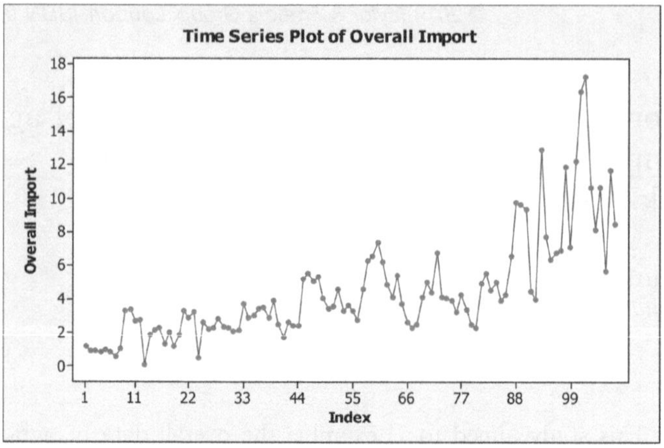

Figure 2. Import value of agricultural products imported via the Chong-Mek border since the fiscal year 2008.

The import value of agricultural products imported from Laos to Thailand was about 90 million USD in 2016. The Chong-Mek border had a total import value of about 113 million USD per year.

Therefore, forecasting the import value of agricultural products imported from Laos to Thailand will help clarify the vision for the development of infrastructures for a fresh food supply chain.

2 METHODOLOGY

Numerous studies have used a wide variety of forecasting techniques. An appropriate technique depends on the pattern of data. In this study, we focus on the following forecasting models: 1) simple linear regression, 2) exponential smoothing, 3) double exponential smoothing, and 4) moving average.

Simple linear regression is a linear regression model with a single explanatory variable (Dudek, 2016: 140). Exponential smoothing is a technique to forecast the demand of the next period based on the average of previous data. The parameter α is used to prioritize recent demands (Tratar, Mojškerc, and Toman, 2016: 163). Moving average is a time series technique to predict future data based on previous data (Barrow, 2016: 6089). In this study, we use forecast accuracy measures such as the mean absolute deviation (MAD), the mean square deviation (MSD), and the mean absolute percentage error (MAPE) (Kim and Kim, 2016: 669).

3 RESULTS AND DISCUSSION

3.1 *Agricultural products imported from Laos to Thailand via the Chong-Mek border*

Various types of agricultural products were imported from Laos to Thailand in each year. Data from the Chonk-Mek customs house indicate that the import values of cabbage, chipped cassava, sweet potato, and green banana shipped from Laos during the fiscal year 2008–2016 were about 6,200 million USD.

3.2 *Forecasting model for imported agricultural products*

We tested several forecasting models to find the best fit of the model for each product. Forecast error indicators, including MAD, MSD, and MAPE, were used to compare the accuracy

Table 1. Import value of agricultural products from Laos.

Fiscal year	Import value (million USD) Cabbage	Chipped cassava	Sweet potato	Green banana
2008	105.41	–	14.44	102.74
2009	94.52	–	32.98	55.63
2010	102.84	–	48.49	53.29
2011	287.09	–	85.09	62.35
2012	352.23	–	206.13	98.58
2013	473.75	51.58	216.30	109.67
2014	375.77	109.91	206.00	76.42
2015	354.34	521.90	247.02	75.38
2016	535.12	1,286.60	161.44	126.89
Total	2,681.08	1,969.99	1,217.89	760.95
			Grand total	6,629.91

Table 2. Forecasting model and import value of cabbage.

Forecasting model	Forecasting import value of cabbage for 2021	MAD	MSD	MAPE	Trends
Simple linear regression	782.52	*58.69*	*4,667.51*	27.64	Yt = 28.7 + 53.8*t
Exponential smoothing	563.33	73.10	10,521.20	*20.70*	$\alpha = 1.1556$
Double exponential smoothing	728.59	83.74	8,483.01	34.22	$\alpha = 0.4304$ $\beta = 0.3574$
Weighted moving average	444.73	111.50	17,035.40	29.60	WMA = 2

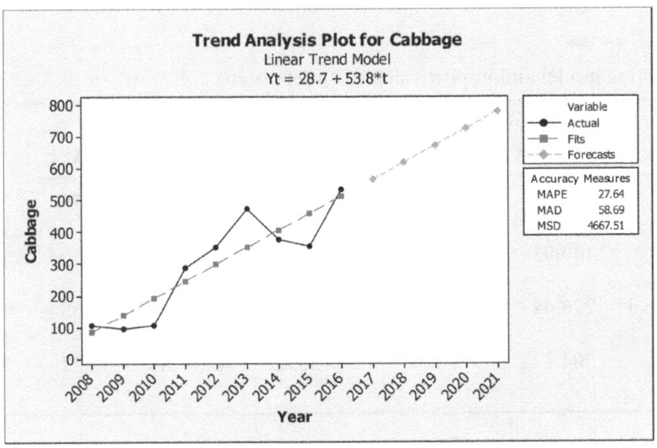

Figure 3. Forecasting import data of cabbage using simple linear regression.

of each model for each agricultural product. The results are summarized in Tables 2–5 and Figures 1–4.

The forecasting models were validated using the import data of agricultural products because of their varieties. Simple linear regression was found to be the best predictor for cabbage and green banana. Exponential smoothing with $\alpha = 1.1266$ was the best predictor for

Table 3. Forecasting model and import value of chipped cassava.

Forecasting model	Forecasting import value of chipped cassava for 2021	MAD	MSD	MAPE	Trends
Simple linear regression	3,168.59	*176.6*	*31,184.1*	137.7	Yt = –537 + 412*t
Exponential smoothing	N/A	N/A	N/A	N/A	N/A
Double exponential smoothing	N/A	N/A	N/A	N/A	N/A
Weighted moving average	904.25	706	568,433	*80*	WMA = 2

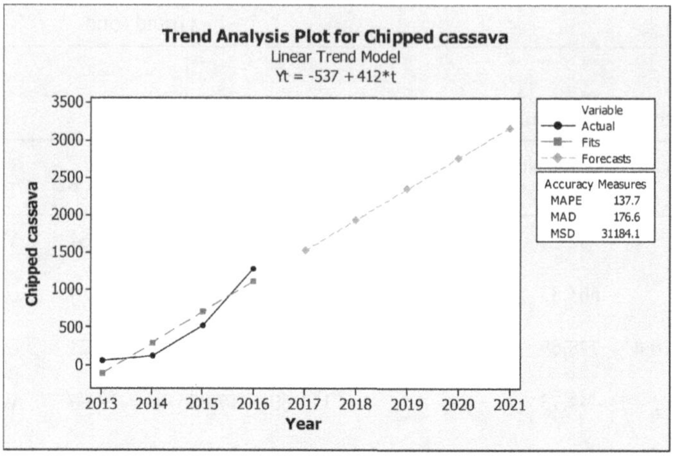

Figure 4. Forecasting import data of chipped cassava using simple linear regression.

Table 4. Forecasting model and import value of sweet potato.

Forecasting model	Forecasting import value of sweet potato for 2021	MAD	MSD	MAPE	Trends
Simple linear regression	386.78	37.01	*2,004.30*	37.76	Yt = –4.4 + 27.9*t
Exponential smoothing	149.93	*36.96*	2,833.48	*30.60*	α = 1.1266
Double exponential smoothing	226.44	39.15	2,945.42	45.68	α = 0.9354 β = 0.0979
Weighted moving average	204.23	55.05	4,648.88	37.27	WMA = 2

sweet potato, with MAD = 36.96, MSD = 2,833.48, and MAPE = 30.60. However, as chipped cassava had very large forecast errors, we only observed its positive trend.

Cabbage, chipped cassava, and green banana showed a positive growth trend during 2017–2021. This is especially the case for cabbage and green banana due to an increase in their domestic consumption. The growth in volume and import value of chipped cassava had been higher since 2013 due to manufacturers' capacity expansion of starch and bio-ethanol in Ubon Ratchathani.

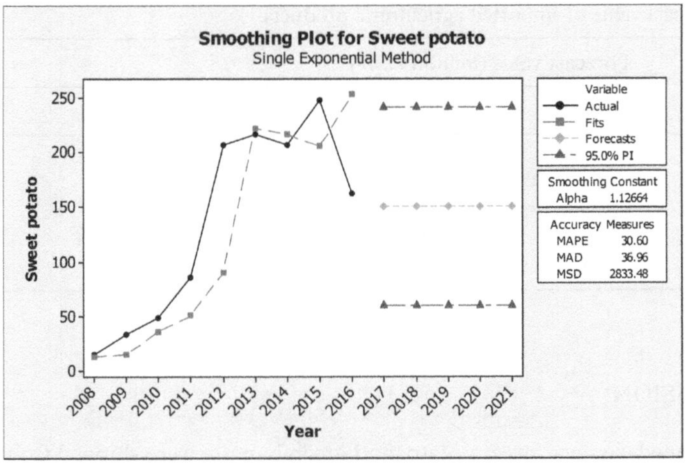

Figure 5. Forecasting import data of sweet potato using exponential smoothing.

Table 5. Forecasting model and import value of green banana.

Forecasting model	Forecasting import value of green banana for 2021	MAD	MSD	MAPE	Trends
Simple linear regression	121.97	21.23	485.63	26.59	Yt = 63.8 + 4.16*t
Exponential smoothing	121.78	22.85	838.44	28.17	$\alpha = 0.8999$
Double exponential smoothing	303.07	80.64	7,059.34	102.37	$\alpha = 1.9149$ $\beta = 0.0788$
Weighted moving average	101.14	28.59	989.40	32.73	WMA = 2

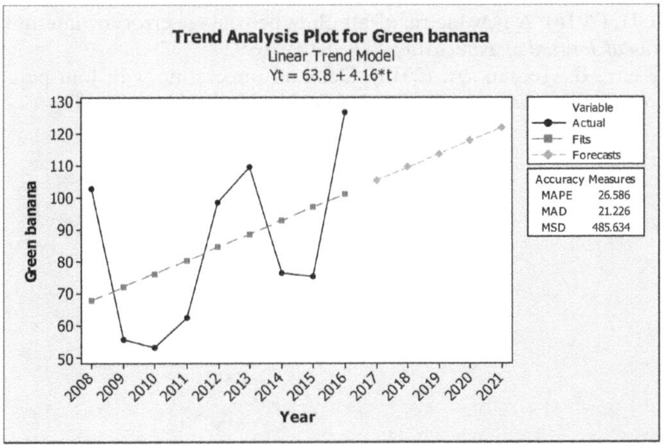

Figure 6. Forecasting import data of green banana using simple linear regression.

Table 6. Forecast value of imported agricultural products.

Fiscal year	Forecast value (million USD)			
	Cabbage	Chipped cassava	Sweet potato	Green banana
2017	567.13	1,521.76	149.93	105.34
2018	620.98	1,933.47	149.93	109.50
2019	674.83	2,345.18	149.93	113.65
2020	728.68	2,756.88	149.93	117.81
2021	782.52	3,168.59	149.93	121.97

4 CONCLUSION

Cabbage, chipped cassava, sweet potato, and green banana were shipped from Laos to Thailand via the Chong-Mek border at an enormous value annually since 2008. We aimed to find the most accurate forecasting models for each of these agricultural products using simple linear regression, moving average, exponential smoothing, and double exponential smoothing. The results indicate that simple linear regression had a small forecast error for cabbage, chipped cassava, and green banana, whereas exponential smoothing was the best predictor for sweet potato. After the selection of the reliable forecasting model, we aimed to forecast the import value for each product during 2017–2021, as given in Table 6.

This result can be used as a basis for policy-makers in both Thailand and Laos. From a logistics and supply chain perspective, facilities for the import and export of agricultural products should be improved or refurbished.

We suggest some future research opportunities to 1) evaluate the accuracy of a forecasting model, 2) improve data collection at the customs house, 3) investigate trans-border mobility in terms of import–export amounts, trans-border tourists, and trans-border vehicles.

REFERENCES

Barrow, D.K. (2016). Forecasting intraday call arrivals using the seasonal moving average method. *Journal of Business Research*, 69(12), 6088–6096.

Dudek, G. (2016). Pattern-based local linear regression models for short-term load forecasting. *Electric power systems research*, 32(2016), 139–147.

Kasemset, C. (2016). *Inventory theory and applications deterministic and continuous inventory model.* Chiang Mai: CMS Press.

Kim, S. and Kim, H. (2016). A new metric of absolute percentage error for intermittent demand forecasts. *International Journal of Forecasting*, 32(3), 669–679.

Tratar F.L., Mojškerc, B., Toman., A. (2016). Demand forecasting with four-parameter exponential smoothing. *International Journal of Production Economics*, 181(A), 162–173.

Urban Studies: Border and Mobility – Kerr et al. (Eds)
 ISBN 978-1-138-58034-3

Social class representation: FoodTruck Culinary Surabaya community

Rahma Rahartika
Universitas Airlangga, Surabaya, Indonesia

ABSTRACT: This study analyses the representation of social classes in the community called FoodTruck Culinary Surabaya (Focus). Social class representation can be assessed by observing the changes in the concept of selling. This food truck community is based on a Western (or global) concept of takeaway, which is more efficient than *cangkruk*. Therefore, it attempts to adapt to local customs while integrating with the global concept. The combination of the global and local concepts is evident in the selected menu of this community, which is a Western or Japanese-style menu adapted to the taste of the people of Surabaya. In addition, the takeaway concept must be adjusted to the local concept of Surabaya (i.e. *cangkruk*), which must provide benches and rental places. This study aims to identify the phenomenon by which local concepts can be integrated with global concepts to attract upper-middle-class consumers. To analyse the phenomenon of this community, the study uses a production approach based on the spatial theory of Henri Lefebvre, which presents spatial practices, representations of space and representational spaces.

The findings of this study provide the evidence of combined global and local marketing strategies, which helps understand the class border of FoodTruck Culinary Surabaya Community. The key conspicuous consumers are the middle to upper classes. Indeed, the food truck community combines the Western (global) concept with the local concept of Surabaya, *cangkruk*.

Keywords: Community FoodTruck Culinary Surabaya, Representation, Social class

1 INTRODUCTION

A culinary business is a huge opportunity in the era of start-up company growth. According to the data from Parama Indonesia, the Indonesian culinary sector has grown by an average of 7–14 percent per year over the last five years (CNNIndonesia.com). The main reason for this phenomenon is that workers who have to work until late at night more often buy food rather than eat food at home. According to the Indonesian Creative Economy (Bekraf), the culinary business makes the largest contribution to the creative economy sector. Three sectors contribute to 30 percent of the creative economy, namely culinary, fashion and crafts. Among these sectors, culinary contributes up to 34 percent. In addition, because per capita income levels have risen, the lifestyle of consumers has also transformed. People no longer eat because of hunger, but due to lifestyle habits (CNNIndonesia.com).

Increasing business opportunities in the culinary field make businesses compete to trigger the development of the creative economy. One of the culinary businesses in Indonesia is selling from a food truck. A food truck is a mobile truck/car modified into a "restaurant". Food trucks first appeared in the United States in the early 17th century, while a new modern food truck was popular in the 19th century after the Civil War in Texas. In 1866, a US Army officer named Charles Goodnight prepared a food truck for a big ride, whose goal was to put all cooking utensils and groceries into it and to immediately cook food when a soldier was hungry (indotrucker.com). Currently, a food truck is equipped with cooking utensils and a mini bar according to the purposes of small owners. The majority of food trucks in Indonesia hawk food and drinks in small or large quantities. Common places that are frequently visited by many people are those on a side street, in a car park or at a particular event.

Communities often found in big cities such as Surabaya gather because of recreational interests, ideas and goals. One of the emerging communities in Surabaya is FoodTruck Culinary Surabaya. Unlike other communities which generally gather in a park, this community prefers to be headquartered at West Surabaya Graha Fair Ground, which is an elite neighbourhood. In addition, the cars are not just any car but with fantastic modification costs, in order to attract consumers.

This community chooses Western menus or Western food that has been adapted to the taste of Surabayan food. The price is quite fantastic for street food. Consumers of roadside food are from the upper middle class. As Belasco (2008: 1) says, food can identify who a person is, which group a person belongs to, and allows us to be what we want. This is also true for food trucks; that is, food sold in the food truck can identify who the consumers are, which class they target and what they show through the concept of selling by a food truck. Another interesting aspect of this community is that there is a shift from the more efficient Western (or global) concept of takeaway to the concept of *cangkruk*. Therefore, this study examines the class border of this community and the mobility of the global concept of local food sold by this community associated with globalisation by applying Henri Lefebvre's production of space theory.

This study used a qualitative research method based on the production of space theory proposed by Henri Lefebvre, in order to present the spatial practices, representations of space and space representation of this community. Production space is able to explain the existence of a new business space commodified by locality, which occurs in a community such as FoodTruck Culinary Surabaya. This research was conducted precisely in the area of West Surabaya, namely the Graha FairGround which is their headquarters from where they sell food, and included observation and interview of some chairmen and members of FoodTruck Culinary Surabaya.

2 DISCUSSION

2.1 *Community as identity*

Humans are social beings who cannot live without others. Interaction is not only common in the family context but also in the social context. In other words, needs and similar hobbies can lead to greater social interaction, called the community. In his book, Cohen (1985: 71) explains that community is the result of a symbolic construction of a set system of values, norms and moral codes that give a sense of identity within an overall limit of its members.

Community is formed not only based on the geographical conditions of the area or where people live, but also due to a common purpose or hobby that people enjoy. As Giddens (in Barker, 2008) says, identity is a project, that is, it is the creation of those who are in a process of thinking that has been overshadowed by the experience in the past and what we expect in the future. This means that the identity is created and explored by an individual or group. As the experience of the food truck community implied, their hobby of modifying cars was combined with selling food in the form of a food truck, which influenced the choices they made that gave the community a new identity. Identity can distinguish one community from another while uniting one individual with another similar individual.

2.2 *The production of space*

As Lefebvre (1991) points out, "space is a social product", meaning that space is always a social space formed by social action, both individually and collectively. This action gives "meaning" to how a spatial space is conceived by those who fill out and turn these spaces. Production of social space with regard to how the spatial practices are realised through the perception of the environment is built over a network that integrates social activities such as work, personal life and free time (Pamungkas, 2016).

The room where the "space" exposes oneself should be used so that both space and commodities have value. This value is closely related to urbanisation, which itself is defined through the ways in which capitalism is created in the space arena, competing commodity production with various interests.

FoodTruck Culinary Surabaya Community also produces space to show their existence. This community uses food trucks to hawk their food. All trading activities are carried out in the car, even the cooking and washing of kitchen utensils. This community does not consider food truck as a restaurant or a "street", but rather as a new space. A food truck can move anywhere like a street vendor, but such a restaurant has a fixed price. This is because the menus provided by the community food truck are not a rigid five-style menu, but are oriented towards a Western-style cuisine combined with the taste of Surabayan food.

2.3 *Food as identity*

As Belasco (2008: 1) says, food can identify who a person is, which group a person belongs to, and allows us to be what we want. In addition, Caplan (1997) also points out that the food consumed shows gender, age, ethnicity and social class. For example, a Balinese Hindu will stay away from food associated with cows, and a Muslim will avoid drinking alcoholic beverages. This will enable us to find which religion a person follows. The selection of food can also show one's class: for example, the upper class would choose to eat in reputable and reliable restaurants, while the lower class would be happy enough to eat at the roadside.

Identity is also manifest through food selection. People's choices are in their hands in terms of where they want to spend the money for a snack with a high nominal rate or whether to choose snacks with large portions. Food can also identify race, not only individually but also collectively. For example, Javanese preferred sweet foods, Madurese favoured salty foods and Padang liked spicy food, thereby indirectly reflecting their mindset. Therefore, food can represent an individual or collective identity.

The food served in this community is mainly Western-style snacks combined with the taste of Surabayan food. This community also chooses where to sell, namely in the area of West Surabaya called Graha Fairground, a well-known elite area. The impact of colonisation is still embedded in the Indonesian society. In fact, they still assume that the West is cool and classy so that they can often enjoy it from the top. Based on this idea, this community dared to show its identity using a new selling method by selecting target customers from the upper-middle class.

2.4 *Spatial practice in Foodtruck Culinary Surabaya Community*

This community used a new selling method by using food trucks. They do not want to be known as street vendors because they rent a place and set up benches to adapt to the culture of Surabaya's community called *cangkruk*. They do not want to be called a restaurant either, even though they rent a place and provide benches. Moreover, the way they adapt themselves to what they are selling is by adapting the Western-style food and beverages and flavours to the taste of their customers (i.e. people of Surabaya).

This community also invites debate about the taxes they are required to pay for such an open business. The food truck cannot be categorised as street vendors nor called a restaurant. Thus, food truck, as described by Munir (2017), is a new business model in the business of food and/or drink which uses a car to do its business. In the case of this new venture, the Law on Regional Tax and Retribution does not provide a tax base for food truck restaurants since they can operate their nomadic business. However, another study has shown that food trucks can be included in the criteria of "like" as described in the Law on Regional Tax and Retribution. Therefore, it can be interpreted that the food truck has taxpayers and a similar tax object to the restaurant tax described in the Law on Regional Tax and Retribution.

2.5 *Representation of space in Foodtruck Culinary Surabaya Community*

Compared with other communities who would gather in a park, FoodTruck Culinary Surabaya Community selects an elite area (Graha Fairground) where they run their business and rent a place. Currently, the food truck community follows various lifestyle events such as bazaars or urban lifestyle events which normally attract many youngsters. Youngsters in Surabaya prefer to spend their time in cafes. Therefore, they always actively participate in these events. Although they target young people, the price of food and beverages is classified as above average so that their customers need to spend more money. The high price is due to the car modification cost, thereby increasing the consumer appeal in favour of a social media lifestyle.

2.6 *Representational space in Foodtruck Culinary Surabaya Community*

Food trucks represent a community that is neither a restaurant nor a street. This community consists of several members which competitively run the business. This is because what they sell is different with creative plating of each member for Western foods and drinks. Moreover, they modify cars to attract consumers' attention. This community aims to target customers by using menu strategies and location choice.

3 CONCLUSION

FoodTruck Culinary Surabaya Community can be identified by their class border-based selection of target consumers. They hawk in the elite area of West Surabaya by selling Western food combined with the taste of Surabayan food. Previously, this community tried to sell on the roadside and then switched to rent a place and set up benches because they had to adapt to the culture of the society of Surabaya, *cangkruk*. They prefer not to be called street vendors or restaurant employers, but food truck employers as food trucks can basically be moved as far as they wish. The food truck community creates its own space, which is represented in each food truck. This representation influences this community's target classes because they modify their cars with a fantastic price and sell the food as in standard-class restaurants.

REFERENCES

Barker, Chris. 2008. *Cultural Studies: Teori dan Praktik*. Yogyakarta: Kreasiwacana.

Belasco, Warren. 2008. *Food: The Key Concept*. Oxford: Berg Publishers.

Bramantio. (2013). Sastra dan Kulinari: Evolusi Gastronomi ke Gastrosofi dalam Tiga Cerpen Indonesia 2. *Jentera, 2*(1). Retrieved from http://ojs.badanbahasa.kemdikbud.go.id/jurnal/index.php/jentera/article/view/391.

Cohen, A.P. 1985. *The Symbolic Construction of Community*. London: Routledge.

Elsy, P. (2012). Bab VI: Dinamika Budaya Masyarakat Kota Surabaya (Anggota TIM). In P. Elsy, *Dinamika & Perubahan Sosial Budaya Kota-kota di Jawa Timur.* Arr-Ruzz Media.

Juniman, OuputTripeni. 2017. Gaya Hidup Masyarakat Menjadikan Bisnis Kuliner.

Lefebvre, Henri. 1991. *The Production of Space*. ebook. Georgetown University Press: New York.

Menjanjikan. *CNN Indonesia.* Retrieved October 20, 2017 from m.cnnindonesia.com/gaya-hidup-masyarakat-menjadikan-bisnis-kuliner-menjanjikan/.

Menurut Henri Lefebvre. *Indo Progress.* Retrieved June 19, 2017 from https://indoprogress.com/2016/01/produksi-ruang-dan-revolusi-kaum-urban-menurut-henri-lefebvre/.

Munir, Ahmad & Andini, Dea Arifka. 2017. "Pengaturan Pajak Restoran atas Foodtruck Menurut Undang-Undang Nomor 28 Tahun 2009 Tentang Pajak Daerah dan Retribusi Daerah". *Mimbar Yustitia* 1(1).

Murdock, G dan P. Golding. 1977. "Capitalism, Communications and Class Relations" in J. Curran, M. Gurevith, and J. Woollacott (eds.), *Mass Communications and Society.* London: Edward Arnold.

Pamungkas, Arie Setyaningrum. 2016. "Produksi Ruang dan Revolusi Kaum Urban".

Supriyadi, H. (2010). *Buku Ajar Ilmu Sosial Budaya Dasar.* Surabaya: Universitas Airlangga.

Urban Studies: Border and Mobility – Kerr et al. (Eds)
© 2019 Taylor & Francis Group, London, ISBN 978-1-138-58034-3

The Bawean ethnic language: Attitude and diglossic community culture

S.W.B. Utami
Universitas Airlangga, Surabaya, Indonesia

ABSTRACT: The Bawean ethnic group is known as the "Boyans". The Baweans have been identified as a group with a deep ethnic consciousness in terms of their language and culture. The purpose of this study was to understand the characteristics of the language and culture of the Bawean ethnic group. This study used a qualitative approach, particularly the ethnography of communication inquiry. The Bawean society is enriched with multi–ethnic groups such as Javanese, Madurese, Bugis, Mandailing, Banjar and Palembang. *Merantau* (migrating) has been a unique local custom of the families living on this island. Historically, early settlers were immigrants from different regions or islands. The adoption of other cultures and languages shows a typical acculturation pattern, which is one of the distinct characteristics of the Bawean society. Therefore, the Baweans regard themselves as a unique society and consider their language to be different from the Madurese language.

Keywords: Migrating, Diglossic, Boyan, Language and Cultural Attitude

1 INTRODUCTION

Bawean Island is one of the small islands in Indonesia, located about 150 km north of Java Island, which can be reached within 3.5 hours by a speedboat from the Gresik port. The Bawean local government is part of Gresik regency.

Bawean people are known to be *perantau* (travellers). More specifically, *perantau* refers to the custom of migrating. Historically, the majority of the Bawean residents were immigrants from various regions in Indonesia. This custom of migrating has endured throughout generations. Consequently, many of its inhabitants travel to other cities, or even to other countries, to work or make a living. This phenomenon—that is, migration to other countries such as Malaysia, Singapore, Vietnam, Thailand and Christmas Island in Australia –has been ongoing since the 18th century (Vredenbergt, 1990; Usman, 1996).

Many Baweans who work in other countries are often identified by different ethnic labels. For instance, *Orang Boyan* is the name for Bawean people settled in Malaysia and Singapore. In both countries, the Baweans live in an area known as the Boyanese village, possessing dual citizenship.

Apart from the migrating and mixed marriage tradition with other ethnic groups, Bawean Island's geographical position as a transit or stopover island has also created a typical sociocultural impact. This is represented in the ethnic and cultural varieties as well as language and speech varieties, forming a unique acculturation pattern.

The history of power and religion in Baweans shows that Javanese and Sumenep (Madurese) are two major ethnic groups on Bawean Island. However, after the spread of Islam, the Madurese ethnic group played an important role (Kartono, 2003: 1) in building religious systems in the Bawean society. Islam has been the only religion practised in Bawean. One of the premises for this assertion is that many Baweans have originated from a Madurese ethnic background. With regard to language, as the Bawean language is similar to the Madurese language, it is often mistakenly referred to as Madurese. The Baweans are often thought of as Madurese by people who live outside Bawean Island. The long migration of the Baweans to foreign locations does not make them forget their native land. This claim is enshrined in the local Bawean proverb that the Baweans who were born on the island will travel but will return and wish to be buried only on Bawean Island.

Furthermore, Bawean people are multi-dialectical and multilingual. This is due to several factors. First, the island's geographical position as a transit or stopover region for many ethnic groups has built a multicultural and multilingual society. Second, the community's high mobility has amplified the variety of cultures and languages on Bawean Islands. Third, access to education and communication media has created an opportunity to be exposed to diverse languages. Such exposure and contact, particularly in relation to language, is defined as diglossia in bilingualism.

These circumstances have fostered the Baweans to become a diglossic community. As Fishman (1972: 92) stated, "*diglossia exists not only in multilingual societies which officially recognize several 'languages', and not only in societies which employ separate dialects, registers, or functionally differentiated language varieties of whatever kind*". In this regard, apart from their local language, the Baweans attach importance to mastering other languages with all their functions and use. It is therefore interesting to investigate the Bawean people's attitude towards their local language, particularly in relation to their speech variety, and their loyalty towards their local language.

This study was conducted in stages, which included observations, questionnaires and interviews about the evaluation of the growing culture and language. Respondents in this study were important authorities and figures such as community figures, cultural figures and village administration staff.

2 DISCUSSION

2.1 *Stopover island*

Bawean refers to the word *pawean* that originated from the word *pawiwahan* which in turn is derived from the old Javanese word *wiwoho* (meeting or encounter). This name projects Bawean Island as a meeting place for many ethnic groups who then resided in Bawean.

Among the islands (Java, Madura, Masalembo and Makasar), the geographical location of Bawean Island has made an open place for both stopovers and settlement. The village Pajinggahan in the Tambak subdistrict was apparently named after boat and ship "transit" from other islands. Some remains of old seaports on Bawean Island can be viewed even today in Kumalasa village.

Its existence as a transit island is also reflected in the existing Bawean legends. According to Bascom (in Dundes [ed.], 1984:5), a "legend is a spoken narrative prose whose story is believed by the community as a true fact". In Bawean legends, the important messengers/figures of cultural teaching generally came from other islands. For example, the Jujuk Campa legend in Bawean tales tells the story about an eminent figure from Campa, Cambodia, which is supported by the remains found in Kumulasa village such as kris, vessel, *sekedub* and graves. According to other cultural findings, Diponggo village is the only village in Bawean that tend to assimilate the Javanese language.

As a transit island, Bawean is highly influenced by cultures from other areas or islands. As Al-Attas (2009) describes, during the colonial period in the 18th century, the Minang, Bugis and other seafaring communities used Bawean Island as their stopover. When appointed as the KPM (Konenklijke Paketwaart Maatschapi) sailing agent, they became influential traders and dominated trade and the economy in the Bawean areas. At that time, most of the Minang ethnic groups migrated to Bawean Island. The strong assimilation of the local culture has led to the non-existence of their unique identities in the Bawean society.

2.2 *Migrating culture*

Despite the availability of jobs in Bawean, the migrating tradition is still upheld by Bawean families. This migrating custom to Malaysia and Singapore began in the 18th century, with many Bawean people becoming citizens of these countries (Usman, 1996: 66).

As Vredenbregt (1964:129–139) explains, in 1849, some Bawean people who settled in the area then called *Malaka* were known as the "Boyans". According to the Singapore statistics, there were a total of 763 Boyan people with 720 males and 43 females. Their number had increased in 1957 and reached a total of 22,167, with 11,580 males and 10,587 females.

Migrating to foreign lands has been the cultural tradition of Bawean. Males (and now more females) are expected to go out of Bawean Island at least once during their lifetime in order to learn by their travel experience. The migrating tradition is reflected in the following proverb:

> *"Ajjɔ moka?-moka?samper? mɔnghitaangarsailange?,*
> *Ben ajjɔ ngokerlange? mɔnghitangoasaeelmolahir ben elmobatin".*
> "Do not take off a woman's sarong before carving the sky, and do not carve the sky before mastering physical and mental knowledge"
> (Interpreted as "Do not get married before going to other lands for a living, and do not travel before mastering self-defence and the Quran").

The word *samper*, in East Javanese dialect *sewek*, indicates a piece of sarong used by women to wrap their lower bodies, from the navel to the ankle. The unwrapping act, such as taking off a woman's sarong, symbolically means to marry a woman, while "carving the sky" expresses gaining experiences in other lands.

Migrating to other lands is a prerequisite for men's marriage. Symbolically, "carving the sky" is an experience that poses a number of strong challenges, for which training in family responsibility of being the head of the family is compulsorily undertaken. Apart from the teaching of migrating to other lands (*merantau*), the knowledge of self-defence and moral teaching are necessary to live safely in other lands as well as to prepare them to be good travellers/settlers.

The "migrating" characteristic of the Bawean ethnic group enables the adoption of other cultures. Multilingualism is welcome in Bawean communities. Many residents are bilingual or multilingual and thus can speak one or two languages other than their mother tongue. Mackey (1962) explains that bilingualism is a practice of using "language" by shifting from one language to another (p. 12). This is demonstrated through the practice of "code switching" among the Baweans. Hence, they chat in the local ethnic language, and then use Malay when meeting with neighbours from other lands and speak in Indonesian on formal occasions. The choice of language domains shows that the Baweans are a diglossic community.

2.3 *Language attitude and cultural identity*

Attitude involves a psychological phenomenon manifested in behaviour or acts based on viewpoints, beliefs or opinions as a reaction to something or some events. According to Anderson (1974), an attitude may refer to language and non-language (political, social, religious) attitude (p. 37). The non-language attitude is related to cultural identity.

There are two types of ethnic attitude: "*revivalism*" is an ethnic attitude towards cultural identity and "*revitalism*" is an attitude related to the wish to access other cultures for the purpose of "catching up with" them (Alqadrie, 2008:197–198).

The results of data analysis indicate that the Baweans' attitude towards their cultural identity is reflected in their reluctance to be identified as the Maduran ethnic group. This is perhaps considered as a unique language attitude given that their language is similar to Madurese, as even the Bawean vocabulary carries the same or similar diction and meaning to Madurese. Some people regard the Bawean language as a dialect of Madurese.

The similarity between Bawean and Madurese is reflected in the use of aspirated sounds [dˀ],[bˀ],[gˀ], [jˀ] → (dh, bh, gh, jh) as follows:

dˀâra → dove
bâdˀhe → place, location
robˀu → collapse
bˀiru → blue
gˀule → sugar
gˀegˀer → fall

j'erre → learn one's lesson, be cured
j'uel → sell

The morphological structure is reflected in the affixes in Bawean /N/,/a-/, /é-/, /ma-/, /-a/, /-na/, /é-a/ that show similar functions to Madurese.

For example:

(N-)	+énom → ngénom	= drink
(é-)	+j'uâl → éj'uâl	= be sold
(a-)	+j'âlân → aj'âlân	= walk
(ma-)	+târros → matârros	= continue
méllé	+(-a) → mélléa	= will buy
d'âlâm	+(-na) → d'ələmna	= the house
(é-)	+pélé + (-a) → épéléa	= will choose

Its similarity with Madurese is further supported by the similarity of vocabulary. Lexicostatistically, 80% of Bawean base words present similarity to Madurese.

As Baker (1992) states, the attitude of the Baweans can be elaborated based on instrumental and integrative motivations. Instrumental motivation aims to achieve social acknowledgement (p. 32). Integrative motivation includes social and interpersonal attributes related to their ethnic identification and activities living together with other ethnicities (p. 32). Such an identity claim is evident in the attitude of the Baweans, as shown in this study, and is apparent within the communities known as the "Boyans" who live in other countries such as Malaysia and Singapore.

There is indeed reluctance on the part the Baweans to be identified as part of Maduran ethnicity. They consider themselves different from the Madurese ethnic group. Based on the interviews and questionnaires, the Baweans in this study had their origin in Madura. In addition, they generally state that they came from Pamekasan or Sumenep. Both areas are "stereotypically" identified as a better, "refined" or a more educated part of Madura. In this context, the Baweans symbolically put priority on virtues such as refinement, proper conduct and honesty, non-violence, obeying the law and harmony.

From the Bawean perspective, the Madurese are known to be strong-willed, rough and dishonest. The Madurese *carok* tradition and their habit of carrying sharp weapons, to kill or steal, are regarded as despicable and thus shunned by the Baweans. The Baweans living in other lands are expected to demonstrate good citizenship behaviours, show honesty and avoid commotion. They would rather shake hands than attack with a sharp weapon when there is a disagreement. This means that they want to resolve disagreement through peace.

The Baweans show themselves to be a people with honesty, enthusiasm and zeal and good behaviours. Such an ethnic identity is shown by the Baweans to distinguish themselves from the Madurese. Ethnic identity describes the attitude of positioning abstract ideas in a specific decision dimension (Baker, 1992: 11). As Azjan (1988) argues, attitude is the tendency to give favourable or unfavourable responses to objects, people or events (Baker, 1922:11).

Language attitude has been a feature that characterises ethnic identity that is commonly known as ethnolinguistic identity. From the perspective of ethnolinguistic identity, the Baweans call themselves Bawean people, similarly to when someone calls himself Indonesian or Japanese to show language ownership. As described by Fishman (1991), ethnolinguistic identity is closely related to ethnic identity (p. 22). In this regard, many elements in culture are verbally spelled out in songs, tales, proverbs and all non-material cultures.

Despite the similarity of language lexicon and structure, and the acknowledgement of their Madurese origin (from Sumenep Madura), the Baweans dislike to be identified as part of Madurese ethnicity. They consider themselves different from the Madurese cultural and ethnic group.

The mention of Sumenep or Pamekasan (Madura) as the place of origin symbolically shows that the Baweans are characterised by "refinement". This shows their strong ethnicity attitude and special cultural identity. In other words, they can be identified as a unique Bawean ethnic group from their attitude and behaviour.

Table 1. Language choice tendencies and domains of language use.

Frequency of language use	Domains of language use				
	Bawean	Indonesian	Indonesian – Bawean	Malay	Arabic
Often	at home in mosque in rice field in the market to neighbours	at school in office on Instagram on Facebook in official meetings	in office while texting on Instagram on Facebook	while texting to family in other countries	while reciting the Quran
Sometimes	to newcomers from Madura	to newcomers	in official meetings	to neighbours from other places	while preaching
Never	to newcomers in official meetings	at home to neighbours	at home to neighbours	to newcomers	in office

As Lebar (1972) explains, the Baweans are also identified as the Babian or Phebian people. Etymologically, the word /*phebian*/ has its origin in /*pawiwahan*/ which represents an assimilation process. In this sense, the Baweans have assimilated through the encounters with diverse people from Madura, Java, Bugis, Minang and Palembang for hundreds of years. This assimilation is a social process that has resulted from the meeting and intense contact between different cultures which finally formed a typically mixed culture (Koentjaraningrat, 1985:254). This factor may explain the reluctance on the part of the Baweans to be identified as Madurese.

The Bawean language has several speech varieties that can be grouped based on the ethnic origins and village names. For example, Daun variety refers to Daun village, which is a mixture of Bugis and Palembang ethnic groups. Pater Selamat variety comprises the villages of Bulu-Lanjang, Sungai Rujing, SawahMulya and Kota Kusuma. Sungai Teluk variety is associated with a mixture of Madurese and Bugis ethnic groups. Kumulasa variety in Kumulasa village can be traced back to the descendants of Champaland. Diponggo variety is closely related to Javanese ethnicity. These associations are often expressed in funny remarks among the villagers in comments such as "during language divisions, all the villagers were listening, except for the Diponggo people, because they fell asleep".

Dialect speakers have no difficulty in interaction. Therefore, a bidialectal language pattern is common in Bawean. In this sense, apart from using speech varieties within the local Bawean dialect in their own village, they also understand dialectal speech varieties from other villages.

The Baweans are not only bidialectal but also bilingual. This is due to the fact that Bawean people migrate to other lands. The long migrating pattern of the Baweans has indirectly placed them in contact with languages other than their mother tongue. This custom has allowed strong influences on their cultural behaviour. This is reflected in the language atmosphere on Bawean Island. Malay, Bawean and Indonesian languages have become key languages of communication.

Technology advancement enjoyed by Bawean communities and improvement in transportation facilities have influenced the high intensity of contact among communities/ethnicities. High mobility to other areas and good communication facilities enable the Bawean people to become bilingual, so that they are able to know and use other languages beside their mother tongue. This is outlined in the following language use.

The domain data show that the Baweans are a diglossic bilingual community. They generally understand that Indonesian possess a more important function than the Malay or Bawean language.

3 CONCLUSION

The Baweans generally feel that they have the same identity in terms of tradition and religion. The migrating tradition has greatly influenced ethnic solidarity and consciousness of the community oriented towards group identity or identification of a people known as the "Boyans".

For the majority of the ethnic groups, it might be rather easier to claim a distinctive language and cultural identity. For example, Javanese ethnicity and Madurese ethnicity can claim their ethnic identity simply from the observation of their language and culture. However, the circumstance is quite different for the Baweans. As a major ethnic group, the Baweans' attitude and representation of themselves ethnically have not reached a similar level of spread in language and culture. The migrating tradition and influences from other languages and cultures may have an impact on the susceptibility of their presently shifting identities. Nevertheless, ethnic consciousness of the Baweans has motivated their desire to regard their local language guidelines as substantially indispensable in order to present themselves as an ethnic group with unique characteristics.

As a diglossic community, the Bawean ethnic group has special attributes. Cultural adoption has enabled the assimilation and acculturation processes to form their cultural identity with ethnic uniqueness. Such acculturation and assimilation have formed a cultural and language identity that is positively valued as their special ethnic characteristics.

REFERENCES

Al-Attas, S.M.N. 2009. *Islamic on Malayu's History and Culture*. Bangi: Kebangsaan Malaysia University.

Alqadrie, S.I. (2008). Identitas Budaya, Identifikasi Etnis dan Keragaman, Kesadaran Etnis dan Hipotesis ekerasan 2020-an. *Industri Budaya Budaya Industri*, Kongres kebudayaan Indonesia 2008. Kenedi Nurhan, ed. 197–198. Kementerian Kebudayaan dan Pariwisata Republik Indonesia.

Baker, C. 1992. *Attitudes and Language*. Adelaide: Multilingual Matters Ltd.

Dundes, A. ed. 1984. *Sacred Narrative*. California: University of California Press.

Fishman, J.A. 1972. *The Sociology of Language*. Rowly-Masshusett: Newbury House.

_______. 1991. *Reversing Language Shift*. Clevedon, UK: Multilingual Matters.

Kartono, Drajat Tri. 2003. Orang Bawean dan Boyan: Perubahan dan Pembentukan Identitas Masyarakat Bawean. *MenciptaBawean, AntologiGagasan Orang-orang Boyan*. 22. Gresik: Boyan Publishing.

Mackey, W.P. 1962. The Description of Bilingualism, in J.A. Fishman (ed.). *Sociolinguistics: A Brief Introduction*. Rowly-Massachusett: Newbury Jouse.

Qushwandhi, Dhiyauddin. 2008. *WaliyahZainab: Puteri Pewaris Syeikh Siti Jenar, Sejarah Agama dan Peradaban Islam di Pulau Bawean*. Gresik: Yayasan Waliyah Zainab Diponggo.

Usman, Z. 1992. *Kisah-Kisah Pulau Puteri (Pulau Bawean)*. Bawean: Next Generation Foundation Perwakilan Bawean.

Vredenbregt, J. 1990. *Bawean dan Islam*. Translated from *Baweanners in hub Moederland en in Singapore*. A.B. Lapian. Jakarta: Indonesian Netherlands Cooperation in Islamic Studies (INIS). 1990. Series INIS. Book VII.

_______.1964. Bawean Migrations Some Preliminary Nates, Bijdragen tot de Taal-, *Land-en Volkenkunde 120*, (1) Leiden: 109–139.

Urban Studies: Border and Mobility – Kerr et al. (Eds)
© 2019 Taylor & Francis Group, London, ISBN 978-1-138-58034-3

Multi-ethnic and religious conflicts in media reported by international online media: http://www.straitstimes.com/asia/se-asia/statue-of-chinese-god-guan-yu-stokes-tension-in-indonesia

P. Wibawanto
Universitas Airlangga, Surabaya, Indonesia

ABSTRACT: An English-language online media site published a news article about multi-ethnic and religious conflicts in Tuban, Indonesia. How did the conflict build on a multi-ethnic and religious background, marking the erection of the Kongco Kwan Kong statue in the backyard of Kwan Sing Bio's Temple in Tuban? This article will answer the following questions: (1) How intolerance marks multi-ethnic and religious conflicts over the erection of the Kongco Kwan Kong statue? (2) How ethnic and religious backgrounds are characterized in multi-ethnic and religious conflicts in Indonesia? We will use the theory of critical discourse analysis by Theodor W Adorno in revealing the ethnic conflict that marks the construction of the Kwan Kong statue, as reported in the news articles. The results indicated that ethnic and religious background marks the erection of the statue of General Kongco Kwan Kong. In addition, the erection of a statue containing religious symbols triggers religious tension in Indonesia. This article shows that ethnic and religious issues can easily develop and thus spark conflict in Indonesia. Nationalists and religious leaders in the country put political pressure on the Indonesian central government to immediately demolish the statue of the Chinese war general.

Keywords: Statue, Ethnicity, Intolerance, Conflict

1 INTRODUCTION

In July to August 2017, the virtual world was busy discussing the erection of the statue of a General or God from China, namely Kwan Kong Kwan Sing Tee Koen in the backyard of the Kwan Sing Bio Tri Dharma Worship Place, Tuban, Indonesia. The statue of the god was built on the complex of the Kwan Sing Bio Tri Dharma Worship Place, Tuban, East Java, Indonesia. A gigantic statue, more than 30 metres high, stands firmly in the area behind the temple. Moreover, the statue attained the record of MURI (Indonesian National Museum for Records) as the tallest sculpture in a temple complex in Southeast Asia, which was inaugurated by the Chairman of MPR RI (Indonesian Congress) Zulkifli Hasan on 16 July. The exact height of the glorious statue of Kongco Kwan Sing Tee Koen or Kongco Kwan Kong in the Tri Dharma Kwan Sing Bio Worship Place is 30.41 metres. It is 7 metres higher than the statue of Kwan Im Goddess in Pematang Siantar, North Sumatra, which was previously recorded as the tallest statue in Southeast Asia, with a height of 22.8 metres.

The construction of the statue of Kongco Kwan Kong, a warlord who is known for his justice and wisdom and enshrined as a God of wisdom, began in September 2015 and took more than a year, costing approximately Rp. 2.5 billion. This tall and majestic statue is a donation of one adherent of the Tri Dharma as a form of devotion and practice in worship. In addition, as an instrument of worship, the existence of the Kongco Kwan Kong statue that carries the machete was also expected to become one of the icons and attractions for the tourists visiting the Kwan Sing Bio Temple, in addition to being used for worship purpose.

Moreover, from the Java Sea, this statue plays the role of a lighthouse, and thus can be a marker for crews and boats. The erection of the statue in one corner of this 6.5-hectare temple is intended to complement the various unique aspects and other sacred attractions in Kwan Sing Bio Temple, Tuban. It also acts as a new icon in the Kwan Sing Bio Temple through the symbol of a giant crab, which is above the gate of the temple area.

The temple is known to exist for more than 300 years; however, its actual construction began in 1971. Kwan Sing Bio Temple is located on Jl. RE Martadinata, Tuban, and has a story believed to be the guidance of Kongco Kwan Sing Tee Koen, the God of wisdom for the people of Tri Dharma. This temple has many unique aspects. It is also the only temple in Indonesia located in the seashore, which is also believed to be the best place of worship. It is believed that anyone who comes here will have his or her wish granted. In addition, it is the only temple in Indonesia that uses the symbol of a crab as its main ornament. In general, temples in Indonesia use a dragon symbol.

The statue of Kong Co Kwan Sing Tee Koen in the Kwan Sing Bio Temple, Tuban, was inaugurated on 16 July 2017. The inauguration of the tallest sculpture in Southeast Asia was conducted by the Chairman of the People's Consultative Assembly of the Republic of Indonesia (MPR RI) Zulkifli Hasan, which was opened with a Barongsai art performance. The north-facing statue looks more dashing with the knightly uniform it wears and the sword it wields. After inauguration, the administrator of the Tri Dharma Kwan Sing Bio Worship Place, Tuban, received an award from the Museum Rekor Indonesia (MURI) for its achievement in making the tallest statue in Southeast Asia in a temple complex.

However, after having been viral on social media since the inauguration, many protests from social media users forced the statue of the God of war to be finally closed. The statue was wrapped in a giant white cloth using a heavy crane. The reason for the closure of the statue was that it did not have a building permit from the local government. Nevertheless, the statue of the Chinese war God Kwan Sing Tee Koen, standing firmly in the backyard of the Kwan Sing Bio Temple, is still visited by many people. The visitors come from all over the country, from Jakarta, Kalimantan, to Aceh. The visitors generally wish to have a closer look at the statue of Kwan Sing Tee Koen.

However, the beauty of the statue can no longer be enjoyed. Problems related to this started from the protests on social media. Finally, the local government and the Kwan Sing Bio Temple agreed to cover the statue so that it is not visible from any angle. The covering was conducted by the vertical team Regional Disaster Management Agency (BPBD) of Tuban Regency by wrapping a giant white cloth around the statue. Even though the wrapping process was planned to be completed in 1 day, the gusts of wind made it difficult for the team to complete the process, and hence took 2 days, and the team worked hard to complete the thorough cloth installation.

1.1 *Geographical location of Tuban*

The total area of Tuban Regency is 183.994.561 Ha, and its sea area is 22,068 km^2. The geographical coordinates of Tuban Regency are 111° 30' – 112° 35 east and 6° 40' – 7° 18' south. The length of the north coast of Tuban is 65 km. The altitude of Tuban Regency ranges from 0 to 500 m a.s.l. Most of the Tuban Regency has a dry climate with varied conditions from dry to very dry, encompassing more than 19 districts, while the wet climate encompasses only one district. Tuban Regency is located on the northern coastline and on the Northern Karst Mountain. The Northern Karst Mountain in Tuban extends from Jatirogo District to Widang District and from Merakurak District to Soko District. Meanwhile, the area of the sea lies in five districts, namely Bancar, Tambakboyo, Jenu, Tuban, and Palang Districts. Tuban Regency is located at the north end and the westernmost part of East Java Province, directly on the border of East Java and Central Java, or between Tuban Regency, East Java, and Rembang Regency, Central Java.

Tuban has a low point of 0 m a.s.l. located in the Pantura Route and a highest point of 500 m located in Grabagan District. Tuban is also traversed by the Solo River, which flows from Solo, Central Java, to Gresik Regency, East Java. Tuban has a majority population of

Javanese and minority ethnicities such as Madurese, Chinese, and Bornean. Tuban's original culture and art are diverse, such as Sandur, Sindir or Langen Tayub, Gemblak, Rodat, and Kentrung Bate, which have become extinct because of the lack of successors. Another culture adopted from the region outside Tuban is Reog, which is commonly found in Jatirogo District.

Tuban is located 90 km west of Surabaya. For a long time, Tuban has been an important area because it is located on the coast and serves as the main port of the Majapahit Empire era. When Islam entered, Tuban became one of the important points. A member of Wali Songo who was also a spreader of Islam in Java, Sunan Bonang or Maulana Makdum Ibrahim, was buried behind the Jami Mosque's town square.

The legendary Kwan Sing Bio Temple is located to the west of Tuban. The place of worship of the people of Tri Dharma, known as the Crab Temple, is one of the landmarks. With a total area of more than 5 hectares, the Kwan Sing Bio Temple is divided into several sections. The first is a place of worship and prayer, which is also the oldest building temple located on the front. Beside it, there is a Chinese learning place, jiamsi clairvoyants, and a secretarial office. In the centre, a hall is built adjacent to a Chinese architectural park, as well as a small lake and a bridge across it. At the very back, which also hosts the most spacious building, stands a multipurpose four-storey building, which is also a place to stay. The crab is a typical symbol of Kwan Sing Bio Temple, Tuban. This is in accordance with the teachings of Tri Dharma, who believes that the crabs were the animals chosen by the gods to protect the people of Tuban. Respect for sea creatures like crabs is demonstrated by the never-ending Confucian, Taoist, and Buddhist pilgrims presenting crabs as an offering to the gods. Another uniqueness related to crabs is visible from the shape of the city of Tuban, which is similar to the form of a crab with two claws. The first claw is located at Tjoe Ling Kiong Temple (Dewi Laut) just north of Tuban Square, while the second claw is located at the Klenteng Kwan Sing Bio Temple in the western suburbs. In the two temples, there are also two freshwater springs. This is unique because both are located in the coastal area. The Kwan Sing Bio Temple is one of the most visited temples in East Java. Not only do the residents of Surabaya and Semarang (as seen by the many cars with H and L license plates) visit this old temple, but also residents from neighbouring countries, such as Singapore, Malaysia, and China, whether with the purpose of traveling or praying.

1.2 *The origin of Tuban's name*

Formerly, Tuban was named Kambang Putih. This was the case from the 11th to the 15th centuries in Chinese writers' writings, from the days of the Southern Song dynasty in the period 1127–1279, and the Yuan dynasty (Mongol) from 1271 to 1368 to the Ming dynasty during 1368–1644. Tuban is referred to as one of the main port cities on the north coast of Java, which is rich and has a large Chinese population. The Chinese call Tuban by the name of Duban or Chumin. The Chinese-Mongolian troops (Tartar Army) in 1292 invaded to attack the eastern part of Java (an event that led to the rise of the Majapahit Empire). They landed on Tuban beach. Also from there, the remnants of the Chinese Army left the island of Java to return to their country (Graaf 1985: 164). However, since the 15th and 16th centuries, medium-sized merchant ships had been forced to drop anchor at sea far enough from the coastline. After the 16th century, finally, the north coast of Tuban became shallow with mud deposits. This geographical situation made the city of Tuban no longer an important port city (Graaf 1985: 163). In fact, to reduce the confusion about the anniversary of Tuban, the Tuban Regent (who was then Drs. Djoewahiri Martoprawiro) set the date of 12 November 1293 as the anniversary of Tuban. A small committee formed by the Tuban Provincial Government provided the reason that the date is set according to the time of Ronggolawe's appointment as Regent of Tuban. Ronggolawe is considered a hero by the people of Tuban and is regarded as the first Regent of Tuban. As with other cities in Java in general, historical sources on the city of Tuban are very difficult to obtain. The written material is full of mixed history and legends, such as the book "Babad Tuban" written by Tan Khoen Swie (1936).

Another version mentions the origin of the city of Tuban from a rock called "Tiban". Tuban is said to be the name of a very famous harbour from the time of the kingdoms to the time of colonisation. However, the name of Tuban city is still confusing today. The number of nicknames given to this city are taken from the characteristics of the city, such as *Tuban Bumi Wali* (Tuban, Land of Saints), because in this region, many Islamic saints, who were regarded as the guardians of Allah, were buried. Another name, *Tuban Bumi Ronggolawe* (Tuban, Land of Ronggolawe), refers to a very famous and brave regent in the Majapahit Empire. Another nickname, *Kota Toak* (City of Palm Wine), is coined from the large number of people who make the fruit of siwalan (*Borassus flabellifer*) into palm wine. Another name, Kota Seribu Goa (*City of A Thousand Caves*), is used because of the presence of thousands of caves in this region. The origin of the city of Tuban is believed by the local community to come from the word "Watu", which means stone, and "Tiban", which means fall. This Watu Tiban can now be found in the backyard of the Kambang Putih Museum, located at Jalan Kartini No 3, Tuban. According to Rony Firman Firdaus, the educational counsellor of Kambang Putih Tuban Museum, Watu Tiban is one of the famous and historic icons in the museum after Kalpataru. Watu Tiban is a kind of stone, which is large and made of andesite or river rock. The stone is located behind the yard of the Kambang Putih Museum. There, there are two "Watu Tiban" of different sizes. It is said that "Watu Tiban" is a historic icon. According to a legend, the two stones are the origin of Tuban's name. However, there are two different story versions regarding the stones. One legend tells that the two stones were brought from Majapahit to Demak by storks. Above the limestone hills, the stones were dropped. The fall of the stones was then called Watu Tiban, which was then abbreviated as Tuban by taking the last syllable of both words.

On the contrary, when viewed from the shape of the stones, it can be concluded that the stones are Yoni. Yoni is the pair of Lingga. The lingga is a symbol of Dewa Siwa, and Yoni is a symbol of Dewi Parwati, his wife. Hindus used to worship Dewa Siwa and Dewi Parwati, and the stone functioned as a medium of worship of Dewa Siwa and Dewi Parwati. The exact date of Watu Tiban's arrival in the backyard of Kambang Putih Museum is unknown. However, since their discovery, the stones have been already in the backyard of Kambang Putih Museum and never moved. The stones are not only an icon, but are also considered to be sacred to a number of societies who consider them mystical. Residents also often come to the museum for praying or pilgrimage with certain purposes.

Another version mentions that the name Tuban is derived from the word *me-tu ban-yune* ("the water came out"). It comes from a story that tells of someone from a foreign land who had supernatural abilities and wanted to challenge Sunan Bonang, carrying his magic books. However, while sailing in the northern sea of the Java island, his boat was tipped over by the waves and stranded on a beach. When he woke, in front of him stood an old man in a white robe. He asked Sunan Bonang who was in front of him. Without a word, Sunan Bonang stuck his rod to the ground, and then water sprouted into the air, along with the books belonging to person who challenged Sunan Bonang. Seeing all of his books drenched in the water released by Sunan Bonang, the person realised that he had faced Sunan Bonang, whom he wanted to challenge. Seeing that Sunan Bonang was so powerful, finally the person refrained from challenging him, and instead he studied with Sunan Bonang. The rod mark of Sunan Bonang is still located on the edge of Boom Tuban beach, just north of the Tuban Square. Although its position is now somewhat in the middle of the sea, the well still lets freshwater out. The well is known as the Srumbung Well. It was from that event that Tuban's name was finally coined, that is, from *me-tu ban-yune* ("the water came out").

2 CRITICAL DISCOURSE

In this study, we used the theory of critical discourse analysis by Adorno in expressing social and ethnic conflicts that marked the erection of a statue of Kwan Sing Tee Koen in Kwan Sing Bio Temple, Tuban, which was reported on an international news website (http://www.straitstimes.com/asia/se-asia/statue-of-chinese-god-guan-yu-stokes-tension-in-indonesia).

Discourse is the most complex and most complete element of culture. The linguistic supporting units include phonemes, morphemes, words, phrases, clauses, sentences, and paragraphs, up to the whole article. However, discourse is basically also a pragmatic language element (Mulyana, 2005: 1). Ismail Marahimin in Sobur (2012: 10) defined discourse as the ability to progress (in discussion) according to the regular and proper sequence and the communication of thought, whether oral or written, which is official and orderly. Writing is a discourse. The term discourse is used to encompass not only conversation or chat but also public speaking, writing, and formal efforts such as scientific reports and plays or acts (Tarigan in Sobur, 2012: 10). Meanwhile, according to Samsuri in Sobur (2012: 10), discourse is a complete recording of the language of communication events, which usually consists of a set of sentences that have a relationship of understanding with one another. Communication can utilise spoken language and also written language. Discourse always presupposes the speaker/writer, what is being said, and the audience/reader. Language is a mediation in the process. Discourse according to Tarigan in Sobur (2012: 11) includes four purposes of language use, namely self-expression, exposition, literature, and persuasion. Thus, discourse is treated as a series of words or sequences of speech expressing a thing (subject) that is presented regularly, systematically, in a coherent whole, and shaped by segmental and non-segmental elements of language. Webster in Mulyana (2005: 2) expands the meaning of discourse as: (1) communication of words, (2) expression of ideas, and (3) written treatises, lectures, etc.

3 METHODOLOGY

This research used the descriptive-qualitative method to describe, expose, and explain data or objects naturally, objectively, and factually. This research was a qualitative research with the researcher as the first instrument and data analysed in the form of words or sentences, and not in the form of numbers as in field research. This research can be said to be qualitative because it contains photographs, writings, phrases, or symbols that represent people, actions, or events in social life (Newman, 2007: 323). The primary data in this research was the news article (http://www.straitstimes.com/asia/se-asia/statue-of-chinese-god-guan-yu-stokes-tension-in-indonesia) on an Internet-based media. The secondary data is data obtained from sources of scientific information relevant to research, such as articles, journals, and other books that support the research. Data collection was done by way of *close reading*. Furthermore, we focused on the events behind the protests of religious groups and then analysed them from the perspective of discourse analysis, so that the researcher found the picture in the textual discourse. The author used a qualitative research design. The data in this study was obtained through the literature review technique because the data source was the text contained in online news media. Afterwards, we performed an analysis of the text used as the source of research data. The source for these research data was an online news article (http://www.straitstimes.com/asia/se-asia/statue-of-chinese-god-guan-yu-stokes-tension-in-indonesia). Then, we applied discourse analysis theory. The results of data analysis were then presented informally using words (Kesuma, 2007: 73). Data were also analysed contextually, depending on the context in which discourse was expressed. Therefore, this study will answer the following questions: (1) How does intolerance mark multi-ethnic and religious conflicts over the erection of the Kongco Kwan Kong statue? (2) How are ethnic and religious backgrounds characterised in multi-ethnic and religious conflicts in Indonesia? (3) How is nationalism characterised in ethnic and religious conflicts over the erection of the General Kong Kong Kwan Kong statue.

4 DISCUSSION

The erection of Kongco Kwan Kong Statue or the statue of a Chinese General, Kwan Sing Tee Koen, in the backyard of the Tri Dharma Kwan Sing Bio worship place, Tuban, which was inaugurated by MPR Chairman Zulkifli Hasan in mid-July 2017 drew protests on social media. Accounts on social media on behalf of (Islamic) religious groups protested and demanded the statue of the Chinese deity to be demolished. Tensions in the virtual world had even ignited a strike by certain groups outside the Parliament in Surabaya after protests made in cyberspace. The existence of the ethnic Chinese deity on social media was due to the fact that the tallest god statue in Southeast Asia was compared to the statue of General Sudirman in Jakarta, triggering the debate. Meanwhile, local authorities assessed that the statue is illegal because it has no construction permit. Initially, the statue as high as 30.4 metres was massively publicised on various Facebook groups and compared with the statue of General Sudirman, the height of which is only 12 metres. In addition, the existence of the tallest god statue in a temple complex in Southeast Asia is now starting to be polemicised by the local government. The management of the temple as the responsible party was accused of not having permission to build the statue, which cost Rp. 1.5 billion. However, to be able to move it, the local authorities need to be careful because these issues are sensitive and can be used by irresponsible parties to trigger bigger issues, such as SARA (ethnic, racial, and religious) issues. Thus, to follow up on the issue, the local government together with the Indonesian Council of Ulama (MUI) of Tuban held a closed meeting with the Regional Leadership Forum of Tuban after the hectic protests on social media. According to local authorities, the process of erecting the statue was banned by Tuban Regency and stopped, but the management of the temple Kwan Sing Bio was still desperate to continue the inauguration of the statue. The regency government did not issue a permit because the status of the management of the temple remains unclear. In addition, the district government solved this problem based on the recommendation of Majelis Ulama Indonesia. Meanwhile, the Kwang Sing Bio Temple had not commented on the statue of Kongco Kwan Sing Tee Koen after the controversy on social media. It is well known that the statue of the Chinese war god Kwan Sing Tee Koen was inaugurated by the Chairman of the MPR RI, Zulkifli Hasan, and was the topic of debate on the Internet. This is due to the fact that this statue also obtained the MURI record as the tallest statue in a temple complex in Southeast Asia. Not only was there an expression of admiration and pride, social media chatter also led to ethnic issues that could lead to ethnic, religious, and group issues, which are prone to be misused by irresponsible parties. In the late 1900s, Gus Dur, the fourth President of the Republic of Indonesia, had often called for Pluralism, which is the notion of openness and freedom of religion. Since then, Confucianism has officially become the sixth religion in Indonesia. The Chinese minority, since the Gus Dur era, has felt supported and was growing rapidly, culminating in the enactment of the Chinese New Year as a national holiday. However, the Chinese minority still appears to be under pressure, both politically and religiously, marked by cases of a religious and political nature.

5 CONCLUSION

The results of this study indicate that ethnic or religious backgrounds marked the erection of the statue of General Kongco Kwan Sing Tee Koen in Kwan Sing Bio Temple, Tuban. The erection of the tallest statue in a temple environment in Southeast Asia using religious symbols triggered religious tensions in cyberspace and in Indonesia. Multi-ethnic and religious conflicts had represented the controversy over a sense of nationalism in Indonesia since the erection of the statue. This study shows that ethnic and religious issues could easily develop and become social conflicts, such as that published in the online news media (http://www.straitstimes.com/asia/se-asia/statue-of-chinese-god-guan-yu-stokes-tension-in-indonesia) to divide the integrity of ethnic communities. Nationalists and religious people put political pressure on the Indonesian central government to immediately demolish the statue of the Chinese war general. It triggered tensions and political constellation in Indonesia, which originated from ethnic and religious issues developed in cyberspace. In fact, there was a strike. Previous racial and ethnic issues also briefly coloured the political constellation in the Jakarta Governor's election arena, where Ahok, the prominent Governor of Jakarta who participated in the Jakarta regional head election, was protested and defeated by the issue of

Islamic blasphemy. After being declared defeated in the arena of Election of DKI Jakarta, Ahok, who is of ethnic Chinese and Christian background, was sentenced to 2 years' imprisonment by a local court authority. The media text says that recent religious tensions have taken place in Indonesia due to a sense of nationalism, ethnicity, and religion. The issue of a religious majority and minorities is still a dominant issue to divide the unity of the nation. In fact, since the era of President Gus Dur, the people of Indonesia had championed pluralism, where citizens glorify freedom of religion, diversity, and tolerance. However, the rising issue of intolerance today undermines pluralism itself. In the realm of politics and government, constitutionally, the central government cannot deliver the pressures on the statue's tearing down. Although a wave of protests was born from the majority religion, it seems that the central government has no strong foundation for putting pressure on the Chinese minority over the controversy.

REFERENCES

5antoso, L., Alfian, I.N., 50elistyarini, T.D., & Meyrasyawati, D. (2013). *Agama dan relasi sosial: konstruksi santri tentang multikulturalisme dan relevansinya terhadap model pembela.* Surabaya: Universitas Airlangga.

Amalia, Y. (2018). *Ethnic Diversity in Indonesia: Do We Stand a Chance to be Truly Multicultural?* (t. R. Conference, Performer) France, Le Havre University.

Hapsari, N.F. (2014). The Discourse of Islam in Indonesia Portrayedin a Daily Mail's Article "Gang-Raped Indonesian Woman May Be Caned Publicly and a New York Times' Article" Indonesia: Shariah Official Urges that Gang Rape Victim May Be Caned. *Proceeding "Language Phenomena in Urban Society".*

Hapsari, N.F. (2015). Dunia Siber dan Norma Maskulinitas Anak Muda Pasca Reformasi di Indonesia. *15*(2).

Kesuma, Tri Matoyo Jati. (2007). *Pengantar (Metode) Penelitian Bahasa.* Yogyakarta: Carasvati books.

Mulyana. (2005). *Kajian Wacana: Teori, Metode dan Aplikasi Prinsip-prinsip Analisis Wacana.* Yogyakarta: Tiara Wacana.

Newman, W.L. (2007). *Basic Social Research: Qualitative and Quantitative.* New York: Pearson Education.

Sobur, Alex. (2012). *Analisis Teks Media: Suatu Pengantar Untuk Analisis Wacana, Analisis Semiotik, dan Analisis Framing.* Bandung: Remaja Rosdakarya.

Urban Studies: Border and Mobility – Kerr et al. (Eds)
© 2019 Taylor & Francis Group, London, ISBN 978-1-138-58034-3

Demystifying Nusantara

Agung Bahroni
Universitas Airlangga, Surabaya, Indonesia

ABSTRACT: The territory of Indonesia is inseparable from the historical knowledge of a territorial concept of Nusantara. The concept of Nusantara itself refers to the territory in the reign of the Majapahit Kingdom. The myth saying that the territory of Nusantara is the territory of Indonesia in ancient times continues to grow. In this study, we aimed to reveal what is behind the myth of Nusantara. By utilising the mythology theory of Roland Barthes, this study focused on the following two aspects: (1) how the construction of the myth of Nusantara was formed and (2) the myth demystification and its meaning. This myth emerged in some Javanese manuscripts, particularly Pararaton and Negarakretagama. This myth continued to spread until the period of the Indonesian National Revolution, which represented Nusantara as the historical background of Indonesia. The new meaning emerged from the term Nusantara as the territory of Indonesia to unify the Indonesian society with nationalist ideology.

Keywords: Demystification, myth, Nusantara, territory of Indonesia

1 INTRODUCTION

Indonesia is a country with an extensive history. Unfortunately, the Indonesian society often forgets some of its history. The extensive history of Indonesia spans from the pre-colonialisation era to the present times, and apparently has a significant impact on the lives of the Indonesian people. Nowadays, there is a heated battle over identity contestation in relation to history. The legitimacy of history ownership is an issue, which is politically contested today.

When discussing the history of Indonesia, the term "Nusantara"[1] should not be ignored, which is very entwined with the lives of people in the Indonesian society. This term has a generally similar meaning in the Indonesian society. The term is also inseparable from the "borderline" concept in the history of Indonesia, particularly in the pre-colonial time. The study regarding the borderline in Southeast Asia is highly complicated if the study is in the context of pre-colonial time. The borderline concept itself was, in fact, conceived by the colonials to compartmentalise the territories of sovereignty. Previously, the borderline did not exist, since the Southeast Asian society was, in fact, a trading community, which had extremely high mobility.

If associated with the territory, Nusantara, for Indonesian society, is the territory of Indonesia before the colonial period. The territory, in the minds of the Indonesian people, is closely related to the territory of the Majapahit Kingdom. This meaning, in fact, came from a Javanese manuscript written in the Majapahit era, i.e., *Kakawin Negarakretagama*. The statement is strongly rooted in the minds of the Indonesian people since it obtained legitimacy in the form of the insertion of the assertion in history in schools.

The term Nusantara, at that time, became an extremely popular term. For example, the term is utilised by an Islamic organization in Indonesia, i.e., Nahdlatul Ulama, which uses the term "Islam Nusantara" as the depiction of the Islamic organization, which adopts a local culture in religious life practice. Moreover, there is also the term "the culture of Nusantara" (*Budaya*

1. Italicised since it comes from Old Javanese and capitalised since the term describes a cultural area.

Nusantara) to describe the culture in Indonesia. The term also often utilizes Indonesian traditional arts as the reference.

The numerous studies about the history of Majapahit render the statement about Nusantara more enduring. Before discussing this even further, it is worth noting that the description of territory in this study is not in the domain of history about Nusantara. Yet, it is in the domain of myth about the term Nusantara that is applied and lasts in the Indonesian society. Myth, according to Roland Barthes, is a meaning towards a particular form. The myth itself is part of the study of sign introduced by Saussure 40 years ago with the term semiology (Barthes, 2004: 155).

Barthes emphasized that meaning should return to semiology, although semiology was incapable of explaining all the different aspects. Yet, all of them had similar statuses. This is because the whole aspect makes up the science of value. They are not satisfied with the value. They are not satisfied by discovering the facts. They define and explore the facts as a sign of something else. Myth has an imperative yet conflicting character, rooted in the concept of history and directly emerging from things that are coincidental (Barthes, 2004:177).

In this study, the term Nusantara and its meaning in terms of the Majapahit territory is a myth supported by some established histories. The myth of Nusantara is viewed from the theory of mythologies by Roland Barthes to reveal the myth about Nusantara. The emerging assumption is that the myth of Nusantara appeared as a form of legitimacy for the territory of Indonesia after its independence, by which, of course, the Indonesian revolutionary figures affected the formation of the myth. Thus, the research questions applied in this study were: (1) How was the myth construction of Nusantara formed? (2) What are the demystification of the myth and its meaning?

2 DISCUSSION

2.1 *Nusantara as a myth*

Many studies on Nusantara have been conducted, particularly in the field of history. The object that is mostly used in those studies is a manuscript from the past (philology). The results of the studies are also the source of past cultural knowledge. This study is not a historical study. This is a study regarding the myth of the term Nusantara, although the object of this study also uses manuscripts written in the past, i.e., *Kakawin Negarakretagama.*

To consider Nusantara as a myth, the initial step is examining the denotative and connotative meanings proposed by Roland Barthes. The denotative meaning of Nusantara can be noted from the meaning possessed by the term. On the basis of the word sequence, Nusantara is split into two words, "nusa" and "antara". Nusa means "island" and antara means "other", which are combined to mean other islands outside Java island (Zoetmulder and Robson, 1982). In KBBI (2006:485), Nusantara is the term for Indonesia that means "consisting of many islands", a land of hundred thousand islands.

Meanwhile, the connotative meaning is the emerging cultural meanings or the emerging meanings due to a cultural construction, so there is a shift. Nonetheless, it is still attached to the term Nusantara. Connotatively, Nusantara is interpreted as another term for Indonesia, which refers to a concept of the territory of the Majapahit Kingdom in the past with reference to *Kakawin Negarakretagama.*

In the proposed myth, Roland Barthes offers a new definition that becomes an initial step, which is used to read the myth (inoculation). In this step, the term Nusantara is defined as a term used in a cultural area in the pre-colonial time that is given by the society in colonial time. Thus, in the pre-colonial time (Majapahit), the term Nusantara was assumed non-existent in this writing, due to deletion of history, which is the next step that will be proved by re-reading *Kakawin Negarakretagama.*

2.2 *Reading Kakawin Negarakretagama*

Nagarakretagama means "a country with a sacred tradition (religion)". The name *Nagarakretagama* itself is not contained in the poem *Nagarakretagama*. In the *pupuh* 94/2,

Prapanca called his work *Deçawarnana* or explication of villages. However, the name given by the poet has been forgotten by the public. The poem is called *Nagarakretagama* until now. The name *Nagarakretagama* is included in colophon published by Dr. J.L.A. Brandes: *Iti Nagarakretagama Samapta*. The name *Nagarakretagama* is apparently an additional copyist of *Arthapamasah* in the month of Kartika in the Saka year of 1662 (October 20, 1740 A.D.). *Nagarakretagama* is copied in Balinese scripts in *Kancana*.

Desawarnana/Nagarakretagama is written in the form of a poem (*kakawin*). Each poem consists of four lines, and each line consists of 8–24 syllables. This poem's manuscript consists of 98 *pupuh*, which are divided into two parts, each of which consists of 49 *pupuh*.

In fact, the term *Negarakretagama* is never mentioned in the original manuscript. Meanwhile, the adaptation of the manuscript of *Arthapamasah* has been found; even so, it is yet to be published. The renaming of the manuscript *Desawarnana* to *Negarakretagama* is only based on the unfamiliarity in using the word. According to Prof. M. Yamin (112) in Sapta Parwa, the name *Desawarnana* is a summary of *Wilwatikta-negara-kretagama*, which means the history of Majapahit: *Kretagama* = good deeds/event/history and *Wilwatikta-negara* = the country of Majapahit.

In the original manuscript, it is mentioned that the manuscript is entitled *Desawarnana* or the explication of villages visited by King Hayam Wuruk.

> *rin çakadri gajaryyamaçwayujamasa çubhadiwaça purnnacandrama, nka hingan/ rakawin/pamarnnanakhadigwijayaniranarendra rin praja, kwehnin deça riniñci donikaminustakamanarana* ***deçawarnnana****, pangil/panhwatasanmatanrpatimengeta rin alawas atpaden lanö.* (XCIV: II)
>
> "The year of *Sakagununggajahbudi* and *janma (1287), the month of Aswina, the day of full moon. To be ready a poem of adoration about the journey of traveling around the Country. To be arranged the entire villages in a sequence, worthy to be called* ***Desawarnana****. To intend that Your Majesty recall while reading the sentences of wisdom.*" (XCIV: 2)

The *pupuh* 94 above explains that in 1287, in the month of Aswina, a poem (*kakawin*) about the journey of the Great King Hayam Wuruk, who went around villages, was written. Then, the poet named the poem *Desawarnana*, so that the King remembers that the one who wrote the manuscript was Mpu Prapanca (pseudonym), who accompanied everywhere.

> *lwir nin nusapranusapramukhasakahawat/ksonirimalayu, nan jambi mwan palemban karitanitebalen/darmmaçraya tumut, kandiskahwasmanankabwarisiyakirkan/kamparmwani pane, kampeharwathawemandahilini tumihaṉ parllak/mwanibarat.* (XIII:1)
>
> *hi lwaslawansamudramwanilamuribatan lampun mwan ibarus, yekadinyan watek/ bhumi malayusatanahkapwamatehanut. len tekan nusa tañjun nagara ri kapuhas lawan ri katinan, sampit/mwan kutalinga mwan i kutawarinin/sambas mwan i lawai.* (XIII:2)

"In detail, the islands of subaltern Countries, at the beginning: M'layu, Jambi, Palembang, Toba, and Darmasraya. Also following: Daerah Kandis, Kahwas, Minangkabau, Siak, Rokan, Kampar and Pane." (XIII:1)

"Vast with Samudra and Lamuri, Batan, Lampung also Barus. Those are the Malay Countries to comply. The countries in Tanjungnegara island; Kapuas-Katingan, Sampit, the City of Lingga, the City of Waringin, Sambas, Lawai". (XIII:2)

Majapahit is extremely well known for the *Sumpah Amukti Palapa* (the Palapa Oath) by Patih Gajah Mada who unified Nusantara from Sabang to Merauke. This is proved from the data above at *pupuh*13, stanzas 1 and 2. Majapahit's subordinate countries from the mainland to the sea area are described in detail. The territory of Majapahit is extremely vast. This statement is also a questionable statement since in *Negarakretagama*, there is no term Nusantara, and Gajah Mada declared the Palapa Oath.

2.3 *The term Nusantara*

In the historiography of Javanese manuscripts, the term Nusantara appears in several manuscripts, which have become the master copy of the historical reconstruction of Java island, such as *Babad Tanah Jawi* and *Pararaton*. Both manuscripts are of *Jawa Pertengahan*. *Pararaton* narrates the progress of the Singhasari Kingdom, which was initiated by the leadership of Ken Arok up to the Majapahit. Moreover, the kings' lineage of Majapahit is also the main content of *Pararaton*. This manuscript is arranged based on other various sources at the time it was written (Krom, 1921: 101). *Pararaton*'s year of writing can be observed from the last event written in this manuscript, i.e., the eruption of a mountain in 1481 A.D., and it can be concluded that this manuscript was written not long after that in the following year (Djafar, 2009: 20). *Pararaton* should also be regarded as the existence of a historical construction established by the author based on his own interest. Likely to be dated from the year of writing at the end of the Majapahit era, the content narrates the historical journey from Singhasari to Majapahit.

The term Nusantara was first found in *Pararaton*, particularly in the Palapa Oath declared by Gajah Mada, whose content is as follows:

> *Sira Gajah Madapatih Amangkubumi tan ayun amuktia palapa, sira Gajah Mada: Lamun huwus kalah ring Gurun, ring Seran, Tanjung Pura, ring Haru, ring Pahang, Dompo, ring Bali, Sunda, Palembang, Tumasik, samana isun amukti palapa.*

In the citation, it is "narrated" that Patih Gajah Mada has declared an oath that until today it is still the robust background for the usage of Nusantara. In his oath, Gajah Mada would never eat *palapa* (the information on this food is not clear) before being able to conquer several islands mentioned and unifying them into a unity called Nusantara. The researcher's suspicion is that given when this oath is written in *Pararaton*, from the year of writing, it can be concluded that the author was not living in the same era, meaning the author did not directly witness the oath declared by Gajah Mada.

Further in the historiography at the time after the fall of Majapahit, the term Nusantara is mentioned as representing the territory of Majapahit on various islands by referring to *Pararaton* as the reference as well as several other sources. The term Nusantara is not only present in Javanese manuscripts, but also spread, and was quite popular at that time. As stated in the manuscript discovered in Bangka, the term Nusantara is mentioned approximately 2,592 times (Evers, 2016:5).

In the colonial period, the term Nusantara started to reappear and became popular when approaching Indonesia's year of independence. Several figures began to consider the name for the country after it became independent from the Dutch colonizers. The socialist figure of Indonesia, Tjokroaminoto, once mentioned the term Nusantara as a possibility for the soon-independent country's name when leading the Islamic Union (*Sarekat Islam*). Furthermore, Ki Hadjar Dewantara, who established the Student's Garden (*Taman Siswa*), who employed Javanese ideology in his pedagogy, also proposed Nusantara for the soon-independent Indonesia (Evers, 2016:5). However, the proposal was not accepted due to the term's ownership status, which was also possessed by Malaysia, and Indonesia became the name after colonial time (Bastian: 1884).

2.4 *The everlasting Nusantara*

After independence, the term Nusantara was still often raised by many figures of the Indonesian revolution. One of them was M. Yamin, who often mentioned and perpetuated the term in some of his works. Two of M. Yamin's studies are often used as historical education references in Indonesia; thus, the myth of Nusantara is more everlasting. *Sapta Parwa* is the work of M. Yamin, which explicates the statecraft of Majapahit, which is detailed in seven separate volumes (1962). The next work which perpetuates Nusantara is *Gajah Mada* (1953), which narrates the journey of Patih Gajah Mada from various sources of manuscripts and in the work; the illustration of Gajah Mada's figure makes its appearance for the first time. The work of Gajah Mada, indeed, perpetuates the Palapa Oath that still persists to today.

Meanwhile, Sapta Parwa, which seems too complicated with its seven separate volumes, informs that the territory of Indonesia named Nusantara is the territory of Majapahit. M. Yamin is sufficiently smart to lead the statement by inserting some archaic manuscripts that relate to the history of Majapahit. From some of the existing sources, the main source to utilise is *Negarakretagama*, which was previously discussed in this study. The works of M. Yamin related to Nusantara and Majapahit were quite acceptable at the beginning of Indonesian independence due to the high sense of nationalism at that time. Furthermore, the term Nusantara is the reference of the territory of Indonesia due to the requirement of the modern world, which obligates the existence of borderlines in a country. In a meeting, M. Yamin also once asserted that the territory of Indonesia after independence is the entire territory that the Dutch once ruled. Furthermore, the term Nusantara is the historical background of Indonesia. Nationalism is the appropriate ideology to perpetuate the myth of Nusantara.

This term becomes the joint property of a united Indonesian society. The feeling of owning a similar historical background in colonial time leaves a gap for Nusantara to invade as a myth in the minds of the Indonesian society. As stated by Benedict Anderson (2008), nationalism is a shadow that is shared by a community; in this case, the Indonesian society possesses Nusantara as a shadow of unity. Several studies using the term Nusantara with the concept of Nusantara as the territory of Majapahit that has now become Indonesia render the eternity of this myth in the academic world.

In 2015, there was a petition from the Indonesian President Joko Widodo to rename Indonesia as Nusantara. The strong rationale for this petition was that the name Indonesia was the name given by outsiders (part of colonial times) and that it was not part of the name of the authority, nation, region, or shared spirit, which commonly becomes the reference for a country's name. The term Nusantara has been used as a synonym to address the archipelago of Indonesia for a long time. However, this petition was a failure since the term Nusantara was also considered by Malay (Malaysia) and it was feared that it would generate a new conflict (Evers, 2016:6).

3 CONCLUSION

Nusantara is an alternative term for Indonesia in the modern world, and nowadays its usage is highly popular. It is a sign that can be read with the theory of mythologies by Roland Barthes. The myth of Nusantara is unravelled by providing another definition to this term as a cultural area of the Malay Archipelago in the pre-colonial period. The deletion of the existing history behind the myth appears in the absence of the term in the manuscript that existed in the Majapahit era, i.e., *Negarakretagama*. The manuscript of *Pararaton* is also relevant in the deletion of history in terms of the myth of Nusantara. *Pararaton* is the pioneer of the emergence of the term Nusantara, which was narrated in the discussion of the Palapa Oath declared by Gajah Mada. Yet, the validity of this manuscript was doubtful since the manuscript was not written in the Majapahit era.

Myth can evolve if there is a process that can perpetuate the myth. In the case of the myth of Nusantara, this process began in the years approaching Indonesian independence as the term was offered by some revolutionary figures regarding the name that would be employed after being free from the colonizers. Moreover, the study conducted by researchers, particularly M. Yamin, who formulated that the territory of Indonesia was the territory of Majapahit, referred to as Nusantara in the past, is the process of perpetuating the myth of Nusantara. A new meaning emerges after the myth of Nusantara when it is revealed that Nusantara becomes a territorial concept used to unify the Indonesian society with nationalism as the main ideology.

REFERENCES

Anderson, Benedict. 2008. *Imagined Communities: Komunitas-komunitas Terbayang.* Yogyakarta: Pustaka Pelajar.

Anwari, Ikhsan & Anwari, Mujahidul. (2015). Sistem Perekonomian Kerajaan Majapahit. verleden. 3. 104–115.

Barthes, Roland. 2004. *Mitologi*. Yogyakarta: Kreasi Wacana.
Bastian, A. 1884. *Indonesien: oder die Inseln des Malayischen Archipel*. Berlin: Dummlers, 5 vols.
Djafar, Hasan. 2009. *Masa Akhir Madjapahit*. Jakarta: Komunitas Bambu.
Evers, Hans-Dieter. 2016. *Nusantara: History of Concepts*. Malaysia: JMBRAS. Vol.89. Part 1.
Kamus BesarBahasa Indonesia, 4th ed. 2008. Jakarta: Gramedia Pustaka Utama.
Krom, N.J. 1921. *De Samenstelling van de Pararaton*. TBG.
Perkasa, Adrian. 2016 Playing with Flames: Counter State Ideas in the Preservation of Majapahit's Heritage, *SEASUK Conference*. London.
Perkasa, Adrian. 2017 Living Heritage of Ruins? Contesting the Paradox in Trowulan's Heritage Zoning. In Hsiao, Hsin-Huang Michael. Yew-Foong, Hui. Peycam, Phillipe. *Citizens, Civil Society and Heritage-making in Asia*, Taipei: ISEAS-Yusof Ishak Institute.
Sunardi, S.T. 2002. *Semiotika Negativa*. Yogyakarta: Buku Baik.
Yamin. M. 1962. *Tata Negara Madjapahit: Sapta-Parwa*. Jakarta: Prapantja.
Zoetmulder, P.J. with Robson, S.O. 1982. *Old Javanese-English Dictionary*. The Hague: Nijhoff.

Urban Studies: Border and Mobility – Kerr et al. (Eds)

Commercial activities and development of the towns in the west side of Banda Sea Indonesia, early twentieth century

L.O. Rabani
Universitas Airlangga, Surabaya, Indonesia

ABSTRACT: This article analyses why cities developed on the west side of the Banda Sea in the early 20th century. These cities are Baubau, Raha, Kendari, Bungku, Kolonodale, and Luwuk. The six cities are on the east coast of Sulawesi island. Four factors support the development of the cities namely: communities' and governments' active trading and shipping activities; secondly, the geographical factor in the global spice trade route encouraging integration between cities on the western side of the Banda Sea with other cities; thirdly, the city's economic base being driven by people living in a strong maritime tradition; fourthly, inter-regional connectivity that can only be done by sea. Thus, the character of the city leads to the typology of a new city, the maritime city. The study of maritime cities in Indonesia is very important because it has a vast sea with thousands of islands.

Keywords: commodities, integration of the sea trade networking, maritime town, development, and west side of Banda Sea

1 INTRODUCTION

The State of Indonesia is a country that has a large area of sea. Lapian said that the sea has a liaison role between one island and another (Lapian, 2002). Elaborating further into his statement, the sea not only connects islands but also connects human beings.

The history of Indonesian trade as proposed by J.C. van Leur is that commerce and shipping in the archipelago took place between islands and connected each other. This relationship brings consequences for the interplay of cultural and social aspects. The community kinship relationship between one island and another is also strong, so it is not surprising that communities come from other islands to several islands like in Sulawesi and other big islands (Leur, 1953). For example, on the Island of Sulawesi, there are settlements of Malays (Makassar city), Javanese people, and Bugis, Butonese, and Bajo people on the north and east coasts of Sulawesi Island. Their existence proves the inter-island relationship is intensive.

Intensive inter-island relations involving many communities are one of the characteristics of the ongoing process of a region's urbanity. The areas of the encounter are often characterised by an ever-expanding morphological concentration of people and settlements, as well as increasing population growth. Such an area usually becomes a city when the needs of its citizens are fulfilled (W.F. Wertheim, 1999),[1] at least when it is near clean water sources. The question is what kind of cities are formed by activities that use the sea as the main (connecting) road?

To answer that question, this paper describes the development of cities in the growing and developing east coast of Sulawesi Island or when viewed from the perspective of maritime history, the west side of Banda Sea is the geographical space of this study. It was at the beginning of the twentieth century when cities in Indonesia and the east Sulawesi coast showed their established urban properties. The discussion begins by describing the geographic area of research (space of town) on the east coast of Sulawesi Island that stretches along the west

1. Wertheim says that the nineteeth century was the earliest period of cities in Indonesia showing its urban nature.

side of the Banda Sea, then, identifying the urban areas and economic activities taking place on the eastern coast of Sulawesi Island, and the final part of the paper discusses the tendencies of certain city typologies such as what grows and develops in geographical spaces such as the eastern coast of Sulawesi Island.

2 THE CONTEXT OF GEOGRAPHY OF THE WEST SIDE OF BANDA: HISTORICAL AND GEOGRAPHICAL BACKGROUND

In August 2015, the researcher travelled overland from the city of Luwuk to Kolonodale (now the capital of North Morowali) (Poelinggomang: 2008)[2] and collected information from various sources. It was concluded that there were no roads connecting Luwuk and Kolonodale. Travel overland to Kolonodale is only possible through Baturube. The City Planning Agency of Banggai Regency, Luwuk City, confirmed that it was, indeed, true that there were no roads to Kolonodale. This means that the sea is still functioning as a liaison that is relied upon between one town and another through the media of motorboats or other sea transportation. The eastern coastal areas of Sulawesi referred to in this paper include Luwuk, Kolonodale, Bungku, Kendari, Muna, and Buton in Southeast Sulawesi, as well as the surrounding islands.

The particular areas that have not developed into "big towns", will be seen as supporters of the "big city" region on the east coast of Sulawesi. The western side of the Banda Sea stretched from the eastern coast of Sulawesi Island to the Maluku Islands (as a spice producer). Furthermore, Bitung as a copra producer is connected to the Sulu Sea in the Philippines as one of the centres of sea cucumbers. The above commercial activity centre is attached to the Port of Makassar which has been used as a "free port" since 1847 by the Dutch colonial government to compete with British controlled Singapore (Poelinggomang 2002, p.hlm. 27).

Slowly the control of power of traders was consolidated by the Dutch colonial government. Previously, the market was always free with traders who were under British control, without taxes. When the Dutch colonial government imposed a free-port policy, it turned into a taxed trade as a result of the half-hearted free trade liberalisation. The Dutch monopolised the spice trade in the port of Makassar, taxes on the arms trade, and tax obligations for boats (Poelinggomang, 2002).

In its development, the policy affected the surrounding area, particularly commodity producers with traders entering the port of Makassar. The taxation and monopoly system on commodity trading in the "free port" of Makassar by the Netherlands had implications for the strengthening of new economic centres on the East coast of Sulawesi in line with the exploitation of forest, plantation and mining commodities.

The above conditions seem to be the turning point in strengthening trading activities on the East coast of East Sulawesi.[3] Early identification and information derived from the Dutch Colonial government reports in the nineteenth century to the early twentieth century recognised that the area was very fertile and could develop into a plantation area, especially in Banggai, Kolodale, and Bungku. Buton was relatively infertile, but, in the 1920s, found asphalt to support the economic development of rulers and communities in the region. Muna became the producer of kapok, tea, rattan, and pearl. Kendari produced sea Lola, Teripang, and fish from Bajo and Bugis fishers (Ligtvoet, 1878). The Bugis merchants became a container and then traded products such as sea cucumbers and Lola. There were products sold to Makassar, some of which were sold to Chinese traders (Sutherland and Nas, 1985). The product surplus that occurred in some parts of East Sulawesi required transportation. Transportation facilities were filled by people who live on Buton island and in the surrounding districts.

2. Kolonodale is a small town in the bay of Tolo, the west side of the Banda Sea.
3. The cities included in this paper are the cities of Luwuk, Kolonodale, Bungku, Kendari, Muna, and Baubau. The discussion of these cities will not be considered one by one but will be discussed as a unity which is bound by the same maritime activity.

3 TRADE AND SHIPPING COMMUNITIES ON THE EAST COAST OF SULAWESI ISLAND

Commerce activities and shipping in the archipelago has been going on for a long time and was done inter-island. J.C. van Leur, in his dissertation at the University of Leiden (1934), found that the shipping and trading activities carried out by the people of Nusantara actively occurred long before the presence of the Europeans (Leur, 1934). European trade and freight companies were unable to shut down shipping and trading activities in the archipelago. Malay, Javanese, Madurese, Bugis, Makassar, Mandar, Buton, and Bajau traders intensively traded and did shipping in some parts of Indonesia, particularly in Sulawesi, the Maluku Islands and beyond.

There was an increase in trade and commodity transportation to the world market from the archipelago in the 20th century by the Europeans (UK) and also the Dutch. Dutch shipping companies that transport products such as KPM served as the backbone for transporting commodities to the world market in Asia, Africa, and Europe.[4] These ships also transported passengers to West Asia such as Muslims for the hajj and Westerners to Indonesia, encouraged by tourist books and travel literature (Majid, 2008).

There were two severe disruptions to the world trade in the early twentieth century: World War I and the economic crisis in 1930. Before World War I, the Dutch shipping companies that began operations in the late nineteenth century were keen to open trade routes in various regions in the archipelago (*Handelsvereeniging Makassar Jaarverslag over 1921*, 1921).

Based on the historical record, shipping and trading on the east coast of Sulawesi has lasted for a long time. Despite this, it has failed to catch the attention of researchers. One of the causes of the lack of interest of researchers in the area is the ability of language and access to limited sources of history. Also, researchers focus on large historical sections, such as the Dutch empire or history in Indonesia.

The report on piracy activities on the intensive west side of the Banda Sea was an early indication that there was productive activity on the west side of the Banda Sea, reflected in the colonial government (Netherlands) report contained in several BKI numbers in the late nineteenth and early twentieth centuries (Peterson 2015). This and the engagement in intensive activities with local communities had an impact on the slow changes towards modernity in the European mind, yet fast in responding to global change. Two colonial government posts in Buton and Kendari bay, up to the end of the 19th century and the first years of the twentieth century, had not been able to overcome the piracy in the region.

The turnover of trading value in this illegal economy was much more profitable because of the highest market prices, lack of taxes, and the traders having the opportunity to sell at the highest rate. Such an economic system could flourish on the east coast of Sulawesi for three reasons, namely adequate navigation capabilities and the abundance of islands scattered in the region, making it shelters and hiding places from a variety of distractions. Also, the limited vessels and patrols owned by the Dutch colonial government and the vastness of the territory to be controlled was an obstacle to the enforcement of policies in colonial trade.

Therefore, unofficial trade and voyages occurred widely from the beginning of the twentieth century, especially on the eastern coast of Sulawesi and surrounding islands. The signing of Korte Verklaring (1918) implied that the Dutch colonial government had most recently conquered the western region of the Banda sea and its surroundings through pax-Neerlandica (Velthoen 2002). The situation indicated the limitation and high rejection of a Dutch Colonial presence on the west side of Banda. With the recognition of the Dutch colonial government as rulers by the local kingdoms on the west side of the Banda Sea through the short agreement of 1918, the territorial administrative arrangement was also made by the government, primarily by forming Afdeling East Sulawesi in 1911–1924. In 1925–1942, Afdeling Buton and Laiwoei were formed again. The agreement for these areas, as is universal in the colonial government bureaucratic system, is to facilitate monitoring and taxation in new areas by government employees, usually the controllers.

4. KPM operates in the transport of passengers and commodities towards the end of the 19th century by routes to port and harbour areas and the commodities export.

Commodity data and trade potentials reported by controller employees as contained in the memory of the handover of the position (memorie van over gave) can be said to vary. These commodities are mostly marketed to British merchants and or bought by Arabs and Chinese for further trade in Singapore.

The traders of the eastern coast of the island of Sulawesi, in addition to having local commodity produce, also buy merchandise from other areas for trading such as from the islands of Maluku, Halmahera, Jailoli, Galela, the Kei Islands and the Arafura Sea as well as in Nusa Tenggara, Iron Ore Islands, and in the waters of Northern Australia. The commodities were collected and traded to areas that offered favourable prices such as Singapore, Makassar, and Java.

The influence of trade and sailing on the east coast of Sulawesi in this article is seen from a social and economic viewpoint. Both sides are connected to the developmental context of towns on the eastern coast of Sulawesi Island that grows and develops in line with trading activities and shipping connectivity. Both commercial activities intensify into areas that are economically profitable. Economically, the eastern coast of Sulawesi produces the commodities needed by the global market, especially spices and marine products.

The eastern coastal community of Sulawesi was a haven for the survival of the people because it met basic needs such as clothing, food, and shelter. The west side of the Banda Sea had fertile soil that can develop farming and plantation systems. With the availability of fertile land, the community could maintain its survival with abundant food. History also notes that the west side of the Banda Sea became a destination for the Sultanate of Buton, Bone, and Laiwoei kingdoms when the situation at the local level was volatile, especially during the Makassar War and the Dutch war against Bone in the 1820s. The Mori Kingdom, located in the Gulf of Tolo, also experienced an impact from the chaos in its neighboring country, particularly in terms of the influx of residents from outside its territory. As a result, the population of the Mori Kingdom became heterogeneous and developed residential community space. That is, part of the process which turned the area into a town is in progress. That's what underlies the development of cities on the east coast of Sulawesi.

4 DEVELOPMENT OF CITIES ON THE EASTERN COAST OF SULAWESI ISLAND

The growing and developing cities on the east coast of Sulawesi Island are the cities of Luwuk, Kolonodale, Bungku, Kendari, and Baubau. The development of Luwuk city towards more modernity began when the colonial government built the road and port infrastructure in Lalong bay. As a fertile area, Luwuk became a source of rice, copra, wood, candlenut, cocoa, and sea cucumber. These commodities made Luwuk a city which developed with a colonial design, along with the towns of Baubau, Kendari, and Kolonodale. Growing cities on the east coast of Sulawesi are areas that historically have had local authorities at the same level as the kingdom. The town of Baubau flourished in the territory of the Sultanate of Buton (Zuhdi, Ohorella and Said, 1996). The city of Kendari thrived in the realm of Laiwoei (Melamba, Salahuddin and Janu, 2011). Bungku and Kolonodale cities thrived in the Mori kingdoms (Poelinggomang, 2008), and the town of Luwuk grew in the territory of Banggai (Bosscher and Matthijssen, 1854).

These cities are in one geographical area that is interconnected. In certain circumstances, they are dependent on one another. The city of Kendari, in the 1830s, became the post of the Dutch colonial government post under J.N. Vosmaer. The town centered on the bay side of Kendari was meant to control pirate activity in the waters and become a trading post. The consequence of the function was that then Kendari was developed into a city that met the needs of its citizens, Europeans, and traders. Kendari port was built, and the road network was also developed to economically profitable areas, especially to Wawotobi as a rice centre.

The above description gives a strong indication that the mainland (producer) of commodities did not mean anything in that period (early 20th century) if not connected to the eastern coast of Sulawesi as a place that moved the commodities to the global market. The bustling towns were Baubau, Luwuk, and Kendari (Vosmaer, 1839; Velthoen, 2010). These towns are crowded because, in addition to being navigable by overseas shipping companies, KPM also had a safe berth. The three towns are located in the bay to provide security and care for the merchants.

Some of the cities on the east coast of Sulawesi have relatively low accessibility levels, especially in terms of shipping and trading. Those towns are Bungku and Kolonodale. Both cities in the early period of the 20th century were highly dependent on other cities that had adequate shipping and trading infrastructures. The people of Buton, Bugis, Makassar, Madura, Mandar, and Bajau became connectors between towns on the east coast of Sulawesi. Various commodities produced by cities on the east coast of Sulawesi are transported and marketed to other cities. The implications of the presence of other communities on the eastern shore of Sulawesi, especially in areas with relatively low accessibility encourage more extensive city development and, at the same time, introduce new things within the social space.

Commodities, boats, and traders are significant contributors to spur the pace of development in cities on the eastern coast of Sulawesi. The colonial government contributed to the design and layout of the town and some of its infrastructure (Liebner, 2004) The exchanges involving local people and coupled with government intervention by controlling economic resources as well as politically controlling a region helped spur more focused urban development, although it did not go as expected. Trade and shipping that went beyond the control of the government continued until near the end of the colonial period. Not all trading ran through official channels. Both lines continued to contribute to the development of the city because local rules still bound personal ones agreed with local authorities, often undocumented, only from recurring (tradition) memories and habits. Thus, it appears that cities on the eastern coast of Sulawesi experience significant developments when trading and shipping activities are strengthening and, at the same time, the colonial government is developing new sectors of the economy, namely commodity trading.

5 CONCLUSION: TYPOLOGY OF A MARITIME TOWN FOR INDONESIA ON THE WEST SIDE OF THE BANDA SEA

This working paper finds substantial maritime activity as factors affecting the development of cities on the eastern coast of Sulawesi. The sea conducts commodity trading and socioeconomic relations. The sea and shipping became the primary medium connecting communities with ports and markets, especially in terms of commodity exchange. Items are traded by traders to meet the needs of the townspeople. These things can only be acquired by sea and ship in the cities on the east coast of Sulawesi.

The people in those cities also rely on sea access for mobility. Indonesia, which has a large area, has presented some typology and characteristics of the city, such as central cities, royal cities, colonial cities, port cities, and coastal cities. Three typologies of the towns that I have mentioned last are still under debate, even though they are conceptually evident. The researcher's perspective on the three typologies of the city becomes the determinant by which a city's characteristic is appropriate for the city area studied.

The context of cities on the eastern coast of Sulawesi Island, as the preliminary findings of this research show, leads to the typology of a maritime town. The maritime city is growing because of the dominance of maritime activities that affect the development of the town. Almost all the connectivity that supports the development of the town is generally mediated by the sea. The maritime town seems to be a binder or fusion of some typologies of coastal cities, port cities, and coastal cities. The concept of the maritime town has not been known in Indonesia, perhaps because of the abundance of fertile land making the nation more agrarian. Therefore, in a particular period, a city may have different characteristics in each phase of its development. The east coast cities of Sulawesi are evidently maritime cities in terms of typology.

ACKNOWLEDGEMENT

I would like to thank Prof. Dr. Bambang Purwanto, M.A, my supervisor for substantive discussions during the process of writing this article as part of my dissertation at Gadjah Mada University. Prof. Bambang Purwanto is a professor in history at Gadjah Mada University,

Yogyakarta. His current research focuses on social and economic history in Southeast Asia, especially Indonesian history of the colonial and revolution era.

REFERENCES

Bosscher, C. and Matthijssen, P.A. (1854). Schetsen van de Rijken van Tomboekoe en Banggai. *Tijdschrift voor Indische Taal-, Land-, en Volkenkunde* Vol. II.

Hafid et al., A. (2006). *Sejarah Kota Kendari*. Kendari: Pemkot Kendari dan FKIP Univ. Haluoleo.

Handelsvereeniging Makassar Jaarverslag over 1921 (1921).

Lapian, A.B. (2002). 'Some Explorations in the History of The Maritime World of Southeast Asia'. In: *Dalam Sunaryo Purwo Sumitro (peny.), Dari Samudra Pasai Ke Yogyakarta, Persembahan Kepada Teuku Ibrahim Alfian*. Jakarta: Sinergi Press-Yayasan Masyarakat Sejarawan Indonesia.

Leur, J.C. va. (1934). *Eenige Beschouwingen Betreffende Den Ouden Aziatischen Handel*. Leiden University.

Leur, J.C. va. (1953). *Indonesian Trade and Society, Essays in Asian Social and Economic History*. Leiden: Foris Publication.

Liebner, H.H. (2004). 'Tradisi Kebaharian di Sulawesi Selatan: Tinjauan Sejarah Perkapalan dan Pelayaran'. In: *Dalam Dias Pradimara (Peny.), Kontinuitas Dan Perubahan Dalam Sejarah Sulawesi Selatan*. Yogyakarta: Ombak.

Ligtvoet, A. (1878). Beschijrijving en Geschiedenish van Boeton. *BKI*I:1–.

Majid, M.D. (2008). *Berhaji Di Masa Kolonial*. Jakarta: CV. Sejahtera.

Melamba, B., Salahuddin, A.A. and Janu, L. (2011). *Kota Lama Kota Baru Kendari. Kajian Sejarah Sosial, Politik, Dan Ekonomi*. Kendari: Teras bekerjasama dengan LAT dan Lembaga Pengembangan Pengkajian Sejarah dan Kebudayaan Sultra.

Peterson, E.J. (2015). *Insulary and Adaptation Investigating the Role of Exchange and Inter-Island Interaction in the Banda Islands, Indonesia*. Washington.

Poelinggomang, E.L. (2008). *Kerajaan Mori, Sejarah Sulawesi Tengah*. Jakarta: Komunitas Bambu.

Poelinggomang, E.L. (2002). *Makassar Abad XIX, Studi Tentang Kebijakan Perdagangan Maritim*. Jakarta: Kepustakaan Populer Gramedia.

Schoorl, J.W. (1994). 'Power, Ideology and Change in the Early State of Buton'. In: Schutte, G.J. (ed.) *State and Trade in the Indonesian Archipelago*. Leiden: KITLV Press, pp. 17–59.

Sutherland, H. and Nas, P.J.M. (1985). Eastern Emporium and Company Town: Trade and Society in Eighteenth Century Makassar. In: *The Indonesian City: Studies in Urban Development and Planning*. Dordrecht: Foris Publication.

Velthoen, E. (2002). *Contested Coastlines: Diasporas, Trade and Colonial Expansion in Eastern Sulawesi 1680–1905*. Murdoch University.

Velthoen, E.J. (2010). 'Pirates in Periphery: Eastern Sulawesi 1905'. In: Kleinen, J. and Osseweijer, M. (eds.) *Pirates, Ports, and Coasts in Asia, Historical and Contemporary Perspectives*. Singapore: ISEAS, pp. 200–221.

Vosmaer, J.N. (1839). Korte beschrijving van het Zuid-Oostelijk-Schiereiland van Celebes. *Verhandelingen van het Bataviaasch Genootschap van Kunsten en Wetenschappen* Vol. 17.

Zuhdi, S., Ohorella, G.A. and Said, D. (1996). *Kerajaan Tradisional Sulawesi Tenggara: Kesultanan Buton*. Jakarta: Depdikbud.

Wertheim, W.F. (1999). *Indonesian Society in Transition, Social Change Study*, Yogyakarta: Tiara Wacana.

Borders and mobility in literature and culture

Urban Studies: Border and Mobility – Kerr et al. (Eds)
© 2019 Taylor & Francis Group, London, ISBN 978-1-138-58034-3

Remixed Javaneseness: Lyrics of levelling adiluhung non-adiluhung

E.D. Riyanto
Universitas Airlangga, Surabaya, Indonesia

ABSTRACT: In this paper, we respond to Nancy Florida's work published in 1978, which describes how adiluhung as a trait of Javanese literature was a deliberate construction, which separated it from the non-adiluhung. Here, we investigate whether the separation is still valid in contemporary Java as seen in its popular culture by analysing the lyrics of the songs produced by the Jogja Hip Hop Foundation, especially Kulonuwun and Gangsta Gapi. In this paper, we present remix as the theoretical tools, followed by the nature of Javanese language registers. The lyrics of the songs will be analysed based on their high or low register. In contrast to the separation above, the main impact of the remix in those lyrics is the levelling of adiluhung and non-adiluhung elements of the Javanese culture.

Keywords: adiluhung, Java Hip Hop, levelling, non-adiluhung

1 INTRODUCTION

This paper is a revisit of Nancy Florida's work published in 1978. The title "Reading the Unread in Traditional Javanese Literature" reflects the idea in the article that there were some unread works in the 19th century Javanese literature. Those works were intended to be unread, or should not be read, to make way for the construction of adiluhung Javanese literature. Thus, Florida concluded that adiluhung is a trait of the Javanese literature, which was a deliberate construction carried out by Dutch philologists and Javanese rulers. The deliberate act was carried out by limiting the access to those "unreadable" works for the public. This construction was quite successful, which makes Javanese culture renowned for its refinement. Stumbling on some not-adiluhung works in a Sunanate library in Solo, Florida deconstructed the formulation.

Almost three decades on from the publication of Florida's article, now I would like to present the levelling of adiluhung and non-adiluhung in remixed Javaneseness in the form of Java Hip Hop produced by the Jogja Hip Hop Foundation (JHF). Founded in 2003, JHF has successfully presented Javaneseness in its Java Hip Hop (Riyanto, 2016).[1]

This paper is an attempt to illustrate how a remix of Javaneseness was produced and presented as seen in the lyrics of Kulonuwun (Excuse me, let me in) and Gangsta Gapi (Gangsta Sh*t) as parts of the discography of the Jogja Hip Hop Foundation. These are rare but powerful examples of a Java Hip Hop product, in which the Javanese language stratifications were levelled.[1]

2 REMIX, THE TWO AND OTHER VIEWS

Technology is embedded in the process of remix. In this case, a process of remix happens when an artist "remixes, or quotes, a wide range of 'texts' to produce something new"

1. Previously, JHF had another name: Ki Jarot (Marjuki, Jahanam, Rotra). Marjuki was the leader, Jahanam consisted of Balance and Mamox, and Rotra consisted of Anto and Lukman. Thus, there were five members of JHF. However, since 2017, there have only been four members because Lukman has decided to leave the group.

(Lessig, 2009: 93). Using technology, the remix "happens at different layers", for example it "may quote sounds over images, or video over text, or text over sounds" (ibid.). Although Lessig did not specifically discuss Hip Hop, it is known that sampling is the core of Hip Hop music production and the development of digital technology has accelerated the sampling process in an unprecedented way. Sampling is similar to "quoting" mentioned by Lessig. Tonya M. Evans stated that digital sampling has been "an essential and integrated component to create Hip Hop Music" (2011: 856). Using digital technology, all digitalised sounds can be captured and it "lets the engineer appropriate any sound and bend and twist it to fit onto a new record" (ibid.). Evan continued that the technology "can take any 'sample' of recorded sound, convert it into a series of numbers and manipulate it in virtually limitless ways by changing the numbers".

Thus, the advance of technology plays a vital role in the process of remix. Furthermore, as Evan suspected, instead of an artist, an "engineer" is central in the process of remix. Knowledge of technical bits is more important than artistic creativity. Musical artistry would not be able to be expressed without mastering the technology. The technology has mediated the artists and the expressions. Luckily, the technology has been produced in a much more custom-friendly way so that it becomes easier for most people to use it. The explosive progress of computer hardware and software (such as Fruity Loops, Native Instrument Traktor Po, Mixvibes Cross, just to mentioned a few) has made it much easier to become an "engineer".

However, to understand the remix production in the Javanese context, it needs the other view, which can be traced from Church's proposition that remix is more a recurrent than a new phenomenon. As Church stated, "historical antecedents to remix can be located in the rhetorical tradition" and "societies have not always recognized a specific concept of intellectual piracy" (2013: 24). Furthermore, a "pre-literate culture of orality" produced "cultural narratives and texts in a spirit of collective remembrance", in which "collective interactivity" was dominant in order to "keep cultural records alive" (ibid. 26, 28). From this viewpoint, the train of cultural transformation as the result of Gutenberg's invention is seen as a "new disruption" of a long history of oral traditions. This is because the Gutenberg invention that generated copyright culture has not totally destroyed the oral traditions. I argue that the aforementioned "historical antecedents" and "pre-literate culture of orality" cannot be applied to the Javanese context.

I argue that remix in the Javanese context tends to be a continuation of a long tradition rather than an interruption of an establishment. Two considerations can be mentioned here. The first is the common phenomenon of remix, and the second is the fact of the rarity of copyright issues, which should have been raised because of the great number of remix products.

It might be concluded that remixes have been done for centuries in the Javanese music tradition and are now becoming increasingly popular in practice. Sutton (2010) discussed the centuries-old history of mixing in Indonesian music, especially Java, in which he referred to both remix and hybrid. He defined hybrid as characterised by a process in which "at least somewhere in the process of creation/production, and perception/reception, the mix, however fused or separately perceivable the constituent elements, must be evident as mix" (Sutton 2010: 183). Sutton stressed that this "mix" should be "evident", "foregrounded", "dominant", and becomes a "trait" (ibid.). Sutton explained the hybrid in Javanese music history starting from gamelan to campursari (literally meaning the mix of the essence) and Jazz. In those three cases, real instruments were involved and became a mark of the hybrid, e.g., Javanese gamelan combined with European drum and guitar. More importantly, in contrast to the above notions of contradiction between remix and copyright, Sutton's article constitutes a piece of evidence that remix is not contradictory to the issue of copyright.

In the 1980s, campursari revived and gradually became very popular in the Javanese musical landscape. Supanggah asserted that campursari represented the "new image of Javanese society", showing "a societal shift toward the modern" (2003: 18). Supanggah indicated a problem in the process of modernity by being called pseudo-modern, for example, when Javanese people treated technology not as a means to "make their lives happier" but more "as a status symbol" (ibid.: 4,5). While it is clearly mentioned that campursari took benefits from many existing musical pieces,[2] Supanggah did not mention that there was any indication of copyright issues.

Similarly to Supanggah, who perceived signs of cultural shifts in the Javanese society, Nancy Cooper analysed the campursari by focusing on the cultural meanings of the genre. She scrutinised the contention in choosing between Javanese and Western tuning systems, which reflects "a modernity defined more completely by powerful...outside actors and forces" against "a modernity drawn from local histories and cultural habits" (Cooper, 2015: 55, 59). It can be assumed that this "genre" has deliberately intended to accommodate any possibilities of mixing. However, instead of discussing copyright, Cooper explained about the tensions between Western and Javanese values.

Unlike Cooper, who put more weight on cultural tensions, Sutton, who analysed the tradition of mixing from colonial to contemporary periods, viewed the phenomenon as "a key locus of musical creativity in Indonesia" (Sutton, 2010: 193). Sutton continued that such endeavours and the surrounding contexts deserved close attention. In this paper, I follow Sutton's suggestion but, at the same time, present the issue Sutton avoided, which is the "dichotomous notions" such as "high-brow vs. low-brow...standard vs. deviant" (ibid.: 181).

Thus, in this paper, we fill the gap because what I am discussing here is the mixing of elements originating from different layers of Javanese culture. Furthermore, the thing I discuss here is the lyrics rather than the instrument. Thus, here I discuss the issue avoided by Sutton by scrutinising the same Javanese language constituting high and low registers. This language stratification reflects cultural stratifications of the Javanese society. At the same time, I also avoid the tendency to contradict the local versus global, because what I am discussing in this paper is the tensions inside Java's own local cultural elements.

3 JAVANESE LANGUAGE STRATIFICATION

Javanese people constitute the largest population in Indonesia. The Javanese language is spoken by more than 70 million, and it has become a marker of identity (Suyadi, 2014: 244). Among the traits of the identity is the high stratification of the community as reflected in the language itself (ibid.). "Javanese language recognizes the undhak-usuk system or speech level" (Septianingtias, V. et al.: 27).

Basically, the Javanese language is stratified into two main registers: the high register, kromo, and the low register, ngoko. There is one in between, madya. There are also many "in-betweens" producing more variations, up to nine categories according to some linguists (Wibawa et al.: 45).

As the highest register, kromo inggil (high kromo) "is used when addressing, or talking about, someone with special respect" (Quinn, 2011: 364). Meanwhile, the "ordinary" kromo is used when someone is talking "to people who are socially distant ... older, or of higher social status, or simply not well known" (ibid.). Technically, the distinctions are performed in the appropriateness of the usage of the language or unggah-ungguh, which is "intricate and elaborate" and "probably unique among the major language of the world" (ibid.). The main feature of unggah-ungguh is to show politeness, with the main idea being "to manage 'face' in many different ways so that one's partner of communication" will not be ashamed because of losing face (Sukarno: 61).

These stratifications are in line with Florida's notion of the separation of adiluhung from no-adiluhung. Krama as the highest and most refined register is part of the adiluhung culture. The separation is reflected in the notion of the "right and true" and the "face management" as quoted above. There should not be any misplacement in terms of when, where, and how to speak correctly. Otherwise, it would be a shame for the speaker as it shows his/her inability to put himself/herself in the most appropriate position.

The difference between Florida's account and the language stratification is that the first deals mainly with the literature, while the second is about the day-to-day usage of the

2. To mention only a few examples used by Supanggah in the article: Ki Nartasabda "PrahuLayar" and "AjaLamis" and Gesang's "CapingGunung", "Ali-Ali", and "Yen IngTawang Ana Lintang".

language. In essence, both are the same, that is, separation and stratification. In practice, one is about written literature and the other is more about the daily performance of the spoken language.

In contemporary music contexts, early accounts of the use of local languages such as Javanese in Indonesian underground music showed an agreement with the above differentiation and separation of adiluhung versus non-adiluhung. This separation had been found to be a hindrance for the creative process of the musicians. Local languages had been reported as being used as everyday vernaculars in "the everyday social life of underground fans" in Indonesian cities such as Bandung, Yogyakarta, and Surabaya (Wallach, 2003: 65). However, those languages were seen as having disadvantages because of their limitation in being local and "their association with 'backward' village life" (ibid.).

Although Wallach did not mention specific languages, he did refer to a "more refined register", which can be translated as the kromo register in the Javanese language. For Wallach (ibid.), there are three disadvantages of the local languages, which caused the reluctance of the underground music artists to use them. The first was that this register was considered "inappropriate" because of its association with "elders and traditional culture". This association is in opposition to "modern and youth-oriented" culture in underground music. Second, the language (especially, the florid language of the Javanese court) was "inextricably associated with gamelan and other traditional music". It seemed that Wallach was trying to contrast this traditional music with modern underground music. Finally, the spirit of regionalism in local language is seen as incompatible with the national consciousness of underground music lyrics. Wallach then concluded that "the Indonesian underground seems quite a long way off from achieving any kind of synthesis between 'Indonesian' and 'Western' music" (ibid.: 80).

Wallach published his book in 2003, the same year as the founding of JHF. At that time, the first Hip Hop album in Java had been around for about 7 years. It is possible that the album was undetected by Wallach because it was "too local". However, I would like to discuss this locality more in this paper. Indeed, I focus on how layers in a locality are deconstructed as seen in the remix produced by JHF.

4 THE CHANGING TONE AS THE LEVELLING STRATEGY

The above speech levels, through which the language is divided into low and high registers, have been exploited by JHF. The Rotra song "Kulonuwun", released in 2007, is a mixture of high and low Javanese registers in one six-stanza song. This signifies Hip Hop music, which is strongly characterised by anti-hierarchical expressions and subject positions. The levelling is presented somewhat bluntly in a piece of Hip Hop music. It successfully attracted many listeners by presenting the high register as an elevated form of self-identification, which is then combined with low and coarse language showing a threat to those who defied the "I" presentation. It is part of the tradition of Hip Hop to present and challenge identity.

The three lines in the first stanza of the song reflect two contrasting registers, combined with a moderate register (kromomadya) in the middle. Ndereklangkung, nyuwunsewunjalukdungalanrestu, jirolu!/Akutaknyobamelumlebu/Nek oraentukmengkotakgajulmatamu, preksu! (Please excuse us, please apologise and bless us, one two three!/I will try to enter and join (the performance)/If I am not allowed, I will hit your eyes, damn you dog!).

These three lines are directed to three different audiences. The first is to the spectators, the second is to the singer himself, and the last is to any possible competitors. In rap performance, this is common, especially in the form of a rap battle. Thus, this deployment of different registers to change tones for different audiences fits in with the tradition of rapping in Hip Hop culture.

The second to sixth stanzas are all in low register (ngoko). Those are intended to challenge the competitors, for example, in stanza 2 line 2, Akuora urus kowengomongoporasah do kemlinthi, akuorawedi (I don't care what you say, you don't think that you are tough, 'cause I am not afraid) and stanza 6 line 1 Ayo dadisiji. Ngadepimusuh-musuh sing soyokemaki (Let's unite, to face the ever-growing bullies).

The second part of the song consists of five stanzas. All of them use high register and encourage peace and happiness, for example, stanza 9 line 2 …mbotenparengcrahsuloyo (you are not allowed to create conflicts) and stanza 10 line 1 supadoskawontenannacaktentrem, ademayem, atimarem, mesam-mesemlansumringahugibungah (in order to make a situation of peace, calm and ease, content, smiles and spirited, and also happy).

It may be said that these lines were reflecting the spirit of living in harmony as an important character of the Javanese society. Thus, although it is a rap song and the existence of "enemies" is clearly mentioned, the song ends with the encouragement to be peaceful. There is a clear sign of conflict avoidance.

5 THE REVERSE LANGUAGE (WALIKAN)

To show politeness, Javanese employs indirect strategies of communication. It has become an important part of the Javanese language conventions. It prevents a Javanese from any frontal and direct conflict engagement. For Lukman, as a member of JHF, this politeness is the key marker of being a Javanese and he strives to maintain it.[3]

Javanese indirectness emerges in the slang system known as walikan or reverse language. This Javanese linguistic tool has already been available for almost half a century (Jackson and Rahmat, 2013). In some ways, this can be seen as a "new register". The employment of this "new register" creates a sense of sameness and belonging for speakers (ibid.:147).

This reverse language is a proof that the Javanese language has provided resources to be exploited by its constituents in facing the test of time. In this case, the resource is the Javanese characters, which consist of 20 letters in 4 lines with 5 letters in each line. By exchanging the corresponding letters from the first line to those in the third line and the second line with those in the fourth line, Yogyakarta people created a new and exclusive language. It is exclusive because to understand the language, knowledge of Javanese characters is necessary. With this requirement, this reverse language marks the boundary between in-group and out-group members of certain circles in the Javanese society.

The use of Basa Walikan in Java Hip Hop can be seen, for example, in the naming of the group (Rotra), which means Jogja. Rotra is a member of JHF. Other examples are in the titles of the songs, and in the lyrics, too. The titles of songs using this reverse language are "JagalPabu" (Dog Butcher), "Watch Out, Dab" (Watch Out, Mate), and "Gangsta Gapi" (Ga-pi = ta-hi or human faeces). In the lyric of "Watch Out Dab", the word saciladh appears, meaning bajingan or bastard.

It is a common view in Javanese that words such as "dog", "poo", and "bastard" are considered offensive. Worse still, Javanese stress politeness so that it is almost impossible to use those words in front of other people. However, the slang system of walikan provides a solution for that. Walikan provides a cover for the obscenity so that the content of the expression is still delivered. Only those who are able to uncover the walikan can understand the message. Thus, this has produced a condition of exclusivity; that is, only those in a limited circle of recipients can understand.

By using this reverse language, JHF was successful in presenting those words and subject matters, which are normally taboo. Those taboos should be hidden from display in "adiluhung" Javanese culture. Using Florida's term, JHF has made those "unreadable" subjects readable. As a result, JHF has shown that, in the Javanese culture, there are some "minor" and "not so-civilized" elements such as swearing (saciladh = bastard) and dog hunting to consume dog meat (as seen in the song of Jagal Pabu). Yet, this reverse language shows that JHF still maintains its Javaneseness by presenting these taboos indirectly.

6 CONCLUSION

Unlike Wallach's notion that underground music, such as Hip Hop, was not able to use local language because it was either too raw or too refined, JHF has proved its capability in using both high and low registers of the Javanese language in its lyrics. JHF had defied the

3. Interview with Lukman, 8 March 2016.

traditional separation of kromo—ngoko by using both registers in one piece of its song. JHF has also crossed the boundary of taboo by presenting the obscene words (such as dog and poo) and themes (such as dog meat eating) in public. All of these were possible because they were remixed with Hip Hop music.

From its remix techniques, as seen in the lyrics, JHF was able to mix the high register and low register in one cultural product, such as in the song of Kulonuwun. In other cases, JHF used tone-changing strategies combined with the deployment of reverse language. These remix techniques have proved to be successful in producing popular Java Hip Hop songs.

The impact of this remixing is the deconstruction of the separation between the adiluhung and non-adiluhung. Once it was strongly preserved, the social stratification as reflected in the language has become more fluid. It shows that socio-culturally, Javanese people are changing.

REFERENCES

Church, Scott H. (2013) *All Living Things are DJs: Rhetoric, Aesthetics, and Remix Culture.* Nebraska: The Graduate College at the University of Nebraska.

Evans, Tonya M. (2011) Sampling, looping, and mashing... oh my! How hip hop music is scratching more than the surface of copyright law. *Fordham Intellectual Property, Media & Entertainment Law Journal*, Summer, Vol.21(4), p. 843–904.

Florida, Nancy. (1987) Reading the Unread in Traditional Javanese Literature, Indonesia. *Southeast Asia Program Publications at Cornell University,* No 44 (Oct), pp. 1–15.

Jackson, Nicholas and Rahmat. (2013) Decoding Basa Walikan—A Preliminary Analysis of Yogyakarta 'Reverse' Language. *International Journal of Indonesian Studies*, Vol 1. pp. 141–151.

Lessig, Lawrence. (2008) *Remix making art and commerce thrive in the hybrid economy*. London: Bloomsbury Academic.

Lessig, Lawrence. (2009) *Remix*. Bloomsbury Publishing. From Pro Quest Ebook Central, http://ebookcentral.proquest.com/lib/monash/detail.action?docID=591059.

Pike, Andy. (2015) *Origination: The Geographies Of Brands And Branding.* Chichester: Wiley Blackwell.

Quinn, George. (2011) Teaching Javanese Respect Usage to Foreign Learners. *Electronic Journal of Foreign Language Teaching.* 2011, Vol. 8, Suppl., pp. 362–370.

Riyanto, Edi. (2016) Hip Hop With Attitude, *Inside Indonesia,* 126: October-December.

Septianingtias, V. et al. (2014) Javanese Speech Level In Bargaining And Declining Strategies At Sarinongko Market Of Pringsewu Of Lampung Province: a Sociopragmatic Study. *International Journal of Language Learning and Applied Linguistics World* (IJLLALW). Volume 5 (1), January; 26–42.

Sukarno. (2010) The Reflection of the Javanese Cultural Concepts in the Politeness of Javanese. *k@ta*, Volume 12, Number 1, June: 59–71.

Supanggah, Rahayu. (2013) Campur Sari: A reflection. *Asian Music*, Spring-Summer, 34, 2.

Sutton, R.A. (2010) Gamelan Encounters with Western Music in Indonesia: Hybridity/Hybridism. *Journal of Popular Music Studies,* 22(2): 180–197.

Suyadi, M. (2014) The Use of KramaInggil (Javanese Language) in Family Domain at Semarang and Pekalongan Cities. *International Journal of Linguistics,* Vol. 6, No. 3.

Wallach, J. (2003) Goodbye My Blind Majesty. *Music, Language, And Politics In The Indonesian Underground.* In M. T. Carroll and H.M. Berger (eds) *Global Pop, Local Language*.. Jackson: University Press of Mississippi: 53–85.

Wibawa, Aji P. et al. (2014) Augmented Javanese Speech Levels Machine Translation. *Ecti Transactions On Computer And Information Technology,* Vol.8, No.1 May. pp. 45–55.

Urban Studies: Border and Mobility – Kerr et al. (Eds)
© 2019 Taylor & Francis Group, London, ISBN 978-1-138-58034-3

The expression of cultural values in Sundanese manuscripts of the *Mandala* period

H.M. Lyra, D. Indira & T. Muhtadin
Universitas Padjadjaran, West Java, Indonesia

ABSTRACT: This article reveals and describes the ancient Sundanese culture and their cultural values in order to enrich the people's knowledge about the spiritual contents of the ancient Sundanese culture. The terminology used in the Sundanese manuscript of the *Mandala* period was referenced from Darsa (2013), who mentioned that the ancient Sundanese literature belongs to Sundanese manuscripts inherited by the intellectuals from formal educational society. The methodology used in this study was a qualitative method with a descriptive design. The cultural values expressed in the ancient Sundanese manuscripts of the *Mandala* period are based on five categories, namely the role of cultural values in the relationship between human and God, human and nature, human and society. human and the other human, and human and himself. In fact, one category is interrelated with another. However, it has been claimed that the categorization of these cultural values is artificial and intentionally made.

Keywords: cultural value, manuscripts, Mandala, ancient Sundanese

1 INTRODUCTION

1.1 *Sundanese manuscript*

This article is an effort of the authors to address the despondency of the translator of the book *Tiga Pesona Sunda Kuna* (2009), who states that only a few intrinsic contents of the ancient Sundanese culture are noticed by society. The book was begun by the Dutch scholar J. Noorduyn and continued by his partner A. Teeuw. For the purposes of this study, two manuscripts were chosen from this book.

A Sundanese manuscript is a form of cultural handwriting in which both physic and content relate to the life of Sundanese culture, generally written by Sundanese who have ever lived in Sundanese regions (Darsa, 2012). The authors try to unravel the understanding of what the ancient Sundanese culture is in order to reveal and describe its values. Two ancient Sundanese manuscripts are found in the book of *Tiga Pesona Sunda Kuna*, namely *Para Putera Rama dan Rawana* (The Sons of Rama and Rawana) and *Pendakian Sri Ajnyana* (The Ascent of Sri Ajnyana) manuscripts.

The cultural values expressed in the Sundanese manuscripts of the *Mandala* period are described using the theories of Warnaen (1987), Djamaris (1993) and Soelaiman (2003). Moreover, the general description and functions of the Sundanese manuscripts of the *Mandala* period are described using the theories of Suryani (2008) and Darsa (2013).

1.2 *Methodology*

The methodology used in this research was a qualitative method with a descriptive design. This study describes the data of Sundanese manuscripts of the *Mandala* period, which were delivered after being transcribed into Latin. This qualitative study suggests that the determiner elements are distributed in the language itself (see Djajasudarma, 1993). The general description of Sundanese manuscripts of the *Mandala* period was based on a referential technique.

2 RESULTS AND DISCUSSION

2.1 *General description of Sundanese manuscripts of the* Mandala *period*

The terminology used in the Sundanese manuscripts of the *Mandala period* was referenced from Darsa (2013; 99–100). He described that the ancient Sundanese literature belongs to Sundanese manuscripts inherited by the intellectuals from formal educational society such as *Mandala*. In other words, *Mandala* is a formal educational institution in the monarchy era. The Sundanese manuscripts of the *Mandala* period generally have several characteristics:

1. The materials used were made up of palm leaves such as *lontar* (the leaves of Palmyra palm) and *nipah* (the leaves of Nipa palm), as well as bamboo strips.
2. The stationery used was *péso pangot* (a kind of knife) to slit open, *paku andam* (a kind of nail) and *harupat 'tulang injuk'* (a kind of black sugar palm fibre) to write with ink.
3. The letters used to record or write the language in the manuscripts were the ancient Sundanese letters (XV–XVIII century) and the letters of *Buda/Gunung* (Buddhist letters).
4. The language used was generally the ancient Sundanese language, that is, a temporary dialect of Sundanese which is generally used to express the pre-Islamic literature influenced by Sanskrit and ancient Javanese languages.
5. Physical characteristics include the following:
   a. The string used for binding the pages was made of plant fibre such as *haramay* (the fibre of the banana trunk), *lulub* (the bark fibre of the hibiscus tree) and *areuy* (a kind of overgrowing plant).
   b. *Regula* (the use of uncoloured shadowing lines to arrange writing in an orderly way).

The majority of the Sundanese manuscripts of the *Mandala* period are categorized into religious manuscripts because they mainly describe the highest divine truth, the divine power of universe, the final destination of life, the way of worshipping, etc. (Munandar, 2010). In that era, the main religion followed in Java Island was Hindu.

2.2 *Expression of cultural values in Sundanese manuscripts of the* Mandala *period*

The cultural values expressed in the Sundanese manuscripts of the *Mandala* period are based on five categories, namely the role of cultural values in the relationship between (1) human and God, (2) human and nature, (3) human and society, (4) human and the other human and (5) human and himself. The following sections discuss these five relationships in detail. According to Warnaen *et al.* (1987), the categorization of these cultural values is artificial and intentionally made. In fact, one category is interrelated with another. The role of cultural values in the relationship between one human and another emphasizes loyalty in relation to, for example, social norms, religious views and himself.

2.2.1 *The role of cultural values in the relationship between human and god*

Based on the available data, Sundanese people's belief in the *Mandala* period was inspired by Hinduism. The manuscripts of *Pendakian Sri Ajnyana* (the Ascent of Sri Ajnyana) explicitly indicate the naming of Gods, heaven, *siwa* (siva/a God in Hindu) and *brahmaloka* (the world of *brahma*). Furthermore, the manuscripts of *Para Putera Rama dan Rawana* (The Sons of Rama and Rawana) reveal the naming of *sanghiang* (the article on sacred objects and places), *hiang* (the article on supernatural existence such as God and Goddess) and *batara* (the article on virtuous ancestors). Each of these namings was influenced by the Hindu religion.

The role of cultural values in the relationship between human and God highlights religious devotion, praying and surrendering to God. The religious devotion in this context refers to realization of God (1a). The realization of the human's religious devotion to God is realized by praying. Praying is considered as a wish of hope or a compliment. Through praying, human tries to get closer with God. The human believes that God is a place to ask and wish, God is the ruler of the universe and Almighty God and a person can only try as hard as he can (1b), (1c), (1d):

dosa a(ng)geus kanyahoan (1a)
sin his has been known
'his sin has been known'

candra wulan aditia (1b)
moon full and sun
'full moon and the sun'

deungeun sanghia(ng) angkasa (1c)
with *sanghiyang heaven*
'with the sacred heaven'

kalawan hiang pertiwi (1d)
and *hiang* earth
'and the Goddess of Earth'

2.2.2 *The role of cultural values in the relationship between human and nature*

There are cultures that view nature as a great whole so that human essentially surrenders to it without even fighting. However, there are also cultures that consider nature as an entity that must be fought, in order that human must try to conquer nature. Furthermore, there are also cultures that consider the need for the human to live in harmony with nature (Kontjaraningrat in Djamaris, 1993).

Based on the description presented above, the role of cultural values in the relationship between human and nature involve unification and utilization. Unification considers nature as a living organism that possesses power. This can be seen in cosmology stories which describe the mystical power of nature. Nature gives signs of premonition that will befall the human (2a–d). In the Sundanese manuscripts of the *Mandala* period, the mystical power of nature dominates the storyline. The value of utilization considers nature as a gift of God that must be utilized wisely. Nature as a source of life provides wealth. The human utilizes anything in nature to preserve his life:

Sabuat ta hujan poyan (2a)
that moment fall rain heat
'at that moment heated rain falls'

téka metu angin rebut (2b)
along with wind storm
'along with the storm'

téka ceudeum téka ceukreum (2c)
till cloudy black pitch
'till pitch black cloudy'

ketug lini tujuh kali (2d)
quake earth seven times
'seven times earthquake'

2.2.3 *The role of cultural values in the relationship between human and society*

According to Djarmaris (1993), in the life of the human, problems cannot be solved by relying on himself but by joining together with others. The role of cultural values in the relationship between human and society indicates agreement (3a–c):

--- *nyaho diulah pamali* (3a)
which understand deed forbidden
'understanding forbidden deed'

geus janji beunang matingtim (3b)
already agree result agreement
'already agree with the result of agreement'

tuwawa beunang rarasan (3c)
arranged with full consideration
'arranged with full of consideration'

2.2.4 *The role of cultural values in the relationship between human and the other human*

The other human is separated from the society. The relationship between one human and the other human focuses more on the interindividual relationship. Human must respect and honour the right of others. The role cultural of values in the relationship between human and the other human expressed in the Sundanese manuscripts of the *Mandala* period includes the values of friendliness, love and affection, keeping promises, loyalty, respecting parents, wisdom and willingness to sacrifice.

Sundanese people are widely known for their friendliness expressed by smiling, speaking and behaving properly. The friendliness highlighted in the Sundanese manuscripts of the *Mandala* period is expressed in the forms of greeting and allowing guests to sit and rest as well as by serving them:

1575 .... (4a)
Raden Laksamana lalu duduk
'*Raden Laksamana* then sit'

kemudian dijamu sirih pinang (4b)
afterwards served *sirih pinang*
'afterwards served with *sirih pinang*'

selanjutnya ditanya maksudnya (4c)
later asked intention his
'later asked his intention'

Love and affection are realized not only in love but also in social life. These feelings are shown by someone to others. In the story of *Para Putera Rama dan Rawana* (The Sons of Rama and Rawana), love and affection were shown by the old man *Hayam Canggong* to Sita (Rama's wife). These feelings of the old man *Hayam Canggong* were displayed by his acts of giving home to Sita and how she was taken care of by him. Similarly, love and affection were also expressed by *Patih Sombali* by taking care of an infant (King Manabaya) who was found near the dead bodies of Manodari and Rawana:

heman ku na p(n)ten tandang, (5a)
so love to princess crown
'so love to crown princess'

katineung basana haat, (5b)
remember will kindness her
'remembering her kindness'

According to Widagdho in Djamaris (1993), loyalty can be realized by love, affection, respect or a union. These feelings are expressed willingly, as found in the manuscripts of both *Para Putera Rama dan Rawana* (The Sons of Rama and Rawana) and *Pendakian Sri Ajnyana* (The Ascent of Sri Ajnyana). In the manuscript of *Para Putera Rama dan Rawana* (The Sons of Rama and Rawana), the loyalty of *Patih Sombali* was shown to *Manondari* (his king Rawana's wife). Meanwhile, in the manuscript of *Pendakian Sri Ajnyana* (The Ascent of Sri Ajnyana), the loyalty was shown by a woman to a man:

Karajeun, lamun dék mangkat (6a)
very well should you want go
'very well, should you want to go'

na aki mungku tinggaleun (6b)
now old man not will silent
'now this old man won't be silent'

mo nyorang moha ka sia (6c)
not will despise to you
'won't despise you'

Parents can refer to the concept of mother–father or people who are older than us. This concept appears in the cultural value of 'respecting parents' in the Sundanese manuscripts of the *Mandala* period:

Ambu, aing meré nyahu (7a)
mother I ask leave
'mother, I ask permission to leave'

pahi deung tuang ayah' (7b)
as well to father
'to you as well father'

The cultural value of willingness to sacrifice in the manuscript of the *Mandala* period is explicitly stated in the sentence "*menyatakan hendak turut berkorban dan berkorban demi sang Rawana*" (willingness to sacrifice for Rawana):

awaking sok béla tapa, (8a)
I only can help pray
'I can only help praying'

ngabélaan sang Rawana berkorban demi sang Rawana (8b)
sacrifice for the *Rawana*
'sacrifice for the Rawana'

2.2.5 *The role of cultural values in the relationship between human and himself*

The relationship between human and himself is always described by personal characteristics that are recognized as his typical behaviour. This typical behaviour can only be found through the impression or interpretation of consistency shown by certain observable behaviours (Warnaen, 1987). This behaviour shows good and bad characteristics. From these data, the authors classify the cultural values which view human as being himself:

Saurna Bujanggalawa (9a)
'*Bujanggalawa* answers'

Ambu ulah kita borang (9b)
mother not afraid
'do not afraid, mother'

saha nu ngadeuleu-deuleu (9c)
who which see sarcastic
'whoever see sarcastically'

ingkeun pilacaneun aing (9d)
let that which become enemy my
'let it be my enemy'

saha oge lakina' (9e)
who he
'whoever he is'

3 CONCLUSION

The cultural values expressed in the Sundanese manuscripts of the *Mandala* period are based on the five categories described in this article, namely the role of cultural values in the relationship between (1) human and God, (2) human and nature, (3) human and society, (4) human and the other human and (5) human and himself. These five categories are inherited in the life of the human. As this finding has not been reported in any of the previous research in the literature, the authors suggest that the transcribed and translated Sundanese manuscripts become the object of multidisciplinary research.

REFERENCES

Darsa, Undang Ahmad. (2012) *Sewaka Darma dalam Naskah Tradisi Sunda Kuno Abad XV-XVII Masehi.* Dissertation. Bandung: Graduate School of Universitas Padjadjaran.

Djamaris, Edward. (1993) *Nilai Budaya dalam Beberapa Karya Sasra Nusantara*. Jakarta: Ministry of Education and Culture.

Ekadjati, Edi S. (1995) *Nusa, Bangsa, dan Bahasa.* Bandung: Faculty of Humanities Universitas Padjadjaran.

Muhtadin, Teddi, et al. (2011) *Kekayaan Batin Kaum Intelektual Sunda Abad 16: Kajian Ungkapan Bahasa dalam Naskah Sunda Kuno di Kabupaten Garut. Laporan Penelitian Muda* (LITMUD). Bandung: Universitas Padjadjaran.

Munandar, Agus Aris. (2010) *Tinjauan Napas Keagamaan Hindu-Buddha dalam Beberapa Naskah Sunda Kuno (Abad ke-14-16M)* page 27 in Jumantara Journal. Jakarta: National Library of Republic of Indonesia.

Noorduyn, J. and Teeuw, A. (2009) *Tiga Pesona Sunda.* Jakarta: Pustaka Jaya.

Setia, Edi, et al. (1990) *Fungsi dan Kedudukan Sastra Lisan Melayu Serdang*. Jakarta: Pusat Pembinaan dan Pengembangan Bahasa (Language Center) Departemen Pendidikan dan Kebudayaan.

Soelaiman, Darwis A. (2003) *Nilai Budi Bahasa dalam Budaya Melayu-Aceh. Warisan Budaya Melayu Aceh.* Banda Aceh: Research Center of Melayu-Aceh, 26–36.

Suryani, Elis. (2008) *Merumat Warisan Karuhun Orang Sunda yang Terpendam dalam Naskah dan Prasati.* Jatinangor: Alqaprint.

Warnaen, Suwarsih, et al. (1987) *Pandangan Hidup Orang Sunda Seperti Tercermin dalam Tradisi Lisan dan Sastra Sunda.* Bandung. Ministry of Education and Culture.

Urban Studies: Border and Mobility – Kerr et al. (Eds)
© 2019 Taylor & Francis Group, London, ISBN 978-1-138-58034-3

Criticisms of the depiction of freedom of characters in Dewi Lestari's novel entitled *Supernova: Kesatria, Putri dan Bintang Jatuh*

M.N.A.T. Gemilang
Universitas Airlangga, Surabaya, Indonesia

ABSTRACT: Dewi Lestari's novel entitled *Supernova: Kesatria, Putri, dan Bintang Jatuh (KPBJ)* depicts problems related to phenomena typical of social realities in real life such as homosexuality, love affairs, and prostitution. Despite the fact that those things are not in accordance with the social values or even highly prohibited, humans actually have the right and freedom to choose their own life path. Therefore, the researcher aims to analyse the novel from the perspective of freedoms of the characters. This article uses Jean Paul Sartre's theory on existentialism to analyse the freedom of the characters. According to Sartre, humans are creatures with freedom that have the right to take life decisions on their own but are still liable to the consequences of the actions they take. The research is conducted using a qualitative approach and descriptive method.

From the perspective of a society that adheres to the social norms and values, homosexuality, love affairs, and prostitution are considered complex issues of a highly delicate nature. In a country like Indonesia, where religious and cultural norms and values are highly regarded, the complexity of these issues is far more undeniable. This hypothesis often obscures the limit of morality and immorality, freedom and oppression.

1 INTRODUCTION

Literary works are essentially inseparable from real life phenomena. In fact, literary works are the depiction of real life experiences in the form of social reality. One of many novels involving the idea of existentialism is *Supernova: Kesatria, Putri, dan Bintang Jatuh* written by Dewi Lestari. In this novel, Dewi Lestari invites the readers to think about the essence of being a human. Various characters with different behaviours are presented to explain how humans have the freedom to make decisions for every action they commit.

Existentialism emphasises freedom as the most essential trait of humans. Existentialism refers to concrete entities, individual, and dynamic, stating that each person has the ability to learn from what they experience in actuality. Existentialism also dictates that humans are prone to avoid dealing with abstract things leading to various speculations. The core concepts of existentialism are related to freedom and existence. In this research, the author aims to focus on the theme related to freedom.

There was previous research on a similar topic conducted by Wahidin Hendrayana entitled *Nilai-nilai Eksistensilisme dalam Trilogi Novel Paulo Coelho.* This research uses the study of existentialism as taught by Muhammad Iqbal, while this research is conducted applying the analysis of existentialism as taught by Jean Paul Sartre by using a qualitative approach and descriptive method.

The research questions are formulated as follows 1) How is the freedom of the characters in *Supernova: Kesatria, Putri, dan Bintang Jatuh* depicted in relation to social interactions? 2) How does the sense of freedom affect the life decisions taken by the characters in *Supernova: Kesatria, Putri, dan Bintang Jatuh* considering the responsibilities they have to take for every action they commit?

2 DISCUSSION

2.1 *Jean Paul Sartre's existentialism*

According to Misiak and Sexton (2009: 83), there is no philosophical figure more prominent than Sartre (1950–1980), especially when novels, dramas, and articles tend to popularise Sartre's ideas of existentialism. Those who believe in existentialism often neglect the common social norms, since the concept of freedom has become their rigorous measure for any actions they commit. This is in accordance with Sartre's remark (in Graham, 2005: 127) stating that:

> The existentialism of a humanism in relation with its meaning is based on individuals and their personal values as well as the realization of humans' real identity as a living creature. He understands life as a quest for self-identity and emphasizes a belief that the only way for realizing one's self is knowing that humans' personal freedom constitutes their real existence, and that they are held responsible for every action they commit while their life is determined by a complete nothingness. Sartre always sees humans' attempt to escape their responsibility as "unacceptable belief", and he also believes that humans can never escape from their burden of personal loneliness.

Sartre (in Misiak and Sexton, 2005: 84) states that the freedom and ability to make a choice is humans' most essential trait. Freedom is not merely a causality or attribute owned by humans since humans themselves constitute freedom and thus all choices are theirs to take. It is then inevitable that humans cannot avoid any consequence resulting from their decisions, and hence freedom is the only thing in life they can always depend on. Sartre also believes that humans are meant to be free and thus they will always be what they choose to be themselves. If they attempt to escape from their freedom, they will suffer from nausea, anxiety, sadness, and desperation.

2.2 *Main themes in Jean Paul Sartre's existentialism*

1. Freedom

Humans have the freedom to choose and make decisions on their own. This is in accordance with Sartre's ideas (in Dagun, 1999: 106) which always leads to the concept of freedom defining humans as a form of freedom itself. In Sartre's opinion, humans are creatures whose existence had long existed before essence. What is the definition of humans? What is their essence and to what purpose do they exist? In Sartre's belief, these kinds of question will finally lead to absurdity because it is always possible for humans to say either yes or no. From the previous explanations, there are several types of freedom suggested by Sartre:

a. Freedom in Human-to-Human Interaction

According to Sartre (in Abidin, 2014: 85), each individual chooses his/her existence, because he/she is the holistic and interrelated entity, so that it eventually becomes quite possible for him/her to express his/her range of choices in each aspect of his/her actions.

Sartre does not reject the ideas proposed by psychology, but instead, tries to integrate life as a whole. This is in accordance with the concept of existentialism as an idea which is not meant to lead a person towards inhumanity, as argued by Sugiharto (2008: 174). In fact, existentialism tries to take part in answering a number of fundamental issues related to life. Existentialism is neither a group for promoting anti-social behaviour nor a group for anti-humanism. The ideas of existentialists such as Sartre, Heidegger, Jaspers, and Marleau-Ponty focus on explaining the fact that humans are indeed personal creatures. However, this personal nature does not restrict itself to a concrete and rigid identity. The personal nature of existentialism is meant to make humans more open to the world and any other entities and phenomena outside their "selves".

In Sartre's opinion, humans' relationships and the basic functions of their physical body play an important role because humans' existence is represented not only by their consciousness but also their physical bodies. Therefore, their existence is observable as long

Table 1. Indicator of the problems.

No	Pokok permasalahan	Indicator
1	Criticisms of freedom	– Homosexuality – Love affairs – Prostitution
2	Freedom in humans' relationships	– Indifference – Love – Hate – Sexuality
3	Freedom and its relationship to responsibility	– Responsible freedom

as they exist as a whole entity. If the "self" within my physical body is under others' control, then "I" have jeopardised my freedom, while humans can never exist without their physical entity as well (Muzairi, 2002: 167).

Several forms of relation between humans' entity exist and can be noted from the act of indifference, love, hate, and sexuality. Some of those relations can explain Sartre's opinion stating that a human's existence is always interrelated with that of others as well as reciprocal (in Muzairi, 2002: 180).

b. Freedom and Responsibilities

According to Koeswara (1987: 14), freedom is a concept that is the most essential element of existentialism, since existentialism always emphasises freedom as the most essential trait of humans. Existentialists see freedom as something that is strongly related to responsibility in taking decisions. Humans are both free and held by their responsibilities to make decisions and to make their choice when committing an action in order to form the life and existence of their "selves". For this reason, Nietzsche refers to freedom as humans' capacity to decide what they will become and how they act on that decision. Sartre restated Nietzsche's argument in his well-known remark: "I am everything I choose to be".

Humans' existence in this world begins with their subjectivity. In existentialism, the most fundamental basis is the acknowledgement of "my self just the way I choose it to be" commonly referred to as mineness. At this point, existentialists place humans' subjectivity in the highest position of importance. As noted by Sartre (in Sugiharto, 2008: 175), humans' subjectivity can be understood in two categories which are humans' individual freedom and humans' capability in dealing with their "selves".

Sartre noted that humans have to be responsible, and responsibility in terms of existential subjectivity itself is not simply understood as a responsibility which is restricted only to shallow individualism, but rather the responsibility towards others as well. If one understands existentialism thoroughly, the subjectivity aforementioned is not a restriction to one's personal self, but rather the ability of an individual to responsibly make a subjective judgement while taking responsibility in terms of others (Sugiharto, 2008: 175–176).

In order to make the research more well-structured and focused, some indicators regarding the object of the study are presented as follows:

The researcher used data coding as a way of referring to the object of the study to ease the process of analysis using SKPBJ/SATA/2012:16 with a description of SKPBJ (*Supernova: Kesatria, Putri dan Bintang Jatuh*), SATA (stands for the indicators of the issue), –1 (showing the order of the data), 2012 (referring to the year the book was published), 16 (number coming after the year is the page of the book.

2.3 *The freedom of the characters and its relationship with human interaction depicted in the novel supernova: Kesatria, Putri, dan Bintang Jatuh*

Homosexuals are a pair of lovers consisting of two males. One of many common ways to express human-to-human relationships is through love. Despite the fact that their relationship is often considered unorthodox, humans basically have the right to have their own will and act on it based on their conscience. The gay couple in the novel even wrote a masterpiece for their love story to be remembered. This can be seen from the extract from the novel below:

> "*Ten years from now, I will have to write and have a literary work of my own. A masterpiece. One piece of writing or a research that will bridge all discussions concerning science*".
> "*Ten years? Why does it have to take so long anyway?*".
> "*Time flies, my friend*".
> "*Fine, you have ten years and so do I*".
> "*One masterpiece. A novel with a broad dimension that will affect people's hearts in a way they have never imagined*". (SKPBJ/SC_2/2012: 12).

They choose to write a masterpiece rather than doing negative things. It has been proven that they are a gay couple with no intention of openly expressing their social desire. This is expressed in an extract from the novel as follows:

> "*... Neither are we a gay couple that express lust in public sphere as commonly assumed by most people. We are best friends. Life partners*". (SKPBJ/SC_3/2012: 112).

From the extract above, it can be understood that Ruben and Dimas choose to have a real existence through an unusual love relationship. Apart from Ruben and Dimas, freedom in a relationship is also depicted by the married couple Arwin and Rana. Arwin really loves Rana, despite the fact that Rana has a love affair with Ferre. Arwin accepts that Rana has treated him that way, and this is depicted by the extract from the novel below:

> "*... I'm sorry for not being able to be the one you've always wanted. But I really love you as a wife...or not. You will always be Rana that I admire. And, I am sure that there is nothing like my feelings to you. And I wish you knew this is true*". (SKPBJ/SC_4/2012: 229).

The decision taken by Arwin shows that he chooses to be a wise man and have such a relationship and express his abundant love for his wife, rather than getting becoming angry and hating his wife for betraying him. The decision taken by Arwin results in harmony within his family. In expressing his existence and making his choices, de does not lose his wife, as explained by the extract below:

> *... Now it's Arwin's turn to be a little surprised as his wife embraces him and never lets him go. It does not feel like a hug for goodbye, but rather it feels like he's welcoming his wife back*. (SKPBJ/SC_5/2012: 230).

By making his choice, Arwin has regained his beloved one. This shows that freedom in making choices will lead to a "pride" as noted by Sartre in his explanation about existential humanism: humans should take "pride" in being the only centre of reality that preserves their autonomy from forceful powers coming from outside their selves (Abidin, 2014: 189). Other than love, humans' freedom is also shown in sexuality. This is an illustration of common human symptoms related to Sartre's existential phenomenology concerning the physical body, "my physical body for others" in which we have the conscience about our physical body and its functions as well as its relation with "others". A character named Diva is a beautiful woman working as model with a side job as a high class prostitute. This is explained in the extract below:

> "*... If you want me to wear it in front of you, there is going to be an extra fee. I don't want you to run out of money*". (SKPBJ/HS_7/2012: 95).

The extract above shows that Diva is a character who is really consistent when it comes to her body. Diva is a smart and diplomatic figure who is academically rigorous but prefers to be a prostitute. This proves the fact that existentialism is a sign of humans having the freedom to choose an attribute attached to the self. This is in accordance with an extract from the novel below:

"Div", he called her gently, "sometimes I think you are far smarter than my CEO. And I wonder why you choose this profession instead. With a knowledge such as yours, you can even get a position better than mine."

That woman smiles sarcastically. "The fact that I am smarter than her is the reason why I do not want to work like you. Is there any difference between my profession and your profession? I already told you that we are both do the trading, only the commodity is different. What you sell is a thing that should not be sold, in my opinion. My mind has to be free. I never do trading recklessly, anyway." (SKPBJ/HS$_{-8}$/2012: 79).

It can be understood that every person has their own freedom, because humans are destined to be free. The extract with Diva's remark explains the fact the existentialism does not lead people to anti-humanism. As Sartre explained, humans are personal creatures who do not restrict themselves in a rigid identity since existentialism is supposed to make them more open to the world and others.

2.4 *Freedom of the characters and their responsibility in making decisions and committing actions depicted in the novel supernova: Kesatria, Putri, dan Bintang Jatuh*

Humans always make all necessary effort to turn what they want into a reality, and they can seek freedom while still taking the responsibility for every decision they make and every action they commit by the process of thinking. This is shown by the extract below:

"… Actually, I can say that there is no one winning or losing today. You now look so cute despite the fact that you will be fat, short and have lots of pimples on your face".

"Don't forget to keep playing once you got home. Do not wear those shoes of yours and certainly, no need to use your mom's lipstick…" (SKPBJ/KBJ$_{-1}$/2012: 85).

In reality, there are so many people who take pride in such a thing, but not Diva. Diva was one of the judges in that event. Unlike the other judges, she decided to talk that way so that she ended up irritating the other judges. From this point, it is clear that each individual has the liberty to choose certain actions they want to commit in their life. In addition, Diva also demonstrates an act of freedom while still taking responsibility as noted in an extract below:

… She pays the tuition for his children's school, and even pay for all the extracurricular courses they take. She also gives books. Mr. Ahmad's wife was even sent to a sewing course and asked to open a small library for the surrounding neighborhood. Of course, they are all done in Diva's expense. (SKPBJ/KBJ$_{-2}$/2012: 174).

The uniqueness of the character named Diva lies in the fact that, despite her profession as a prostitute, she always cares for her surroundings. Besides, Diva also makes some decisions and commits some actions that are categorised as courageous and free. Despite the fact that she is actually a prostitute, she is really good at putting together an argument and understands idealism introduced by Marx and other philosophers so that her customers are often made curious. This motivates one of her customers to suggest her buying a university diploma as explained by an extract below:

…. "Just get rid of it, Sir", said she, "just wait until I build a school of my own. School should be a place for learning, not for searching a diploma". (SKPBJ/KBJ$_{-3}$/2012: 97).

One of Sartre's most famous remarks is "I become whatever I choose myself to be" (Koeswara, 1987: 15). From the extract above, it can be understood that Diva has a strong belief and principles leading her to ignore the opinion of others and stick to what she always believes. It shows that existentialism is depicted by Diva, in terms of responsibility for every action she commits. She has personal responsibility.

2.5 *Criticisms of freedom depicted by characters in the novel supernova: Kesatria, Putri, dan Bintang Jatuh*

This research presents several phenomena that commonly occur in social reality such as: homosexuality, love affairs, and prostitution. From all these phenomena, there are several criticisms on freedom depicted the characters of Ruben, Dimas, Arwin, and Rana. Freedom can be expressed indifference, hate, love, and sexuality. However, this novel only depicts humans' freedom from the perspective of love and sexuality. Ruben and Dimas represent a critic on humans' freedom in love. They are a gay couple, and this is often considered as a taboo. However, their actions represent the fact that humans have the freedom and rights to form their selves and commits certain actions based on their conscience.

Criticisms of freedom are also shown by Arwin's love to his wife, Rana. Arwin knows that Rana has a love affair with Ferre but he decides to be wise in dealing with his problem. Despite all the betrayals his wife has done to him, he does not express his feeling in anger, just to avoid losing his wife. Arwin shows his unconditional love to his wife. He expresses his love in a choice he takes and take a pride for it. Criticisms of human freedom are also represented by Diva in terms of sexuality. It can be said that Diva has neglected all the common norms, considering the fact that humans have their freedom. All the things she does explain the concept related to "my physical body for others", stating that our physical body is closely related to our interactions with others. Existentialism dictates that humans are free to choose attributes that are attached to their selves.

Freedom in making decisions is also represented by Diva. Despite the fact that she has a side job as a prostitute, she takes responsibility such as telling the kids at the modelling event to keep playing like any other normal kids, paying the tuition for her driver's children, criticising her customers regarding their opinions on education, and planning to build a school where education is the sole purpose, not a diploma. This has proven that the most essential elements of existentialism is not only freedom but also a sense of responsibility.

3 CONCLUSION

From the explanation above, there are several conclusions that can be formulated based on the idea that existentialism is meant to be a way of finding the real self. Those who believe in existentialism have more freedom than those who don't. For that freedom, they are usually held responsible for every action they commit. Freedom in terms of human relationships shows that existentialism is neither a group based on anti-social behavior nor anti-humanism, nor a group that isolates itself from others. Existentialism is openness to the world and others. Human freedom can be expressed in love, indifference, hate, and sexuality. Freedom as a form of responsibility cannot be understood in a shallow way, but rather it has to be understood thoroughly.

REFERENCES

Abidin, Z. (2014). *Philosophy of Human, Understanding Humans through Philosophy*. Bandung: PT Remaja Rosdakarya.

Anwari, I.R. (2014). Karya Sastra: Antara Propaganda Pemerintah dan Media Kritik Sastrawan Masa Pendudukan Jepang 1942–1945. *Seuneubok Lada.*

Dagun, S.M. (1999). *Filsafat Eksistensialisme.* Jakarta: PT. Rineka Cipta.

Graham, H. (2005). *Humanist Psychology.* Yogyakarta: Pustaka Pelajar.

Koeswara, E. (1987). *An Introduction to Existential Psychology.* Bandung: Rosda Offset.

Lestari, D. (2012). *Supernova: Ksatria, Putri dan Bintang Jatuh.* Yogyakarta: Bentang Pustaka.

Manuaba, I.B. (2002). Martabat Manusia dan Keadilan Sosial dalam Novel-novel Pramoedya Ananta Toer.

Misiak, Henryk, & Sexton, V.S. (2009). *Phenomenological, Existential, and Humanist Psychology.* Bandung: PT. Refika Aditama.

Muzairi. (2002). *Jean Paul Sartre's Existentialism.* Yogyakarta: Pustaka Pelajar.

Saraswati, R. (2012). Aspek Religiositas dalam Dua Novel Eric-Emmanuel Schmitt: Monsieur Ibrahim and the Flowers of the Koran dan Oscar and the Lady in Pink. *ATAVISME.*

Sugiharto. (2008). *Humanism and Humanities: Their Relevance to Education.* Yogyakarta: Jalasutra.

Urban Studies: Border and Mobility – Kerr et al. (Eds)
© 2019 Taylor & Francis Group, London, ISBN 978-1-138-58034-3

Translation ideology recommendation for translating cultural issues in children comics from English into Indonesian: Crossing the borders between language and culture of SLT and TLT

Nurlaila, M. Nababan, Djatmika & R. Santosa
Universitas Sebelas Maret, Surakarta, Indonesia

ABSTRACT: In this study, we aim to propose a recommendation of translation ideology selection to be implemented in translating cultural issues in children comics. To achieve this aim, we measure translation ideology tendencies of children comic translators. The data used in this study are texts related to cultural issues in children comics, translation techniques applied to translate the issues, and translation quality scores of the translation of the issues. Document analysis is used for data collection by comparing the English version of the issues and their translated Indonesian version, noting the translation techniques used and measuring the frequency of use of these techniques. Questionnaires are used to collect translation quality scores. Validation is conducted by focus group discussion involving three translation experts and four doctoral program students majoring in translation. The result of this study shows that 15 translation techniques are used with a frequency of 1,680 to translate 1,617 data of cultural issues. Three out of the 15 techniques, with a frequency of 470, are categorized into source language-oriented techniques, while the rest, with a frequency of 1,210, are categorized into target language-oriented techniques. It indicates that the translation ideology tendency used by translator is domestication. This tendency contributes to the high level of translation quality of the cultural issues. As a result, domestication is recommended for translation of cultural issues in children comics from English into Indonesian.

Keywords: Cultural issues, Children comic, Translation ideology recommendation, English, Indonesian

1 INTRODUCTION

There are three interrelated theories involving translation techniques, translation methods, and translation ideologies, which are applied to solve verbal text problems arising in translation. The three theories are applied in different levels of text. Translation techniques are used to solve problems in the micro level of text, whereas the translation method is used for the macro level of text (Molina & Albir, 2002: 507–509; Newmark, 1988: 81). The micro level of text can be defined as small units involving word, phrase, clause, and sentence (Newmark, 1988: 65; Porter, 2005: 300). Thus, translation techniques are used to solve problems in the aforementioned four levels, while translation methods are more global. They are applied for paragraphs or for the whole text. On the contrary, translation ideologies are beyond the text. They are chosen before starting translation processes and determining translation methods (Venuti in Baker, 2001: 240). Translation techniques, methods, and ideologies are interrelated, as the choice of translation techniques is determined by translation methods specified and translation methods are determined by the translation ideology chosen.

According to Zohar (1990), translation ideologies are closely related to polysystem, which denotes positions of the literature in a country: primary and secondary positions. A country lacking local literature and fulfilling its literature procurement mostly by import will lead imported literature to occupy the primary position and local literature to occupy the secondary position. By occupying the primary position, the literature will become central and a model that leads the direction of other literatures. Meanwhile, if its literature procurement has been fulfilled by local literature, which will occupy the primary position and the imported literature will only be complements and occupy the secondary position. The relationship between positions of imported literature with translation ideologies, which will be applied to translate them, is that if they occupy the primary position, then they will tend to be translated by using foreignization ideology, whereas if they occupy the secondary position, then they will tend to be translated by using domestication ideology (Zohar, 1990; Venuti in Baker, 2001).

By using the foreignization ideology, the result of the translation will introduce foreign cultures and languages. The application of this ideology will lead to the use of source language-oriented methods for the whole text involving word-for-word translation, literal translation, faithful translation, and semantic translation (Newmark, 1988: 45). In the level of words, phrase, clause, and sentences, three techniques can be used, including pure borrowing, naturalized borrowing, and literal translation (Molina & Albir, 2002: 509–512). The application of this ideology, methods, and techniques will make target readers realize that target language texts (TLT) they read are translated versions of foreign literature.

On the contrary, by applying the domestication ideology, a source language text (SLT) will be translated by conforming to the local literature trend in order to be accepted in markets. The result of this translation will sound similar to local products. The application of this ideology will lead to the use of target language-oriented methods for the whole text involving adaptation, free translation, idiomatic translation, and communicative translation (Newmark, 1988: 45). In the level of words, phrase, clause, and sentences, there are 17 techniques to be used, including adaptation, amplification, calque, compensation, description, discursive creation, established equivalent, generalization, linguistic amplification, linguistic compression, modulation, particularization, reduction, substitution, transposition, and variation (Molina & Albir, 2002: 509–512). The application of this ideology, methods, and techniques will make target readers feel that they read local literature. Its relation is shown in Figure 1.

The objective of this study is to propose a recommendation of translation ideology selection to be implemented in translating cultural issues in children comics. The recommendation is constructed based on the empirical data found in the translation of Donald Duck comics from English into Indonesian. We will first identify the translation ideology applied in the translation of cultural issues in Donald Duck comics. Then, we will test the quality of the translation. The ideology that produces a qualified translation of cultural issues will be suggested to be applied. The recommendation of the use of the translation ideology for cultural issues can guide translators and ease them to translate cultural issues in children comics.

The ideology discussed in this study is two ways of translation orientations: whether translators tend to translate a text by orienting to target language (TL), which is called domestication, or tend to translate a text by orienting to source language (SL), which is called foreignization (Venuti in Baker, 2001: 240). Another study discussing the translation ideology was conducted

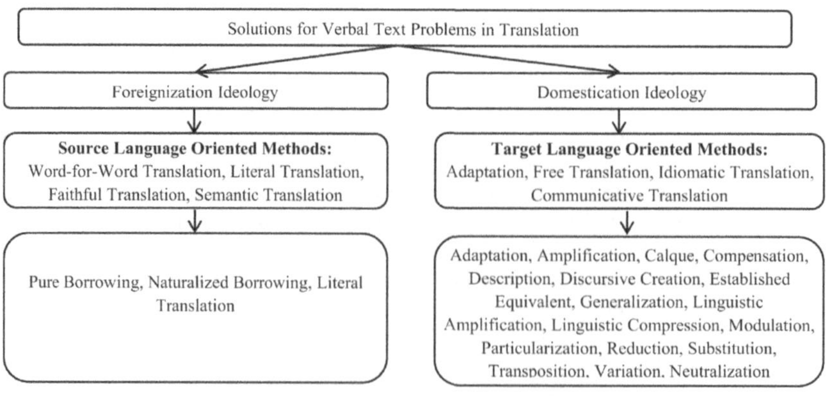

Figure 1. Relation of translation ideologies (Venuti in Baker, 2001), methods (Newmark, 1988) and techniques (Molina & Albir, 2002; Berezowski, 1997).

by Sharifabad, Yaqubi, and Mahadi (2013) aiming to measure the use of domestication and foreignization ideology on English-Persian Translation of News Phrasal Verbs. The result of that study shows that domestication ideology tends to be used, mainly, to translate culture-specific terms and words, which are difficult to find their equivalent in TL. Another researcher, Siregar (2015), conducted a study to identify the translation ideology used in Stephen R. Cove's The 8th Habit from English into Indonesia. The result of that study shows that the foreignization ideology tends to be used (50.70%) rather than the domestication ideology (49.30%).

In the following, we show that the foreignization ideology tends to be used rather than the domestication ideology. The difference between this and the aforementioned studies is that this study aims to identify the use of the translation ideology for translating cultural issues on children comics. This study does not focus on all texts in the comics, but rather focuses on words, phrases, clauses, and sentences, which are categorized as cultural issues. Other words, phrase clauses, and sentences outside the category are categorized into contexts that cannot be separated from data in determining translation techniques as a micro part, which must be identified before identifying the translation ideology.

This study uses children comic as data source as comic is a favorite literature for children as it provides simple texts to read and is full of pictures (Boltz, 2007: 9). Besides that, even though Indonesia produces local comics, the procurement of children comics in Indonesia is still fulfilled by importing. Thus, a study describing translation ideology currently applied by translators and recommending the ideology suitable for translating children comics is needed in order to help translators provide qualified translated comics for children.

2 METHODOLOGY

This qualitative study uses the case study method, the data source of which is five series of Children Comics "Donald Duck", with their Indonesian Translation compiled in one bundle of comic entitled "Komik Terbaik Disney Karya Carl Bark" and six respondents who give translation quality scores through questionnaires. The data obtained from the comics are 1,617 cultural issues in the form of words, phrases, clauses, and sentences contained in the Donald Duck Comic panels and their translation. These data are collected by document analysis and taken by purposive sampling. The researcher put the words, phrases, clauses, and sentences in a two-column table to compare the English version of Donald Duck comic (SLT) and its translation in Indonesian (TLT). Both the SLT and TLT are compared to identify translation techniques as micro units of translation ideology.

3 DISCUSSION

The analysis of 1,617 data shows that they are translated by using 15 translation techniques, with the frequency of 1,685. The translation techniques used are pure borrowing (269), naturalized borrowing (208), literal technique (4), established equivalent (317), adaptation (271), neutralization (197), deletion (189), generalization (71), explicitation (63), discursive creation (61), reduction (13), implicitation (12), particularization (7), modulation (2), and description (1). All of the 15 translation techniques are classified into two categories: source language-oriented techniques and target language-oriented techniques. Three out of the 15 translation techniques are classified into source language-oriented techniques: pure borrowing, naturalized borrowing, and literal technique with a total frequency of 481.

The remaining 12 translation techniques with a total frequency of 1,204 are classified into target language-oriented techniques, including established equivalent, adaptation, neutralization, deletion, generalization, explicitation, discursive creation, reduction, implicitation, particularization, modulation, and description. It indicates that the translation ideology tendency of cultural issues translation in Donald Duck Comic is domestication, as shown in Table 1.

The quality of translation is measured by the following three aspects: accuracy, acceptability, and readability (Nababan, Nuraeni, & Sumardiono, 2012: 50). Questionnaires scores show that the use of the 15 translation techniques to translated 1,617 data of cultural issues

Table 1. Frequency and orientation of translation techniques in Donald Duck comics.

No	Technique orientation	Translation techniques	Frequency	Total	Ideology
1	Source language	Pure borrowing	269	481	Foreignization
2		Naturalized borrowing	208		
3		Literal	4		
4	Target language	Established equivalent	317	1204	Domestication
5		Adaptation	271		
6		Neutralization	197		
7		Deletion	189		
8		Generalization	71		
9		Explicitation	63		
10		Discursive creation	61		
11		Reduction	13		
12		Implicitation	12		
13		Particularization	7		
14		Modulation	2		
15		Description	1		

resulted in a high level of quality. The scores show that the translation of 1,617 data of cultural issues is accurate (2.6 out of 3), accepted (2.4 out of 3), and easy to be understood by target readers (2.9 out of 3). Discussions will be presented in the following paragraphs completed with examples taken from cultural issues data in five series of Donald Duck comics.

Pure borrowing, naturalized borrowing, and literal technique are classified into source language-oriented techniques as their characteristic, which tends to keep SLT forms. By using pure borrowing, SLT terms can be directly taken to be used in TLT without any change. Pure borrowing can be applied in some cases such as an SLT personal name *Edgerton*. In the Donald Duck comics, the name is borrowed to be used in the TLT without any change. A personal name can also be translated using naturalized borrowing, that is, by taking SLT names and adjusting them to the pronunciation and linguistic rules of the target language. One example is an SLT personal name *Donald*, which is translated into *Donal*. A double consonant "ld" pronunciation is not appropriate in Indonesian so that the dominant consonant pronounced, *l*, is taken. Both of the techniques lead to the high level of accuracy as they do not change, add, reduce, or omit the meaning of SLTs. However, the use of pure borrowing may reduce the acceptability and readability scores as it presents foreign texts in the TLT.

Literal technique is applied to translate the SLT terms word by word and out of context. This technique leads to the low scores of accuracy, acceptability, and readability as the meaning of SLT is not transferred in the TLT. In addition, its translation results are not commonly used in the target language and are not familiar for target readers. One example is an interjection *oh, my stars!* that is translated into *oh, demi bintang—bintang*. Even though the meaning of each word in that sentence is equal, the meaning of the whole words as an interjection is not delivered in the TLT. Its literal meaning is not recognized as an interjection in the target language and is not common to be used in utterance.

Established equivalent, adaptation, neutralization, deletion, generalization, explicitation, discursive creation, reduction, implicitation, particularization, modulation, and description are classified into target language-oriented as their characteristics tend to produce a translation, which is accepted and easily understood by the target text reader. By using established equivalent, cultural issues of the source text are translated into the target text equally in terms of meaning and form. Thus, this technique produces high-quality translation. One example is an SLT idiomatic expression *oh, why didn't I keep my big mouth shut!* that is translated into a TLT idiomatic expression having an equal meaning *oh, mengapa aku tidak menutup mulut besarku*. Another example is an SLT abbreviation *P.S.*, which is translated into a TLT abbreviation having an equal meaning *N.B.* Both the abbreviations are used to add additional information in the end of letter.

By using adaptation, SLT cultural issues are translated into their closest equivalents whose functions and forms are similar and commonly used in the target language even though their meanings are not exactly equal. This technique reduces the accuracy but increases the acceptability and readability of translations. One example of cultural issues translated using adaptation can be seen in the sentence *I'll paddle the first kid that laughs*. This sentence is categorized as cultural issues describing a speaker's habit in the SLT in terms of punishing children. That habit is not common in target culture. Instead of paddling a kid, people in the target culture usually pull kid's ear for punishment. Thus, that sentence is translated into *yang pertama tertawa akan kujewer!* (back translation: I'll pull the ear of the first kid that laughs!). Another example of the use of adaptation technique can be seen in the translation of a measurement unit term *feet* into *meter*. *Feet* is an imperial measurement term, which is *kaki* in Indonesian. Even though *feet* is equal to *kaki*, the metric measurement term *meter* is preferred to be used in TLT. In the target language, *meter* is more familiar and widely used than *kaki*.

By using neutralization, SLT cultural issues are translated into neutral languages in the TLT. This technique reduces the acceptability but increases the accuracy and readability of translations. One example is a word *unca'*, which is a dialect of *uncle*. This word is translated into *paman*, which is not considered as a dialect in the target language. This technique reduces acceptability scores of this dialect, as its translation is not considered as dialect in the target language.

By using deletion, SLT cultural issues are fully omitted in the target text. This technique leads to low scores of accuracy, acceptability, and readability as it does not communicate the SLT information in the TLT. One example of cultural issues translated by deletion is *Wistful Union,* which is written on a letter in the SLT. This name of shipping company is deleted or omitted in the target text so that the reader cannot get information about the name. Another example can be seen in slang terms *put up your dukes* in sentence *put up your dukes and let's go*. That sentence is translated into *kita mulai sekarang!* (back translation: *let's go!)*. The slang term in that sentence is omitted in the TLT.

By using generalization, SLT cultural issues in forms of hyponyms are translated to their hypernyms. One example of cultural issues translated by generalization is *rhubarb juice*, which is translated into *juice*. *Juice* has a broader sense than *rhubarb juice*. Another example is a term of addressee *boys* in the sentence *no, boys,* which is translated into *tidak, nak*. The word *nak* (back translation: *kids*) has general meaning as it can refer to either male or female. It is translated into more general terms as Indonesian does not differentiate gender for addressing children. The use of this technique reduces accuracy scores as it does not communicate SLT information precisely in the TLT. However, this technique increases acceptability and readability scores.

By using explicitation, SLT cultural issues are translated by making their functions, forms, and meaning explicit. This technique produces translation with high scores of accuracy, acceptability, and readability. One example of cultural issues translated by explicitation is *kids* in a sentence *anyway, I had sense enough to leave the kids at home!* uttered by Donald to his nephews. The word *kids* can refer to a broad sense, that is, to all children. However, in this example, it refers to Donald's nephews. Thus, it is translated by explicate the addressee meant by Donal into *para keponakan* (back translation: nephews).

By using discursive creation, cultural issues of the source text are changed into the target text, in terms of their meanings. This technique leads to low scores of accuracy, acceptability, and readability as the meaning of SLT is changed in the TLT. In other words, the TLT does not present the information contained in the SLT. One example can be seen in an SLT idiomatic expression *we're up to our ears* which in the target text means *kita sudah sampai di tempat tujuan*. However, the SLT is translated into *dan sekarang kita sendirian di hutan belantara* (back translation: and now we are alone in the jungle). The TLT is not the translation of the SLT. It is a creation of translator in producing a text, which is not in the SLT.

By using reduction, SLT cultural issues are omitted partially in the TLT. This technique reduces accuracy scores as it reduces the information of the SLT. One example is interjection in the source text *oh, my stars,* which is translated into *oh*. In that case, the phrase *my stars* is omitted in the TLT.

By using implicitation, SLT cultural issues are stated implicitly in the text structure. This technique produces translation with high scores of accuracy, acceptability, and readability. One example of the use implicitation can be seen in the pronoun *'em* in *ran 'em down the hose*, which is translated into *alirkan lewat selang* (back translation: *ran … down the hose!)*. The pronoun *'em*, which in the target language means *mereka* or explicitly means *bola tersebut* (back translation: the balls), is implicate in the TLT. This technique does not reduce, add, change, or omit meaning of the source text.

By using particularization, SLT cultural issues in the form of hypernym are translated into their hyponym. One example is *rice*, which is translated into *beras*. *Rice* is a hypernym term referring to *padi, gabah, beras*, and *nasi* in the target language. This technique produces translation with high scores of accuracy, acceptability, and readability

By using modulation, SLT cultural issues are translated by changing their point of view in the TLT. The meaning of SLT will not be changed, reduced, added, or omitted. Thus, this technique produces translation with high scores of accuracy, acceptability, and readability. One example is *look at him sweat!*, which is a kind of dialect of *look at his sweat!* This expression is translated into *dia berkeringat* (back translation: he is sweaty).

By using description, SLT cultural issues are translated by more than one word to explain more information about the issues. This technique produces translation with high scores of accuracy, acceptability, and readability. The example can be seen in the verb *voodooed*, which is translated into *melakukan mantera voodoo* (back translation: doing voodoo spell). The verb *voodoo* means affecting someone using black religious cult ("voodoo," n.d.). To express the meaning, it is translated in the form of description in the TLT.

4 CONCLUSION

The frequency of use of the target language-oriented techniques is higher than that of the source language-oriented techniques. It indicates that the translation ideology tendency to translate cultural issues in five series of Donald Duck comics in the Bundle "Karya Terbaik Carl Bark" is domestication. The result of cultural issues translation using this ideology tendency is considered accurate, accepted, and easy to be understood by target readers. Thus, this ideology is recommended to be used to translate cultural issues in children's comics.

REFERENCES

Berezowski, L. (1997). *Dialect in translation*. Wroclaw: Wydawnictwo Uniwesytetu Wrocławskiego.

Boltz, R.H. (2007). What We Want: Boys and Girls Talk about Reading. *School Library Media Research*, *10*(December), 1–18. Retrieved from: http://proxy.lib.uiowa.edu/login?url=http://search.ebscohost.com/login.aspx?direct=true&db=eric&AN=EJ851693&site=ehost-live.

Molina, L., & Albir, A.H. (2002). Translation Techniques Revisited: A Dynamic and Functionalist Approach. *Meta: Journal Des Traducteurs*, *47*(4), 498. https://doi.org/10.7202/008033ar.

Nababan, M.R., Nuraeni, A., & Sumardiono. (2012). Pengembangan model penilaian kualitas terjemahan. *Kajian Linguistik Dan Sastra*, *24*(1), 39–57. Retrieved from: https://eprints.uns.ac.id/11853/.

Newmark, P. (1988). *A TEXTBOOK OF TRANSLATION*. New York: Prentice Hall International.

Oittinen, R. (2000). *Translating for Children* (Vol. 11). New York: Garland Publishing Inc. https://doi.org/10.1017/CBO9781107415324.004.

Porter, E.S. (2005). *Idiom of the Greek New Testament* (2nd ed.). London: Continuum.

Sharifabad, E.D., Yaqubi, M., & Mahadi, T.S.T. (2013). The Application of Domestication and Foreignization Translation Strategies in English – Persian Translations of News Phrasal Verbs. *Theory and Practice in Language Studies*, *3*(1), 94–99. https://doi.org/10.4304/tpls.3.1.94–99.

Siregar, R. (2015). Translation Ideology in the Translation Process of Stephen R. Covey's The 8th Habit into Indonesian. *International Journal of Comparative Literature & Translation Studies*, *3*(4). https://doi.org/10.7575/aiac.ijclts.v.3n.4p.54.

Venuti, L. (2001). Strategies of Translation. In M. Baker (Ed.), *Routledge Encyclopedia of Translation Studies* (pp. 240–243). New York: Routledge.

Voodoo. (n.d.). Retrieved November 15, 2017, from: https://en.oxforddictionaries.com/definition/us/voodoo.

Zohar, I.E. (1990). Polysystem studies. *Poetics Today: International Journal for Theory and Analysis of Literature and Communication*, *11*(1). Retrieved from: www.tau.ac.il/%0A~itamarez/works/books/ez-pss1990.pdf.

Urban Studies: Border and Mobility – Kerr et al. (Eds)

The director's responses and the shaping of Indonesia's identity in the European film festival funding

R. Ihwanny & M. Budiman
Universitas Indonesia, Jakarta, Indonesia

ABSTRACT: The Indonesian cinema is on the rise once again and this time it has gone further by going international. Sincere appreciation needs to be extended to the deserving independent film directors who pioneered the rise of Indonesian cinema and put the country's name on the map of world cinema. More and more Indonesian films are shown in the international film festivals and they often won awards. The European film festivals became one of the important factors behind these recent successes. Their involvement is shown in the fund they provide to the Indonesian directors. Unfortunately, it raises the issue of a hegemonic practice. The concern here is that the directors have to comply, albeit reluctantly with what the film festivals want to say. The focus of this article is to analyse the responses and strategies of the Indonesian directors with regard to the hegemonic practices associated with the European film festival funding. Interview with the directors and reading of the funded films revealed two important findings; firstly, how Indonesia's identity is shaped from two different perspectives, namely the Indonesian directors as recipient of the fund and Europe as the funder. Secondly, the practice of hegemony should not be seen solely as domination by one of the parties and subordination of another. The directors in this regard are active agents who constantly strategizing and negotiating, despite their position as subordinates who must rely on the funding from the European film festivals.

Keywords: film festival funding, shaping of identity, Indonesian cinema, hegemony, responses

1 INTRODUCTION

In the last ten years, there has been a spectacular resurgence of independent cinema in Southeast Asia (Baumgärtel, 2011: 58). A region that was never considered as a serious contender in world cinema has now gained international attention,

> A region where for a long time was more or less a blank space on the map of world cinema has made itself heard in the international art house and festival circuit. (Baumgärtel, 2011: 58).

Liz Shackleton in her article entitled 'Indonesia Spreads its Wings', which was published in 2012, named Indonesia as one of the promising creative regions in Asia and mentioned two Indonesian films that debuted in a prestigious international film festival; *Postcards from the Zoo* (Edwin) and *Peculiar Vacation and Other Illnesses* (Yosep Anggi Noen). These two films were supported by the funding programme from the European film festivals. Shackleton's article suggests the importance of a funding programme, not only for the directors who received international acclamations, but also for Indonesia.

Indonesian films, which are eligible for receiving financial assistance from the European film festivals, are those that belong to the art film genre (art house) or commonly referred to as independent films. Even though for the most part they are successful in the international film festivals and some even distributed and screened in European movie theatres, most of them will never see the light of day in their home country (Baumgärtel, 2011: 59).

Mainstream films are certainly the bread and butter of the domestic film industry, whereas independent films have to be content by being in the periphery, despite being established in the international world (Aartsen, 2011: 22).

Independent films from Indonesia are able to survive because of one crucial factor, i.e. the funding programmes from Europe, as stated by Lalu Roisamri, "Most films are financed by local private investors, although for art house films you still need to turn to Europe" (Shackleton, 2012). The domestic infrastructure is simply not strong enough to support an independent film industry, which is evident from the difficulties these directors have to face in securing funds for their films. Not to mention the insufficient number of theatre screens to show their works because most of the theatres are showing films from Hollywood, and the prevailing issue on censorship.

The funds from the European film festivals programmes, therefore, becomes the saviour for the Indonesian independent film directors. Not only do they get the necessary financial support, but more importantly, they also gain international recognition that legitimizes their presence and puts Indonesia at the centre stage. Mubarak and Ageza (2015) stated the simple fact that Indonesian films are often selected to be shown in various international film festivals, but question the type of representation that each of these festivals wants to show by inviting films from Indonesia.

Mubarak and Ageza (2015) mentioned a contradiction, wherein on one hand these film festivals have pronounced *en masse* that they are here to maintain the diversity in the world's independent films; however, at the same time, they are also erasing any trace of diversity by "farming" through the politics of taste of the funding programme. Here we see a blatant practice of hegemony in operation. The European film festival funding retains the prerogative to fund films that meet their criteria of taste, which in turn create uniformity instead of diversity.

The funding programme undoubtedly has a positive impact; however, that does not mean that it is not problematic. The issue here is the politics of taste of the programme, wherein only a certain type of film or film with a specific theme is eligible for funding. Therefore, there is a nagging suspicion that these directors will be producing films that conform to their patron's taste.

Film is an important medium for propagating images of a culture and as such, as it becomes an arena for the struggle between representations. Nadine Chan (2010: 38) states that cinema, like any other modes which mediate, represent, and distribute cultural images, is a carefully policed minefield where the struggles of representation are played out.

2 METHODOLOGY

The study focuses on Indonesian directors who received financial assistance from the Hubert Bals Fund. In the last ten years, ten Indonesian directors have received financial assistance from the programme, including, among others, Ismail Basbeth and Yosep Anggi Noen.

According to Aeron Davis (2008: 53), there are three approaches that can be used to investigate cultural production. They are broadly categorised as the political economic approach, the textual analysis and the sociological/ethnographic work. This research focuses on the textual and sociological/ethnographic approach, using the films and interviews as the source of data. Interview with the directors was conducted to analyse their position and strategies in dealing with the dynamics of hegemonic practices in the funding programme, while the textual approach is used to expose how Indonesia's identity is shaped in the funded films. The director's position is analysed using Stuart Hall's theory on hegemony responses, while previous studies on funded films are used to establish the patterns that lead to the politics of taste in film festivals and the shaping of Indonesia's identity.

3 DISCUSSION

Halle (2010: 314) states that the dynamic of orientalism here supports the production of stories about other peoples and places that it, the funding source, wants to hear, or in other words, a film receiving financial assistance must tell a familiar story or something that the international/European audiences want to hear.

Therefore, what are the stories these audiences would like to hear from the "third world?" The answer to this question may be surmised from a number of scholarly works that have analysed films from the "third world", which are funded by the Hubert Bals Fund. Miriam Ross and Daniel Steinhart discuss Latin American films funded by the Hubert Bals Fund; meanwhile, Lindiwe Dovey discusses films from Africa that are benefitted from the fund. Additionally, more films funded by the European film festivals are also analysed by other scholars, such as Deborah Shaw who discusses films from Latin American countries that benefitted from the fund provided by Cannes and Berlin film festivals. Zainab Taymuree discusses films from Iran, which are funded by the Berlin film festivals, and Irid Neidhart discusses films from Middle East countries supported by the Berlin film festivals.

Some of the conclusions drawn from those studies are as follows: (1) the ever—present shadow of orientalism that leads to a practice of self-exoticization and continuation of West-East stereotyping; (2) a perpetuation of the "first world" – "third world" stereotype by exploiting poverty; and (3) the types of films eligible for funding are those that conform to the European agenda. These conclusions provide the answers to the question of what the European audiences want to hear, and explain the politics of taste within the European film festival programme, but mainly it presents a specific type of the "third world" identity to the international eyes.

Orientalism and the dynamics of the "first world" – "third world" work in the same dimension, a binary opposition between superior and inferior, i.e. the West and the East. Old style orientalism is still used and even repackaged in a new form (Said, 2010: 443). Instead of appearing as scholarly books and imaginative records, the "new style" of orientalism appears in modern media, including films, posters, and news (Said, 2010: 443). This new style is not only defined by its forms but also by its executors. If in the past the East was constructed by experts, scholars, or professor of Orientalism from the West, nowadays in the era of film festivals, it is the Eastern directors who are advocating orientalism; it is they who coined the term self-exoticization. In the language of Said, the modern East is taking an active role to "easternize" itself (Said, 2010: 508).

In addition to orientalism and the "third world" themes, other genres of films may also receive financial support provided that they conform to the European agenda, i.e. films that depict oppositions against an oppressive regime, opposition against censorship, injustice, or films that act as mouthpiece for humanitarian values, freedom, and advocating the rights of the marginalised. The European agenda is still within the framework of West-East, where the West through their film festivals is providing a safe haven to be creative. The festivals are providing space for discussing issues on controversial topics or for directors who are being censored in their own countries. It is intended to demonstrate Western superiority over the East, where the West puts a high value on the freedom of speech and expression.

The funding programme of the "first world" film festivals provides the West with the right to represent the East. Ultimately, the films supported by the fund become an extension of orientalism, or what is referred to by Said as the "new style" of orientalism. McKenzie Wark said that the European audiences are raised by 'orientalist' media literacy (van Heeren, 2012: 2). European film festivals are operating under the shadow of orientalism, because they have to offer something that is familiar to the European audiences. Cindy Hing-Yuk Wong (2011: 87) states that many festival films from outside Western Europe also tackle social and political issues that are taboo, especially if these are issues that are less problematic within a Western audience framework.

The potential issue that may arise from the funding programme is how to make a film that tells a local story and yet can be understood globally. Authenticity of such a story then becomes a problem. Tamara L. Falicov (2010: 14) said that there is an expectation that a foreign director's work is understandable and can cross borders and thus be accessible to what is considered as a universal, typically educated audience. The fact that the fund provider retains the prerogative to select the film to be funded; therefore, the authenticity in this regard is the European version, which once again points to orientalism and the "third world" stigma.

Another Trip to the Moon (*Menuju Rembulan*) is the title of film by Ismail Basbeth who received a script and project development support from the Hubert Bals Fund. The film was entered for the Hivos Tiger Award competition at the Rotterdam film festivals. The official website of Rotterdam film festivals states; "*In the oldest legends and traditions—also in those of Islam in Indonesia—the role of the woman is usually a subordinate one, but Ismail Basbeth reverses the traditional roles in his film*". A European agenda that depicts a strong willed woman who defies tradition is neatly embedded within this statement.

The film tells the story of Asa; the daughter of a seer and she lives with her female friend (Laras) in the forest. The film focuses on their daily life, from when they wake up, when they are taking a bath in the river, catching fish, and hunting rabbits. Unfortunately, one day Laras was struck by lightning and died. Soon after, Asa saw strange creatures whenever she visited Laras' grave. Afterwards, Asa's mother performed a ritual to bring Asa home. Shortly Asa met a human canine, and both of them left the forest and returned home. She and the human canine got married and he turned into human, and they had a child. However, as the time passed, she became disillusioned with domestic life, so she decided to leave her home and return to the forest.

Another Trip to the Moon is steeped with a nuance of orientalism. Mystical nuances appear strongly in the various rituals portrayed in the film. Exoticism is broadly explored through Asa and her friend while they were still living in the forest; from their primitive clothing; their unusual sleeping positions; being half-naked while taking a bath in the river, all of these with a backdrop of beautiful forest with shady trees and crystal clear water. A sense of exoticism is manifested in Asa's sensuality and the beautiful panorama of the forest. Exploration of the exotics continued. When Asa and the human canine returned home and got married, in a sense they were returning to civilization, with the call to prayer beckoning in the background and a scene when Asa took a bath outside the house. They lived in an exotic Javanese house, which was referred to by the official website of the Rotterdam film festivals as a "*beautiful classical Javanese house*". The clothing worn by Asa, her mother, and the other female characters did not resemble any modern attire. When it was time, Asa gave birth at home assisted by two women complete with some ritual offering, and not in a hospital under the care of a specialist. The fact that Asa and the human canine return to civilization does not mean that they leave their primitiveness in the forest.

Peculiar Vacation and Other Illnesses (*Vakansi yang Janggal dan Penyakit Lainnya*) by Yosep Anggi Noen received a post-production support in 2012. The film was a contender in the Locarno film festivals in the same year. The Rotterdam film festivals refer to this film as "*Another side of Indonesia. With ordinary locations around Yogyakarta (and hence not the touristy Borobudur), an original story told in fragments about an unthinkable love. A striking red sofa represents the protagonists' desire and the director's sense of humour*".

It tells the story of Ning, a wife who worked in a second-hand clothing store, and then she got a job at a furniture store. One day, Ning and her male co-worker, Mur, were told to deliver a sofa from Yogyakarta to a remote village in Temanggung. This film is a road movie that follows the two-day car trip of Ning and Mur, and the places where they stop to take a break along the way. The travel scenes are interposed by scenes showing the activities of Ning's husband, when he watched TV at home, selling gasoline at a roadside stall, or when he hired a prostitute.

The film focuses on the working class in Yogyakarta. Through the two-day car trip, the director is exploiting the day-to-day existence of the lower middle class living in Yogyakarta and its surrounding area, and the presentation is wrapped by scenes depicting the natural beauty of Indonesia. The exploitation of Indonesia's natural beauty is a manifestation of the director's self-exoticization; meanwhile, the simple existence of the lower middle class family who is struggling to make ends meet fulfils the aspect on the "third world" stereotype.

The two films were shot in location where the actual daily struggles occurred. The locations selected for the film are cities or regions that are not so developed and far from modernity, yet they are endowed with beautiful landscapes. Both films have not managed to free themselves from the trapping of self-exoticization and exploitation of poverty. Anshari (2016: 80) states that Selecting a location outside Jakarta can be seen as a progressive move because most Indonesian films are Jakarta centrist; however, it also creates its own complication because the film is trying to sell exoticism in its attempt to highlight the difference between local and modern existence.

The funding programme is the solution; however, along the lines it also becomes problematic. A film brings a strong symbolic message and it has a great influence, specifically when it comes to an image of a country (Herold, 2004: 6). Considering the fact that the films funded by the programme are screened at international stage, the identity presented in the films becomes an important issue. Indonesia as presented in these films is Indonesia with beautiful panorama; however, at the same time, it is also Indonesia that still struggles with poverty and local traditions that are far from the touch of modernity, or in other words, Indonesia is a "third world" country that has yet to catch up with the advanced civilization of today.

Stuart Hall (1999: 515) claims that there are three subordinate positions in hegemonic practices, i.e. the dominant hegemonic, negotiated, and oppositional. The first means accepting the hegemony in full and straight (Hall, 1999: 515). The subordinates in this case are passive; they identify themselves with the hegemon and accept the position without questions (Sturken & Cartwright, 2009: 72–73). The second position means the subordinates acknowledge the legitimacy of the position, while, at a more restricted, situational level, they make their own ground rules, they operate with exceptions to the rule (Hall, 1999: 516). Meanwhile, the last position means the subordinates are in a position that fully opposes the hegemon.

While the directors are showing that they are under a dominant hegemonic position with regard to the films they produced, which implies a compliance with the taste of the film festivals, they are also expressing a negotiated position in the interest of the films they produced. The directors are showing the flexibility of their positions in response to the hegemonic practices in the funding programme, unlike real subordinates who are in complete submission to the wish of the hegemon.

Ismail Basbeth was taking a negotiated position when explaining why the fund was not used as intended. In his own words, Basbeth was acknowledging the position of the fund provider by asking for their permission beforehand; however, at the same time he was operating under the framework of rules that he had created for himself. Basbeth was in a negotiated position when he used the fund that should be used for the script and project development for something else,

Basbeth was nominated to receive financial assistance for a script and project development; however, his script was already completed when he was announced as a recipient of the fund. Therefore, Basbeth negotiated so the fund could be used for other purposes.

Meanwhile, Yosep Anggi Noen negotiated for the location of his film's international premiere. As a recipient of the Hubert Bals Fund, the premiere of Noen's film should be at the Rotterdam film festivals as stipulated by the agreement. However, after considering various aspects, Noen chose to hold the premiere of his film at Locarno film festivals. Noen showed how important it was for a director to make a practical decision with regard to his work, even though they have to negotiate with their patron. On one hand, Noen was showing that he was under a dominant hegemonic position because of the fear that the fund would be withdrawn; on the other hand, he was showing the courage to renegotiate his position with regard to the location where his film would be screened first.

Noen also shows that a director should not only limit his focus on the creative aspects of the film, but they should also consider other aspects such as promoting their works. He has maturity in calculating what is best for his film. Following a thorough deliberation, he decided to go against the terms and conditions set forth by the fund provider. Both of them show that hegemony does not always go full and straight and the subordinates in the relationship are active agents that are capable to penetrate the hegemonic leadership. James Scott argues that the subordinates are the active agents, who through their experiences are able to penetrate the hegemon's ideology (Eklöf, 2003: 10). The negotiation conducted by the directors is one form of their ability to contest the hegemonic leadership.

Basbeth and Noen are two newcomers in the national and international film scene. Their films, which are funded by the Hubert Bals Fund, are their first full feature films. Nonetheless, they have demonstrated that they are mature enough to play at international level. The maturity is shown when Basbeth explains what he knows with regard to the taste of film festivals.

Basbeth realized that the European/international audiences already have their own opinions about Indonesia. This implies Basbeth's deep understanding about the taste of the international market. Noen also exhibits a similar maturity. He understands the taste of the film festivals and ponders using scenes that would please the crowd, but ultimately uses his own reasons to do it.

Basbeth and Noen's thorough knowledge about the subject clearly shows their maturities despite their status as new directors. They did not simply leap into the film industry empty handed and without any knowledge. Their maturities ultimately become the crucial instrument in the negotiations they have to take. Their positions as subordinates in the hegemonic practices of the funding programme have made them quite flexible in attending their interests.

Sturken and Cartwright (2009: 75) state that one weakness in Stuart Hall's theory is that it limits responses of the subordinates to only three positions, while in reality they are a series of negotiated positions. The directors mentioned are always in the position of a 'constant negotiation' between the dominant hegemonic, or operating under the framework of their own making for the sake of their own interest or the objective they want to achieve. The word negotiation becomes the real focus and must be highlighted when analysing the dynamics of these directors.

4 CONCLUSION

Film festivals are no longer about film screening, as they have gone even further by initiating funding programmes for the films. Financial assistance was specifically extended to directors from the "third world" by festivals such as the Rotterdam festival, which eventually raises some suspicions among scholars about the politics of taste in the film festivals that promote films from the "third world", which are thickly embedded with orientalism views and stereotypes about the "third world". These suspicions were evident in the films supported by the Hubert Bals Fund.

Indonesia is one of the countries eligible for assistance from the Hubert Bals Fund, and in the last ten years, ten Indonesian directors have received financial assistance from the programme, including, among others, Basbeth and Noen. Orientalism and stereotypes about the "third world" are clearly evident in the works of these two directors. This situation clearly has a great impact in the shaping of Indonesia's identity at the world stage, considering the fact that the films supported by the fund are screened and distributed in the international film festivals and international market. Indonesia's representation in these films matches the common perception held closely by the West about a "third world" country, i.e. a beautiful but poor country and still has a long way to go from achieving a modern living.

Even though the directors are dominated by the film festivals' taste, they are not the subordinates that are completely submissive to the wish of the hegemon. Their flexibilities in this regard are shown by their capacity to renegotiate the rules set forth by the hegemon. These directors are the active agents who will not hesitate for a minute to challenge the hegemonic leadership. Although these fund recipients are newcomers in the industry, their flexibilities in dealing with the hegemonic practices in the funding programme show that they are quite mature and ready to do battle in the international film industry. These directors are proof of a different hegemonic practice in operation, where the subordinates are positioning themselves in a series of negotiations.

REFERENCES

Aartsen, J. (2011). *Film world Indonesia the rise after the fall*. Utrecht: Universiteit Utrecht FTV MA. Available at http://dspace.library.uu.nl/bitstream/handle/1874/205138/thesis_filmworldindonesia_jv_aartsen.pdf?sequence=2. (Accessed 16 May 2017).

Anshari, I.N. (2016). *Pendatang dan representasi lokal dalam film Indonesia.* In Herlina S, D. and Andini, K (eds.). *Laut bercermin (The mirror never lies): Sebuah catatan dan tafsir film.* Yogyakarta: Rumah Sinema, 71–81.

Baumgärtel, T. (2011). Imagined communities, imagined worlds: Independent film from South East Asia in the global mediascape. *Transnational Cinemas*, 2, 57–71.

Chan, N. 2010. Slumdog millionaire and the troubled place of cinema and nation. *Spectator*, 30(2), 37–45.

Davis, A. (2008). Investigating cultural producers. In Pickering, M (ed.). *Research methods for cultural studies*. Edinburgh: Edinburgh University Press Ltd, 53–67.

Dovey, L. (2015). *Curating Africa in the age of film festivals*. New York: Palgrave Macmillan.

Eklöf, S. 2003. *Power and political culture in Suharto's Indonesia: The Indonesian Democratic Party (PDI) and decline of the New Order* (1986–98). Leifsgade: NIAS Press.

Falicov, T.L. (2010). Migrating from South to North: The role of film festivals in funding and shaping global South film and video. In Elmer, G., Davis, C.H., Marchessault, J., & McCullough, J (eds.). *Locating migrating media*. Lanham: Lexington Books, 3–22.

Hall, S. (1999). Encoding decoding. In During, S (ed.). *The cultural studies reader second edition*. London: Routledge, 507–517.

Halle, R. (2010). Offering tales they want to hear: Transnational European film funding as neo-orientalism. In Galt, R. and Schoonover, K (eds.). *Global art cinema: New theories and histories*. New York: Oxford University Press, 303–319.

Herold, A. (2004). EU film policy: Between art and commerce. *EDAP: European Diversity and Autonomy Papers,* 3, 1–21.

Mubarak, M. and Ageza, G. (2015). *Agenda politik selera dalam festival film dunia*. Available at http://cinemapoetica.com/agenda-politik-selera-dalam-festival-film-dunia/. (Accessed 20 October 2017).

Ross, M. (2011). The film festival as producer: Latin American films and Rotterdam's Hubert Bals Fund. *Screen*, 52(2), 261–267.

Said, E.W. (2010). *Orientalisme: Menggugat hegemoni Barat dan mendudukkan Timur sebagai subjek.* Yogyakarta: Pustaka Pelajar.

Shackleton, L. (2012). *Indonesia spreads its wings*. Available at http://www.screendaily.com/features/in-focus/indonesia-spreads-its-wings/5045560.article. (Accessed 9 June 2016).

Sturken, M. and Cartwright, L. (2009). *Practices of looking: An introduction to visual culture second edition*. New York & Oxford: Oxford University Press.

Taymuree, Z. (2014). Self-exoticization for the film festival. *Avicenna: The Stanford Journal on Muslim Affairs*, 4(1), 26–29.

Van Heeren, K. (2012). *Contemporary Indonesian film*. Leiden: KITLV Press.

Wong, C.H. (2011). *Film festivals: Culture, people, and power on the global screen.* New Brunswick, New Jersey & London: Rutgers University Press.=.

Urban Studies: Border and Mobility – Kerr et al. (Eds)
© 2019 Taylor & Francis Group, London, ISBN 978-1-138-58034-3

Children in Indonesian cinema during colonialism: The border of cross-identity

S. Wibawa
Universitas Airlangga, Surabaya, Indonesia

ABSTRACT: This essay will examine how the image of children in Indonesian cinema embodies cross-cultural identity framed by dissonance ideas of Indonesia as a nation during colonialism. I employ the discourse of national cinema to explore how the idea of a nation in the cinema emerged while Indonesia, as a state, did not exist. National cinema has been transformed from a classical definition of the limited singular identity of a nation into a complex discourse within the nation. Furthermore, the nation in national cinema is not seen merely from the perspective of the banality of a national symbol, but as the discourse that appears in the cinema and within the nation. In this context, film can be placed as a historical site that reveals the dynamic discourse of the nation. Thus, since the first Indonesian cinema during colonialism has placed children in its main narrative, I argue that Indonesian film during colonialism utilises children's images to develop a national consciousness to construct the identity of Indonesia within a colonial framework. From this perspective, children are constructed to convey the cross identity of the Dutch East Indies as the Dutch colony and Indonesia as an emerged-imagined nation. The image of children, thus, becomes a cultural agent that crosses the border of identity and culture within the change of Indonesia as a nation.

Keywords: Children, Indonesian cinema, colonialism, cultural, identity

1 INTRODUCTION

A bold question relating to Indonesian cinema is "How do they project Indonesia-ness?". Relating to this is the question of how children's depiction in Indonesian cinema represents the nation. As an independent state, Indonesia declared its proclamation in 1945 against Dutch colonialism. However, cinema cultures in Dutch East indies started in the early 1900s while film production started in the 1920s, marked by the first local film production in 1926. This film, *loetoeng Kasaroeng*, is acknowledged as the first local film made in Indonesia, by a Dutch filmmaker with local casting, setting and narrative. Indonesia's film culture started as part of the colonial film industry with a historical screening in 1900 (Masak, 2016). Until 1942, before the Japanese occupation, more than one hundred films were produced, predominantly by Chinese film businesses (Setijadi-Dunn & Barker, 2010; Masak, 2016). A question arises then about how these colonial films can be integrated after the nation's independence. This historical complexity, at some stage, affected the way cinema projected this social practice as suggested by Graeme Turner (2006).

Turner argues that film is seen as a social practice representation that is related to production and consumption of the communities and determined by the dominant power, ideology, or the objectives of the film production. A film, or films, continuously develops a reality of a discursive site of ideological framing that discloses the dynamic discourse of the nation through its narratives which includes children's figures in the film narrative. Applying Ben Anderson's definition of a nation contributes an understanding that Indonesia as a nation is formed as "an imagined political community – and imagined as both inherently limited and sovereign". In this context, Anderson's idea provides a basic idea of national cinema as a place for "imagining the nation" as an "imagined community" (Abe, Tortellini & King, 2008; 2). Therefore, there is a discursive space to identify how the films that were produced during Dutch colonialism embody ideas of Indonesian national identity. A study in early cinema by Nanna Verhoeff (2006) argues that, as a concept, the nation in national cinema is embodied within three semantic layers: "the national, the nationalist and the nation-ness" (2006, p. 19). Verhoef argues that the national signifies a film narrative's place of origin which includes nation-ness as "a sense of belonging and of cultural ownership" that produces the nationalist and the value of nationalism (p. 160). Verhoef asserts that film narratives provide a "constant interaction between nationality of films to identity in films" (pp. 160–161). These embedded values in cinema narrative have been part of an ongoing discourse on national cinema.

Critical discourses on national cinema (Higson, 1989) suggest that a cinema does not merely depict a "homogenous national culture and identity" (Higson, 1989, p. 44). In an Indonesian context, cinema during the Dutch period can be seen as a "dialectical engagement" with transnational cinema in a way to identify a construction of national identity through a range of discourses of actions, practices and rituals (Choi, 2011). However, while several scholars have discussed the discourse of Indonesian national cinema (Said, 1991; Heider, 1991; Sen, 1994; Robert, 2000; Barker, 2011; Paramaditha, 2014), there is a lack of specific discourse on Indonesian cinema during colonialism. Therefore, this paper intends to fill the gap on how Indonesian-ness was projected during the Dutch period in Indonesia.

A study by Christopher A. Woodrich (2014) asserts that a subliminal identity of Indonesian nation-ness has also been established in early Indonesian films during Dutch colonialism. By examining four films written by Saroen, a famous Indonesian journalist-cum-film writer, Woodrich argues that these films depict a positive hope of Indonesian independence. Woodrich argues that Saroen contrasted a spirit of a dynamic city against uninteresting old village life to portray a Dutch East Indies as the Indonesian modern nation in the future. By reading Saroen's films, Woodrich endorses that Indonesia as a nation has been developed while it was a Dutch colony. Then, how does the first fiction film that was also made during colonialism depict these issues?

Loetoeng Kasaroeng reflects Verhoeff's sense of a nation's belonging. The film is attributed with being the first film made in Indonesia with local casting and story. Kristanto (2005) indicates that this film features children, the first film that presents children in its main narrative. In this context, Veronique Benei (2008) suggests a relationship between children as citizens and a state in a way placing children to represent the nation's values (p.72). Furthermore, since the first fiction film in Indonesia casts children in the main role, and borrowing Benei's thesis, I argue that Indonesian film during colonialism utilises children's images to develop a national consciousness to construct the identity of Indonesia within a colonial framework. From this perspective, children are constructed to convey the cross-identity of the Dutch East Indies as the Dutch colony and Indonesia as an emerged-imagined nation. The image of children, thus becomes a cultural agent that crosses the border of identity and culture within the change of Indonesia as a nation.

Examining early Indonesian films and identifying which films use children in the cast is challenging. Firstly, there are no copies of those films. This widens the research gap on Indonesian cinema before and after independence (Sen, 1995; Heider, 1991; Woodrich, 2014). This study relies on synopses and second sources such as film posters, news and references. Secondly, in general, film reviews or synopses only focus on famous casts, or the director and the film company. At some level, it is hard to define the cast of children figures, though it's possible and has been done. Thirdly, the semantic issue and language differences in the written reviews produce another challenge to identify which films actually present children in the cast. The keyword "anak" in Indonesian language cannot be easily translated as a term "child" into English. "Anak" possibly means the child figure; however, it could be referring differently to a son or daughter of someone. However, after carefully examining some synopses and film references, several films that were produced during the periods of Dutch colonialism that present a children's cast in the main narratives were identified, such as: *Loetoeng Kasaroeng* (1926), *Poei Sie Giok Pa loey Tay* (1935), *Anaknja Siloeman Oeler Poeti* (1936), and *Rentjong Atjeh* (1940).

2 DISCUSSION

Loetoeng Kasaroeng[1] (1926) is adapted from a famous folktale from West Java, Indonesia. The story is about a love journey of Purbasari, and Purbararang. *Loetoeng Kasaroeng* was made by L Heuveldorp of the Java Film Company. This company was invited by the Dutch government to produce a documentary. However, after getting support from the mayor of Bandung city, the company produced *Loetoeng Kasaroeng*. The mayor insisted the film should present Sundanese culture, thus a folktale was chosen to be the story, starring children of the Mayor. *Poei Sie Giok Pa loey Tay* (1935) is a film about a boy named Poei Sie Giok. His mother, Tjoei Hoa is a famous figure in their martial arts community and trained him in martial arts. One day, Sie Giok beats another student from a martial art school owned by Loei Lo Ho. Lo ho is then also killed in another fight thus his wife takes revenge for him against Tjoei Hoa. This film presents cultural hybridity by mixing local context with the Chinese origin of the story. Another example is *Anaknja Siloeman Oeler Poeti* (1936)[2]. This film is about a boy who looks for his father while his mother is the white snake ghost. People around this boy keep abusing him which makes him run away to the forest where his mother lives. *Rentjong Atjeh* (1940)[3] features a story of children whose parents were killed by a pirate called Bintara. These children, Maryam, Rusna, and Daud, survived the attack. Maryam is held in a pirate's ship while Rusna and Daud escape to the jungle. Years later, they come back and take revenge against Bintara with help from Panglima Ali. These films, through depicting children in various roles, also project Verhoeff's idea of a nation's senses of belonging by featuring local story adaptation, language and resistance to dominant colonialism construction in terms of values of family.

These films develop their narrative through adaptation of Indonesian or local oral tradition such as legends, folktale or myths (Heider, 1994). *Loetoeng Kasaroeng* is taken from Sundanese folktales, while *Rentjong Atjeh* adopts a local story of an Acehnese pirate. *Poei Sie Giok Pa loey Tay* and *Anaknja Siloeman Oeler Poeti* mix mainland Chinese stories with a local context. This culturally hybrid story become a standard film narrative for Chinese film companies in Indonesia. A study by Setijadi-Dunn and Barker (2010) finds that, during Dutch colonialism, films made by Chinese producers managed to draw attention from Indonesian indigenous audiences and the Chinese diaspora.

> As a result, early local films were diverse, cosmopolitan, and projected an image of an Indonesia that is complex, idiosyncratic, and unique, yet connected to global flows and modern practices. We argue that such an image of Indonesia is different from later insular indigenist imaginings of Indonesian belonging. (Setijadi-Dunn & Barker, 2010:27).

They note that most diasporic Chinese filmmakers tended to create a film that mixes local Indonesian or Chinese origin stories with Indonesian stamboel theater narrative then visualised with the Mandarin cinematic style. This formula, an idiosyncratic mixed with fantastical taste, was a popular genre especially among lower class audiences (Setijadi-Dunn & Barker, 2010). Indonesian stamboel theatre was adopted from European theatre performance but was then modified to a local style. At first, this performance was exclusively for the European and Chinese business community. However, due to a growing demand, it also became a popular performance among local Indonesians. It used to adopt Arabian nights stories, but then adopted an Indonesian context and began to be delivered in Malay language to accommodate local audiences. Furthermore, this performance had its popular place among Indonesians; the local audiences had more chance to express their feelings because the performance involved audiences in informal interaction. This shifting of origin from European-Asian dominance to the local Indonesian context articulates a shifting int political identity that differentiates colonial European culture from local Indonesian.

---

1. https://www.imdb.com/title/tt0347429/
2. https://www.imdb.com/title/tt1847537/
3. https://www.imdb.com/title/tt1852087/

As this shifting to stamboel style became closely related with local Indonesians, then it was seen as a way to approach local audiences in cinema by Chinese filmmakers. This formula, used by most Chinese filmmakers, created a culture replica made from a mixture of stories and images from local culture, the Dutch, Chinese and other cultures at that time as the early cinematic vision of Indonesia. Karl G Heider argues that this formula developed the idea that these films "were responsible for creating the image of the common Indonesian culture" (1994:170). These Chinese and Dutch made films developed the early image of Indonesia as a nation as a globally mixed culture with local taste and context.

Loetoeng Kasaroeng was produced by Europeans. The story does not feature Indonesia as a nation-state. However, *Loetoeng Kasaroeng* is a well-known Sundanese folktale. The casts are all Indonesian. As the film was sponsored by the Mayor of Bandung city, most of the actors are the *priyayi*s. They have been trained by a local school teacher, Kartabrata. The main roles were taken by children of Wiranatakusumah, the regent of Bandung city. Despite the casts and stories, local elements in *Loetoeng Kasaroeng* can be identified through its promotion poster. The poster shows a girl in a traditional salute style. She wears traditional clothes and sits on a box crafted with traditional ornamentation.

This local-classic visual also appears in a poster of *Rentjong Atjeh*. The cast wears an Acehnese costume set up as a focal point. "*Rentjong*" is the name of Aceh's traditional sword. Interestingly, while local cultural elements appear dominantly, a picture of a man holding a rencong in his mouth looks but does not feel so local. The pose looks similar to a global well-known character, Tarzan. This strengthens the ideas of globally mixed culture: a story of a pirate in Aceh, with a local cast featuring a narrative of revenge using animals in the forest and a pose like Tarzan.

While *Loetoeng Kasaroeng* adopted a local folktale, *Rentjong Atjeh* modified the local context with a global taste. *Sie Giok Pa Loey Tay* and *Anaknja Siloeman Oelar Poeti* adapted Chinese mainland origin stories into a localised context. Shaolin school branch and kung fu martial arts in *Sie Giok Pa Loey Tay* represent Chinese cultural elements but the setting and context are in Indonesia. Similarly, the white snake ghost is well-known in Chinese myths rather than Indonesia. This localised context, and also cast, as discussed previously, works well as a marketing strategy attracting local audiences. The text and context of these films are Indonesian; however, by nation-origin, they were made by non-Indonesian filmmakers. Thus, Indonesia presents in these films a cultural artefact. The film text develops Indonesia as an imagined nation through the form of a local narrative. In the context of that time, while Indonesia was a Dutch colony, this cultural artefact, in a way, challenges colonialism's dominant narrative. Locality elements develop the sense of cultural belonging and ownership into the imagined nation of Indonesia. By representing local cultures and values, the sense of this imagined community of nationalism can be easily shared. Thus, in this context, the use of Indonesian language as the standard language plays a significant role in developing nationhood and Indonesian-ness.

As suggested by Wimal Dissanayake (1994), nationhood's discourse encourages local narrative's involvement. However, nationhood also aims to unify and, at some level, homogenise the diversity in a nation. Generally, in unifying diversities, a national language plays an important role. Language is utilised to identify an imagined community as a nation and, at the same time, to bind the people in the community (Hobsbawn, 1996). Michael Billig (1995) argues that language plays a significant role in an ideological deployment and nationalist consciousness that shares the imagined idea of nation. Thus, the language turns out to be the principal identification of national identity (Smith, 1994). In 1908, the Indonesian language was recognised as the language of a nation and in 1928, the Indonesian language was declared the national language. (Elson, 2005). Furthermore, the Indonesian language has been formed as political language to unify the East Indies as an independent nation (Foulcher, 2000). Since then, the Indonesian language has been placed as an element of national identity which emerges from the first Indonesian film (Heider, 1994).

All of the films in this paper highlight the Indonesian language, usually called Malay, as its official language for the film. A promotional poster of *Poei Sie Giok Pa loey Tay* uses Chinese characters and words. However, it underlines "Bitjara Melayu" (speaking Malay) to inform potential audiences that this film uses the Indonesian (Malay) language. The words "Rame dan Heibat" also becomes the focal point in this poster to promote the film. These words in English literally translate to "crowded and great"; however, contextually they can be interpreted as "exciting and great". This emphasis refers to the idea that, in the beginning, films in the Dutch East Indies were intended to be part of colonial lifestyle. Watching films in cinema became exclusively for Westerners and rich Chinese especially those who spoke and understood English or Dutch. Thus, by using Malay language as a root language for Indonesian, these films constructed a communal sense of national identity: an Indonesian.

This communal identification then develops an early notion of resistance against the dominant colonial culture at that time. In these films, one of the notions emerges from how these films depict the image of children within a family system. The Dutch colonial construction of an idealised family was developed within the ethical policy period. This colonial metaphorical view places a family with a white Caucasian as a father figure who guides his brown-skinned Asian children (Gouda, 1993). This family figure represents the power of the Dutch colonial empire where European controlled non-Europeans in their colonies. Family metaphors have been politicised to justify their colonial system.

As in this context, the Dutch government placed Indonesians as children in the family; therefore, the children need to be supervised. Akiko Sugiyama (2008) argues that the Dutch government conducted policies to develop families' social prosperity by standardising a comprehensive social nuclear family model. This model consists of a monogamous type family—in which neither polygamy nor child-marriage that were common practice at that time were acknowledged—and women's roles remain in the domestic sector.

Meanwhile, children are less important figures under the aristocratic shadow of the colonial father figure.

By contrasting this colonial concept of the family model, *Loetoeng kasaroeng, rentjong Atjeh Poei Sie Giok Pa loey Tay* and *Siloeman Oeler Poeti* depict a notion that challenges the mainstream colonial model by placing children as independent individuals in the family. In *Loetoeng Kasaroeng*, the main narrative focuses on the life of two sisters, Purbasari and Purbararang, who independently decide their own life while the parent figure is missing in the story. Similarly, children are depicted surviving in their world without their parents in *Rentjong Atjeh*. This film features orphaned children who have lost their parents. They survive and come back to take revenge on the pirate who makes them suffer and has killed their parents. These films depict the image of fragmented families where children are surviving without their parents. This image contrasts against the dominant colonial construction of family; a family that placed children as the weakest members needing a father figure as a family leader to guide the children (Gouda, 1993).

The absence of a father figure also appears in *Poei Sie Giok Pa loey Tay* and *Anaknja Siloeman Oeler Poeti*. *Poei Sie Giok Pa loey Tay* projects the mother figure as the idealised model member of the family. She trains her son, Poei Sie Giok, in skilled martial arts to fight against his competitor. The mother figure in this film takes the role of mother as "child's educator" as socially constructed but also taking over the position of family leader that is traditionally held by the father figure. The colonial construction of the family model positions the mother figure in a subordinate role as serving her husband and as "a child bearer and child educator" (Locher-Scholten, 2003: 39). *Anaknja Siloeman Oeler Poeti* depicts the absence of a father figure while his son demands a mother figure in his life. The boy in this film prefers to see his mother even though he realises his mother is not a human. Despite her physical appearance, the mother figure in this film is depicted as a lovable individual for her son. This depiction suggests that the idea of the mother-child relationship is a metaphor for nation-building (Pollard, 2005).

The mother's dominant role in these films can be seen through a poster of *Anaknja Siloeman Oeler Poeti*. The poster emphasises the mother who is holding a baby. A man, assumedly the father, sits beside them. While the colonial perspective of the father figure shows similarities with traditional Javanese culture, that places the father figure as a dominant aristocrat, these films suggest an alternative depiction of the family model against the colonial framework: a dominant mother and missing father figure.

3 CONCLUSION

At the beginning, I argued that the image of children in Indonesian cinema embodies a cross-cultural identity framed by dissonance ideas of Indonesia as a nation during colonialism. I employ the discourse of national cinema to explore how the idea of a nation in the cinema emerged while Indonesia as a state did not exist. National cinema has been transformed from a classical definition of the limited singular identity of a nation to a complex discourse within the nation. Furthermore, the nation in national cinema could be seen not merely from the perspective of the banality of a national symbol, but as the discourse that appears in the cinema and within the nation. In this context, the film can be placed as a historical site that reveals the dynamic discourse of the nation. This discourse develops a national consciousness to construct the identity of Indonesia within a colonial framework. From this perspective, children are constructed to convey the cross identity of the Dutch East Indies as the Dutch colony and Indonesia as an emerged-imagined nation. The image of children, thus becomes a cultural agent that crosses the border of identity and culture within the change of Indonesia as a nation. I have identified that the national consciousness is developed through several ideas. Firstly, the sense of Indonesian belonging is developed by presenting localised stories, context and culture. *Loetoeng Kasaroeng* and *Rentjong Atjeh* introduce Indonesian origin stories, while *Sie Giok Pa Loey Tay* and *Anaknja Siloeman Oelar Poeti* adapted Chinese origin folk tales into an Indonesian context. This localised context features local landscapes, creating a symptom of national consciousness through Nana Verhoeff's layers of nation. Despite the fact that some films adopted Chinese origin folktales, the context of these films localises Indonesia as an origin geographically and culturally. This can be identified as developing a sense of cultural belonging and ownership of a nation. National consciousness is also constructed through the use of Indonesian language, or Malay as an early form of Indonesian language, as a recognisable element of nationhood. All of the films' narratives are delivered in Indonesian language. Indonesian language is deployed as a unified language among a diverse audience at that time. The use of Indonesian language can be identified as cultural ownership to recognise representation of "nation" against the dominant colonial culture. This dominant culture was also challenged by juxtaposing an alternative image that opposes a colonial construction of the family model which placed the family under the father figure's domination, subordinating the mother figure and children as the weakest family members. In contrast, these films depict the image of children as independent individuals aligned with a strong but lovable mother figure who acts the role of family leader.

REFERENCES

Billig, M. (1995). *Banal nationalism*. London: Sage.

Cohen, M.I. (2006). *The Komedie Stamboel: popular theater in colonial Indonesia, 1891–1903* (Vol. 112). Ohio: Ohio University Press.

Dissanayake, W. (1994). *Colonialism and nationalism in Asian cinema*. Bloomington: Indiana University Press.

Elson, R.E. (2005). Constructing the nation: Ethnicity, race, modernity and citizenship in early Indonesian thought. *Asian Ethnicity, 6*(3), 145–160. doi:10.1080/14631360500226556.

Gouda, F. (1993). The gendered rhetoric of colonialism and anti-colonialism in twentieth-century Indonesia. *Indonesia* (55), 1–22.

Gunning, T. (2008). Early Cinema as Global Cinema: The Encyclopedia Ambition. In R. Abel, G. Bertellini, & R. King (eds.), *Early Cinema and the "National"*. New Barnett UK: John Libbey Publishing Ltd.

Heider, K. (1994). National cinema, national culture. *Colonialism and Nationalism in Asian Cinema*, 1162.

Heider, K.G. (1991). *Indonesian cinema: National culture on screen*. Honolulu: University of Hawaii Press.

Higson, A. (1989). The Concept of National Cinema. *Screen, 30*(4), 36–46. doi:10.1093/screen/30.4.36.

Hobsbawn, E. (1996). Language, culture, and national identity. *Social research*, 1065–1080.

Kristanto, J. (1995). *Katalog Film Indonesia 1926–1995*. Jakarta: Pt Grafiasri Mukti.

Locher-Scholten, E. (2003). Morals, Harmony, and National Identity: “Companionate Feminism” in Colonial Indonesia in the 1930s. *Journal of Women's History*, 14(4), 38–58.

Masak, T.P. (2016). *Sinema pada Masa Soekarno*. Jakarta: Fakultas Film dan Televisi Institut Kesenian Jakarta.

Pollard, L. (2005). *Nurturing the nation: The family politics of modernizing, colonizing, and liberating Egypt,* 1805–1923. California: Univ of California Press.

Said, S. (1991). *Shadows on the silver screen: a social history of Indonesian film.* Jakarta: Lontar Foundation.

Sen, K. (1995). *Indonesian cinema: framing the New Order*. New York: Zed Books. Setijadi-Dunn, C., & Barker, T. (2010). Imagining “Indonesia”: Ethnic Chinese Film Producers in Pre-Independence Cinema. *Asian Cinema*, 21(2), 25–47.

Smith, A.D. (2013). *Nationalism: Theory, ideology, history*. Cambridge: John Wiley & Sons.

Sugiyama, A. (2007). Ideas about the family, colonialism and nationalism in Javanese society, 1900–1945. In L.Y. Andaya (ed.): ProQuest Dissertations Publishing.

Turner, G. (2006). *Film as social practice*. London: Routledge.

Verhoeff, N. (2006). *The West in early cinema: After the beginning*. Amsterdam: Amsterdam University Press.

Woodrich, C.A. (2014). *Between the Village and the City: representing Colonial Indonesia in the films of Saroen.* Vol. 1: International Indonesia Forum.

Urban Studies: Border and Mobility – Kerr et al. (Eds)
© 2019 Taylor & Francis Group, London, ISBN 978-1-138-58034-3

Mimicking East Asian popular culture products: Temporality of urban global culture in Indonesia

Shuri Mariasih Gietty Tambunan
English Studies Program, Faculty of Humanities, Universitas Indonesia, Indonesia

ABSTRACT: The vibrant flow of popular culture products from South Korea and Japan has transformed the way 'East Asia' is perceived in Indonesia. Recent cultural phenomena as a result of this new global force are copycat (or mimicking) products, such as I-Pop's boy/girl band and K-Dramas remakes (*sinetron* or soap operas copying plots from the original K-Dramas). Instead of condemning the process of standardization, commodification and massification implying that the Asian globalization has created a homogenous popular culture scene, this research will go beyond the "economically reductionist explanations of globalization." By looking at how the cultural industry in Indonesia is, in this moment, being strained to accommodate these foreign products, this research aims to show the cultural dynamics of today's increasingly globalized environment through a Cultural Studies perspective. Homi Bhaba's thoughts on mimicry, Fredric Jameson's pastiche and Raymond Williams' arguments on the evolution of culture will work as the conceptual foundations of the analysis. The main question to be explored will be: while mimicking the new global, i.e. East Asia pop culture, how does this force us to revisit the complexity and temporality of cultural globalization in Indonesia within the Asian context?

Keywords: East Asia, Globalization, Mimicry, Pastiche, Popular Culture

1 INTRODUCTION

On 13 November 2017, a remake of a 2002 renowned *sinetron* or Indonesian television soap opera was premiered in SCTV, a local television station. On the next day, "*Siapa Takut Jatuh Cinta*" or "Who's Afraid to Fall in Love?" was on the top list of Google Indonesia's Trending Search (Asih, 2017). The popularity of the remake is a fascinating and also problematic spectacle because the "original" *sinetron* in 2002 was an adaptation (or some would argue that it was a copycat) of *Meteor Garden*, a Taiwanese television series from 2001. *Meteor Garden* was a "trans-medium production" (Chua, 2012), because it was produced in Taiwan, while re-making a Japanese*Manga s*tory into a television drama. *Meteor Garden* was then made into Japanese (*Hana Yari Dango* in 2005) and Korean (*Boys Over Flowers* in 2009) versions, which also gained popularity in many parts of Asia.

These cultural phenomena could be considered as a reaction toward to emergent transnational flow of cultural products in Asia, especially from East Asian countries, such as Japan, South Korea, Taiwan and even China to other parts of Asia. As a new global force, the intra-Asian cultural traffic invites scholars to re-evaluate how we understand cultural globalizations. Moving on from the cultural imperialism model where globalization is considered as one dimensional force creating a homogeneous culture with a one-way flow, scholars, such as Koichi Iwabuchi and Chua Beng Huat, have argued that cultural globalization in Asia has created a regional dynamic that works as a network instead of a center-periphery dichotomy. Furthermore, as products move from one Asian country to another, in the receiving country, the products are not merely consumed; however, they are being re-produced as explained in the beginning of this article.

The dominant discourse that has been built around these re-produced products is how Indonesia is a "copycat" nation with no creativities as we keep on making cheap copies of East Asian products, such as television dramas (*sinetron*) copying the plot and narration from Japanese/Taiwanese/Korean dramas or I-Pop consisting of boy/girl bands copying K-Pop artists. This argument lies on a similar premise with the cultural imperialism model as it condemns the process of standardization, commodification and massification implying that the Asian globalization has created a homogenous popular culture scene. By choosing several case studies (K-Dramas remakes and I-Pop), this research aims to move beyond the "economically reductionist explanations of globalization." By looking at the case studies from a Cultural Studies perspective which takes into consideration the power relation and agency of the actors involved in the meaning-making process in the cultural industry, the question to be explored in this article is: while mimicking the new global, i.e. East Asia pop culture, how does this force us to revisit the complexity and temporality of cultural globalization in Indonesia within the Asian context? As the recent phenomenon occurs in 2017, I would argue that this copycat/mimicking occurrence is a part of the Indonesians' everyday reality and this research's main goal is to make sense of this pragmatic observation by examining several conceptualizations, Homi Bhaba's thoughts on mimicry, Fredric Jameson's pastiche and Raymond Williams' arguments on residual culture.

2 DISCUSSION

2.1 *Crossing the borders: Culturally modified narrative elements in television dramas' remakes*

The transnational flow of these dramas as they are dispersed to many parts of Asia has constructed the thematic and visual element driving a distinct textual and genre formulation. This was also the case for Taiwanese television dramas as they were rapidly exported to many parts of the region in the early 2000s, especially with the popularity of *Meteor Garden* as explained earlier in this article. At that time, "The rating for MG reached 5,1 with a 29,9 share and this was considered as a very high rating. This has also influenced other television station's policy in importing East Asian television series" (Merdikaningtyas, 2001: 3). Soon after, the Japanese and Korean versions were made and broadcast all over Asia.

These three similar, yet different, dramas reflect how the textuality of television dramas in Taiwan, Japan and Korea, are constructed in connection to one another. "By 'textuality', I mean the different ways in which something can function as a text for its readers" (Couldry, 2000:71). Thematically, the three dramas have the same story line, focusing on the romantic involvement of the characters as they become entangled in familial issues. Visually, the Taiwanese, Japanese and Korean versions depict urban consumerist lifestyle, emphasizing the good-looking characters in a city setting. In the matter of style and format, dramas from these countries are commonly done in a one hour-long format consisting of a number of episodes. This kind of inter-textual reworking alongside has shaped the textual and visual elements of these television dramas as they flow across Asian countries, reaffirming a particular

(Taiwan) (South Korea) (Japan)

Figure 1.

formulation of textuality. The fabricated thematic and visual elements echo how the dramas from Japan, Taiwan and Korea have formulated the textuality, which is heavily influenced by their transnational disposition. As an illustration, the early development of Korean and Taiwanese television dramas came as a response to a regional market need. For that reason, there are a number of similarities in the thematic content and visual aspects as these dramas are produced based on dramas, which were circulated earlier. The circulation network has indeed shaped these comparable textual elements. Furthermore, on the consumption level, as audience members moved from Japanese dramas to Taiwanese and then Korean dramas in the year 2000, they were seeking for familiar formulae.

In 2002, a year after *Meteor Garden* became a huge hit in Asia, a local television station in Indonesia broadcast *Siapa Takut Jatuh Cinta* which was a local television drama copying the characters, plot and other narrative elements from the Taiwanese television drama. Since then, a number of local television dramas or *sinetron* start copying other East Asian television dramas. What is stimulating from this phenomenon is the assumption of familiar formulae as mentioned earlier. The industry, in this case the television station, aims to gain as much profit as possible by taking advantage of the momentum. Creating a *sinetron* with similar narrative elements with *Meteor Garden* was a sure win and surely enough, *Siapa Takut Jatuh Cinta* (2002) was a success and was even considered as the moment in Indonesian television when Oriental-looking actors and actresses were establishing their repertoire in Indonesian television. Within the period of *F4's* popularity, the four main male actors in Meteor Garden, there were sudden recognition and acceptance of "oriental" looking actors and actresses in television programs and films. These "oriental" looking actors were often seen as similar or identical to Chinese Indonesians, who have been at the brunt of long-term historical discrimination in Indonesia. "Oriental" looks became the mainstream discourse of East Asian-ness in Indonesia. "Oriental artists, mostly from Chinese descendants in Indonesia, became very popular and fans adored them including the artist's hairstyle and fashion style" (Juliastuti, 2004, p. 95). Juliastuti emphasizes the irony of this, as the Chinese ethnicity was a discriminated minority before the abolition of the Assimilation Law. Other remakes, such as *Buku Harian Nayla* (2006) which was copying characters and plots from the Japanese television series *One Litre of Tears* and *Benci Jadi Cinta* (2006), a remake from a Korean series *My Girl*, are other reproduced texts which carry thematic similarities and visual aspects as these dramas from the "original version".

In the 2017's version of *Siapa Takut Jatuh Cinta*, the narrative elements are not completely similar with *Meteor Garden* (2001) even though it keeps the basic foundation of the story: a poor (economically) struggling girl who fell in love with a rich (snobbish) boy. The visual representation is still quite similar with the Taiwanese, Japanese and Korean versions portraying a high-class and urban lifestyle with good-looking characters. These textual and visual elements have become the formulae of these types of television dramas as they move from its origin country to other Asian countries formulating its textuality. *Siapa Takut Jatuh Cinta* (2017) retains this textuality even though upon further investigation, the *sinetron* formula still dominates the narrative elements. For example, in the mise-en-scène, the exaggerated multiple close-up shots used to emphasize on the dramatized shots with dramatic music are still used constantly throughout the episode. Another example is the notion of piety in which the main female character, Laras, in her troubling times when she was evicted from her house and lost her scholarship to the prestigious college, she went to the mosque and pray (episode 9). Afterward, her problems were solved even though it was not all solved at once, the idea of praying as a part of the solution could be found in most of Indonesian local soap operas or *sinetron*. The textual reproduction offers the audience familiar formula (from the East Asian television dramas) in their meaning-making process of the original Japanese/Taiwanese/Korean television dramas, while conforming to a specifically *sinetron* narrative elements.

2.2 *Crossing the border: Reconstructing a sense of authenticity in Indonesian boy/girl bands*

On 22 November 2012, the Tourism Minister, Mari Elka Pangestu, speaking at the Indonesia Creative Products Week (PPKI), stated that "The domestic music industry is threatened by the entry of Korean Pop (K-Pop), and now is time for us to create I Pop." In this sense, I-Pop has become a reaction to the popularity of K-Pop. Jung (2011) argues that in Indonesia, K-pop's popularity has lasted longer because, on one side, the Korean entertainment industry seeks out market possibilities to sell and distribute their products. On the other side, for the Indonesian media industry, there is a desire to be globalized and South Korean popular music products offer a new kind of global as an alternative to the global West. K-pop has also shaped up the mainstream discourse of how Indonesians imagine East Asia.

Many scholars have formulated how one should understand this global phenomenon of K-Pop. Shim (2008) argues that K-Pop is all about the "idol-making system and global marketing strategies." It is not about the product itself, such as the music or the singer's capability in singing, but K-Pop is an industry in which idols are made and products are disseminated through a strategically structured global distribution and marketing. These strategies integrate the production process consisting of training of talents and even the management system. Training here involves not only the training of singing and dancing, but also on how to be multi-skilled entertainers by, for example, learning other languages to appeal to fans in different countries, emphasizing its global characteristics. Furthermore, K-Pop's "cross bordering characteristics" as argued by Siriyuvasak and Shin (2007) are significant because as a culturally specific product, for example how it is mostly presented in Korean language, K-pop needs to have cross-bordering characteristics to guarantee that it will be able to enter other countries. One strategy is to mix the language in order to creating a sense of identification for different consumers in different countries. The specificity of Korean-ness, such as the Korean lyrics, needs to be repressed and even reduced. One could argue that by, for example, creating lyrics that could be understood by consumers from different countries, K-Pop is creating a "sense of authenticity" and it is actually "denationalizing K-Pop."

I-Pop is reproducing K-Pop in the Indonesian context. The development of I-Pop could be seen from two different stages. The earliest ones happened in 2010–201 with the popularity of *SM*SH* and *Cherrybelle*. They are 'look and sound-alike' groups that mimic its Korean counterparts and they actually receives a lot of resistance from the Indonesian consumers. Both have been repackaging K-pop's image and music in order to look and sound like the South Korean idol bands but with a speed-up idol making system. The second stage of I-Pop development was in 2012 to 2013, when *S4* (boyband) and *SOS* (girlband) were formed as the products of *Galaxy Super Star* talent show, which was collaboration between YS Media Entertainment (Indonesia) & Rainbowbridge Agency (a South Korean-based K-pop Artist Incubation Company). Both boy/girl bands were trained for 8 months in a training camp in Seoul and they underwent a physical transformation as they dyed their hair and changed their fashion style. Their music videos were made in South Korea by a Korean video director with Korean models. They also performed in South Korea and learned Korean language so they were able to sing in Korean. *S4* and *SOS* mimic K-pop in a distinct 'Korean' way in the sense that the members trained in Seoul, transformed physically to look precisely like K-pop idols, sing some songs in Korean and use Korean language to greet their fans

I-Pop has reworked the idol making system and standardized practices of the South Korean popular music scene as it mimics the 'new' global in K-pop. I have argued elsewhere (Tambunan, forthcoming) that the examples from the second stage of development of I-Pop reflect an intensification of Korean-ness articulating K-pop elements. Compared to earlier I-pop groups, *S4* and *SOS* amalgamate the actuality that they are "made in Korea" Earlier waves of I-pop have reproduced K-pop products by mimicking the songs, dance movements and also the idol making system. However, for its later predecessor, it is more significant to enhance the Korean-ness of their products.

Another form of textual reworking of idol bands happened in late 2011. In a talent show competition, *Boyband Girlband Indonesia* (Nov 2011-Feb 2012), there were 8 finalists and the winner was Sunni, a group of 8 teenage girls wearing hijab from Gresik, East Java. In July 2015, a girldband which claimed that they represent 'Hi-Pop' (*Hijabers* Pop), Noura, was formed. All four members were originally *Hijabers* models and they went through auditions. In several interviews, the claimed that they were singing pop songs influenced by

(SUNNI)

(NOURA)

Figure 2.

Islamic values. Their single, "*Kekasih Halalmu*" or "Your Halal Lover," exemplifies a different kind of textual reworking if compared to the previous boy/girl bands.

From the analysis, I would argue that the idol making system in which each boy/girl band is fabricated through multiple media activities resonate the idol making system, even though the Indonesian one is done within a shorter time period. There is a significant distinction between the two generations of I-pop. *S4* and *SOS* are products that actually accentuate "made in Korea". As entertainment companies fabricate different ways of mimicking K-pop, it also reflects the elements that constitute K-pop as a new global product. Furthermore, Hi-Pop embodies a more culturally modified textual reworking as the formation of both girl bands that are chosen as examples in this research could be further contextualized within the context of Islamic reawakening in the Reformasi period (Budiman, 2011). By utilizing distinct K-Pop features and culturally modified elements, I-Pop and Hi-Pop are results of a cultural fusion and intertextual reworking of K-pop that could not be simplified as solely products of plagiarism.

3 CONCLUSION

From a Cultural Studies perspective, I would argue that to make meaning out of these cultural phenomena as products of plagiarism distorts the complexity of cultural globalization in Asia. By investigating the chosen case studies, one could argue that by copying the elements, be it the narrative elements in the television dramas and the industrial characteristics in the boy/girl bands, the Indonesian popular culture industry diminish creativity and submit to the new global force. However, through the aforementioned radical contextualization, I would argue that more substantial abstraction is needed to explain this phenomenon. Mimicry, proposed by Homi Bhabha, is a helpful concept to understand how mimicking East Asian television dramas or K-Pop could be a kind of performance that exposes the artificiality of these popular culture products as "symbolic expressions of power." From this conceptualization, there is agency in mimicking the "colonizer" while challenging the power relation. Could we conclude that there is agency in these textually reworked products? It is also problematic to look at the East Asian countries as "colonizers" because this would lead us back into the center-periphery dichotomy and the one-way flow of cultural globalization in the cultural imperialism model.

I would also argue that the intricacies of the textual reworking could be understood as what Raymond Williams explains in his book *Marxism and Literature* on how the complexity of culture could and should be understood not only its materiality and social characterizations but also from its "dynamics interrelations at every point in the process." I would argue that one of the main characteristics of urban culture is "temporality" because of the speed-up everydayness in the cities. Anything that regain popularity and become what Williams would argue as the "dominant" culture does not exist in a long period because they will be taken over by the other "dominant" cultural forms. In this case study, research findings reveals that the basic formula of Indonesian *sinetron* is in a sense dominating the process of textual reworking that no matter how similar the remakes are with the "original version" it would never be completely similar. On the other hand, when for example I-Pop has transformed significantly over the last 7–8 years in Indonesia, the evolution of K-Pop as the dominant culture which at some point diminish its dominance and then revive its popularity only by textual reworking and transformations. Additional research needs to be done to further explore this complexity of the cultural process of East Asian popular culture products in relation to the Indonesian counterparts.

REFERENCES

Asih, R. 2017. Siapa Takut Jatuh Cinta Memuncaki Google Trend Indonesia or Who's Afraid to Fall in Love is on Top of the Google Trend List in Indonesia. *liputan6.com*. Retrieve November 14, 2017 from http://showbiz.liputan6.com/read/3162026/siapa-takut-jatuh-cinta-memuncaki-google-trend-indonesia.

Budiman, M. 2011. The Middle Class and Morality Politics in the Envisioning of the Nation in post-Suharto Indonesia. *Inter-Asia Cultural Studies* 12(4): 482–499.

Chua, B.H. & K. Iwabuchi. 2008. Introduction in Chua Beng Huatand Koichi Lwabuchi *(Eds.). East Asian Pop Culture: Analysing the Korean Wave* (1–12). Hong Kong: Hong Kong UP.

Chua, B.H. 2004. Conceptualizing an East Asian Popular Culture. *InterAsia Cultural Studies* 5(2): 200–221.

_________. 2012. *Structure, Audience and Soft Power in East Asian Pop Culture.* Hong Kong: Hong Kong UP.

Couldry, N. 2000. *Inside culture: Re-imagining the Method of Cultural Studies.* London: SAGE.

Iwabuchi, K. 2002. *Recentering Globalization: Popular Culture and Japanese Transnationalism.* Durham: Duke UP.

Jameson, F. 1991. *Postmodernism, or, the cultural logic of late capitalism.* Durham: Duke University Press.

Juliastuti, N. 2004. Budaya Televisi, *Meteor Garden,* dan Remaja Perempuan or Television Culture, *Meteor Garden* and Female Teenagers. *Jurnal Perempuan* 37: 87–97.

Merdikaningtyas, Y.A. 2001. Perjalanan Meteor Garden di Indonesia or Meteor Garden in Indonesia. *Indosiar.* Retrieved November 10, 2017 from http://www.indosiar.com/gossip/6297/function.reguire-once.

Shim, D. 2008. The Growth of Korean Cultural Industries and the Korean Wave. In C.B. Huat & K. Iwabuchi, *East Asian Pop Culture: Analysing the Korean Wave*. Hong Kong: Hong Kong University Press: 15–31.

Siriyuvasak, U. & S. Hyunjoon. 2007. Asianizing K-pop: Production, Consumption and Identification Patterns among Thai Youth. *Inter-Asia Cultural Studies 8(1):* 109–136.

Tambunan, S.M.G. (forthcoming). Cultural Borrowing/Appropriating South Korean Popular Culture: An Analysis of I-Pop and K-Drama Remakes in Indonesia. *Cultural Dynamics in a Globalized World: Proceedings of the Asia-Pacific Research in Social Sciences and Humanities*: CRC Publisher.

Williams, R. 1977. *Marxism and literature* 1. Oxford Paperbacks: Oxford and New York.

Urban Studies: Border and Mobility – Kerr et al. (Eds)
© 2019 Taylor & Francis Group, London, ISBN 978-1-138-58034-3

Representation of nostalgia for home in diasporic poetry: An analysis of selected poems of mahtem shiferraw

S. Elias
Universitas Airlangga, Surabaya, Indonesia

ABSTRACT: Nostalgia for home has been a common subject in African diasporic literature. In the past four decades, the African continent has witnessed an increase of movements of people leaving their home to new places around the world. These movements, whether voluntary or forced, have resulted in displaced communities in the world, diasporic communities being one type of those communities. Deploying a discontinuity hypothesis, this paper (1) examines the representation of nostalgia for home in selected poems in *Fuchsia* (2) interrogates the contribution of nostalgia on redefining the identity of the poet. Shiferraw's poems capture and communicate the traumatic experience of leaving home through the use of memories coloured by nostalgia of time, place and people. On the one hand, nostalgic feelings may evolve without one yearning to go back home, though home remains a stimulus of a reflective past and, on the other hand, nostalgic feelings raise memories that significantly re/define a poet's identity in the host homeland.

Keywords: Diasporic Literature, Home, Nostalgia

1 INTRODUCTION

Since its advent in the 17th century, the concept of nostalgia has undergone several reviews to suit the academic needs in a wide range of fields. For instance, it has been viewed as a mental illness and an incurable modern condition at the end of 18th century and early 19th century (Davis, "Nostalgia, Identity and the Current Nostalgia Wave"), as longing for bygone times, of course, as a mode of selective past memories (Casey), as imaginative enrichment and, more importantly, the emotional aspects inherent in nostalgia during the Romantic Era (Boym) and as "a form of melancholia caused by prolonged absence from one's home or country" (Simpson, 1989) as cited in (Feldbrügge, 2010: 27) to mention a few.

In the 20th century, nostalgia was reviewed and redefined to reflect contemporaries of the academia (Feldbrügge). Subsequently, nostalgia was used to signal sentimental yearning for any object, event or place in the past (Davis, "Nostalgia, Identity and the Current Nostalgia Wave"). Today it has been extended to include "the recalling of memories that give rise to certain complex emotions resulting to positive mood" (Andersson, 2011:13). In this paper, I treat nostalgia as a longing for bygone times or a mode of selective past memories of place, events, time and people with respect to selected poems of Mahtem Shiferraw.

Neelime (2014) argues that one may recollect and mourn through memories of the past of the ideal home, and nostalgia feelings can occur without one yearning to go back home. Yet, Swamy (2013:5) still restricts the concept by stating that "the sense of nostalgia is often seen among the dislocated and displaced people in most of the diaspora writings" and it contributes to identity re/invention as they try to adjust to new settings (D'Costa, 2006). Conforming to the foregoing view, the African continent has witnessed an increase in the movement of people to other countries in the world due to cultural, political, economic and social upheavals and so has formed larger displaced community in the world over the past four decades. The surrounding circumstances that necessitated the departure of these African diasporic communities from their homelands have influenced the way communities feel about and associate with their homelands. In this view, Rubenstein (2001:6) acknowledges that "culturally displaced or exiled people may mourn their separation from home/land, community, language, and/or cultural practices that contribute to identity" and D'Costa (2006) adds that displaced or exiled people deploy "nostalgia to re/centre their identity about their homeland".

Rijsdijk (2015) whose focus was to examine *Nostalgia and the (re) construction of South African identity in District 9* sees that authors who seek continuity with their past always do so through nostalgia. Mahtem Shiferraw in *Fuchsia* has not strayed far from what Rijsdijk (2015) sees in authors. In her debut book *Fuchsia* she creatively presents the migrant's experiences through memories in terms of imagination and nostalgia for place, time and people. It is from nostalgia that she also manages to construct her 'otherness' while immersed in the new multicultural experiences. The poems in her book connect dots of her past experiences from childhood to adulthood of different places and times. It is in this light that the paper (1) examines the representation of nostalgia for home in selected poems in *Fuchsia* (2) interrogates the contribution of nostalgia in terms of re/defining the identity of the poetess.

Mahtem Shiferraw was born in Eritrea, grew up in Ethiopia and now lives in the United States of America. She holds an MFA in Creative Writing from Vermont College of Fine Arts. She is a poet, story composer, visual artist and an activist. She was nominated for Best of Net in 2013 and her story 'The River' was entered on the final list for the Glimmer Train Press Open Fiction.[1]

Fuchsia is her debut collection of poetry published by the University of Nebraska Press in 2016. The book won the Sillerman First Book Prize for African poets and secured a place in the African Poetry Book Series. The poems in *Fuchsia* are crafted with the use of metaphorical colours which evoke detailed memories and life in a different space and time. The poems in the book carry the readers through the convoluted experience of the author while in Africa and as well as when a migrant in America. It is a book of displacements, nomadic life, memories of pleasure, grief, and loss and identity in search of the contemporary diasporic communities. Poems in it interrogate the surrounding environments that might have lead to her departure to America and the later effects on her identities and sense of belongingness as a part of a displaced community but also towards other migrants who have had the same experience. Kwame Dawes, in the foreword of the book, explains that poems in *Fuchsia* invite us to spaces of 'home' through time and geography (Shiferraw, 2016: ix).

As this paper involves textual analysis, I first use close reading to analyse the representation of nostalgia in selected poems of Shiferra as depicted through the use of words, symbols and metaphors without considering external factors as advocated under new criticism. This is followed by a contextual method. I use this method to find the association between the selected poems and the contexts in which they were produced. It is in this section that the relevance of Shiferraw's biography becomes a consideration. Close reading and contextual methods lead to a thematic analysis method. Guided by the topic under discussion I explore the thematic representation of nostalgia in selected poems in Shiferraw's book *Fuchsia*.

2 DISCUSSION

Shiferraw, in her poem "Fuchsia", recalls and longs for memories of her childhood and family while at home in Addis Ababa. Through the persona, Addis Ababa is described as a place where they dwell and are comfortable, secure, and emotionally and culturally attached. This reiterates what Webster (2010) conceives about home. Webster points out further that when there is a significant change in terms of security, comfort and emotional attachments, it is common to displaced individuals. Thus, Rubenstein (2001) stated that displaced individual will mourn their separation from home, people, language and cultural practice. On the surface of the poem "Fuchsia", through the persona, readers are taken through an emotional selection of pleasant past memories of the poet and her Ethiopianness while at home in Addis Ababa with her families. "When I was little, growing up in, Addis Ababa, my father bought the fattest sheep

1. http://www.glimmertrain.com/pages/finalists/2013_09_sep_fopn_hm.php

from street vendors for holidays…." (1–3). In the foregoing lines of the poem, the poet recalls and longs for her pleasant past during her childhood that she cannot have access to at the moment. She keeps watching as her parents butcher the sheep and prepare varieties of meals from it as per traditions like *berbere* and other spicy cuisine. The other part of the sheep (head) is shared with the neighbours for them to make soup. The momentous memories of her past trigger emotional and complex moods of adjustment in the new place as immigrant.

> insides, and knife slashing between slabs of organs, all to be eaten differently—bones of the rib cage deep fried, bleeding texture of kidneys minced into bite-sized shapes and soaked in onion and pepper oil, small blades of the stomach dutifully cut into long strips, and mashed with spiced butter and berbere. Even the skin, bloodying fur, will be sold to passing vendors, its head given away to neighbors who will use it for soup. (1–3).

Notably, through the persona, the pleasant past of the poet signals the communal living of their entire community in Addis Ababa. Communal living is reflected as they share the other part of the sheep (head) with the neighbours for them to make soup. Unlike in the new land where the poet dwells at present, communal living is a typical African way of living in many African societies.

However, sensual imagery of violence comes into the poem as it comes to an end and memories of death are inscribed metaphorically through the use of colours. The lines "If you ask how to say 'burgundy' in Tigrinya, you will be/told, it's the color of sheep-blood, without the musty smell/ of death attached to it. It's also the color of my hair, dipped" (1–3) are allusive of memories and symbolic representations of violence in Addis Ababa.

In contrast to what Hofer (1688), Daniels Eugene B (1985) and Sedikides, Wildschut and Baden (2004) state when they argue that nostalgic feelings revolve around going back to one's tangible home place, I argue that nostalgic feelings may evolve without one yearning to go back to a home place as in "Fuchsia". Nostalgic feelings in "Fuchsia", that the poet depicts through the persona, are made of selective pasts that give rise to emotional imagery of security, belongingness and the comfortable life of the poet and of the entire family in Addis Ababa, but they are presented from a distant angle of multicultural experiences where she currently lives. However, Davis in *Yearning for Yesterday: A Sociology of Nostalgia*, argues that nostalgic emotions of this kind that may not require one to go back to a physical home place but protect the individual (poet) from discontinuity with her past during moments of adjustment in the foreign environment. This conforms with Havlena, who argues that nostalgia may be stimulated by memories of hard times or times of turmoil. On the surface of the poem "Synesthesia", the poet metaphorically presents violent, painful and fearful memories of home. Through the persona's voice, colours are crafted to carry the readers through convoluted memories of violence, troubles, wounds, restlessness, sadness and the post-violence scenery of the poet's home. The following lines provide an exemplification of the violent and painful memories of the poet.

> White sometimes comes back at odd hours. White are stranger's eyes drenched in sadness. White is the uniform of doctors, the smell of alcohol and something mad. White is absence. Purple comes back as shoes, American shoes. Sky and blood under a quiet shadow. The shadow of a young tree planted in memory of a murdered teacher in high school. And the milky paste of over-ripe figs spurting prematurely, spiking insides. Purple is warmth in mid-July, when rain hails on corrugated tin roofs and the leaning green arms of lonely corn plants.
>
> Yellow is crying; it's a bell, a cathedral in Asmara? A school? Or the shriek of a mass funeral. Yellow is dead. But listen to black. Listen to black notes, black heart, listen. Black is art. Not of the artist, the art of being. The painful art of memory. Here's to remembering (30–31).

The foregoing fragment of the poem alludes to violence and instability in the poet's African homes. As pointed out earlier, the poet has Eritrean-Ethiopian origins; the two nations went through a two-year long war from May 1998 to June 2000 causing deaths and more than 100,000 causalities. According to Tronvoll, the dispute over the border has continued to be the main source of instability in the Horn of Africa until today. Through the persona's voice, the poet presents to readers the painful memories of people, place and time. The poet emotionally recalls the time when causalities of violence are being attended to by doctors in white uniform, ism in the post-violence landscape. Consequently, the poet's neighbour's houses in *Abbay* were empty. As depicted in the fifth death of people in Asmara,[2] there is insecurity as people are being murdered in the place and an absentee stanza in the poem "Synesthesia" through the persona; there is grief on people's face as is shown in the first verse of the poem: "Blue are the waters embedded in my grandmother's eyes. Blue is" (30–31). Through the persona's voice, the poet orientates the readers through chaotic incidents in her African home from a different space and time.

Accordingly, Addis Ababa and Eritrea are momentous places in the past of the poet that connect to her present life experience while living in another place. The nostalgic feelings for the homeland in Africa not only describe who the poet is, but also protect her from discontinuity with her past. This confirms what Davis asserts (1979), in hypothesising that nostalgia allows people to retain their identity when they enter new life experiences and, therefore, nostalgia functions to protect personal identity against threats of discontinuity (Bassett). Alternatively, Andersson (2011: 29) states that "people who face existential disruptions, like moving to a new location, changes in an intrapersonal relationship, occupational crises (e.g. layoffs), loss of family members etc., will be more nostalgic about the past than those who maintain greater continuity in their lives". She chooses Addis Ababa to associate with memories of her past that define her "otherness"[3] as an Ethiopian/Eritrean and Afro American while in America. It is from her past that she is lead to "otherness" in the new land. Following this assertion, the poet confirms her "otherness" identity in the poem, "Talks about Race" (20). The lines "I am 'other'; it is such/an indistinguishable form, beyond the construct of the proper self" (20) defines her entrance in the foreign place. Though, in the poem "Talks about Race", the poet also interrogates her Africanness because she also questions her identity while she is in her homeland. Her past, as represented through nostalgia, adjusts her to fit in the new location, but also redefines her identity. This correspond to Lowenthal (1989) who affirms that "We are at home in [the past] because it is our home – the past is where we come from".

Through the persona's voice, the poet mourns her Eretria nativeness which is revisited through memories of Tigrinya language.[4] According to Rubenstein (2001), it is common for culturally displaced people to mourn for their separation from language and community. However, in this particular scenario, her past Eritreaness includes unpleasant imagery with death attached to it.

> If you ask how to say "burgundy" in Tigrinya, you will be told, it's the color of sheep-blood, without the musty smell of death attached to it. It's also the color of my hair, dipped (1–3).

Shiferraw's poems, namely "Fuchsia" and "Synesthesia", capture and communicate the traumatic experience of leaving home through the use of memories coloured by nostalgia of time, place and people. Consequently, the experience that the poet goes through follows Andersson who argues that nostalgic feelings lead to complex and emotional memories that raise a certain mood as the author struggles to fit himself or herself into a foreign culture.

3 CONCLUSION

As it can be deciphered from the foregoing discussion, nostalgic feelings are reflections of past experiences of disrupted individuals. Also, as is the case in the selected poems, the poet chooses

---

2. Capital city of Eritrea.
3. Afaf Ahmed Hasan Al-Saidi (2014), in Post-*colonialism Literature the Concept of self and the other in Coetzee's Waiting for the Barbarians: An Analytical Approach*, describes the concept of the "other" as the foreignness: a person who does not belong to a group, does not speak a given language, does not have the same customs; unfamiliar, uncanny, unauthorised, inappropriate, and the improper.
4. Tigrinya language is a widely spoken language in Eritrea and the first language of the poet. The poet has been identifying herself with respect to the two nations because of the historical explanations that abide in the two nations. Eritrea was part of Ethiopia before 1993 when it became independent.

to present both the pleasant and unpleasant past life experiences in Africa. However, the discussion appears to rebel against the traditional sense of nostalgic feelings as revolving around going back to ones' homeland. Rather, nostalgic feelings in the selected poems have been a mode of presenting selective memories of the past that render continuity of the past with the present experience of the migrant as she adjusts herself to fit her new experience in terms of culture and place. On the other hand, the representation of nostalgia significantly contributes to re/defining the poet in the new place she currently lives. Her Africanness (past) shapes her current identity in the new place and this agrees with Bassett and Andersson in the sense that nostalgic memories protect personal identity as well as threats of discontinuity and pressures of disruptions.

REFERENCES

Bassett, Jonathan F. 2006. "An Experimental Test of the Discontinuity Hypothesis: Examining the Effects of Mortality Salience on Nostalgia." *Journal of Articles in Support of the Null Hypothesis* 4.1: 1–8.

Boym, S. 2001. *The Future of Nostalgia*. New York: Basic Books.

Buikema, Rosemarie. 2005. A Poetics of Home: On Narrative Voice and the Deconstruction of Home in Migrant Literature. *Migrant Cartographies: New Cultural and Literary Spaces in Post-Colonial Europe* 1986: 177–187.

Cahyo, P.S.N. (2017). Cultural Studies: Perlintasan Paradigmatik Dalam Ilmu Sosial. *KOMUNIKATIF*, 3(1), 19–35.

Casey, Edward. 1987. "he World of Nostalgia". *Man and World* 20: 361–384.

Christinawati, D., Manuaba, I.P., & Hum, M. (2005). Makna Resistensi Kultural dalam Puisi-Puisi Indonesia Mutakhir.

D'Costa, Alzena. 2006. Anglo-Indian Nostalgia: Longing for India as Homeland: 24–25.

Daniels Eugene B. 1985. Nostalgia and Hidden Meaning. *American Image* 42: 371–383.

Davis, Fred. 1977. Nostalgia, Identity and the Current Nostalgia Wave. *Journal of Popular Culture* 11.2 (1977): 414–424.

_____. 1979. *Yearning for Yesterday: A Sociology of Nostalgia.* New York: NY: Free Press.

Feldbrügge, Eingereicht von Astrid. 2010. Nostalgia, Home and Be-Longing in Contemporary Postapartheid Fiction by Zakes Mda and Ivan Vladislavi Ć. *Dissertationsschrift Eingereicht von Astrid Feldbrügg.*: n. pag.

George, Rosemary Marangoly. 1996. *The Politics of Home. Postcolonial Relocations and Twentieth-Century Fiction.* Berkely: University of California Press.

Havlena, Susan L. Holak & William J. 1992. Nostalgia: An Exploratory Study of Themes and Emotions in the Nostalgic Experience. *In NA—Advances in Consumer Research Volume 19, Eds. John F. Sherry, Jr. and Brian Sternthal, Provo, UT : Association for Consumer Research.* N.p., 1992. 380–387.

Lowenthal David. 1989. Nostalgia Tells It like Itwasn't. *In The Imagined Past,*. N.p., 1989.

Neelima, C. 2014. The place of nostaligia in diaspora writing: home and belonging in the fiction of bharati mukherjee. *International Journal of English and Literature (IJEL)* 4.6: 33–40.

Rijsdijk, Ian-Malcolm. 2015. *Nostalgia and the (Re) Construction of South African Identity in District 9*: n. pag..

Rubenstein, Roberta. 2001. *Longing and Belonging Nostalgia and Mourning in Women's Fiction.* New York: Palgrave.

Safran William. 1991. Diasporas_in_Modern_Societies_Myths_of_H(1).pdf.: n. pag.

Sedikides, Constantine, Tim Wildschut & Denise Baden. 2004. *Nostalgia: Conceptual Issues and Existential Functions*. N.p.

Shiferraw, Mahtem. 2016. *Fuchsia*. Lincoln: University of Nebraska.

Simpson, J.A & E.S.C. Weiner. 1989. *Oxford English Dictionary*. 10.

Swamy, Mrs. G. Serwani Venkata. 2013. Immigrant Identity, Nostalgia for Home and Home Land: A Perception in Chitra Banerjee Divakaruni'S the Vine of Desire. 1(3): n. pag.

Tronvoll, Goitom Gebreluel & Kjetil. 2013. Ethiopia and Eritrea: Brothers at War No More. *Aljazeera.* Retrieved November 3, 2017 from www.aljazeera.com.

Webster, Magda B. 2010. *Home As The Intersection Of Place And Cultural Identity : The Idea of Home in Diaspora and Art as a Response to an Immigrant ' S Nostalgia.* _____.

Urban Studies: Border and Mobility – Kerr et al. (Eds)
© 2019 Taylor & Francis Group, London, ISBN 978-1-138-58034-3

Returned participants' Perception of the Sarjana Mengajar di Daerah Terluar, Terdepan dan Tertinggal (SM-3T) program

Yuni Sari Amalia, Cahyo S. Budiono & Rizki Andini
Universitas Airlangga, Indonesia

ABSTRACT: The Ministry of Education and Culture has implemented a program of Sarjana Mengajar di daerah terluar, terdepan dan tertinggal (graduates educating in the frontier, outermost and disadvantaged regions), well known as SM-3T, since 2010. This program is a one-year social service program for rural and remote schools to prepare future professional teachers. The purpose of this study is to describe the SM-3T returned-participants' perceptions of the influence of the program on their professional skills as teachers. This is a qualitative study for which data were collected through interview and documentation. The result of this study shows that the program has no significant influence on the participants' professional skills in teaching. However, they think that the program influences participants' personal, social and pedagogic competencies. They feel that the program made them tougher, more independent, and more caring to others. This study also reveals that the participants felt insufficient preparation for dealing with the challenges in the targeted regions (frontier, outermost, and disadvantage regions) so that they could not perform well and learn optimally during the program.

Keywords: Teacher education, Teaching strategies, Teacher competence

1 INTRODUCTION

Good quality education can be achieved through the interaction of multiple factors, the most important of which is the quality of teachers and teaching (Anderson 2002; Lewin and Stuart 2003; Verspoor 2006). Unfortunately, the quality of teachers in Indonesia is still not satisfying and contributes to the low performance of students in several international tests such as PISA and TIMS. A study conducted by the World Bank (2010) shows teachers' low competencies in terms of knowledge, pedagogic skills, and general academic intelligence. The other test conducted by the Ministry of Education and Culture in 2015 also indicates the low pedagogic and professional competencies of teachers with an average score for teachers being 53.02 out of 100 (Kememdikbud, 2015).

Dealing with the issue, the Ministry of Education and Culture has issued several policies such as PLPG (Pendidikan dan latihan Profesi guru), PPG (Program Profesi Guru), and SM-3T (Sarjana Mengajar di daerah Terdepan, Terluar, Tertinggal). SM-3T graduates educating in the frontier, outermost, and disadvantaged regions are designated as pre-service teachers to teach in remote Indonesian areas for a year. The program aims to prepare future professional teachers who are then recruited to PPG, a professional teacher education program. It also aims to help the regions to overcome the problem of the lack of teachers and improve the learning quality in the targeted regions (Rustad, 2013).

The SM-3T program provides an opportunity for bachelor degree graduates to be volunteers in the targeted regions in eight provinces such as Aceh, Riau Archipelago, East Nusa Tenggara (NTT), East Kalimantan, West Kalimantan, North Sulawesi, Papua, and West Papua. During the voluntary program, the participants are challenged to: (1) improve the quality of the learning process in the targeted area based on the needs of the region's condition and the participants' expertise; (2) promote learning innovation activities at school; (3) help the school to conduct extracurricular activities; (4) promote and help the school to implement good practices in school management; (5) conduct social tasks and society empowerment (Rustad, 2014).

According to Rustad (2014), the SM-3T the program is designed to help the targeted regions to deal with education problems such as the low quantity and quality of teachers in the region. Furthermore, the program is expected to cultivate the characters of care, empathy, responsibility towards the nation's progress, loving the homeland, and defending the country and make the participants tough and persistent in dealing with challenges in the disadvantaged regions. Finally, the program is designed to build 'chemistry' toward the profession as a teacher who are ready to be stationed in any disadvantaged regions and to get the participants ready to be future professional teachers.

Before joining the SM-3T program, the candidates are required to join the pre-condition activity to give them a general orientation on the condition of social, cultural, and infrastructural aspects of the targeted regions. This orientation is held for 12 days providing academic and non-academic activities. The academic pre-condition includes: (1) training on how to run education tasks in challenging conditions (low number of teachers, low ability of the students, and low access of facilities); and (2) education leadership and management at school. The non-academic pre-condition includes: (1) mental training and survival, (2) training in social skill, (3) concept of nationalism and defending the country, and (4) scouting and first aid.

The effectiveness of the Ministry of Education and Cultures' policy to prepare the future teachers through the SM-3T program is an interesting topic for exploration. This study aims to describe the development of professional teaching skills through the SM-3T program and the returned participants' perceptions of the SM-3T program in developing their professional teaching skills.

2 LITERATURE REVIEW

2.1 *The effective teacher*

Because one of the main goals of the SM-3T is to prepare quality future teachers, it is important to present studies on 'who the effective teachers are'. Crosser (2002) defines effective teachers as good planners. They anticipate everything and anything that may occur in their classrooms. Furthermore, many studies show that classroom management is the key element that an effective teacher has to possess (Crosser, 2002). Classroom management is not only about how to manage the students' behaviour but also everything that is involved in the class, including the learning environment (Duncan, 2010).

John Dewey (1916) believes that an effective teacher promotes active learning to stimulate students' thinking on his own. Teachers are supposed to recognise and respect that the students are able to form their own opinions, and conclusions and ideas. In evaluating their students' progress, effective teachers also employ authentic assessment. Wiggins' (1991) describes authentic assessment as, "It involves engaging problems and questions of importance and substance in which students must use knowledge (and construct meaning) effectively and creatively." Therefore, when teachers employ authentic assessment, they customise instruction, deeply involve students, and assess the assessment support changes in the curriculum, teaching, and school organisation.

Effective teachers also have an ability to adjust learning to the individual needs of their students. Baglieri and Knopf (2004) suggest that teachers should design lessons based on their students' needs, clarify their goals, and consistently update whether the goals are being attained (McTighe & Brown, 2005). Effective teachers also need to know how to motivate their students. According to Shalaway (1998), teachers should develop the students' internal motivation in order to make the students approach learning tasks seriously.

2.2 *Stages of teacher development*

Katz (1972) mentioned four developmental stages in teacher development, i.e., survival, consolidation, renewal and maturity. These stages are related to the length of the teaching experience.

The survival stage is indicated by teachers' self-questioning about their own competence and willingness to teach.

Stroot et al. (1998) explain that the teaching style used by the teachers in this stage is the teacher-directed method which tends to stick to the lesson plan. The teachers have difficulties in dealing with a more student-centered approach and control in the lesson is dominantly directed by the teacher.

The consolidation stage is characterised by the increased focus on the students' individual needs. The teachers give more attention to the children with special needs. Even though they still struggle in their teaching practice, they begin to be more open to professional development. They begin to have confidence to develop discussions with other teachers at the same stage of development (Stroot et al., 1998).

The teachers in the renewal stage face boredom in using patterns of teaching. Even though the patterns are effective, the teachers begin to seek new ideas for teaching. Although they have expertise in some management strategies, they want to learn new methods to improve their performance in their jobs (Stroot et al., 1998).

The maturity stage is indicated by the mastery of some effective teaching strategies. The teachers are more reflective and keep their interest in discovering new ideas and resources. Their teaching focuses on deep issues and concerns in inspiring the students and the impacts of their teaching (Stroot et al., 1998).

3 METHOD

This research is descriptive research. The subjects of this research were two participants in the SM-3T program of Surabaya State University who were stationed in Sumba Timur, Nusa Tenggara Timur province and Memberamo Tengah, Papua province. The data was collected through in-depth interviews and documentation. The questionnaire was filled in by the SM-3T participants. Narrative offers a rich description of the data (Amalia et al, 2015) and is considered most relevant for this study. Thus the data was analysed using a narrative description method.

4 RESULTS

Subject 1: Paramita Ananda, S.Pd. (Paramita – pseudonym).
Paramita was stationed in Nusa Tenggara, Timur province bordering Timor Leste and Australia to the east. Because Paramita is a graduate from Bahasa Indonesia department, she is supposed to teach in the Junior or Senior high school. But, she was assigned to teach elementary school students in SD Masehi Billa, Pindu Hurani village. It is about 123 km from East Sumba city centre. The village is situated in a hilly area on the edge of Laiwangi Wanggameti national park.

Paramita attended the pre-conditioning program in Surabaya. Unfortunately, it was designed to prepare the participants to deal with general problems commonly faced in the disadvantaged areas. It was not specifically designed to deal with the specific challenges in the targeted areas in which the participants would be stationed.

During the program, Paramita taught Bahasa Indonesia for 4, 5, and 5 graders. The students only use and access Bahasa Indonesia in the school, thus their competence in Bahasa Indonesia is relatively low. The students' competences are also divergent. Some of them even have difficulty reading and writing. She needed to learn techniques to teach reading which she had not learnt in the university or pre-conditioning program. Fortunately, her partner at the school, the other participant in the SM-3T program, had an elementary education program background. She was very helpful and spent a lot of time discussing techniques for teaching reading.

Generally, the teaching conditions were not ideal. Paramita had to teach from Monday to Saturday, 07.30–12.30, sometimes this was cut short in the rainy season as the students are allowed to leave early as a precaution as many have to traverse a river to their homes.

Her responsibilities included: planning lessons, teaching, conducting extracurricular, cocurricular, remedial, enrichment, and social activities. Extra sessions were impossible due to the students' obligations to help their parents. Because of the students' low and divergent competencies and the lack of any requests for her to submit lesson plans, Paramita devised her own plan which diverged from the standard promoted by the Ministry of Education to account for the students' needs and used it as a guideline. Unfortunately, violence was still prevalent in the school as a form of punishment, which Paramita struggled with as it was an accepted norm even amongst the students.

Even though Paramita felt that she could not apply and improve her professional competencies during the program, she thinks that she benefited from experiencing changes in the students' competencies and attitudes. She understood the importance of giving respect to the students to build positive interaction in enhancing the students' attainment.

Subject 2: Taruna Gilang, S.Pd. (Taruna – pseudonym).
Taruna was an Indonesian teacher at Al Azhar Junior High School in Surabaya. Taruna was assigned to Kelila village, Kelila district, Memberamo Tengah District in Papua Province, a district that is accessible by land and air but difficult to reach because of mountainous terrain and rocky roads. In his pre-conditioning training, he found the non-academic training useful, although the 2013 curriculum taught in the academic training proved useless as the target school still used the 2006 curriculum. He was stationed at the only public high school, Kelila State High School.

SM-3T teachers came to the district on 21 August 2015. Housing and infrastructure were immediate problems as, not only was the school being renovated, but the teachers' housing was not yet finished. The teachers had to stay in the principle's house before moving to a priest's house and, although there were compromises as a solution for the lack of classrooms (sharing with the gradeschoolers), problems with personnel forced the teachers to teach outside the classroom at one point.

The teachers concluded that the largest issue they had to face was with the school culture especially dealing with teachers' values and motivation. Attendance of the teachers was unsatisfactory, with attendance of three out of six days considered good. The SM-3T teachers not only substituted for absent teachers but felt as if they were the main teaching force.

Taruna taught Bahasa Indonesia for class X, XI, XII. Due to the renovation and the inconsistency of 12th grade attendance, he did double classes with inconsistent numbers of students. Most of the students like Bahasa Indonesia especially in speaking sessions. In teaching, Taruna used the school text book as the main reference and simplified and adjusted the learning materials for the students' ability. The students' competencies in Bahasa Indonesia were low and divergent. There were students who had 'relatively good' ability and there were also students who still had difficulties in reading and writing. In general, their ability in reading and writing were like students in sixth grade in Java. This was also applied in teaching Economics, a subject he was not trained in yet was forced to teach due to a lack of teachers.

Similarly to Paramita's experience, after school activities were difficult to conduct due to family obligations, the exception being sports, and the principal did not request a lesson plan, prompting Taruna to create a lesson plan adapting to the situation as well.

An accepted value in the village was that inviting people to have a meeting means providing them with money and souvenirs, which made meetings costly and unaffordable. The teachers also have problems educating people about healthy lifestyles (one of biggest issues in Papua) and have difficulties in conducting social activities. What they decided to do was to help and educate people through conversations on any occasions when they could meet people.

In preparing the 12th grade students for the national examination, Taruna had to train the students on how to darken the answers with pencils and treat the exam paper. According to Taruna, preparing for the academic aspect of the national examination would be useless because they will never achieve the standard. Because the pressure from the parents for their children to pass the exam was so high, the teachers helped them to pass the national examination.

One of the benefits Taruna took from this program was that he can now use anything around him as media for learning and teaching. Even though they do not have technology, the students could keep learning competencies for their life.

5 DISCUSSION

The most difficult task for the participants of SM-3T was how to deal with bad school culture, especially the teachers' work ethics. It is a challenging even impossible task for the SM-3T participants to change the conditions. The way teachers think about their attendance at school, for example, would be difficult to change without the principal's leadership and their willingness to change (Amalia, 2009; Listiyono, 2006).

According to Paramita and Taruna, non-academic contents for the debriefing program are useful during the SM-3T program. And some contents in the academic part are irrelevant or not useful during the program e.g., 2013 curriculum. Based on Paramita's and Aldi's explanations about the debriefing or pre-orientation program there is a recommendation that the contents of the program needs to be revised, especially in terms of the academic content.

Both Paramita and Taruna valued their resources and luxuries in Java more after the experience, and it also made Paramita and Taruna more patient in handling their students. They felt that the students in the outermost areas in Indonesia are more challenging and more divergent in ability.

The challenge to deal with local people and maintain good relationships also improved Paramita's and Taruna's social skills. During the program, both of them were required to have many conversations with locals, deliver educational information to the parents and join traditional ceremonies. These experiences improved their respect and tolerance for other people. In their role as teachers, their experiences during the SM-3T program gave them an understanding that the success of their students can be achieved through orchestrating all of the stakeholders' roles including parents, education department of the Surabaya municipality, and community. In addition, the SM3T program also improved the participants' pedagogic skills.

Unfortunately, according to Taruna and Paramita, the program does not give enough opportunity for the participants to sharpen their professional competencies. Professional competencies in this case are basic competencies in terms of subject matter which the teacher has to possess (based on the regulations issued by the Ministry of Education). Paramita thinks that the program did not require her to learn more and more details about topics or basic competencies in Bahasa Indonesia. It is true that Paramita gained new pedagogic skills during the program, but they were all appropriate for elementary schools. As a Bahasa Indonesia teacher, she is supposed to teach in Junior or Senior High School. Even though Taruna has a task to teach at Senior high school, he thinks that, in terms of professional skills, the program did not challenge him to explore more. He thought that his competencies were good enough for teaching the students in the school.

Before participants are stationed in the outermost areas, a better preparation program is needed. The program should be adjusted to the challenges in each area. It cannot be a 'one size fits all' program that will endanger the achievement of the goals of the programs. Considering the participants' need to upgrade their competencies to deal with challenges in the targeted areas, the duty-reward scheme for this program also needs to be reviewed. For the 2015 program, the participants have to do voluntary service for one year and they will get a scholarship for one year to get a professional teacher certificate. It would be better if the next program is done the other way round—the participants get one year's scholarship first to get better competencies and they do the voluntary service afterwards.

6 SUMMARY

The SM-3T program is designed to help the targeted regions to deal with educational problems and to get the participants ready to be future professional teachers. The most difficult task for the participants of SM-3T is how to deal with bad school culture, especially the teachers' work ethics.

The SM3T program improved the participants' personal, social and pedagogic skills. Unfortunately, according to Taruna and Paramita, the program did not give enough opportunity for the participants to sharpen their professional competencies.

Before participants are stationed in the outermost area, a better preparation program is needed. The program should be adjusted to the challenges in each area. It cannot be a 'one size fits all' program that will endanger the achievement of the goals of the programs. Considering the participants' need to upgrade their competencies to deal with challenges in the targeted areas, the duty-reward scheme for this program also needs to be reviewed.

REFERENCES

Amalia, Y.S., Johnson Mardones, D., Johnston-Parsons, M., Shen, W., Shin, Y, S. & Swanson, J. (2015). Dialogue in narrative inquiry. In P. Smeyers, D. Bridges, N. Burbules, & M. Griffiths, (Eds). *International handbook of interpretation in educational research methods*. Thousand Oaks, CA: SAGE Publications.

Amalia, Y.S. (2009). Teachers' Perceptions Toward the Indonesian Government's Instruction for a Decentralized and Competence-based Curriculum. Thesis. University of Illinois at Urbana-Champaign, US.

Anderson, Stephen E. (ed.). (2002). *Improving schools through teacher development: Case studies of the Aga Khan Foundation projects in East Africa*. Lisse: Swets and Zeitlinger.

Arimbi, Diah A. (2017). Politicizing piety: Women's rights and roles in the Tarbiyah movement in Indonesia. *Religious Studies and Theology, 36*(2), 228–244.

Crosser, S. (2002). Managing the early childhood classroom. *Young Children, 47*(2), 23–29.

Dewey, J. (1916). Thinking in education. *Democracy & Education*. New York: The Macmillan Company.

Duncan, A. (2010). Teacher preparation reforming the uncertain profession. *Education Digest,* 75(5), 13–22.

Katz, L.G. (1972). The developmental stages of preschool teachers. *Elementary School Journal, 73*(1), 50–54.

Kemendikbud. (2015). 7 *Provinsi raih Nilai Terbaik Uji Kompetensi Guru 2015*. Retrieved on 15th October 2017 from http://www.kemdikbud.go.id/main/blog/2016/01/7-provinsi-raih-nilai-terbaik-uji-kompetensi-guru-2015.

Lewin, Keith M. and Janet S. Stuart. (2003). "Research Teacher Education: New Perspectives on Practice, Performance and Policy". *MUSTER Synthesis Report*. Sussex: University of Sussex and Department for International Development (DFID) Educational Papers.

Listiyono, S. (2006). Tanggung jawab pendidikan dalam masyarakat (tidak) sadar multikultural. *Jurnal Ilmiah Kebangsaan dan Keindonesiaan, 1*(2), 44–67.

McTighe, J., & Brown, J.L. (2005). Differentiated instruction and educational standards: Is détente possible? *Theory into Practice, 44*(3), 234–244.

Rustad, Supriadi, et.al. (2013). *Pedoman Pelaksanaan Program Sarjana Mendidik di Daerah Terdepan, Terluar, dan Tertinggal (SM-3T)*. Jakarta: Direktorat Jenderal Pendidik dan Tenaga Kependidikan, Dirjendikti, Kementrian Pendidikan dan Kebudayaan.

Rustad, Supriadi, et al. (2014). *Pedoman Pelaksanaan Program Sarjana Mendidik di Daerah Terdepan, Terluar, dan Tertinggal (SM-3T)*. Jakarta: Direktorat Jenderal Pendidik dan Tenaga Kependidikan, Dirjendikti, Kementrian Pendidikan dan Kebudayaan.

Stroot, S., Keil, V., Stedman, P., Lohr, L., Faust, R., Schincariol-Randall, L., Sullivan, A., Czerniak, G., Kuchcinski, J., Orel, N., & Richter, M. (1998). *Developmental states of teachers: Peer assistance and review guidebook*. Columbus, OH: Ohio Department of Education. Retrieved from http://www.utoledo.edu/colleges/education/par/Stages.html.

Verspoor, Adriaan. (2006). "Schools at the Center of Quality." *ADEA Newsletter,* Special Issue—Biennale 2006: 3–6.

Wiggins, G. (1991). *"Task" design ideas, principles and guidelines*. Geneseo: Center on Learning, Assessment and School Structure.

World Bank. (2010). Transforming Indonesia's teaching force volume ii: From pre-service training to retirement: producing and maintaining a high-quality, efficient and motivated workforce. Retrieved on 15th October 2017 from http://siteresources.worldbank.org/EDUCATION/Resources/278200-1099079877269/547664-1099080063795/Transforming_Indonesia_teaching_force.pdf.

Urban Studies: Border and Mobility – Kerr et al. (Eds)

Capital reconversion practices by Srintil in the novel *Ronggeng Dukuh Paruk*

T.W. Iswara
Universitas Airlangga, Surabaya, Indonesia

ABSTRACT: *Ronggeng Dukuh Paruk* (1982), one of the novels written by Ahmad Tohari, describes the lush secluded village with its traditional *ronggeng* culture. This novel describes the practice of conversion of cultural capital ownership as a dancer to attain the *ronggeng* status. The aim of this study is to reveal the process of struggle by Srintil to achieve the legitimacy of a dancer, who only has cultural capital in general. There were two research focuses: to analyse the role of cultural capital in the life of Srintil and Srintil's struggle for reconversion in social, economic, and symbolic capitals to achieve legitimacy. This study was qualitative. The data used for this study were characters' speech and narration in the novel. Therefore, in this study, we use the capital forms of Pierre Bourdieu. The results showed that Srintil's proposal not only has capabilities for integrated cultural capital, but she also converts it to other capitals such as social capital, so that she can establish relationships with officials from her village to the capital. In economic terms, Srintil earned money from the bukak klambu ritual. Finally, she attained legitimacy as *Ronggeng*, namely in the form of symbolic capital.

Keywords: Srintil, Modal form, Legitimacy, Pierre Bourdieu

1 INTRODUCTION

The novel *Ronggeng Dukuh Paruk* describes the social life of the main woman character, Srintil, who is a beautiful dancer, just like Cinderella in any ordinary story. It is important to note that this story contains capital, where the struggle is performed by Srintil. As an artist, she wants to show her struggle to attain symbolic capital in her village. Capital can be defined more broadly and includes material things that are symbolic and significant in accordance with culture. The basic capitals of a species are social capital, cultural capital, economic capital, and symbolic capital (Bourdieu, 1986: 242).

Moreover, the story also indicates that it is generally well known that economic capital is the most effective capital that is most easily converted into other capitals. However, Srintil, has the ability, by means of cultural capital attempts, to convert her capital to other capitals from economic capital, then social capital, and finally symbolic capital as a mark of legitimacy as a Ronggeng dancer who is renowned by the people not only around her village, but also in other villages.

2 DISCUSSION

In his classic work, "The Form of the Capital" (1986: 82), Bourdieu developed the concept of capital as a flexible, interdisciplinary-based human science. There are four forms of capital: economic, cultural, social, and symbolic. Linguistic capital is a practical competence of the speaker to produce a highly valued expression in the relevant market (Thompson, 1991: 18). Each capital really has differences that can be converted and transferred to other capitals. However, the level of modal subspecies differs due to various fields, such as culture, politics, and education, and various forms, such as language, ethnicity, and symbol.

The social space in this novel will be classified into two social aspects: the social space of Srintil as a villager and Srintil's social space as a *ronggeng* dancer. Srintil's social space before she becomes a *Ronggeng* dancer at Dukuh Paruk is represented by herself, whereas her social space after she becomes a dancer is represented when she proceeds to perform reconversion. This social space shows the position of Srintil and other characters as social agents in the lower classes in their first struggle.

Srintil lives in a small village that is well known for poverty and ignorance that is passed on from one generation to the next generation. Some villagers truly believe in the strong bond of brotherhood, even the familial bond as the descendants of Ki Secemenggala. With this condition, Srintil struggles to become a *ronggeng* dancer because the last *ronggeng* dancer in Dukuh Paruk died when Srintil was a child. During shows, *ronggeng* dancers sing poems while dancing, accompanied with a rebab, violin, or gong. At that time, *ronggeng* was a kind of work for a woman who makes herself and her society proud.

"Dukuh Paruk without *ronggeng* is not Dukuh Paruk. Srintil, my granddaughter, will restore the former glory of this dukuh. Sakarya said to himself. The soul of Ki Sacamenggala will burst into laughter in his grave when he found out one day that there is another *ronggeng* in Dukuh Paruk" (Tohari, 2016: 15).

The above statement shows how the *ronggeng* is a noble profession to be shared in the community and how it increases the trust of the Dukuh (hamlet, a subdivision of a village). As the highest belief in Dukuh Paruk, they believe in the grave of Ki Secamenggala. Everyone will obey all the rules associated with it. According to their belief, Ki Secamenggala is the ancestor who will give peace in their lives. Without *ronggeng*, Dukuh Paruk becomes similar to any other village. *Ronggeng* makes them feel happier than before. Dukuh Paruk is a field in which Srintil wants to enact change in terms of social space and contest to become a *ronggeng* dancer.

The highest position is possessed by Kartaraja as the creator of *Ronggeng* and his wife who helps him prepare all kinds of dancer needs. The positions of Sakum and Sakarya are equated because they share a role in supporting Srintil. In addition, Srintil's position is the lowest because she tries to adjust to the culture and learns how to become a *ronggeng*.

Cultural capital plays a role in the life of Srintil. Srintil has wanted her dream of becoming a *ronggeng* dancer to come true since the age of 11. In addition, Srintil is the only girl in Dukuh Paruk who has the ability to become a dancer. Therefore, "*Ronggeng*", as understood by Srintil, is a high symbol in Dukuh Paruk at that time. With that symbol, she also earned an honourable title. *Ronggeng* in the world of Dukuh Paruk is a picture as well as a symbol of passion and excitement (Tohari, 2016: 114). In the culture, Srintil has the ability to present cultural products in Tatar Pasundan or the Javanese culture.

"When Srintil sang a difficult song, which she had never learned before, Kartareja's intention was settled. He had to believe that Srintil had been bestowed the *indang* (blessing).

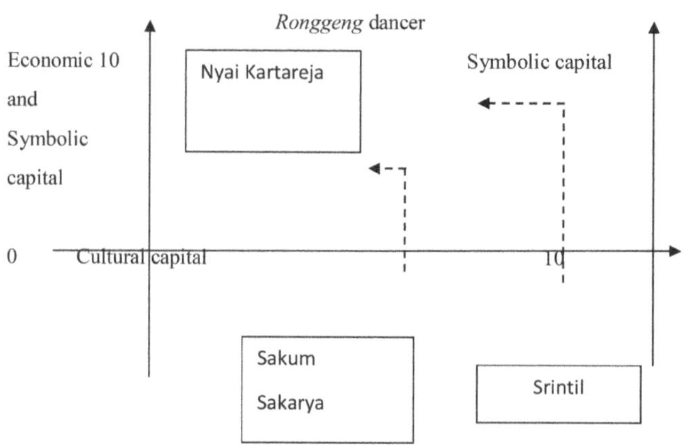

Figure 1. Srintil's social space as a villager.

Kartareja fully believed that Srintil was born in Dukuh Paruk with a task to become a *ronggeng*" (Tohari, 2016: 20).

The above statement shows how lively Srintil dances. She was able to sing the difficult Kartaraja as the creator of the *ronggeng*. Srintil generally shows herself in performances as a *ronggeng* dancer in general. Although that is her first appearance before being inaugurated by a legitimate *ronggeng* ritual, Srintil is able to sing her song with flexible and *kenes* movements. In Javanese, *kenes* is a lively and fascinating trait in children's attitudes, with a lot of talk and stylish fun.

Cultural capital emerges as a capital show of capital. Srintil as a *ronggeng* dancer makes everybody interested in her performance, especially as a dancer, which is shown by her lentik (smooth) movements.

"During the dance, Srintil's face went cold, her charm gripped every spectator. Many people were touched and amazed to see how Srintil threw her *sampur*. Srintil was able to flick her fingers, the most difficult movement performed by a *ronggeng*" (Tohari, 2016: 20).

A dancer has the ability to throw *sampur*, a Pasundan word, which in the Javanese language means a long shawl worn by women. However, Srintil, in her teenage years, was able to do it. In the objective cultural capital, to support her work as a dancer, Srintil has a variety of cultural products. One of the identity markers as a *ronggeng* dancer is the kris. Traditional tools are a form of sharp call as a jaran goyang (rocking horse). As a *ronggeng*, Srintil also uses a susuk (implant), which, in the Javanese culture, is a skill made from a golden needle and diamond, which is inserted into the skin, lips, and forehead with a spell to make the wearer look beautiful and attractive. In terms of institutionalised capital, Srintil has no educational background. All the villagers are illiterates.

2.1 *Srintil's struggle in reconciliation practices*

First, Srintil will make her first appearance after Dukuh Paruk had lost its *ronggeng* culture a few years earlier. "I never thought Srintil could dance well," he said. "If she allows me to take her on my lap" (Tohari, 2016: 20). This expression shows how Srintil's talent gained remarkable appreciation and response. Second, Srintil must fulfil the last requirement as *ronggeng*, namely bukak-klambu. In traditional hamlets, bukak klambu is a type of open competition for every man considered as a candidate for the virgin *ronggeng*. The man will give the *ronggeng* some money determined by the *ronggeng* shaman, so he will be able to enjoy her virginity at night.

"I have decided the time to be next Saturday", said Kartaraja in the morning in front of many men in the market and you ask for a gold ringgit?

Why? Too expensive? Remember carefully, has there ever been such a beautiful *ronggeng* like Srintil? (Tohari, 2016: 52)

Bukak klambu identifies the highest number that says Nyai Kartaraja "Gold ringgit" equals the price of a large buffalo. Basically, this requirement is too difficult for ordinary men who only work in the market, but there are two men who complete the requirements: Dower and Sulam, both from Pecikalan. However, Sulam has a high social status as a son of a tribal chief. Gold ringgit can be categorised as economic capital. In addition, due to Nyai Sakarya's deception, she gets two people who reach the conditions she determines.

Third, the agent who practiced the conversion of Srintil tries to become a *gowok* (a woman hired to teach sex to men) after a *ronggeng* dance performance at Alaswangkal. Her social capital gives her influence in other jobs, such as the new experience to become a *gowok*. "A woman who teaches humans how to be a real man. Nyai Kartaraja surely believes that finding the answer to Srintil's question is as easy as blinking the eye" (Tohari, 2016: 331).

In addition, as a dancer, Srintil has rich knowledge to serve men or to get them interested in her, through which she gets more money for her course in seven days. This indirectly gives her an advantage in terms of economic capital, the amount of which can be transformed into the other forms of capital.

Srintil's struggle requires some efforts and strategies. She performs reconversion of the first capital she owns. With cultural capital as a *ronggeng* dancer, she is able to rise to a social class position higher than before. She tries to show and explore her capital and identity very

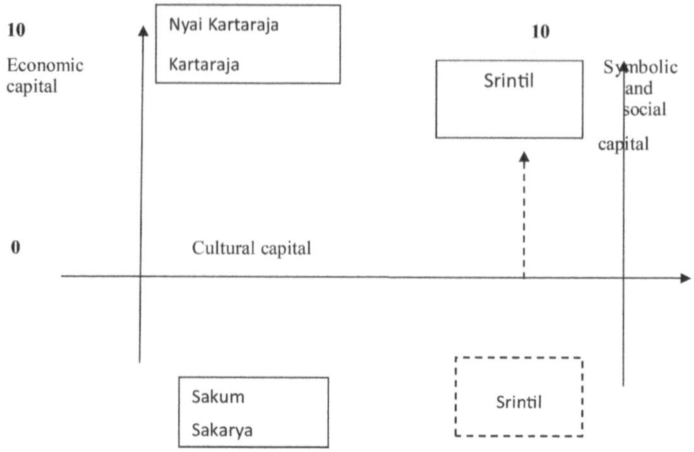

Figure 2. Social space of Srintil as a dancer.

well with the help of *Ronggeng* shamans, Nyai Kartaraja, and his husband. Srintil indirectly affects them to move to a higher class.

3 CONCLUSION

After analysing *Ronggeng Dukuh Paruk* (1982) using the capital model theory by Pierre Bourdieu focusing on the practice of reconversion, mainly using the forms of capital, as a dancer, Srintil is able to collect cultural capital from all her dance skills. To perform the reconversion, Srintil shows her struggle. First, she will make her first appearance after the previous *ronggeng*'s death a few years ago. Second, Srintil must fulfil the last requirement as a *ronggeng*, namely bukak-klambu. With the show, she gets a lot of money in economic capital. Third, she tries to do a new job as a *gowok*, which gives her additional capital as a social, economic, and symbolic capital as another ability to be a *gowok*.

Finally, with two figures, the writer shows the movement of Srintil positions, how agents occupy different social spaces from fighting for symbolic capital and after the agent attains symbolic capital. The symbolic capital of Srintil is the legitimacy as a *Ronggeng* dancer at Dukuh Paruk. Although Srintil dies by the end of the story, she has undergone a struggle in her life as a *ronggeng* dancer and has brought Dukuh Paruk alive.

REFERENCES

Boudieu, Pierre. (1986). *The Form of Capital*. In J. Richardson (Ed.). *Handbook of theory and research for the sociology of education*. New York: Greenwood.

Bourdieu, Pierre. (1991). *Language and Symbol Power*. Cambridge: Polity Press.

Bourdieu, Pierre. (1993). *Sociology in Question*. London: SAGE.

Bourdieu, Pierre. (1993). *Thefield of Cultural Production: Essay on Art and Literature*. Columbia: Columbia University Press.

Manuaba, I.P. (2002). *Martabat Manusia Dan Keadilan Sosial Dalam Novel Pramoedya Ananta Toer Laporan Penelitian Dosen Muda.*

Ritzer & Goodman. (2012). *Teori Sosiologi Klasik—Post Modern Edisi Terbaru (Trans: Nurhadi).* Yogyakarta: Kreasi Wacana.

Tohari, Ahmad. (2016). *Ronggeng Dukuh Paruk*. Jakarta: PT Gramedia Pustaka Utama Jaya.

Thompson, John B. (1991). *Language and Symbolic power* (Editor's introduction). Cambridge, MA: Harvard University Press.

Wulandari, R.A., & Handayani, D. (2008). *Kajian Filologis dan Kajian Pragmatik Serat Patiwinadi.* Study. Fakultas Ilmu Budaya Universitas Airlangga from http://journal.unair.ac.id/filerPDF/05%20vol, 207.

Borders and mobility in language and multilingualis

Urban Studies: Border and Mobility – Kerr et al. (Eds)
© 2019 Taylor & Francis Group, London, ISBN 978-1-138-58034-3

Morphological system of Javanese verbs in the border area of East Java (Tapal Kuda)

A.S. Rohmah, Mahdar & W.A. Sari
Universitas Airlangga, Surabaya, Indonesia

ABSTRACT: In this study, we aim to describe the morphological structure of Javanese verbs in the border area of East Java (Tapal Kuda). The Javanese language in the border area of East Java (Tapal Kuda) is the result of various ethnic interactions in social roles and cultural attractions. In the border area, two languages, Javanese and Madurese, interact. The Javanese language in the border area of East Java is the result of interaction of various ethnicities in social roles and cultural attractions. It produces a combination of Javanese, Madurese, Balinese and Chinese existing in the eastern part of East Java. However, the most prominent variation is the combination of Javanese and Madurese. This study is uses a descriptive qualitative method. The source of data is Javanese speech in various situations including seven districts, namely Pasuruan, Probolinggo, Lumajang, Jember, Situbondo, Bondowoso and dan Banyuwangi. For data collection, we used tapping, recording and interviewing techniques. The data were obtained from the speech of Javanese speakers in that region using questionnaires and observation. In this study, we used structural theory. The collected data were analysed by using the synchronous approach. In Javanese verb morphology in the border area of East Java, the affixation is more influenced by standard Javanese and Malangan Javanese, but the reduplication is more influenced by Madurese and Suroboyoan Javanese.

Keywords: Javanese, Verbs, Morphology, Border area

1 INTRODUCTION

Labelling of a language dialect is often associated with the geographical location in which the dialect is used. This can be seen in the Java language, which is recognised in several dialects based on its geography. For example, the Javanese language used in Malang is called the Javanese language of Malang, the Javanese language used in Surabaya is called the Javanese language of Surabaya, and the Javanese language used in Tuban is called the Javanese language of Tuban (Basuki et al., 1995).

Sudaryanto (2001:3) states that the extent of use of the Javanese language enables very prominent differences and various geographical dialects. Soedjito et al. (2006:2) mention that the Javanese language has several geographical dialects such as Banyumas, Tegal, Solo, Surabaya, Samin and Using.

According to Kisyani (2004), the Javanese language in the northern part of East Java and Blambangan consists of the following two dialects: (1) Using dialect and (2) East Java dialect, covering the Southern Banyuwangi subdialect, Bojonegoro subdialect, Gresik subdialect, Lamongan subdialect, Mojokerto subdialect, Pasuruan subdialect, Rowogempol subdialect, Sidoarjo subdialect, Surabaya subdialect and Tengger subdialect.

Furthermore, Sutarto (2004) divides the area of language use in East Java into four major cultural regions: the cultural areas of Java Mataraman, Arek, Madura Island and Pandalungan. In addition, there are six cultural areas that are not too large, but smaller, namely Java Panoragan, Osing, Tengger, Madura Bawean, Madura Kangean and Samin.

Pendalungan culture in the Tapal Kuda area has unique characteristics in using not only geographical Javanese dialects but also the Javanese language structure, especially morphology. Geographically, Tapal Kuda covers seven districts, namely Pasuruan, Probolinggo, Situbondo, Bondowoso, Jember, Banyuwangi and Lumajang. These districts are inhabited by Madurese and Javanese tribes. The tribe of Madura is even a majority in some places, especially in the north; most of them cannot speak Javanese despite living in Java.

The morphological system of Javanese verbs in the border area of East Java (Tapal Kuda) is an interesting research topic because the Javanese verbs used are unique compared to other Javanese verbs used in other regions.

Paryono (2011) conducted a study related to the language structure in Banyuwangi, Jember and Lumajang concerning the structure of the Javanese language in Pasuruan, Situbondo and Bondowoso. When combined with the previous study, this study will complete the former to produce complete results on the structure of the Java language within the Pendalungan culture in the "Tapal Kuda" area.

2 METHODOLOGY

The data sources of this study were obtained from the speech of Javanese speakers in Pasuruan, Probolinggo, Situbondo, Bondowoso, Banyuwangi, Jember and Lumajang in the form of questionnaires and loose speech when making observations. The questionnaire was made up of a list of Swadesh basic vocabulary questions in the form of words and morphology. In this research, there were two main informants and three supporting informants in each area of study. The data collected were valid and reliable linguistic data, because both of these conditions were the main factors of proper analysis (Sudaryanto, 1990: 34). The data collection techniques used in this study were recording and interview or cooperation with informants. In the data processing stage, we transcribed the recorded and tapped, classified and analysed data. In data classification, all data corpuses were grouped by type and then analysed.

3 DISCUSSION

A word is the smallest unit that can stand alone as a sentence (Crystal, 1997: 189). The word formation process is distinguished as a root and derived word. The root word has no morphological process, while the derived word is formed from the word that already exists. The word formation process is also called the morphological process. According to Kridalaksana (1988:56), the morphological process is classified into: (1) derivation zero, (2) affixation, (3) reduplication, (4) shortening, (5) reverse derivation and (6) blend.

3.1 *Affixation to form a verb*

Affixation is a process that converts lexemes into complex words (Kridalaksana, 1992: 28). The affix forms of verbs in Javanese Pendalungan in Pasuruhan, Probolinggo, Situbondo, Bondowoso, Banyuwangi, Jember and Lumajang include prefixes, suffixes and confixes.

3.1.1 *Prefix*

The prefix is a bound morpheme. A prefix can be identified from the beginning of a syllable. The prefixes of Javanese in the Tapal Kuda area in Banyuwangi, Jember and Lumajang are N- (m-, n-, ny-, ng), / me/ di- /, / ke- /, / se /, / sira / and / sun- /. Word formation with prefix / N- / can cause various variations of morphemes, such as / m- /, / n- /, / ny- / and / ng- /. The variation of the allomorph occurs due to the influence of the initial letter of the word attached. For example: /N-/ + cicil: nyicil 'to pay instalment'. Another example is /N/+ gawa: nggawa 'to bring'. The variant of verb forms is due to the phonological condition of the basic forms concerned so that, in fact, they are a morpheme, morpheme / N- /.

3.1.2 *Suffix*

The suffixes found in Javanese Pendalungan in Pasuruhan, Probolinggo, Situbondo, Bondowoso, Banyuwangi, Jember and Lumajang are / -e /, / -i /, / -an /, / -en /, / -o /, / - no / and / -ono /. For example: takbatek + /-e/: takbateke 'will be pulled out soon'. Another example is from the suffix /-o/, which means to command someone to do something as in its base word: gawa + /-o/: gawao 'bring it'.

3.1.3 *Confix*

A confix is a morphemic process with an affix flanking the basic form. There are two confixes or ambifixes in Javanese Pendalungan in Pasuruhan, Probolinggo, Situbondo, Bondowoso, Banyuwangi, Jember and Lumajang forming verbs, namely / N - no / and / di-no/. For example: /N—no/ + turu: nurokno 'make someone sleep'.

3.1.4 *Combination of prefix and suffix*

The combination of prefix and suffix is a combination of the basic form with prefix and suffix, which is not simultaneous, meaning that if one of the prefixes and suffixes are omitted, the affixes together and the basic form still have meaning. The combinations of prefixes and suffixes that form verbs in Javanese verses of Pendalungan include / N-i /, N-no / and / di-i /. An example of a combination of prefix and suffix is /di-i/ is /di—i/ + donga: didongani 'being prayed'.

3.2 *Reduplication to create a verb in the Tapal Kuda area*

In this study, it was found that, there is some form of reduplication or repetition in Javanese Pendalungan in Pasuruhan, Probolinggo, Situbondo, Bondowoso, Banyuwangi, Jember and Lumajang. Reduplication of the basic form in Javanese Pendalungan in the Tapal Kuda area uses two forms, namely repetition of the whole or full verb and partial repetition (Lutfi, 2004). The basic form that forms the verb in the reduplication in the Java language Pendalungan can be observed in the following data.

3.2.1 *Whole repetition*

Certain reduplications in Javanese Pendalungan in Pasuruhan, Probolinggo, Situbondo, Bondowoso, Banyuwangi, Jember and Lumajang can express (1) the meaning of repetitive activities such as mlayu-mlayu: 'to run around', (2) activities done casually, not seriously like ndeleng-ndeleng: 'look around', (3) stating mutual meaning like in batek-batekkan: 'pull each other' and (4) expressing the meaning of 'bluffing'or 'pretend', such as manten-mantenan: 'pretending to be a bridegroom'.

3.2.2 *Repetition changes the sounds to create a verb*

The repetition of sounds used by Javanese speakers of Pendalungan in Pasuruhan, Probolinggo, Situbondo, Bondowoso, Banyuwangi, Jember and Lumajang usually conveys the meaning of deeds done repeatedly. For example: mocal-macul, which means 'plough continuously' and ngrobak-ngrabuk, which means 'fertilise continuously'.

3.2.3 *Partial repetition to create a verb*

Reduplication of certain parts of Javanese Pendalungan in Pasuruhan, Probolinggo, Situbondo, Bondowoso, Banyuwangi, Jember and Lumajang can express (1) the meaning of repetitive activities such as dung-tedungan: 'lying down' or on-alon: 'go slowly', (2) activities that are not done seriously, like adha-khanda: 'to tell' or adon-jandon 'to talk', (3) expressing mutual meaning, like lep-salebhan: 'pass each other', or leg-leregan: 'glancing at each other', (4) meaning of 'bluffing' or 'pretend' as in dhi-masedhi: 'pretending to be sad' or mes-malemmes: 'pretending to be weak' and (5) reduplication preceded by additional words ndak or gak mean a hope. This can be seen from the phrase ndakna-nyana: 'unpredictable' or gakka-sangka meaning 'unpredictable'.

This study on the morphology system of Javanese verbs in the Tapal Kuda area uses structural theory. Structural linguistic theory places the form and meaning in a balanced position, but it develops into various versions. There are versions that emphasise form, and there are also versions that emphasise meaning. According to Soeparno (2002, 47–52), structural theory has characteristics such as (1) language based on behaviouristic behaviour; (2) speech language; (3) language of sign system (signifier and signified); (4) language as habit factor; (5) grammatically based on the context; (6) grammatical levels are neatly enforced; (7) pressure analysis on the morphological field; (8) language as a syntactic and paradigmatic row; (9) descriptive language analysis and (10) analysis of language structures based on direct elements. Study using structural theory basically explains that each language consists of a collection of linguistic units (linguistic units), which are systematic and can be described. It means that the units in one level have certain structures that can be described to know the characteristics of each unit in relation to the unit itself or with other units in other levels. Principles of structural analysis based on structural linguistic theory are used as the basis for the study of morphological research.

In the process of formation of Javanese verbs in the Tapal Kuda area, affixation and reduplication play a very important role in the morphological process. The affixation occurring in the formation of the Javanese language in Tapal Kuda includes the prefix / N- / and its variations, the prefix / N- / into / ny- /, prefix / N- / am / / / -, prefix / N- / into / m- /, the prefix / N- / changing to / n- / when it joins the base form beginning with the consonant / d / and / t /, and the / N- / prefix / changing to / m- /, but not changing initially when attached to the basic form beginning with the phoneme / b /. The prefix / N- / changes to / n- /, but does not change its prefix when attached to the base form beginning with the phoneme / d / and / j /. Other prefixes are prefix / ke- / and prefix / di- /.

Similarly, the suffix in the formation of a Javanese verb in the horseshoe includes suffix / -e /, suffix -i, suffix / -en /, suffix / -o /, suffix / -no / and suffix / -ono /. As for confixes or ambifixes in Javanese Pendalungan in Pasuruhan, Probolinggo, Situbondo and Bondowoso, Banyuwangi, Jember and Lumajang verbs, there are two, namely / N - no / and / di-no /, while combinations of prefix and suffix that form the verb are combinations of prefix and suffix / N-i /, prefix and suffix / di-no / and prefix and suffix / di -i /. The suffix / di-i / /, which forms verbs in the base form, does not change; despite joining any type of word, the shape is fixed / di-i /. Judging from the process of Java language affixation in Tapal Kuda, it resembles the standard Java language.

The formation of verbs through reduplication is through repetition intact, repetition of sounds and partial repetition. Whole repetition involves a certain reduplication, a reduplication whose basic shape gets a suffix / -an / and a reduplication whose basic form gets the prefix / -an /. Partial reduplication includes some partial reduplication, reduplication of the basic form, which gets the suffix / -an /, a reduplication whose basic form gets the suffix / -an / and a reduplication preceded by an additional word ndak or not. Judging from the process of reduplication in the formation of Javanese verbs in the horseshoe, it resembles the Java language Surabaya.

Viewed in terms of its relationship, morphemes can be sorted out in terms of relationship structure and position relationships. A morpheme is seen in terms of its structural relations, which can be divided into three morphemes: (1) additives (additive), (2) replacement (replasif) and (3) subtractive. In terms of their position relations, morphemes can be distinguished from sequential, inset and simultaneous morphemes (Samsuri 1991: 186–187).

4 CONCLUSION

The process of formation of a Javanese verb in the border area (Tapal Kuda) is most often through the process of affixation and reduplication. The affixation process is achieved by giving a prefix, insertion, suffix or combination, while the reduplication process can be intact repetition, partial repetition, repetition with affixed combination and repetition with phoneme variation.

The formation of verbs through the affixation process in Javanese in Pasuruhan, Probolinggo, Situbondo, Bondowoso, Banyuwangi, Jember and Lumajang is influenced by standard Javanese and Malangan languages, while the formation of verbs through reduplication is influenced by Madurese and Surabayan Javanese.

REFERENCES

Arikunto, S. 2006. *Prosedur Penelitian Suatu Pendekatan Praktik*. Jakarta: PT Rineka Cipta.

Baihaqi, M.L. 2004. *Struktur Fonologis dan Morfologis Bahasa Jawa di Pesisir Utara Jawa Timur*. Surabaya: Balai Bahasa Surabaya.

Basuki, I.A., et al. 1995. *Ciri Struktur dan Leksikon Bahasa Jawa Malang*. Surabaya: Departemen Pendidikan dan Kebudayaan Kantor Wilayah Propinsi Jawa Timur.

Chaer, A. 2003. *Psikolinguistik (Kajian Teoritik)*. Jakarta: Rineka Cipta.

Crystal, D. 1971. *Linguistics*. Harmonsworth: Penguin Book Ltd.

Handayani, D., Hum, M., & Ratnawati, S. (2011). Penggunaan Bahasa Tabu Dalam Konteks Tuturan Bahasa Jawa Pada Masyarakat Jawa Timur (Doctoral Dissertation, Lembaga Penelitian Universitas Airlangga).

Kisyani. 2004. *Bahasa Jawa Di Jawa Timur Bagian Utara dan Blambangan*. Jakarta: Pusat Bahasa. Departemen Pendidikan Nasional.

Kridalaksana, H. 1992. *Beberapa Prinsip Perpaduan Leksem dalam Bahasa Indonesia*. Yogyakarta: Kanisius Publisher.

Mahsun. 2005. *Metodologi Penelitian Bahasa*. Jakarta: Rajagrafindo Persada.

Mayani, A.L. 2004. *Perbandingan Fonologis, Semantis, dan Leksikal Antara Bahasa Jawa Dialek Surabaya dan Bahasa Jawa Dialek Standar*. Surabaya: Balai Bahasa Surabaya.

Montolalu, L.R., et.al. 2005. *Tipologi Bahasa dan Bahasa-bahasa di Dunia. Dalam Pesona Bahasa: Langkah Awal Memahami Linguistik*. Jakarta: Gramedia Pustaka Utama.

Paryono, Y. 2011. *Struktur Bahasa Jawa di Jember, Lumajang, dan Banyuwangi*. Surabaya: Balai Bahasa Jawa Timur.

Poedjosoedarmo, S. 1979. *Tingkat Tutur Bahasa Jawa*. Jakarta: Pusat Pembinaan dan Pengembangan Bahasa, Departemen Pendidikan dan Kebudayaan.

Samsuri. 1982. *Tata Kalimat Bahasa Indonesia*. Jakarta: Sastra Budaya.

Samsuri. 1991. *Analisis Bahasa*. Jakarta: Erlangga.

Sartini, N.W. (2009). Menggali nilai kearifan lokal budaya Jawa lewat ungkapan (Bebasan, saloka, dan paribasa). dalam Jurnal Logal, 28–37.

Sartini, N.W. (2015). Perilaku Bahasa Diaspora Orang Bali di Jawa Timur: Kajian Sosiolinguistik. Jurnal Keilmuan Bahasa, Sastra, dan Pengajarannya, 1(1), 54–62.

Soedjito, dkk. 1986. *Pemakaian Bahasa Jawa di Pesisir Utara Jawa Timur Bagian Sempit*. Jakarta: Pusat Pembinaan dan Pengembangan Bahasa, Departemen Pendidikan dan Kebudayaan.

Sofyan, A. 2007. Beberapa Keunikan Linguistik Bahasa Madura. *Jurnal UGM*. Retrieved 20 October 2017 from http://jurnal.ugm.ac.id/jurnal.humaniora/article/view/126/1077.

Subroto, E. 2007. *Pengantar Metode Penelitian Linguistik Struktural*. Surakarta: UNS Press.

Sudaryanto. 2001. *Aneka Konsep Kedataan Lingual dalam Linguistik*. Yogyakarta: Duta Wacana University Press.

Wardhaugh, R. 1986. *An Introduction to Sociolinguistics*. Oxford: Basil Blackwell Ltd.

Wibisono, B. & Haryono, A. 2016. *Wacana Perkawinan di Tapal Kuda*. Jember: Penerbit Tapal Kuda.

Verhaar, J.W.M. 1982. *Pengantar Linguistik*. Yogyakarta: Gadjah Mada University Press.

Urban Studies: Border and Mobility – Kerr et al. (Eds)
 ISBN 978-1-138-58034-3

Javanese *Unggah-Ungguh* level used in some rubrics of the *Jaya Baya* and *Panjebar Semangat* magazines

B.D.Y. Puteri
Faculty of Humanities, Universitas Airlangga, Indonesia

ABSTRACT: In this global community, people do not use standard language anymore. The use of native language in local magazines can be seen in several Javanese magazines. Interestingly, they use not only Javanese, but also other languages in their rubrics. This paper discusses *unggah-ungguh* grade, linguistically, as it is used in two Javanese magazines; they are *Jaya Baya* (7th January 2017) edition and *Panjebar Semangat* (8th January 2017) edition. Qualitative data in this research are specific phrases, clauses, and sentences used in both magazines which are combined from Javanese and other languages, especially Bahasa. This research aims to classify what they are by using a correlational and descriptive method. This research finds some acculturation at three levels of the Javanese language which are *Basa Ngoko*, *Basa Madya*, and *Basa Krama*. They can be viewed from the utterances and splitting of stem words, phrases, clauses, and sentences.

Keywords: Acculturation, Linguistically, Javanese, *Unggah-ungguh*

1 INTRODUCTION

Cultural linguistics views language and culture as two sides of a coin; one cannot exist without the other. Correlated to the above, Duranti (1997: 27) even explicitly says that describing a culture is the same as describing a language. For this, it can be interpreted that a culture can develop and spread throughout the world, in one way through language.

In connection with the above, Wierzbicka (1991) formulates assumptions about language and culture which are (1) different communities and people are talking or acting speech differently (depending on the context of the situation, the language used, etc.), (2) the above differences are primarily in terms of speech or acting the speech and occur systematically and usually on a large scale, (3) the above differences also reflect the existence of different culture values, and (4) differences in ways inside speaking or acting as well as differences in communication styles, especially through the use of different languages, can be explained and understood as reflecting differences in cultural values and cultural priorities.

The assertion put forward by Wierzbicka mentioned above is actually a real picture of the empirical and theoretical relationships between language and culture based on three keywords, namely: (1) a community or *guyub*, both *guyub tutur* and *guyub culture*; (2) how to interact, and (3) cultural values. Different people or people show different ways of interacting and also show different cultural values, especially in implementing them through a language.

Concerning the above, the relationship between language and culture is presented conceptually-theoretically, which is not only named variably, but is primarily interpreted differently. Studies which focus on the relationship between language and culture can be examined through an anthropological linguistics approach to discover and determine the meaning behind its use (Foley, 1997).

Concerning the above, in this research paper, we examine the use of propriety strategies (*unggah-ungguh* and *unda-usuk*) in the rubrics contained in the Jaya Baya Magazine and Panjebar Semangat Magazine which are two very prominent Javanese magazines. The reason why the researcher chose this topic or topic is because the use of the Javanese courtesy strategy (*unggah-ungguh* and *unda-usuk*) in writing magazine rubrics is very rarely found in scientific journals. The concept of cultural or a Javanese cultural context discussed in this paper is reflected in the Java proprietary strategy called *unggah-ungguh*. In Javanese language, especially Javanese language that is used in Central Java and East Java, there is a context or aspect related to social culture or culture that represents the existence of some level of ethics or politeness strategy in the use of this Javanese language. This ethics or courtesy strategy in the use of Javanese is called *unggah-ungguh*. There are several versions of *unggah-ungguh*. The most commonly used type is the level of *unggah-ungguh* that divides *unggah-ungguh* into 3 levels of *unggah-ungguh*, namely: (a) Basa Ngoko which is the language or speech used by: people of the same age, parents of young people, the manager of staff, and those of the same social class (eg: "*Gee, adhimas ki rada ngece*" which means: "*Gee, this sister is a bit mocking*"), (b) Basa Madya or Krama Madya which is the language or type of speech located between Basa Ngoko and Basa Krama and is usually used by leaders with leaders or between persons of the same social class or equivalent (eg "*Napa sampeyan ajeng tumbas*?" meaning "*Do you want to buy*?"), (c) Basa Krama or Krama Inggil is a language or type of speech whose level of courtesy is high or, in this case, uses subtle words of speech acts to honour and appreciate others and is used by young people to older people (eg: "*Pak Farid nembe mucal mahasiswa semester gangsal*" which means "*Pak Farid just taught the fifth semester student*"). In this case, some linguists or Javanese linguists add a level of Javanese courtesy strategy with another level of Basa Krama Kedhaton, a Javanese language that uses subtle words of speech acts to honour and respect the priyayi or royal court or palace of Java, especially in Keraton Jogjakarta and Surakarta palace (for example: "*Punapi sira darbe kawasisan ingkang linuwih*?" Which means "*Do you have more skill*?"). Basa Krama Kedhaton is only around the palace or limited to the palace alone (Bayurianto, 2012).

This study includes descriptive research with referring and literature study as the techniques for data collection (Bungin, 2007; Sudikan, 2001). There are two data sources: two famous Javanese magazines in Indonesia, namely *Jaya Baya Magazine* and *Panjebar Semangat Magazine*. The data in this research are in the form of words, phrases, and sentences analysed descriptively and qualitatively. The data obtained is taken from the rubric *Primbon Majalah Jaya Baya,* rubric *Crita Cekak Majalah Jaya Baya,* rubric *Crita Cekak Magazine Panjebar Semangat, Layang Saka Warga,* rubric *Crita Wayang Magazine Jayabaya,* and rubric *Crita Taman Putra Magazine Jaya Baya.*

2 DISCUSSION

Based on the results of the research conducted, there are a number of words, phrases, and sentences in the rubric of Javanese language in two magazines (*Jaya Baya* and *Panjebar Semangat*) representing the level of politeness of Javanese language (*unggah-ungguh*). In accordance with the socio-cultural context that underlies its use, here are some examples of data in the form of words, phrases, and sentences that represent *unggah-ungguh* in Javanese language in the rubrics of *Jaya Baya Magazine* and *Panjebar Semangat Magazine*, especially taken from the rubric *Primbon* of *Jaya Baya Magazine*, rubric *Crita Cekak* of *Jaya Baya Magazine*, rubric *Crita Cekak* of *Panjebar Semangat Magazine*, *Layang Saka Citizen* of *Jaya Baya Magazine*, rubric *Crita Wayang* of *Jayabaya Magazine*, and rubric *Crita Taman Putra* of *Jaya Baya Magazine*.

In this section, researchers describe the data analysis and discussion. To begin with, the initial discussion described is *Basa Ngoko* which, in this paper, is divided into four classifications, namely:

a. The data of *basa ngoko* in the form of stem words, such as: what, yes, *kowe*, and *piye* (data 1.1 to data 1.7). These stem words are words in Javanese that are not given any affixes (either prefix, insertion, suffix, or prefix and suffix). These words belong to a stem because these words are used by older people when talking or speaking to a younger person. In addition, the above stem words are also used by persons of higher rank or social rank to persons of lower social position or a lower stratum. Also, these stem words are most commonly used in conversation or dialogue or speech among people of the same age, a collection of professional people in work, persons of equal rank and social rank, and those who have a very familiar friendship.

b. Data from *basa ngoko* in the form of words that have been used as a sample for research from *Jaya Baya* Magazine and *Panjebar Semangat* Magazine, namely the words "*kepiye*" and "*ayune*". The word *kepiye* (data 2.1) and *ayune* (data 2.2) include the inscribed words in the stem because the two words are: prefixed to make the word *kepiye*, and suffixed with -ne to make the word *ayune*. These words are used by older people when talking or speaking to a younger person. In addition, the above-mentioned words are also used by persons of higher rank or social rank to persons of lower social position or of a lower stratum. Also, the words are most often used in conversation or dialogue or speech among people of equal age, a collection of people in professions in work, persons of equal rank and social stratum, and people who have a very familiar friendship.
c. Data from *basa ngoko* which formed in phrase can be found in the phrase "*karo sapa*" (data 3.1) and "*gara-gara pedhot*" (data 3.2). Included in the classification of phrases because "*karo sapa*" and "*gara-gara pedhot*" each consist of merging two words or more. This phrase is used by older people when talking or speaking to a younger person. In addition, the above phrase is also used by a person whose position or social stratum is higher to a person whose position or social stratum is lower. Also, the phrase is most often used in conversation or dialogue or speech among people of equal age, a collection of people in professions in work, persons of equal rank and social stratum, and people who have a very familiar friendship.
d. Data from *basa ngoko* in the form of *clauses* and sentences can be found in data 4.1 to data 10.9. This data belongs to a *basa ngoko* because it is used by older people when talking or speaking to a younger person. In addition, the above sentence is also used by a person whose position or social stratum is higher to a person whose position or social stratum is lower. Also, the phrase is often used in conversation or dialogue or speech among people of the same age, a collection of people in professions in work, persons of equal rank and social stratum, and people who have a very familiar friendship.

Next, the discussion describes *basa madya* or *krama madya* which, in this paper, is divided into three classifications, namely:

a. Data from *basa madya* or *krama madya* in the form of basic words, such as the phrase "ngebis mawon" (data 2.1) and "kula nuwun" (data 2.2). They are including to phrase classification because "kula nuwun" and "ngebis mawon" contained of merging 2 (two) syllables. These words belong to *basa madya* or *krama madya* because these words are used by younger people when speaking or speaking to an older person. In addition, the above basic words are also used by persons of lower rank or social stratum to persons of higher social position or stratum. Also, these basic words are often used in conversations or dialogues or utterances uttered by children or students to their parents and teachers.
b. Data from *basa madya* or *krama madya* can be found in the phrase "ngebis mawon" (data 2.1) and "kula nuwun" (data 2.2). Included in the classification of the phrase because "kula nuwun" and "ngebis mawon" that each consists of merging two words. This phrase is used to belong to *basa madya* or *krama madya* because these words are used by younger people when talking or speaking to an older person. In addition, the above phrase is also used by persons of lower rank or social stratum to persons of higher social position or stratum. Also, the phrase is often used in a conversation or dialogue or utterance uttered by a child or pupil to his or her parents or teacher.
c. Data from *basa madya* in the form of *clauses* and *sentences* can be found in data 3.1 to data 4.5. Furthermore, data 3.. up to 4.5 used to belong to *basa madya* or *krama madya* because these clauses or sentences are used by younger people when talking or speaking to an older person. In addition, the above sentence is also used by a person whose position or social stratum is lower to a person whose position or social stratum is higher. Also, the sentence or clause is often used in the conversation or dialogue or utterance uttered by the child or pupil to his or her parents or teacher.

Furthermore, *basa krama* or *krama inggil* is the third or the last classification in this paper, which in this paper is divided into three classifications, namely:

a. Data of *basa krama* or *krama inggil* in the form of stem words, such as: *kula*, *piyambak*, *ngaten*, and *pinten* (data 1.1 to data 1.4). These stem words are words in Javanese that are not given any affixes (either prefix, insertion, suffix, or prefix and suffix). These words belong to *basa krama* or *krama inggil* because these words are used by younger people when talking or speaking to an older person. In addition, the above stem words are also used by persons of lower rank or social strata to persons of higher social position or strata. Also, these stem words are primarily used in conversation or dialogue or speech when, in particular, the speaker assumes that the other person is higher in rank or social level than the speaker.
b. The data of *basa krama* or *krama inggil* in the form of the phrase: "*Lepat kula ugi* ..." (data 2.1) and "*nembe siram*" (data 2.2). This is included in the classification of *phrase* because "*kula nuwun*" and "*ngebis mawon*" each consist of merging two words or more. This *phrase* is used by younger people when talking or speaking to an older person. In addition, the above *phrase* is also used by persons of lower rank or social strata to persons of higher social position or strata. Also, the phrase is primarily used in speech or dialogue or speech when, in particular, the speaker assumes that the other person is higher in rank or social level than the speaker.
c. The data of *basa krama* or *krama inggil* in the form of clauses or sentences mainly encountered in data 3.1 to data 3.7 (see pages 16 to 17 of these papers). The above sentence is used by a younger person when talking or speaking to an older person. In addition, the above sentence is also used by a person whose position or social stratum is lower to a person whose position or social stratum is higher. Also, the sentence is mainly used in conversation or dialogue or speech when, in particular, the speaker assumes that the other person is higher in rank or social level than the speaker.

3 CONCLUSION

Based on the results of the research conducted, there are a number of words, phrases, and sentences in the rubric of Java in two magazines (*Jaya Baya* and *Panjebar Semangat* Magazine) representing the level of propriety of Javanese language (*unggah-ungguh*). From the classification results based on Javanese courtesy, the Javanese language rubrics in the two magazines (*Jaya Baya* and *Panjebar Semangat*) representing the level of politeness of Javanese language can be divided into: (a) *Basa Ngoko*, (b) *Basa Madya*, (c) *Basa Krama* or *Krama Inggil*. First of all, (a) *Basa Ngoko* is classified again into: (1) the form of stem words (such as *apa*, *kowe*, *piye*, etc.), (2) the form of formed words (such as *kepiye* and *ayune*) (3) the form of phrases (such as *karo sapa* and *gara-gara pedhot*), (d) the form of clauses and sentences. Then, (b) *Basa Krama* are classified into: (1) the form of stem words, (2) the form of phrases, and (3) the form of clauses and sentences. The last classification (c) *Basa Krama* or *Krama Inggil*, based on this study, are classified again into; (1) the form of stem words, (2) the form of phrases, and (3) the form of clauses and sentences.

REFERENCES

Amalia, Y. & M. (2014). Dialogue in narrative inquiry: Collaboration in Doctoral Study in the USA. In P. Smeyers, D. Bridges, N.C. Burbules, & M. Griffiths, *International Handbook of Interpretation in Educational Research.* London: Springer.

Becker, A.L. & Yengoyan, A.A. (eds). 1979. *The Imagination of Reality: Essays in the South East Asian Coherence Systems*. New Jersey: ABLEX Publishing Corporation.

Barker, Ch. 2004. *Cultural Studies, Teori & Praktik* (Translated by Nurhadi). Yogyakarta: Kreasi Wacana.

Duranti, A. 1997. *Linguistic Anthropology*. Cambridge: Cambridge University Press.

Finochiaro, M. 1974. *English as a Second Language: from Theory to Practice*. New York: Regent Publishing.

Foley, W.A. 1997. *Anthropological Linguistics: An Introduction*. Oxford: Blackwell Publishers.
Frawley, W. 1992. *Linguistic Semantics*. New Jersey: Lawrence Erlbaum.
Handayani, D. (2010). Penggunaan Bahasa Tabu dalam Tuturan Bahasa Jawa di Jawa Timur. *Mozaik*.
Hassan, R. 1989. *Linguistics, Language, and Verbal Art*. Victoria: Deakin University.
Indarti, Y. (2012). *Nasib Bahasa Jawa di Surabaya Faktor-faktor Penyebab Kepunahan dan Pemerintahan Bahasa*. Surabaya.
Kaelan. 2004. *Filsafat Analitis Menurut Ludwig Wittgenstein*. Yogyakarta: Paradigma.
Koentjaraningrat. 1984. *Kebudayaan, Mentalitas dan Pembangunan*. Jakarta: PT Gramedia.
Koentjaraningrat. 1990. *Sejarah Teori Antropologi*. Jakarta: Penerbit Universitas Indonesia.
Koentjaraningrat & Donald K.E. (ed). *Aspek Manusia dalam Penelitian Masyarakat*. Jakarta: PT Gramedia.
Kupper, A. & Jessica, K. 2000. *Ensiklopedi Ilmu-ilmu Sosial*. Diterjemahkan oleh H. Munandar, et al. Cetakan I. Jakarta: Raja Grafindo Persada.
Mbete, A.M. 1997. Linguistik sebagai Realisasi Pola Ilmiah Pokok Kebudayaan Universitas Udayana. *Makalah*. Disampaikan dalam Ceramah Pramagister Program Studi Magister (S2) Linguistik dan Kajian Budaya Universitas Udayana Denpasar 1997. Denpasar: Program Studi Magister (S2) Linguistik dan Kajian Budaya Universitas Udayana.
Mbete, A.M. 2003. Bahasa dan Budaya Lokal Minoritas: Asal-Muasal, Ancaman Kepunahan dan Ancangan Pemberdayaan dalam Kerangka Pola Ilmiah Pokok Kebudayaan Universitas Udayana. *Pidato Pengukuhan Jabatan Guru Besar Tetap dalam Bidang Linguistik pada Fakultas sastra Universitas Udayana, 26 Oktober 2003*. Denpasar: Universitas Udayana.
Mbete, A.M. 2004. Linguistik Kebudayaan: Rintisan Konsep dan Beberapa Aspek Kajian nya, in Bawa, I.W. dan Cika, I.W (ed.), *Bahasa dalam Perspektif Kebudayaan*: 16–32. Denpasar: Universitas Udayana.
Mulyana. 2005. *Kajian Wacana: Teori, Metode & Aplikasi Prinsip-Prinsip Analisis Wacana*. Yogyakarta: Tiara Wacana.
Ochs, E. 1988. *Culture and Language Development*. Cambridge: Cambridge University Press.
Palmer, G.B. 1996. *Toward a Theory of Cultural Linguistics*. Austin: University of Texas Press.
Palmer, R.E. 2003. *Hermeneutika, Teori Baru tentang Interpretasi*, (Translated by Musnur Hery & Damanhuri Muhammed, dari judul asli: *Interpretation Theory in Schleimacher, Dithey, Heidegger, and Gadamer*). Yogyakarta: Pustaka Pelajar.
Rusnaningtias, E. (2012). Pemakaian Bahasa Jawa di Kabupaten Blitar (Kajian Geografi Dialek). Surabaya.
Wierzbicka, A. 1991. *Cross-Cultural Pragmatics, The Semantics of Human Interaction*. New York: Mouton de Gruyter.
Wierzbicka, A. 1992. *Semantics, Culture, and Cognition*. Oxford: Oxford University Press.
Wierzbicka, A. 1994. Cultural Scripts: A New Approach to the Study of Cross-Cultural Communication. *The Third Australian Linguistic Institute*, Australian National University.

Urban Studies: Border and Mobility – Kerr et al. (Eds)
© 2019 Taylor & Francis Group, London, ISBN 978-1-138-58034-3

The role of language in border relations (Desa Tajungan Kec Kamal, Kab Bangkalan, Madura)

D.R. Sugiharti, Miladiyah & Y.S. Amalia
Universitas Airlangga, Surabaya, Indonesia

ABSTRACT: In this paper, we report the findings of a study that investigated the linguistic situation in the border region, Desa Tajungan. This village is the border region of Madura island and Java island (Gresik). The geographical location of this region is Kamal, Bangkalan Madura. It was interesting as the object of the study since the people living in this region speak Javanese even though they live in Madura island. The aim of this study was to investigate the language use when contact is revitalized after a long period with closed borders. Javanese and Madurese have very different vocabularies and structures, which makes communication difficult. The descriptive method is used in this study. This study was carried out through interviews, a questionnaire, and observation. The results showed an ongoing development where the neighbouring language is increasingly notable, and there is a clear link between attitudes, identity, and language use. Javanese people who speak Madurese tend to mix Javanese terms and Madurese words for typically Madurese concepts and institutions, which finds way into Javanese sentences. This code switching often occurred in a local community consisting of speakers of two languages.

Keywords: Attitude, Border region, Code switching, Identity, Neighbouring language

1 INTRODUCTION

Borders are imaginary lines separating two or more political areas or jurisdictions, such as states, provinces, or sub-national territories. In some parts of Indonesia, borders are marked by boundaries. A boundary can be a stone or a monument, large or small. The border area may be in the form of an ocean or land. Madura is a small island located in the eastern region of Indonesia directly adjacent to the island of Java. Border studies are often cross-cultural studies representing a multidisciplinary field: they involve sociology, anthropology, economy, history, and linguistics (Hofstede, 2001). Even though language can be said to be the most recognisable part of culture, it is not often the focus of research in border regions. However, no activity across borders can succeed without language proficiency. In the Russian–Norwegian border region, contact between people in private, business, and official contexts involves the use of a language that is foreign to one or both parties. Knowledge of foreign languages is crucial regarding successful communication and cooperation (Hofstede, 2001). However, language is often a major cause of cultural clashes.

The focus of this study is Desa Tajungan. Geographically, Desa Tajungan is a border area situated in the island of Madura, Kamal, Bangkalan. It is in the coastal strait of Madura, where it borders the Gresik region. Desa Tajungan was formed in 1995; it started by the landing of some fishermen from Gresik on the mainland and made it as a place for their shelters. Therefore, the fishermen make settlements on the land, so as to form a village named Tajungan. According to Amalia (2016), there are some problems faced by people living in urban and border areas, such as poverty, lack of access to information, infrastructure, quality of education, and poor health services compared to urban communities. Therefore, there are circumstances that require the community of border districts to integrate with neighbouring regions or other regions. This phenomenon usually causes people living in the border area to have the ability to master two or more languages. Sometimes it even makes the original language (mother tongue) of people in the region experience a shift.

There is an interesting phenomenon that makes researchers use Desa Tajungan as an object of research. In their daily life, people living in the village use Javanese as a means of communication. Although they live and coexist with the majority of Madurese people in the community, Javanese is the main language of the village. Even the majority of the shrine population believes and admits that they are Javanese and not Madurese.

The main research question of the present study aims to find out how languages are used in communication in various areas of social life between Javanese and Madurese in the border region. The writers have studied how border contact affects the development of the two languages, Javanese and Madurese, and how language use is linked to perceptions of identity among people in this region.

2 METHODOLOGY

A descriptive method was chosen for this study in order to make a systematic and accurate description of the data collected, characteristics, and the phenomenal relationship of the collected data (Djajasudarma, 1993:8). Data for this study were collected from the people of Desa Tajungan who speak Javanese in the island of Madura. Dealing with the characteristics of the observed language, the research method chosen for this study was a descriptive method. Data were collected through observation and interviews using their techniques. The use of both methods aims to obtain complete data to achieve satisfactory results (Sudaryanto, 1993:133). The spoken data collected through both methods were also supported by written data taken from the questionnaire. The collected data were classified and analysed in accordance with the problem discussed, and the result of the analysis was descriptively presented.

3 DISCUSSION

3.1 *Language use*

Javanese language is increasingly prominent in Desa Tajungan, and it is clear here that the Desa Tajungan community is an independent society. Although the Desa Tajungan community lives side by side with the Madurese-speaking villages, it does not affect the existence of their language. There are many examples in everyday life of the use of the Java language. One example is trading activities conducted in the market, where the domination of Javanese is evident. Traders in the market are not only from the indigenous communities of Tajungan village, but also from neighbouring villages whose original language is Madurese. The phenomenon seen here is that traders with native Madura are affected by the language of the Desa Tajungan. Although, in fact, the village has two languages, Javanese and Madurese, Javanese is more dominant. Some of the informants revealed that Madurese is difficult to pronounce. Sometimes, they can understand when their counterparts speak using Madurese, but they find it hard to reply in the same language.

During our observations in the market, for example, Javanese is found to be very dominant. Some informants said that there is a sense of pride when they use Javanese. They argue that Javanese is already inherited from their ancestors, so that they feel the need to preserve the heritage language. In addition to the wish to preserve the culture, they also feel ashamed when speaking Madurese. From the results of interviews conducted by researchers, they consider that Javanese is more prestigious than Madurese.

In addition, Ahmad Sofyan in Rosyida (2016) in his presentation at the Congress of Madurese held in Pamekasan in December 2008 stated that the Madurese people themselves are even reluctant to use their own language, because they feel embarrassed to use Madurese. This opinion is in line with the reality presented by some informants who were interviewed by the authors. Therefore, rarely, people in Desa Tajungan use Madurese language in a public space. In the assessment of Sri Ratnawati in Rosyida (2016), one of the lecturers in the University of Airlangga Surabaya, there is a tendency to leave their mother tongue. The Madurese generations prefer to use Bahasa Indonesia as a language. Madurese is regarded as a lower economic class. Madurese is identical to the language of the uneducated. Those who use Madurese only are backward people, so the tendency is towards prestige and lifestyle.

Madurese is not so prominent in Desa Tajungan, although it is found that some people use Madurese. They may speak two languages, but Javanese is dominant. The following comments illustrate:

> *"Javanese for us is the language of the ancestors that must be preserved. We do indeed live and coexist with people who speak a majority of Madurese but the Javanese language remains the language that we treasure. Even when doing a sale and purchase transactions in the market for example we still use Javanese."* Suratmi (46 years old)

From the above illustration, it is seen that they are prouder when using Javanese. However, in this case, it is further illustrated that even though Javanese is the dominant language, learning the language of Madura is also important.

> *"Our ancestors are Javanese people who are native to Javanese. Although this time we stayed in Madura Island but we could not leave the legacy language. Because our everyday language is indeed that. However, I think when I want to learn and speak Madurese will be more open access to socialise with neighbouring villages. Although the Madurese is used slightly but opponents said (the original Madurese people) will feel appreciated."* Khomariah (16 years old)

Despite the opportunity to learn Madurese is very widespread, very few people in Desa Tajungan are willing to learn Madurese. On the basis of the second illustration, it is really important that people in this village learn or speak Madurese in all social areas of life in order to be understood. In a bilingual family, for example, if someone from the native village of Tajungan marries someone who is not from the village of Tajungan (original Madurese person), in practice, the family uses both languages, Javanese and Madurese, well. Suppose that the father is the original Madurese and his native language is Madurese, while the mother's mother tongue is Javanese, then the mother speaks Javanese to her children and the father speaks Madurese. The children's parents speak to each other in Javanese, but the language of their day-to-day communication is still Madurese.

3.2 *Attitude to language*

Language is not only a tool for conveying ideas or a socially neutral instrument to convey meaning, but it is also related to the identity of social or ethnic groups and it has consequences for social evaluation and attitudes towards language (Ihemere in Rosyida, 2016).

As declared by Purwo in Rosyda (2016), the relationship between language attitudes and use of language can indeed be in a positive or negative direction. There are three positive attitudes of language attributes described by Garvin and Mathiot in Suwito (1989): (1) language loyalty, an attitude which encourages a given language society to defend its language and, if necessary, to prevent and avoid using other languages; (2) language pride, which encourages people to develop their language and to use it as a symbol of identity and a unity of society; and (3) awareness of the norm, which encourages people to use their language well and is the most influential factor in the use of language.

Rusyana (1989) in Rosyida (2016) added related attitudes to languages: the language attitude of a user of a specific language or language community, both bilingual and multilingual, can be either pride or ridicule, rejection or simultaneous acceptance of a particular language as regards the language mastered by every individual as well as by community members. It has to do with the status of language in society. Likewise, the use of language associated with the life of certain groups of people is often stereotypical, because the language is not only a communication tool, but it can also become a social identity.

The attitude of the village people to the exposure of their own language is certainly influenced by various factors that encompass it and also for various reasons. The reason for maintaining the Javanese language as their mother tongue is not merely the desire to preserve the inheritance of their ancestors, in spite of the fact that linguistic states within the village environment create the possibility of a choice of language. Given the choice of language, it affects the pattern of language usage. Malini (2013) in the Rosyida steady-state pattern of language use means that the use of language has a good and consistent effect on the survival of the language (language maintenance), while the faltering pattern causes a shift in the language (language shift).

One informant said:

> *"Initially we are afraid when we move outside the village, because automatically we will surely interact directly with the Madurese people who are Madurese-speaking indigenous people. Even just to say hello we dare not to talk. It is because there is the notion that the Madurese would be glad if the opponents he is talking to also speak Madurese. Therefore, as time goes by we are trying to learn and speak Madurese even though only one or two words, and even we mix both of the languages. However, the Javanese remains the main language of us."* Lina (17 years old)

The above illustration is the result of an interview with a high school student, where it can be seen that there is a willingness to learn and speak Madurese, although they still have a positive attitude to their mother tongue. Therefore, there are not few people from the village, especially students, who mix or do code switching in some Madurese words in their conversations.

Two major concepts in sociolinguistics are code switching and code mixing, which are often found in bilingual or multilingual societies. Holmes (2008: 102) defines code as: "(1) symbol or phrase system used to describe a particular meaning, and human language is a kind of code; (2) the language system in a society; (3) certain variations in the language". More simply, Holmes states that, code is a kind of system that two or more people use to communicate. Code switching is one of the aspects of language dependence in bilingual or multilingual societies. This means that, in a bilingual or multilingual society, it is possible that a speaker uses various codes in his speech according to the circumstances and various aspects that surround them. Romaine (2000) explains that code switching is a situation in which a speaker deliberately changes the language code he is using for a reason. In line with that opinion, Holmes (2008) states that code switching is the use of more than one language by communicants in the execution of a speech act. On the basis of some of these definitions, it can be concluded that code switching is a common phenomenon in bilingual or multilingual societies. The definition also implies that code switching can also occur in a single conversation. In other words, if a bilingual person uses the local language in his daily conversation and changes the Indonesian language when he is in school, then this activity can be categorised as code switching.

One characteristic of code switching is the existence of the aspect of language dependency in multilingual society. This means that in a multilingual society, it is almost impossible for a speaker to use one language in absolute terms without the slightest use of another language. Another feature is expressed by Romaine (2000: 69), who stated that "The use of two or more languages in the switching of code is indicated by: 1) each language still supports its own functions according to the context, and 2) the function of each language is adapted to the situation relevant to the code changes". This means that code switching can be said to have a social function.

People in Desa Tajungan use some Madurese words when they speak Javanese. These are typically Madurese words that have no counterpart in Javanese. The result is code switching. The following examples were taken from interviews (Madurese is in boldface):

Iki aku duwe **ghuk genghuk**
(I have some snacks)
Kenno **ngakan** jam piro?

(What time will you eat?)
Saporanah aku gak iso teko
(Sorry I cannot come)
Billeh kenno teko?
(When will you come?)
Wes iki **kallak dihibi'**
(Just take it by yourself)

3.3 *Language and identity*

In analysing the existing data, the researchers find that identity is the core category. In the context of Indonesian-ness, regional languages indicate the identity of a region. According to Alwi in Rosyida (2016), it serves as a communication tool for the speakers who come from the same ethnic group. According to more details, language has the following functions: (1) as a symbol of regional pride; (2) as a symbol of regional identity; (3) as a means of communication within the family and local communities; (4) as a means of supporting the local culture; and (5) as regional supporters and literature. This is in line with the theory of Paasi, who claims that "language is closely connected to culture and is the supreme expressive component of identity" (Paasi, 1996:47). In this study, there is a big question: "What is the identity of the village community?" They live in Madura region, yet they declare themselves Javanese. The fact is that, geographically, they are in the region of Madura, but people who were interviewed revealed that most of the communities have a common regard for identity across border regions. They reveal that they are indigenous people who live in Madura. Thus, to date, they claim themselves people. Language, culture, and traditions remain Javanese. Even when they go out of town or travel far away from their home, when meeting with new people, they introduce themselves as Javanese people. This could indicate that people who live in Desa Tajungan speak the Javanese language. It is to show their identity as Javanese people.

4 CONCLUSION

It seems that there are clear links between language use, attitudes to language, and the perception of identity. An open border encouraging more mobility has, over time, changed attitudes towards the neighbouring language and culture. There is interest in learning Madurese in Desa Tajungan, as we have seen that many of the participants in this study start learning the language while living at home. One major reason for the increasing interest in learning the neighbouring language is that people see that mastering the language will open doors to employment on both sides of the border. The movement across the border is mostly from Desa Tajungan to the neighbouring village. Cultural contact has caused some changes in the everyday language on the Desa Tajungan side of the border. People of Desa Tajungan who speak Madurese tend to mix Javanese terms with Madurese words for typically Madurese concepts, and this finds way into Javanese sentences. This code switching is often a token of participation in a local community consisting of speakers of the two languages. The role of language policy in the region may have an effect on the linguistic situation.

REFERENCES

Amalia, Y.S. (2016). *"Critical Reading of Our Worlds": An Ethnographic Study of Dialogue as Praxis in a Community-founded School in Indonesia*. Illinois: University of Illinois at Urbana-Champaign.

Amalia, Y. &. -M. (2014). Dialogue in narrative inquiry: Collaboration in Doctoral Study in the USA. In P. Smeyers, D. Bridges, N.C. Burbules, & M. Griffiths, *International Handbook of Interpretation in Educational Research.* London: Springer.

Djajasudarma, T. Fatimah. 1993. *Metode Linguistik: Ancangan Metode Penelitian dan Kajian*. Bandung: Eresco.

Ekawati, Rosyida. 2016. *Madura 2045 (Merayakan Peradaban)*. Yogyakarta: PT Lkis Pelangi Aksara Yogyakarta.
Hofstede, G. 2001. *Culture's Consequences: Comparing Values, Behaviour*. California: Sage.
Holmes, J. 2008. *An Introduction to Sociolinguitics.* Harlow: Parson Education Limited.
Kardana, I Nyoman. 2013. "Marking System of Information Structure in Indonesian Language". *International Journal of Linguistics*, 5(4): 121–122.
Kwary, D.A. (2013). *Creating and Testing the Indonesian High Frequency Word List*. Surabaya.
Malini, Seri, dkk. 2013. *Sikap Generasi Muda terhadap bahasa bali di Destinasi Wisata Internasional Bali. Jurnal Bahasa dan Seni* 41(2).
Romaine, S. 2000. *Language in Society. An Introduction to Sociolinguistics*. Oxford: Oxford University Press.
Rosyidah, U.N. (2013). *Individu dan Ruang Kota: Studi Ekokritik Novel Dilatasi Sketsa Karya Ari Nur Utam*. Surabaya.
Sudaryanto. 1993. *Metode dan Teknik Analisis Bahasa*. Yogyakarta: Duta Wacana University Press.
Sugiarti, E. (2014). *Dinamika Sosial budaya Masyarakat di Perbatasan Indonesia-Timor Leste*. Surabaya.

Urban Studies: Border and Mobility – Kerr et al. (Eds)
© 2019 Taylor & Francis Group, London, ISBN 978-1-138-58034-3

Indirect criticism in the ethnic Madurese community: Its various semantic formulas, lingual markers, and context of use

E. Jauhari
Universitas Airlangga, Surabaya, Indonesia

D. Purnanto
Universitas Sebelas Maret, Surakarta, Indonesia

ABSTRACT: This article studies the speech act of criticism in the ethnic Madurese community (in Indonesia). The study focuses on indirect criticism. The goals are to explain the various semantic formulas that are used to express indirect criticism in the Madurese community, to identify the lingual markers, and to understand the different contexts of use according to the cultural norms of the Madurese community. The data was collected through the methods of Discourse Completion Tasks and interviews. The results of the data analysis show that the semantic formulas that are used to express indirect criticism in the ethnic Madurese community vary widely. In general, each semantic formula has specific lingual markers which indicate the type of semantic formula. Each type of semantic formula is generally used in a different context.

Keywords: indirect criticism, semantic formulas, Madurese community

1 INTRODUCTION

Criticism is a kind of speech act which is interesting to study because its characteristics make it prone to be face-threatening (Brown and Levinson, 1987). Criticism has an important function in the social life of a community, as a tool for social control. Criticising is, in truth, performing an act of control. Performing an act of control means correcting or rectifying a deviant behaviour.

Criticism can be expressed using a direct or indirect strategy. Direct criticism occurs when the speaker's intention is the same as the linguistic meaning or semantic meaning of the utterance. In contrast, indirect criticism occurs when the speaker's intention is not the same as the linguistic meaning (Thomas, 1995). The speaker presents a primary illocutionary act through a secondary illocutionary act (Searle, 1996). This article investigates indirect criticism in the ethnic Madurese community. The goals are to understand the various kinds of semantic formulas, the lingual markers, and the contexts of use in the Madurese community.

Ethnic Madurese are one of the ethnic communities found in Indonesia who live predominantly on the Island of Madura, in East Java. The people in this community are known to be somewhat sensitive about their self-esteem, as reflected in the popular expression: "*angoanpotetolangetembangpotemata*", 'Better the white of bones than the white of eyes". The meaning of this saying is that it is better to die and be buried in the earth than to have no self-respect. It gives an implication of how important the matter of self-respect is for this community, and criticism is a linguistic act that tends not to respect the person towards whom the criticism is directed. Therefore, it is interesting to observe how the people in this community express criticism while taking care to respect the person who is the target of the criticism.

Up to now, there have been no studies on the speech act of criticism in the ethnic Madurese community, although studies on criticism in various other cultures have already been carried out by a number of experts. They include Tracy, et al. (1987), Tracy and Eissenberg (1990), Wajnryb (1993), Gunarwan (1996), Toplak and Katz (2000), Mulac et al. (2000), Nguyen (2005), Hoang Thi Xuang Hoa (2007), Min Shang-chao (2008), and Nguyen (2008). The results of their studies are, of course, highly beneficial as a background to the writing of this article. Nevertheless, the study of criticism in this article is different from the studies of criticism by previous experts because this article focuses on criticism in its function as a tool for social control.

2 METHOD

The data in this article were collected using the method of a Discourse Completion Task (DCT) and interviews. The DCT was distributed to informants in a number of government offices in the towns of Bangkalan, Sampang, and Pamekasan. Since the people completing the DCT were office employees, the material of the DCT was related to office matters. The completed DCT was then used as material for interviewing the informants. The goal of the interviews was to affirm that the data contained in the DCT matched the conditions of the Madurese community in the field. In addition, interviews were held to obtain additional relevant information. The context of the DCT was determined based on the alternations of ±Power and ±Distance (Brown and Levinson, 1987). The data collected were then sorted into two groups based on the strategy chosen, either direct or indirect. Subsequently, the data which showed indirect strategy was analysed further to understand the various semantic formulas used, as well as the lingual markers and contexts of use according to the cultural norms of the Madurese community.

3 RESULTS AND DISCUSSION

The results of the data analysis show that indirect criticism in the Madurese community can be expressed through a variety of different semantic formulas, including: (1) request, (2) prohibition, (3) command, (4) suggestion, (5) rhetorical question, (6) irony, (7) comparative expression, (8) sarcasm, (9) expectation, (10) advice, and (11) admonishment. These various kinds of semantic formulas are explained further in the following section.

3.1 *Semantic formula 'request'*

Criticism is sometimes expressed through the semantic formula 'request' in the Madurese community. This means, the speaker makes a request but the substance of the utterance is not really a request but a criticism. see example (1) below.

1. *Abdhinasadhajasaroju' manabikodhudisplin. Namongabdhinajhughan<u>nyo'on</u>Saleranama rengeconto se sae da' ka baba'anna. Ka'dintojhughanka'angghuykabhaghusansadhajana.* (Uttered by an employee [a subordinate] to the head of the office who orders his employees to behave in a disciplined manner but does not set a good example of discipline himself to his subordinates.)

In the Madurese community, the semantic formula 'request' is often characterized by the use of the word *nyo'on* 'request' or the word *menta* 'ask', which appears explicitly in utterance (1). However, not all utterances containing a request are criticisms. A request which is a criticism is a request that is intended to enforce control over a deviant behaviour.

Criticism which is expressed using the semantic formula 'request' is felt to be soft criticism. This is understandable because the negative evaluation that forms the substance of the criticism is not stated explicitly or overtly but is stated covertly or is disguised in the form of a request. Perhaps this formula can be found in various situational contexts such as the context (+P+D), (+P–D), (–P+D), or (–P–D).

3.2 *Semantic formula 'prohibition'*

Indirect criticism in the Madurese community contains criticism but it is expressed with a semantic formula which is in the form of prohibition. Consider example (2) below.

2. *Pak Fauzi, sampeyanonenglakohkantorkitanekasangatbanyak ben kaduhsegerapanakehaghin. Dedihtolongsampeyan ta' ngurusilakohsampingan mon jam dinas.* (Uttered by the head of an office to his employee who often spends time on other business in working hours.)

Utterance (2) above is a prohibition (marked by the use of the word *ta'* 'don't'); its substance is criticism. Utterance (2) is to control the behaviour of the person to whom the utterance is directed (the employee) so that he does not conduct his own business during working hours. In the Madurese community, the semantic formula 'prohibition' is usually expressed overtly and is often marked with the word *jha, je'*, or *ta'* which appear explicitly in the utterance. Prohibitions which represent criticisms are utterances which intend to prevent or control the addressee.

Criticism in the form of prohibition is categorized as quite harsh for the Madurese community because it is expressed overtly and explicitly. Based on the analysis, criticism using the semantic formula 'prohibition' is mainly found in the context (–P+D) or (–P–D). However, with the aid of modifiers (whether internal or external), the semantic formula 'prohibition' can potentially also be used in the context (+P+D) or (+P–D).

3.3 *Semantic formula 'command'*

In the Madurese community, indirect criticism is sometimes expressed as 'command'. Thus, the substance of the utterance is, in fact, a criticism but semantically it is formulated as a command. See example (3) below.

3. *Pak Fauzi, kalakoankita se utamapanekahnengkantor. Deddihlastarehagihgelluhkalakoan se bedehnengkantor, mangkenmanabiamponlastareh, sampiyanbisangelakonihkalakoan se laen.* (Uttered by the head of an office to a subordinate who often neglectshis office work to manage his own business.)

Semantically, utterance (3) above is a command. A command is marked by the use of a verb without a prefix, in this case*lastarehagih* 'finish'. However, from the point of view of its illocutionary force, utterance (3) above has the intention of encouraging the addressee (Pak Fauzi) to engage in self-correction, or to stop conducting his own personal business during office hours. In this case, the command utterance is considered a harsh criticism. However, if the command is expressed using a passive verb, the harshness of the criticism can be reduced by several degrees. The semantic formula 'command' is largely found in the Madurese community in the contexts (–P–D) and (–P+D). However, if the degree of harshness is reduced by the use of a modifier or the command is expressed in a passive form, the semantic formula 'command' is sometimes also used in the context (+P+D) or (+P–D).

3.4 *Semantic formula 'suggestion'*

In the Madurese community criticism is also expressed through the semantic formula 'suggestion'. The substance of the utterance is criticism but is also expressed through the semantic formula 'suggestion' (See example (4)).

4. *Ta' langkong Pak, sopajakedisiplinannengkantorka'dintokengengterrosejaga, abdhinaghadhuwanusul. Kadhiponapamanabi Bapak menangkapimpinannengka'dintojhughanmarengeco nto se sae da' ka baba'annaepon.* (Uttered by an employee[a subordinate] to the head of the office who often orders the employees to behave in a disciplined manner but does not set a good example of disciplined behaviour himself to his employees.)

Semantically, utterance (4) expresses a suggestion. This is indicated by the use of word *usul* 'suggest' which appears explicitly in the utterance. However, from the point of view of its illocutionary force, utterance (4) above is clearly not only intended to convey the suggestion that is seen in its semantic meaning but is also intended as a criticism. Through utterance (4) above, the speaker (the employee) intends to enforce control over the addressee (the head of the office) who has acted in an inappropriate manner, by ordering his employees to behave in a disciplined manner while never setting a good example of discipline himself to his subordinates. In the Madurese community, the semantic formula 'suggestion' tends to be expressed in a performative manner by using the word 'suggest' (*usul/usulen'usul'*).

Not all utterances expressing suggestions are criticisms. Suggestions that are not intended to carry out control are, of course, not criticisms but only common utterances of suggestion. In the Madurese community, a criticism that is presented through the semantic formula 'suggestion' is regarded as a very soft form of criticism. The hearer does not feel face-threatened because the criticism is expressed as a suggestion. Criticism which is presented through the semantic formula 'suggestion' is primarily found in the context (+P–D) or (+P+D).

3.5 *Semantic formula in the form of a rhetorical question*

Criticism in the Madurese community may also be expressed through the semantic formula of a rhetorical question. In this type of criticism, the speaker expresses his or her criticism by asking the addressee a question, but the question does not require an answer because the answer is, in fact, already known, both by the speaker and the addressee. The main goal of the speaker in asking the question is to perform control (criticise) of the addressee so that s/he carries out self-correction of the deviant behaviour that s/he has been engaging in. Take a close look at example (5) below.

5. *Pak Gun, kher-akhernekahkaulahngatelakbedehbannyakpegawai se korangdisiplin. Apahkaadaannekahepennenggahpak?* (Uttered by an employee [subordinate] to his superior [the head of the office], who has been allowing the employees to behave in an undisciplined manner in carrying out their office duties.)

In utterance (5) above, the speaker (the employee) is not simply intending to ask the addressee (the head of the office) why he is allowing the employees to behave in an undisciplined manner. In addition to that, the speaker's intention is to criticize (control) the addressee for violating certain norms, in this case, as the head of the office, for allowing his employees to behave in an undisciplined manner. Therefore, the response required from the addressee is not simply an answer to the question asked by the speaker but also an act of self-correction. In the Madurese community, this type of criticism is felt to be a soft criticism because the negative evaluation which forms the substance of the criticism is expressed covertly through a question. Due to its soft nature, this type of criticism can potentially be used in various situational contexts such as (+P+D), (+P–D), (–P+D), or (–P–D).

3.6 *Semantic formula 'irony'*

Indirect criticism in the Madurese community is sometimes expressed through the semantic formula of irony. For this semantic formula, when expressing a criticism to the addressee, the speaker utters a statement which says the opposite of what he or she actually means. Example (6) below is expressed through the semantic formula 'irony'.

6. *Musyafak, ta' pa-rapahkalowankantorba'na ta' taurus, se pentengkalakowansampengannaajhalanterros.* (Uttered by the head of an office to one of his employees (Musyafak) who often conductshis own business affairs during office hours and as a result, his office work is neglected.)

If we look at the context, the speaker (the head of the office) in utterance (6) above clearly does not intend to give permission to the addressee (Musyafak) to conduct his own personal business and neglect his office duties. On the contrary, the speaker is performing an act of control so that the addressee does not conduct his personal business while neglecting his office work. On the basis of this, it is clear that utterance (6) above is a criticism that is expressed through the semantic formula irony. In general, this semantic formula does not include specific lingual markers and tends only to be understood semantically. In the Madurese community, this kind of semantic formula is used in the context (–P+D) or (–P–D).

3.7 *Semantic formula in the form of a comparative expression*

The data also shows that criticism in the Madurese community may also be expressed using a semantic formula which is in the form of a comparative expression. In this case, the speaker, when presenting his or her criticism, makes a comparison between himself or herself and the addressee, or between another person and the addressee about the same or a similar type of problem. In making this comparison, the speaker shows that, in doing the same or similar kind of thing, the speaker can do it without violating any norms, unlike the addressee. In the Madurese community, a criticism that is expressed by means of a comparative expression such as this does not have any particular lingual markers. This kind of comparative expression can usually only be understood semantically. In order to better understand this semantic formula, see example (7) below (in the same context as example (6) above).

7. *Pak Musyafak, Pak Ahmad rowa pada andi' kalakowansampenganganyar. Tape Ahmad ta' tomangurusinnengjhamlakoh.*('Pak Musyafak, Pak Ahmad also has his own business. But Pak Ahmad never conductshis own business during office hours'.)

Utterance (7) above contains a comparison between two people (both employees in the same office), Pak Musyafak and Pak Ahmad, both of whom have their own personal businesses in addition to their office jobs. Through utterance (7) above, the speaker (the head of the office) is not intending simply to compare the business of Pak Musyafak with that of Pak Ahmad. By making this comparison, the head of the office, of course, is making a criticism (control) of Pak Musyafak so that he does not let his own personal business interfere with his office work, which Pak Ahmad never does. It is hoped that this comparison will encourage Pak Musyafak to carry out introspection and self-correction. In the Madurese community, a criticism that is expressed using a comparative expression is primarily found in the context (–P+D) or (–P–D).

3.8 *Semantic formula 'sarcasm'*

In the Madurese community, indirect criticism may also be expressed using the semantic formula 'sarcasm'. A criticism which is expressed through sarcasm is usually formulated in such a way that the addressee is not mentioned directly in the utterance but instead the speaker makes a generalisation. In other words, when making the criticism, the speaker does not refer specifically to the person to whom the criticism is directed but rather expresses it as a general criticism. In the Madurese community, the semantic formula 'sarcasm' appears not to have any particular lingual markers and tends only to be understood semantically. Consider example (8) below.

8. *E kantorreyapossa' pongghaba se lebbimelengurusinkalakowansampengannanengjhamkantoretembhengkalakowanotamana. Areyatakekenningtorot!*(Uttered by the head of an office to his employee by the name of Pak Fauzi who often conductshis own personal business during office hours so that his office duties are neglected.)

Criticism (8) above is expressed using a semantic formula in form of sarcasm. When making his criticism, the speaker (the head of the office) does not mention the name of the employee to whom he is referring. Nevertheless, based on the context, the person who is the target of the criticism, namely Pak Fauzi, knows precisely that the criticism being expressed is intended for himself and not for anyone else. A criticism that is expressed with sarcasm appears to have the intention of 'saving face' because the name of the person who is the target of the criticism (the addressee, Pak Fauzi) is not stated explicitly. In the Madurese community, a criticism that is expressed using the semantic formula 'sarcasm' is found in the context (–P+D). However, due to the soft nature of the criticism, it also has the potential to be used in other contexts.

3.9 *Semantic formula 'expectation'*

Indirect criticism in the Madurese community is sometimes also expressed using the semantic formula 'expectation'. In this case, the utterance of the speaker expresses an expectation but the substance of the utterance is not expectation but criticism. In the Madurese community, a criticism expressed with the semantic formula 'expectation' is usually marked by the appearance of the word *ngarep/arep* 'expect/hope' in the utterance. See example (9) below.

9. *Pak Fauzi, sampeyanollebeihgeduenbisnissampingan e luarkantor, tapehabdinangarepbisnissampeyanloksampe' eganggulakoh e kantor*. (Uttered by the head of an office to his employee who often leaves the office during work hours to take care of his own personal business.)

In criticism (9), the speaker expresses his hope that Mr. Fauzi will prioritise his job at the office. A criticism which uses the semantic formula 'expectation' is mainly used in the context (–P+D) or (–P–D).

3.10 *Semantic formula 'advice'*

Indirect criticism is also actualized using a semantic formula 'advice'. In this case, the utterance is presented semantically in the form of advice but, in substance, it is not advice but criticism. Consider example (10) below.

10. *Sabelummahkaulahnyo'onsaporahpak, manabiapa se kaulahdhebunekahkorangnyaman. Kadihnekahpah, manabibapakmentah kami gebeyasekapdisiplin, alangkahbegussahbapakjuganmerrikcontohdisiplin se begus ka kami. Manabientenkaulapekkeranjuranbapaknekahmlaratelaksanaagih.* (Uttered by an employee [subordinate] to the head of the office who often orders the employees to have discipline but does not set a good example of discipline himself to his employees.)

This semantic formula 'advice' can be seen from its head act, which has the lingual marker *alangkahbegussah* 'how much better it would be'. On the basis of this, it is clear that criticism (10) above is expressed with the semantic formula 'advice'. In the Madurese language, there are several lingual forms that can be used to mark the semantic formula 'advice', including *alangkahbegussah* 'how much better it would be', *langkongsae* 'it would be better', *saestonah* '[you] should', *lebbibecek* 'it would be preferable'. In the Madurese community, criticism using the semantic formula 'advice' is viewed as soft criticism and, as such, it can to be used in various situational contexts, such as (+P+D), (+P–D), (–P+D), and (–P–D).

3.11 *Semantic formula 'expectation'*

In the Madurese community, indirect criticism is also sometimes actualized using a semantic formula which is in the form of a reminder or admonishment. See example (11) below.

11. *Pak Fauzi, kalakoanotamalebhipentengetembangkalakoansampingan. Polanakaulengenga 'aghinsopajakalakoansampinganlokaganggukalakoanutama.* (Uttered by the head of the office to one of his employees (Pak Fauzi) who spends time working on his own personal business during office hours.)

It is clear that semantically criticism (11) above is an admonishment. This semantic formula can be observed from its head act (*Polanakaulengenga'aghinsopajakalakoansam pinganlokaganggukalakoanutama*). In this head act, the speaker explicitly uses the word/phrase *ngenga'aghin* 'remind' to convey his criticism. In the Madurese community, a criticism that is expressed with this kind of semantic formula is considered to be quite a soft criticism. Nevertheless, this kind of semantic formula is mainly used in the context (–P+D) or (–P–D).

4 CONCLUSION

People in the Madurese community use various kinds of semantic formulas to express indirect criticism. These semantic formulas include (1) request, (2) prohibition, (3) command, (4) suggestion, (5) rhetorical question, (6) irony, (7) comparative expression, (8) sarcasm, (9) expectation, (10) advice, and (11) admonishment. Each of these semantic formulas generally has its own specific lingual markers which indicate the type of semantic formula. In addition, each semantic formula also has its own different contexts of use.

REFERENCES

Brown, Penelope & Levinson, S.C. 1987. *Politeness: Some Universals in Language Usage*. Cambridge: Cambridge University Press.

Gunarwan, Asim. 1996. "The Speech Act of Criticizing among Speakers of Javanese". Paper presented at the 6th meeting of the South East Asian Linguistics Society. Unpublished.

Min Shang-chao. 2008. "Study on the Differences of Speech Act of Criticism in Chinese and English". *US-China Foreign Language* 6(3).

Mulac, Anthony et al. 2000. "Female and Male Managers' Criticism Giving: Differences in Language Use and Effects". *Journal of Language and Psychology* 19(4).

Nguyen, Minh Thi Thui. 2005. *Criticizing and Responding to Criticism in a Foreign Language: A Study of Vietnamese Learners of English*. A Thesis Submitted for the Degree of Doctor of Philosophy in Language Teaching, The University of Auckland.

———. 2008. Criticizing in an L2: Pragmatic Strategies Used by Vietnamese EFL Learners. *Reference Global*. Retrieved October 25, 2017 from http//www.reference-global.com/doi/pdf plus/10.1515/IP.2008.003.

Searle, J.R. 1996."Indirect Speech Acts" in A.P. Martinich (ed.). *The Philosophy of Language*. Third Edition. New York: Oxford University Press.

Thomas, Jenny. 1995. *Meaning in Interaction: An Introduction to Pragmatics*. London and New York: Longman.

Tracy, K & Eisenberg. 1990. "Giving Criticism: A Multiple Goals Case Study" in *Research on Language and Social Interaction*.

Tracy, et al. 1987. "Good and Bad Criticism: A Descriptive Analysis". *Journal of Communication*.

Toplak, M. & Katz, A. 2000. "On the Uses of Sarcastic Irony". *Journal of Pragmatics*.

Wajnryb, R. 1993. "Strategies for the Management and Delivery of Criticism". *EA Journal*.

Xuan Hoa, Hoang Thi. 2007. "Criticizing Behaviors by the Vietnamese and the American: Topics, Social Factors, and Frequency". *VNU Journal of Science. Foreign Languages*: 141–154.

Urban Studies: Border and Mobility – Kerr et al. (Eds)

Multilingualism and mobility: Defining borders within Surabaya city through the linguistic cityscape

E. Rusnaningtias
Universitas Airlangga, Surabaya, Indonesia

ABSTRACT: Surabaya as an urban metropolis has undergone a lot of changes. Its development has attracted many people from different regions inside and outside Surabaya to come and stay in the city. They have brought with them their desires for their futures, their cultures, and their local languages. This condition results in a multilingual situation in Surabaya as it is reflected in its linguistic cityscape. Although most parts of Surabaya have been transformed into urban areas, some of the more local areas with their indigenous people of Surabaya are still present and some contrasts between urban and suburban or rural areas are still identified. Using Landry and Bourhis' theory, this paper discusses how the languages used in the linguistic cityscape reflect the linguistic repertoire of the speech community and mark its geographical territories. The results showed that there were six languages making up the linguistic cityscape of Surabaya. The use of Indonesian and Javanese languages represented an area with the indigenous people of Surabaya while the use of foreign languages represented new settlements where outsiders came and stayed.

Keywords: multilingualism, mobility, linguistic cityscape

1 INTRODUCTION

As one of the big cities in Indonesia, Surabaya has undergone rapid changes. Many buildings, roads, hotels, parks, real estates, offices, schools, and shopping centres have been built by the Surabaya City Government to transform Surabaya into a business metropolis (Visi dan Misi Kota Surabaya, 2009). Initially, much of the development had been centralised in the city centre. However, since the number of city centre residents has increased and the city centre has become overcrowded, in 1970, the Surabaya City Government began to break the concentration by expanding its area and extending its development to the suburbs (Fikriyah & Wisnu, 2013). Then, the government opened up opportunities for investment and developments of the suburbs as a satellite town. A satellite town is self-contained and limited in size, built in the vicinity of a large town or city to house and employ those who would otherwise create a demand for expansion in the existing settlement, but it is still dependent on the parent-city to a certain extent for population and major services (a Dictionary of Architecture and Landscape Architecture, 2000). However, the development of the suburbs as a satellite town was not exciting, not until a giant developer called Ciputra Group acquired its investments and began its development in the western part of Surabaya in 1990. Thus, West Surabaya appeared to have greater potential compared to East Surabaya.

Ciputra Group developed its real-estate known as Citraland on an area of 2000 ha and started its commercial operation in 1993 (Fikriyah & Wisnu, 2013). With its slogan '*Kota Mandiri Citra Raya*', Citraland has been established as an independent residential area. It provides not only prestigious residences but also facilities and tourist attractions. In 2003, this slogan was changed to 'Citraland—the Singapore of Surabaya: Living in a Modern International City'. With this new slogan, Citraland was transformed to have a more modern and international look. In addition, many of the landscapes including roads and commercial signs, which are found in this area, are written in English or any other foreign language. Such a view gives a more international feel to the region and the inhabitants.

Most of the signs which appear in the public sphere are written not only in Indonesian Language but also in foreign languages, such as English, an indication of a multilingual community. Furthermore, the use of English for road signs or regulatory signs can also be found in the surrounding areas which are mostly inhabited by Surabayanese people who speak their own local language, Javanese language as the vernacular language and Indonesian language as the national language. The fact that a regulatory sign written in English is placed in a bilingual community (Indonesian-Javanese community) seems inappropriate and raises questions, such as who is going to read the signs, who lives in the area, what language is used as a means of communication in the area, and what functions does it serve. Thus, this paper aims to describe how the languages used in the linguistic cityscape reflect the linguistic repertoire of the speech community and mark its geographical territories.

This study is concerned with the linguistic forms, codes or languages chosen to be displayed in the public sphere; and therefore, it focuses on the written forms or texts of the landscape. Texts which are visible and readable in public spaces form the linguistic landscape of the area or region. Landry & Bourhis (1997:25) stated that "the language of public road signs, advertising billboards, street names, place names, commercial shop signs, and public signs on government buildings combined to form the linguistic landscape of a given territory, region, or urban agglomeration." Gorter (2006a) proposes the term 'multilingual cityscape' instead of 'linguistic landscape', as the word 'landscape' relates to the countryside whereas collections of signs with texts are encountered predominantly in urban areas. However, 'multilingual cityscape' excludes the possibility of monolingualism in advance. Even in multilingual contexts, it is an empirical question whether the linguistic landscape is monolingual or multilingual. Thus, the term linguistic cityscape is used in this study. Torkington (2009) added that the texts which make up the linguistic cityscape can be monolingual, bilingual, or multilingual, reflecting the diversity of the language groups present in a given territory, or region. Since there are some languages present in the multilingual community and the languages do not always appear together in the same cityscape, it is assumed that particular languages are chosen to be used in the cityscape for particular reasons. As Scollon & Scollon (2003) stated, "no choices are neutral in the social world".

Code choice in the linguistic cityscape functions both as an informational marker and a symbolic marker. As an informational marker, it informs in-group and out-group members about the linguistic characteristics, the language which can be used to communicate and to obtain services within the region, and the borders of the territory of the linguistic group (Landry & Bourhis, 1997). As an example, if a predominance of signs uses Mandarin, it indexes that the group of people living in the region is a Mandarin speech community or a Chinese group of people. Meanwhile, as a symbolic marker, it refers to the value and status of the languages as perceived by the members of a language group in comparison to other languages (Landry & Bourhis, 1997). The use of foreign language, say, Italian in a restaurant name, symbolises foreign taste, Italian taste. In addition, the use of a foreign language, such as English, for commercial signs is perceived as being more modern and prestigious than local languages because the local languages are spoken by only a few people in the world beyond regional or national borders. In this way, the use of English in a linguistic cityscape has more of a symbolic rather than an indexical function (Scollon & Scollon, 2003).

Regarding the producers of the linguistic cityscape, Huebner (2006) and Ben-Rafael *et al.* (2006) classified linguistic cityscapes into two types: governmental linguistic cityscapes (top-down) issued by the national and public bureaucracies, administrations, and institutions; and non-governmental linguistic cityscapes (bottom-up) produced by individual social actors and all commercial enterprises. Furthermore, Scollon & Scollon (2003) distinguished four types of linguistic cityscape discourses: regulatory discourses, infrastructural discourses, commercial discourses, and transgressive discourses which are about calls for social action.

This study discusses linguistic cityscapes from a sociolinguistic view. The use of varieties of languages in linguistic cityscapes indicates that the community is a multilingual speech community. Thus, there are different groups of language users present in the community.

In this study, a mixed method approach (quantitative-qualitative approach) was used. The data were 210 photos collected randomly from seven districts in the western part of Surabaya, so there were 30 pictures taken from each district. The seven districts which constituted the research areas were: 1) Asemrowo District, 2) Benowo District, 3) Lakarsantri District, 4) Pakal District, 5) Sambikerep District, 6) Sukomanunggal District, and 7) Tandes District. By using a quantitative approach, the languages appearing in the linguistoc cityscapes were counted and classified based on the types of text (monolingual/bilingual/multilingual). It was done in order to obtain a general picture of which languages making up the linguistic cityscape. After getting the objective description of the languages used in the linguistic cityscape, the data were analysed based on the types of discourse and the context of the language use in order to find out the functions of the linguistic landscape. Finally, the identities of the regions were analysed and the borders were determined.

2 DISCUSSION

2.1 *The code choices in the linguistic cityscape*

Compared to the other parts of Surabaya, the western part of Surabaya has changed much. It has developed faster and it has caused many people from different regions in Surabaya to move to the western part of Surabaya. The west Surabaya community comprises Surabayanese, the indigenous people of Surabaya who speak Javanese as vernacular, Madurese people who speak Madurese, and other ethnic groups who speak their own varieties. As they are Indonesian, they speak Indonesian language as a lingua franca or national language. However, one of the seven districts in the western part of Surabaya, Citraland is inhabited not only predominantly by Chinese people but also by some foreign people. This situation results in the emergence of different languages and this is visible in the linguistic cityscape found in the western part of Surabaya.

As stated by Torkington (2009), the texts which make up the linguistic cityscape can be monolingual, bilingual, or multilingual, reflecting the diversity of the language groups present in a given territory, or region. In this study, six languages were identified as being used in the linguistic cityscape. The six languages were 1) Indonesian, 2) Javanese, 3) English, 4) Mandarin, 5) Japanese, and 6) Arabic. Three languages appeared as monolingual texts: Indonesian language appeared on 40% of all the signs, English appeared on 10.48% of all the signs, and Javanese language appeared the least, on 5.24% of all the signs. On the other hand, there were eight kinds of bilingual texts: Indonesian-English for 31.90% of signs, Indonesian-Javanese for 6.19% of signs, Javanese-English for 0.95% of signs, Indonesian-Arabic for 0.95% of signs, and consecutively, Indonesian-Mandarin, English-Arabic, English-Mandarin, English-Japanese, each for 0.48% of signs. Lastly, there were three kinds of multilingual texts: Indonesian-Javanese-English for 1.43% of signs, Indonesian-English-Arabic for 0.48% of signs and Indonesia-English-Mandarin for 0.48% of signs.

Based on the frequency of the appearance of the languages in the linguistic cityscape, the Indonesian language was the most frequently appearing language in the linguistic cityscape. It appeared on 173 signs or 82% of the total signs. This position was followed by the English language which appeared on 100 signs or 48% of the total signs. Though Javanese is the vernacular of Surabayanese people, the language only appeared on 30 signs or 14.3%. Meanwhile, the least frequently appearing signs were those in Arabic at 1.9%, Mandarin at 1.4%, and Japanese at 0.48%.

The findings show that even though there are many languages present in the western part of Surabaya, the Indonesian language is still the predominant language used in the linguistic cityscape found in the western part of Surabaya. This is due to the fact that the Indonesian language is a lingua franca for different language groups in Indonesia. As stated in Act Number 24 Year 2009 about Flag, Language, Symbol of State, and National Anthem, the Indonesian language is the official language of Indonesia which functions as 1) the national identity, 2) national pride, 3) unifying language, and 4) a means of communication across regions and regional cultures in Indonesia. Besides, there are also regional languages and foreign languages.

Based on the data analysis, there was only one regional language identified, Javanese language. Moreover, the number of linguistic cityscapes using the Javanese language was limited. This was due to the fact that Surabaya is a big city and it is a melting pot of many different ethnic groups speaking different regional languages in Indonesia. In accordance with this, none of the linguistic cityscapes found in this study used the Madurese language. This was because Madurese is not the vernacular of the majority of inhabitants of Surabaya. It is commonly used only for interactions between Madurese people. Therefore, the use of the Indonesian language is necessary to deliver the message to different language groups. Finally, the use of foreign languages in the linguistic cityscape was also limited, except for the use of English. English was found in many monolingual, bilingual, and even multilingual texts. Moreover, English was predominantly used in a particular district, i.e. Lakarsantri District, where many Chinese people or foreigners live in Citraland, a giant real-estate in this district. Briefly, the linguistic cityscape represents the diversity of languages in a particular speech community.

2.2 *The functions the linguistic cityscape, mobility, and borders*

In order to find out the functions of the linguistic landscape, it is important to take account of the producers of the linguistic landscape, the types of discourse and the context of use. The producers of a linguistic landscape are distinguished by two types: government including the national and public bureaucracies, administrations, and institutions; and non-government including individual social actors and all commercial enterprises (Huebner, 2006). Furthermore, Scollon & Scollon (2003) divided the types of discourse into four categories: regulatory discourses, infrastructural discourses, commercial discourses, and transgressive discourses.

Based on the producers and the types of discourse, it was found that there were four types of linguistic cityscape discourses produced by the government: 4 regulatory signs, 60 infrastructural signs, 6 commercial signs, and 2 transgressive signs. In contrast, there were three types of linguistic landscape discourses produced by non-government or commercial enterprises: 4 regulatory signs, 84 commercial signs, and 50 infrastructural signs.

All of the regulatory signs produced by the government were in the Indonesian language; whereas, 5 out of the 60 infrastructural signs were bilingual, Indonesian-English. This supports the fact that the Indonesian language is the official and national language of Indonesian people. It is used as a means of communication in formal domains, such as regulation and documentation. The regulatory signs were addressed to Indonesian people, for example one was a regulatory sign written '*belok kiri mengikuti lampu*,' which means that if you want to turn left, you should follow the traffic light. This sign was found at one of the sides of an intersection in the Tandes District. The majority of people living in this district are Javanese, but there are also Madurese people, and other ethnic groups. Thus, the Indonesian language here acts rather as an informational marker than as a symbolic marker.

Though most of the governmental signs were in the Indonesian language, there were also bilingual signs. One of the bilingual signs was found in a police station '*Polisi* – Police: the Western Regency'. The bilingual (Indonesian-English) linguistic cityscape serves an informational function or as an identity marker that the police station is near the Western Regency. As stated by Laundry and Bourhis (1997), informative function indicates the border of the territory of linguistic community. Beside the police sign, bilingual text was also displayed on transgressive discourses. There were only 2 transgressive discourses identified in this study, and they were made by the government: 1) '*Sehat tanpa narkoba pasti bisa, ayo*say no drug. *Selamatkan generasi kita*', and 2) '*Anda memasuki zona:* save our students'. These signs were addressed to teenagers and had symbolic markers indicating the value and status of the languages (Landry & Bourhis, 1997).

In line with the signs produced by the government, the signs produced by non-government occured in monolingual, bilingual, and multilingual texts. One out of four regulatory signs produced by the government was written in the Indonesian language; for example, '*kawasantan parokok,'* which means area free of smoking. This sign was found in Sukomanunggal District. The Indonesian language was used to inform visitors that smoking was prohibited in the area of the hospital. In contrast, the other three regulatory discourses were monolingual and used English only: 'Emergency call in case of crime, fire, and accident', 'Reduce speed now', and 'school zone'. These signs were found in Citraland real-estate, Lakarsantri District, where there are some foreign people living in this real-estate. Therefore, English is acceptable and is commonly used as a means of communication in this area. Moreover, it matches with the concept of 'Citraland—the Singapore of Surabaya: Living in a Modern International City'. It is obvious that the use of these regulatory signs has informational or indexical function as it informs in-group and out-group members about the linguistic characteristics, the language which can be used to communicate and to obtain services within the region, and the borders of the territory of the linguistic group (Landry & Bourhis, 1997).

2.3 *Mobility, and borders*

The infrastructural discourses varied in terms of marking the boundaries of the territory. In the first area of this study, Asemrowo District, many of the infrastructural signs either made by the government or made by non-government used the Indonesian or Javanese language, for example '*JalanTambak*', '*Jalan Asem*', '*Jalan Asemrowo*' and '*Klinik Asemrowo*', indicating that the region has long been established and inhabited by the indigenous Surabayanese people. As stated in Surabayan government regulations, PERDA No. 2 Year 1975, in naming streets, recreation facilities, parks, and other public places, the regent should take account of the history, heroism and the characteristics of the area. Asemrowo District used to be characterised by its nature which was full of fish-farms and tamarind trees. However, many of these fish-farms and tamarind trees have been transformed into highways and residences or 'kampoengs', but until now, these kampoengs have not changed much.

The second research area, Benowo District, had started to use some English in its infrastructural discourses resulting in a bilingual linguistic cityscape; for example, '*Citra Permata* Regency' '*Tandes* Central Business District', and '*Cemara* Production'. This showed the move towards urban areas. This also supported the evidence that this district has been the connecting door between Surabaya and Gresik.

The third area is Lakarsantri District where Citraland real-estate is located. This area has been known as an area of urban agglomeration since it was established as a satellite or independent city with its own social and economic activities within the real-estate. The people who live in this real-real-estate comprise different linguistic groups. Moreover, this real-estate also provides amusement and thus it becomes one of the tourist destinations in Surabaya. It can be seen from the signs that make up the linguistic cityscape of this area (Landry & Bourhis, 1997). Many of the signs made by the Ciputra developer were written not only in Indonesian but also in English, for example 'Gwalk', 'Food Garden', 'Food Arcade'. Some of the street names were bilingual Indonesian-English texts because they had to meet Surabayan government regulations about naming streets or places. Thus, an international sense was added to the names, such as '*Citraraya* Boulevard', 'International *Timur* Road', '*Bukit Telaga* Golf' and 'East Emerald Mansion'. Moreover, many commercial shop signs used foreign languages, for instance 'Daily Sweet', 'Nagasaki Ya Japanese Restaurant', etc.

The next district is Pakal District which is located between Benowo District and Lakarsantri District. The dichotomy of traditional and modern is obvious in this region. The linguistic cityscape shows the division clearly. In its suburbs, traditional names were used, such as 'Desa Penggalangan', 'Dusun Kukun' and 'Jalan Kendung Kanoman' while in the areas near Citraland, English has started to be used, for example 'Palma *Clasica*', '*Rosewood*', 'Orange Bakery', etc.

The fifth research area is Sambikerep District. Almost all of the linguistic cityscapes found in this district used the Indonesian language and few signs used English. The signs which used English were 'Pilarland *Sambikerep*' and '*Ruko* West Point'. 'Pilarland *Sambikerep*' was a new residence in this region; whereas, '*Ruko* West Point' was new housing with shops. This indicated that English was a new influence in this region.

The sixth district is Sukomanunggal District. Since its location is near the city centre, many of the commercial signs found were written in English; for example, 'Steak Hut: Steak and Whatever', '*Bulir Padi* Resto', 'Live Cooking Djoejogan' and 'Lucky Mart Farma'. Such types of commercial discourses have symbolic rather than informational function because they symbolise foreign taste or associations of the products with their origins (Scollon & Scollon, 2003).

The last region is Tandes District. Citraland seemed to bring influence to the surrounding areas including Tandes District. The obvious evidence was the finding of one regulatory sign 'Reduce speed now' which was inappropriate for use in an area where the majority of the people speak the Indonesian language and Javanese. Many commercial signs were written in different languages, such as 'De *Bali*', '*Barokah* Bike', '*Klambie* Kids Branded', '*Optik Nirwana* Softlens Centre', '*Jazirah* Fit Centre' and '*Toko Shien Chuan*', marking the emergence of many different linguistic groups in this region.

3 CONCLUSION

The six languages identified in Surabaya's linguistic cityscape confirm that Surabaya is a multilingual speech community comprising different language groups. The Indonesian language was used predominantly in six districts; whereas, English dominated the linguistic cityscape in a more urbanised and polyglossic situation. Furthermore, the local languages found on the signs were limited because they did not serve a function as Indonesian or the English language did. Out of the seven districts, Lakarsantri District was found to be a more urbanised area inhabited by a more multilingual speech community. The multilingualism identified in this region influenced the nearest districts, such as Sambikerep and Tandes Districts. The proximity of the district to the city centre also influenced the varieties of languages used in the region, for example Sukomanunggal District. On the other hand, the older the region was and the further the region was from the city centre, the more likely it was to maintain the language as could be seen in Asemrowo District and some parts of Pakal District.

REFERENCES

A Dictionary of Architecture and Landscape Architecture. (2000).

Ben-Rafael, E., Shohamy, E., Amara, M.H., & Trumper-Hecht, N. (n.d.). *Linguistic Landscape as Symbolic Construction of the Public Space: the Case of Israel. In D. Gorter (ed), Linguistic Landscape: A New Approach to Multilingualism. Clevedon: Multilingual Matters.*

Cenoz, J., & Gorter, D. (2006). *Linguistic Landscape and Minority Languages. In D. Gorter (ed), Linguistic landscape: A New Approach to Multilingualism. Clevadon: Multilingual Matters.*

Fikriyah, S., & Wisnu. (2013). Perkembangan Kawasan Real Estate di Surabaya Barat Tahun 1970–2000 (Kontribusi Citraland dalam Perkembangan Kawasan Real Estate di Surabaya Barat. *Avatara e-Journal Pendidikan Sejarah*, 503–518.

Gorter, D. (2006). *Further Possibilities for Lnguistic Landscape: Enviromental Print, Code-mixing and Language Change. In D. Gorter (ed), Linguistic Landscape: A New Approach to Multilingualism. Clevon [etc]: Multilingual Matters.*

Huebner, T. (2006). *Bangkok's Linguistic Landscape: Environmental Print, Code-mixing and Language Change. In D. Gorter (ed.), Linguistic Landscape: A New Approach to Multi-lingualism. Clevedon: Multilingual Matters.*

Landry, R., & Bourhis, R. (1997). Linguistic Landscape and Ethno-Linguistic Vitality: an Empirical Study. *Journal of Language and Social Psychology*, 23–49.

Scollon, R., & Scollon, S. (2003). *Discourses in Place: Language in the Material World.* London: Routledge.

Torkington, K. (2009). Exploring the Linguistic Landscape: The Case of the 'Golden Triangle' in the Algarve, Portugal. Papers from the Lancaster University Postgraduate Vol. 3. *Linguistics and Language Teaching*. Portugal: LAEL PG 2008.

Visi dan Misi Kota Surabaya. (2009, 05 17). Retrieved 10 20, 2014, from fasilitasumumsby.wordpress.com.

Urban Studies: Border and Mobility – Kerr et al. (Eds)
© 2019 Taylor & Francis Group, London, ISBN 978-1-138-58034-3

Border or beyond: Dangdut jazz's reception and liminality analysis in the ITS Jazz community

F.Z. Putri & B.A. Sansoko
Universitas Airlangga, Surabaya, East Java, Indonesia

ABSTRACT: Music is one of the inseparable cultures in Indonesia. People with different backgrounds in Indonesia enjoy many genres of music. One of the music genres that is strongly identified with middle-lower social class is dangdut, while jazz is mostly identified with the middle-higher social class. However, the appearance of dangdut jazz (dangdut songs arranged as jazz) is a unique phenomenon. It raises two questions: (1) does dangdut jazz make dangdut accepted by jazz enthusiasts and (2) does dangdut jazz blur or emphasise the border between the two? In this study, we used Stuart Hall's reception theory and Victor Turner's liminality theory. The data were collected through interviews by first showing the "Fahmi Shahab—Kopi Dangdut (Jazz Cover)" video to the members of the ITS Jazz community. The results showed that the presence of dangdut jazz influenced jazz enthusiasts to be more open to dangdut music. However, instead of blurring the border, it emphasised the border between the two.

1 INTRODUCTION

Dangdut and Jazz are two cultural products that undergo a complex transformation in the history of Indonesian music. For example, in Indonesia, jazz turned into a music for the upper class, while historically, jazz is the music that pictures the struggle of African-Americans who felt discriminated against by the actions of white Americans (Rahadianto, 2010). Jazz in Indonesia in various colonial regimes, the Old Order and the New Order, continues to be reproduced as high-class music (Baulch, 2011) or known as the music of a gedongan (people who live in nice residences, based on KBBI).

Similarly, dangdut also undergoes a cultural transformation; dangdut was identified as a kind of polite music and was also considered the antithesis of fight rock music that is often accused of being wild and brutal. In fact, dangdut has become a medium that is effective enough to preach and spread religious values, especially Islamic teachings (Binus University, 2017). However, over time, dangdut has undergone a change of colour from being music, whose purpose was to preach, into music that is used as social parody, full of whiny or sensual lyrics, complete with sensual dancing by the singers; for example, Inul Daratista has emerged in the 21st century. Her controversial Goyang Ngebor dance was soon followed by many other types of sensual dance by new female singers. She was even banned by the king of dangdut Rhoma Irama, but survived and was determined to restore dangdut to its roots: mass music, which identified with the lower class (Gunawan, 2003).

The class division also cannot be separated from the colonial era of Indonesia. The colonial policy divided society into different social classes, and in its development, it had an impact on the development of jazz and dangdut as a cultural product. Jazz is more developed in large cities like Jakarta, Bandung, Surabaya, and Makassar. At the time, on a daily basis, jazz was also only played in star hotels (Rahadianto, 2010). Studies conducted by Irawati (1992, in Sutopo 2012:66) and Sudrajat (2003, in Sutopo 2012:66), for example, also confirm that jazz is mostly used by middle and upper classes, in contrast to dangdut, which is more enjoyed by the lower class in Indonesia. Nevertheless, dangdut is elastically able to adapt to the development of global music and has finally been born as a music genre in its own right.

In addition to the social border between the consumers of dangdut and jazz, there is also a border that divides the complexity between the two, with dangdut as simple, easy-listening music, and jazz as a more complex music than dangdut. The grammar of jazz is indeed relatively more complicated than other popular music (although not all jazz styles have intricate grammar, often those referred to as core jazz usually have intricate grammar) (Riwayanto, 2006). This causes jazz to sound strange and makes it difficult to listen among the consumers of dangdut, whereas dangdut sounds too simple and monotonous among the consumers of jazz.

However, a unique phenomenon has occurred, which is the emergence of dangdut songs, which have been transformed into a form of jazz by a variety of jazz enthusiasts in Indonesia. This phenomenon became the basis of this study that aims to determine the meaning of the existence of dangdut jazz on the border of dangdut and jazz. It raises two questions: (1) does dangdut jazz make dangdut accepted by jazz enthusiasts in ITS Jazz and (2) does dangdut jazz blur or emphasise the border between the two?

The writers selected members of the ITS Jazz community as the object of this research. This is because ITS Jazz is the first university-based community of jazz connoisseurs in Surabaya that accommodates students of ITS (Institut Teknologi Sepuluh Nopember), PENS (Politeknik Elektronika Negeri Surabaya), and PPNS (Politeknik Perkapalan Negeri Surabaya) who have an interest in jazz music. ITS Jazz was founded on November 10, 2011 with 300 members, and according to our database, it now has 152 active members. ITS Jazz also actively participates in many jazz events, including Jazz Traffic in 2016 and many other Jazz festivals.

2 THEORETICAL FRAMEWORK

The writers use two theories: (1) the theory of receptions developed by Stuart Hall in 1973, or better known by the term encoding–decoding, and (2) the theory of liminality developed by Victor Turner in 1967.

2.1 *Hall's reception*

The emergence of Stuart Hall's theoretical receptive theory was motivated by the dominance of the structuralist thought in cultural studies and media studies, especially about how meaning formed among the audiences of media content. The dominant thought developed at that time was that meaning in the audience is formed based on the message presented in the contents produced by the particular media. It is then criticised by Hall while at the same time trying to build a model of message creation by the producers and construction of meaning by its audiences. Hall (1980:118) then called this model of meaning and message construction encoding/decoding.

Hall (1980:120) used four stages in the communication process: production, circulation, distribution, and reproduction. These four stages are autonomous from one another in spite of their interrelationships. Furthermore, these four stages can be categorised into further two big stages: (1) production and circulation processes in the production stage of the message and (2) the process of distribution and reproduction in the message reception stage.

Because these four stages are autonomous between one another, the meanings formed at the production stage are not always the same as at the reception stage. Encoded messages composed by the message maker do not always give the intended meaning to the recipient of the message. This is influenced by various factors such as social structure, message receiver's class, and message receiver's ideology. Therefore, Hall (1973) formulates three possibilities that can occur in the meaning when messages are decoded by the audience:

1. Dominant-hegemonic reading. This position is that of the audience's acceptance of the message conveyed intact with the desired meaning of the message maker. Audiences who

take this position believe the reality that occurs in accordance with what is delivered by the message maker.

2. Negotiated reading. This position means that the audience acknowledges a general condition that is true and in accordance with the message delivered. It is just that the audience in this position believes that there are detailed situations that then become an exception to the general conditions conveyed in the delivered message.
3. Oppositional "counter" hegemonic reading. This position occurs when the audience understands the meanings given by the message giver, but decodes the message in a conflicting way.

2.2 *Turner's liminality*

Reviews on liminality need to be preceded since the term was first used by Arnold van Gennep in 1909, an anthropologist, who wrote his observations of the tribes in different parts of the world in his book, *Rites of Passage*. The term rites of passage means the rituals commonly performed by a particular ethnic group or group which determines that a person or several persons have successfully passed from the initial stage to the next stage of life. He explains that every society has rites of passage that represent a moment for a person to move and be in the transition or liminal space itself (Thomassen, 2009: 6).

In contrast to van Gennep's liminality that is specifically applied in tribal or religious contexts, Turner extends the scope of its liminality. Limitations in Turner's sense not only include rituals performed in tribes, sects, or religions to mark the passage of a person from the first stage to the next, but also incorporate cultural performances, such as carnivals, festivals, theatres, films, and music as things that can form into a liminal period (Turner 1988: 101). Turner uses the term liminal for rituals and liminoids (or liminal-like) for cultural appearances. Both are still things that can be betwixt and between conditions and have the function of being a moment for the individual in it to muster a new power, creativity, innovation, or viewpoint in the new way of self-presenting after passing through the liminal phase.

Turner attributes the liminality in cultural appearance to the concept of subjunctive mood, i.e., the mood of a verb used to express supposition, desire, hypothesis, possibility, etc., rather than to state an actual fact. Therefore, there is a characteristic ambiguity that is evident in the area or the liminal period. The structure can be reversed, mocked, dyed, or even cursed (Turner, 1988: 101). Turner (1988: 95) states that the liminal phase is anti-structure in the sense that it is opposite to the existing structure and becomes a neither-here-nor-there existing condition so that it is a realm of pure possibility. There is a chaotic or anarchic condition in it that soon becomes the struggle arena for the formation of meaning, the determination of attitudes, and the dynamics of politics.

3 METHODOLOGY

This study was conducted using a qualitative research methodology. Data collection in this study was first conducted by determining the subject of the study. The study subject involves active members of the ITS Jazz community. Then, the data were collected through in-depth interviews conducted through two media. The first medium was to use phone calls recorded with the subject's consent. The second medium was using the chat feature via an online chat app. Interviews were not conducted by meeting as would be usual because some study subjects rejected this either because they were too shy or because they did not have time to spare.

The results of the interview were used to answer the formulation of predetermined problems through the theory of encoding/decoding and the theory of liminality. The first research question was answered by using the theory of encoding/decoding. This theory was used first by determining the message assumption that the message maker is trying to communicate in producing jazz dangdut content. The message is that jazz dangdut is created to improve the acceptance of dangdut music among jazz lovers. Afterwards, the keywords from the interview were determined and answered to check whether the audience increasingly accepts the existence of a dangdut song, ordinary, or even denies its existence. The answer of the resource persons will be attributed to dominant position, negotiated, or opposition.

In order to answer the second research question, the writers used the combination of encoding/decoding and liminality theory. The encoding/decoding theory was used by determining three message assumptions that the message maker is trying to communicate in producing jazz dangdut content. Once determined, the keywords from the interview results were determined, as well as the content of the answers in one of the three positions that was provided. After that, analysis using the theory of liminality was done by finding the elements of the liminal conditions, e.g., ambiguity, which are outside the structure.

4 DISCUSSION

There were 17 informants to begin with. All of them were active members of the ITS Jazz community and had the ability to either sing or play instruments in jazz. In this study, there was a general interpretation of the meaning of the existence of dangdut jazz. First of all, from the 17 informants, 3 informants stated that they neither liked nor disliked dangdut, 3 informants said that they did not like dangdut, 1 informant said that he only liked old dangdut songs (which are full of religious or spiritual messages) and disliked the latest dangdut songs with their non-educational lyrics, and 11 other informants claimed to love dangdut. However, behind the variation in the answers, when the researchers showed the dangdut jazz video of Kopi Dangdut by Fahmi Shahab, which was covered in jazz by one of the jazz bands, 16 of the 17 informants said they liked the dangdut jazz version, while 1 informant said "nothing special".

4.1 *Encoding/decoding*

4.1.1 *Dominant reading*

Even informants agreed that jazz dangdut made them (and possibly other jazz consumers) be more accepting of dangdut. This is because when listening to the jazz dangdut version, they were curious to listen to the original version of the song (dangdut), and they also agreed that after listening to the original song, there was a sense of interest that arose. In fact, two of the informants claimed that they had already covered the dangdut songs into jazz, and one informant, with the initial Va, said that he had enough dangdut songs sung by Nella Kharisma after seeing a dangdut jazz version of Kopi Dangdut.

W, one of the informants, stated that he had covered the dangdut song into the jazz version with his band several times. The saxophone player said that there is a similarity between the dangdut song and Bossanova jazz, so the interest and acceptance came within him. "Nice, well, it is kinda similar to Bossanova", he said. According to W, there are also many jazz bands that do re-arrangements of dangdut songs into jazz versions, and this was proved when he joined Java Jazz in 2015. "There were about 5 bands who performed dangdut jazz in Java Jazz 2015", W said.

4.1.2 *Negotiated reading*

Three informants expressed a negotiative attitude towards the existence of jazz dangdut songs; even after listening to the dangdut jazz version of Kopi Dangdut, there was only a slight interest amongst them in dangdut. The three informants agreed that jazz dangdut could make jazz consumers have more acceptance of dangdut; however, it would be due to curiosity only. "Nothing special, because dangdut and jazz have different market, we have different taste, so it's not easy for me personally to like and accept dangdut, but if it's just curious, yes I was kinda curious to the original song, but that's it," said one informant, named O.

Other informants named R and Vi also provided similar comments that dangdut jazz would not affect their level of interest or acceptance in dangdut songs. They said that they really liked the jazz version, but they were also curious about the original version (dangdut). R is a drummer, and he said that the drum rhythm in dangdut sounds more simple and, indeed, makes it more comfortable to hear especially in the ears of people who are not used to faster and complex rhythms in jazz. Vi, who is a vocalist, also agreed that dangdut songs are easy to listen to, but she does not like to sing dangdut songs because of its cengkok that does not fit her style.

4.1.3 *Oppositional reading*

Meanwhile, six other informants expressed disagreement about whether dangdut jazz could raise their acceptance towards dangdut. One of the informants said the only thing they listened to was dangdut in its jazz version, and he would not be interested in hearing the original version, let alone being a dangdut consumer. "I'm okay with dangdut, I won't run away if any of my friends play them, and I like this dangdut jazz but I didn't bother to look for its original version. I only liked the jazz version", said one informant named M.

Another informant named Oz also said that it can be very difficult for jazz consumers to accept dangdut simply because there is a version of jazz dangdut. According to him, jazz consumers will still hear the jazz version only. "It's hard, personally, I only liked the jazz version, I don't really care about the dangdut version", said Oz. The same thing was also conveyed by another informant named Da that it would not affect his level of acceptance towards dangdut. Moreover, according to him, there have been more dangdut songs recently that have non-educational lyrics, not to mention the singers who dance or dress sensually to attract a male audience. That impression will never leave Da, even though there is a dangdut jazz version. He will still listen to dangdut jazz, just because the version is already in the form of a jazz arrangement.

4.2 *Liminality*

Turner (1988: 95) states that the liminal phase is anti-structure in the sense that it is opposite to the existing structure and becomes a neither-here-nor-there existing condition so that it is a realm of pure possibility. Researchers suspect the existence of this jazz dangdut to be a new liminal form that comes out from both Jazz and dangdut's structures and creates a new sub-genre or even a new music genre. However, on the basis of the results of the interviews, there is no presumption that dangdut jazz is a sub-genre or new genre. If anyone agreed, it was stated as a possibility only.

An informant named Z said that, at first glance, dangdut jazz looks like something new and unique and can create a new space in the middle of the border between dangdut and jazz. However, according to him, basically, dangdut jazz is just a matter of a playing technique only. "Sub-genre or new genre, I don't think so. It's just a matter of different playing techniques, dangdut in a form of jazz. To sum up, it's only a cover version", said Z. According to another informant named Da, for people who do not know dangdut, they will certainly not realise that it is a dangdut song in the form of jazz, and they will think that the song is originally Jazz. Therefore, dangdut jazz will fail to be unique and will be no different from original jazz.

Meanwhile, four other informants still agreed that there is a possibility of dangdut jazz becoming a new sub-genre because of its uniqueness, and one informant named Na thought it would be something new. However, if it was defined as a new genre, he thought it would be difficult because there would be at least one genre, either jazz or dangdut, that would be sacrificed (e.g., the song would be dominant in jazz or dangdut) in order to make it accepted by jazz consumers or dangdut consumers. Meanwhile, according to the informant named Az, this is also very dependent on market demand. "As a jazz consumer, I think I prefer the jazz version, only maybe the original song came from dangdut", said Az.

Overall, 12 informants expressed disagreement that jazz dangdut was or had the potential to be a new sub-genre or new music genre, while 4 other informants agreed with the possibility of dangdut jazz being a new sub-genre. Therefore, jazz dangdut cannot be regarded as liminal, or entering something new that combines the characteristics of dangdut and jazz without losing any of its elements.

4.3 *Border*

Although most informants claimed to have more acceptance towards dangdut due to the existence of dangdut jazz, we managed to reveal that it did not help to blur the border between dangdut and jazz. A total of 10 informants stated that dangdut jazz actually strengthens the border between dangdut and jazz, because they still tend to prefer to listen to jazz dangdut with a strong jazz harmony. An informant named Az said that,

"Actually I like the idea that dangdut jazz blurs the border between dangdut and jazz because indeed as musicians we will learn about both songs. However, the idea of strengthen the border is also make sense, because I personally think it will be weird if the dangdut jazz still shows its dangdut style. I like dangdut jazz with its dominant jazz style."

The same information was also provided by an informant named Oz, "Dangdut is dangdut, jazz is jazz, I kinda like dangdut jazz because … it's basically jazz haha". This is also supported by the statement of informant R, who said, "I guess people who like dangdut will also dislike dangdut jazz with dominant jazz. It's the same for us. So yeah I guess I agree if it's strengthen the border."

Meanwhile, five other informants stated that dangdut jazz could blur the border between dangdut and jazz, creating possibilities for the development of music itself. The informant named W said, "In my opinion, this dangdut jazz creates two benefits: (1) introducing jazz to dangdut consumers, (2) blurring the border between the two, because it could make jazz consumers to accept dangdut better and develop their music together". Another informant named Di said, "If one's bounded to one genre of music, they would never learn. Music will never be as variative as we know now."

However, two other informants could not give an answer, or rather refused to state their answer, because according to them, music is something universal and has no limits. This is what the informant N said: "Music is universal, everything is based on the tastes of each person, and I believe one's taste is also influenced by their surroundings. In other words, it's fluid. I refuse to say that there is even a border in the first place."

5 CONCLUSION

In conclusion, most of the informants agreed with the message that the existence of dangdut jazz helps jazz consumers to have better acceptance of dangdut songs. Seven informants went with the dominant reading by agreeing that dangdut jazz helped them to accept dangdut, three informants were in a negotiated reading position, whereby, according to them, dangdut jazz does not significantly affect their acceptance level towards dangdut jazz, and six other informants were in an oppositional reading position, whereby, according to them, dangdut jazz has no effect at all on raising their acceptance of dangdut songs.

Dangdut jazz cannot be categorised as liminal, even though there is ambiguity due to the mixed genre between dangdut and jazz. Overall, 12 informants expressed disagreement that jazz dangdut was or had the potential to be a new sub-genre or new music genre, because good dangdut jazz according to them would be a song with a strong sense of jazz only. On the contrary, four other informants agreed the possibility of dangdut jazz being a new sub-genre, yet some still argued that its existence also depended on the market demand.

Although dangdut jazz made most jazz consumers in ITS Jazz have better acceptance of dangdut songs, its existence actually strengthens the border between the two. A total of 10 informants thought that dangdut jazz indeed strengthened the border because jazz consumers will only like dangdut jazz when it has a strong sense of jazz. Five other informants stated that they agreed that the emergence of dangdut jazz would help to blur the border, as it would introduce jazz to dangdut consumers and vice versa. Meanwhile, two other informants refused to state an answer because they felt that music is universal in the very first place and there is no such thing as a border.

REFERENCES

Binus University. 2017. (Kumpulan Sejarah) Sejarah dan Perkembangan Musik Dangdut di Indonesia. *Binus University*. Retrieved November 20, 2017 from http://scdc.binus.ac.id.

Gunawan, F.R. 2003. *Mengebor Kemunafikan—INUL, Seks dan Kekuasaan*. Jakarta: Kawan Pustaka.

Hall, S. 1980. Encoding/Decoding. In S. Hall, D. Hobson, A. Lowe, & P. Willis (eds). *Culture, Media, Language*: 117–128. London: Routledge.

Basundoro, P. (2012). *Sejarah pemerintah kota Surabaya sejak masa kolonial sampai masa reformasi (1906–2012)*. Departemen Ilmu Sejarah, Fakultas Ilmu Budaya, Universitas Airlangga.

Rahadianto, O. 2010. *Warta Jazz*. Retrieved November 15, 2017 from Pengantar Menuju Studi Komunitas Jazz di Indonesia: www.wartajazz.com.

Riwayanto, D. 2006. *Warta Jazz*. Retrieved November 12, 2017 from ALIENASI HARMONI JAZZ DI "TELINGA DANGDUT" (menanggapi artikel Dr. Heru Nugroho tentang 'Memahami Musik Jazz'): wartajazz.com.

Santoso, E.B. (2004). "Kota Surabaya" Sebuah Tinjauan Dalam Perspektif Historis (Doctoral dissertation, UNIVERSITAS AIRLANGGA).

Sutopo, O.R. 2012. Transformasi Jazz Yogyakarta. *Jurnal Sosiologi MASYARAKAT*: 65–84.

Thomassen, B. 2009. The Uses and Meanings of Liminality. *International Political Anthropology*: 5–27.

Turner, V. 1988. *The Anthropology of Performance*. New York: PAJ Publications.

refused to give an answer because they [illegible] that most [illegible] there is no such thing [illegible]

[illegible]

Urban Studies: Border and Mobility – Kerr et al. (Eds)

Culture shifting from wearing sarong to wearing trousers amongst the people of Bangkalan Madura

I. Husna
Universitas Airlangga, Surabaya, Indonesia

ABSTRACT: In this study, we analyse the cultural shift amongst the Maduranese people of Bangkalan, which is an Islamic district. Islamic scholars, often called "Santri", generally wear sarong in daily activities as well as special occasions. It is the primary wear for Santri. In this globalisation era, the tradition of wearing sarong by the Bangkalan Maduranese people is being replaced by wearing trousers on formal occasions, such as weddings. Therefore, here we focus on two important questions: (1) how actually is the practice of wearing sarong in Bangkalan Maduranese people? (2) How does the tradition of wearing sarong mark a culture shift to the wearing of trousers amongst Bangkalan Maduranese people? The study was conducted using an interview method and the concept of "Distinction and Taste" of Pierre Bourdieu.

Keywords: Madura, Santri, Sarong, Trousers, Pierre Bourdieu

1 INTRODUCTION

There have been no accurate data yet on when Madura island was inhabited. According to an expert in the field, it was in ca. 929 AD (Zainalfattah, 1951: 7–13), when the island was discovered. At that time, a princess of a kingdom on the island of Java named Mendang Kamulan, without clear cause, was found to be pregnant. Knowing the condition of her daughter, the king was furious and ordered a Patih (minister) named Pranggulang to kill her. Yet, each assassination attempt failed and eventually the princess gave birth to baby boy named Raden Sagoro. Meanwhile, Patih Pranggulang dared not return to the palace and changed his name to Kiyai Poléng. According to some resources, Raden Sagoro and his mother were then swept away into the sea in a ghitèk (woven wood serving as a boat) and finally stranded on land, which was later known by the name of Mount Gegger (Bangkalan district). This land is called "madu oro", which means the corner of a larger area. The word Madura was coined from "madu oro". Raden Sagoro and his mother were referred to as the first inhabitants of the island of Madura.

The life philosophy of the Madurese people is inseparable from the Islamic values they embrace. It is an undeniable sociological fact that almost all Madurese are Muslims. Their obedience to Islam is an important identity amongst the Madurese people. This is indicated in their traditional clothes of *samper* (long cloth), *kebaya*, and *burgo*' (veil) for women, and *sarong* and *songko'* (cap) for men, which have become symbols of Islam, especially in rural areas (Rifai, 2007: 446). Therefore, Islamic identity is a highly important aspect of Madurese society. In accordance with the teachings of Islam they adopt, the Madurese life philosophy leads them to live life for the achievement of happiness in this world and the hereafter. For this reason, *aèkthèyar* (endeavour, effort) is highly crucial for the Madurese because this approach will increase the likelihood of them achieving all their desires and goals (Munir, 1985: 228).

Bangkalan is one of the districts within Madura, which has a tradition of *pesantren* (Islamic boarding schools). This is due to the role of *pesantren*, which have been existing for many generations. One of the famous *pesantren* is Pesantren Kademangan founded by Mochammad Kholil, who was the teacher of great scholars of the nation such as Hasyim

Asy'ari (founder of Pesantren Tebu Ireng and the first *Rais Am Syuriah* for Nahdlatul Ulama organization) and Abdul Karim (founder of Pesantren Lirboyo). In pesantren tradition, it is mandatory for *santri* to wear sarongs. The obligation to use sarongs is usually expressed in the written rules. Yet, there are also *pesantren* that do not put it in the written rules because the use of sarongs is regarded as a common practice and binding for *santri* who gain knowledge in the *pesantren* environment. Sarongs in the *pesantren* environment are not only used as a medium to cover the *aurat* (private body parts in Islam) and a symbol of politeness, but they are also part of the effort of the *pesantren* to familiarise *santri* with the concept of *neriman (self-content)* so that the *santri* will become humble and independent individuals.

Historically, *pesantren* have documented various historical events in Indonesia. Since the beginning of the dissemination of Islam in Indonesia, *pesantren* have become the main witnesses and important instruments for Islamisation. The development and progress of the Islamic society in Indonesia cannot be separated from the role played by *pesantren*. The huge meaning of *pesantren* in the course of Indonesia as a nation must be acknowledged. Moreover, *pesantren* have been regarded as educational institutions strongly rooted in the original culture of the nation (Asrorah, 1999: 184).

Pesantren represent one of the examples of old culture, which still maintains its existence in the educational world in Indonesia. It is embedded in the minds of *santri* that one form of *jihad fisabillah* (to strive in God's way) is by seeking knowledge in accordance with the rules of the *pesantren* they live in. In the Bangkalan society, people live their daily lives from this perspective, that is, thinking about seeking knowledge to the use of sarongs in their daily activities. In social life, people always harmonise themselves with the surrounding environment in an effort to adjust themselves to improve their quality of life. Therefore, a society is actually an adaptive system as it acts as a container to meet various interests and, of course, to survive. Nevertheless, the community itself also has a variety of needs that must be fulfilled so that it can survive (Koentjaraningrat, 2005: 122). When viewed from the geographic perspective, on the island of Madura, especially Bangkalan, where the researcher observed the community, it is not surprising how the *santri* culture is well received. One of the factors is the weather in Madura, which is hot and arid; thus, the community prefers light and multifunctional clothing articles. *Santri*, who had received an explanation, would become accustomed to wearing a sarong and become examples for the Bangkalan society and an identity for the society of Bangkalan Madura to this day. Bangkalan has already hegemonised *santri*'s ways of thinking so that religiosity is the main value and the centre of their daily activities.

In the modern society, a distinction is often made between urban communities and rural communities, which is essentially gradual and it is rather difficult to provide clear limitations on what is referred to as urban because of the relationship between population concentration and the social phenomenon called urbanism (Soerjono Soekanto, 2006: 23).

2 METHODOLOGY

This study was conducted in Bangkalan, Madura. The location selection was based on the notion that the Bangkalan community was imbued with religious values in their everyday life, including the wearing of sarongs for daily activities. The research samples included the villages of Bates, Rabesen, Tanah Merah, Durjan, Sambiyan, Sen-Asen, Campor, and Bungkak. The data collection methods were: (1) interview, which is a process of questions and answers between the researcher and informants aiming to gain verbal descriptions of their views and (2) observational technique where the researcher uses direct observation and records the use of sarongs in the Bangkalan society.

3 DISCUSSION

The Madurese traditional clothing in Bangkalan is of three types: clothes for teenage nobles, adult nobles, and ordinary people. The clothes for teenage nobles consisted of two parts: top and bottom. The top consisted of odheng, the head cover made of batik cloth with black and brown colours, and the motifs were usually flowers and flames. In Madura, they were better known as stor joan, bera' songay, or acoh biru. Next, there was rasughan totep, which was a kind of suit made of thick fabrics like wool, brightly coloured with plain designs. The shape was similar to the traditional clothes from Central Java. The last item was rompi, shaped like an undershirt, made of satin, coloured in brown or black, plain, and with no certain design.

Meanwhile, at the bottom, clothing consisted of several items as well. The first was ¾ pants made of satin, coloured in grey or black with a plain motif. The second was samper kembeng, which was a long cloth made from batik. The third clothing item was sap-osap, i.e., gloves made of white cotton fabric. The fourth clothing item was setagen, a woven fabric in red, white, and black colours. The next was sabbuk katemang raja, a rather-wide cowhide belt in black or brown colour with plain motifs. The last clothing item was terompah, i.e., cowhide footwear, shaped like sandals or closed slippers.

The clothes for adult nobles also consisted of two parts: top and bottom. The top consisted of several clothing items. The first was odheng Peredan, the head covering made of batik cloth in black and brown colours, and the motifs were usually flowers and flames in Madura better known as stor joan, bera' songay or toh biru. The second piece was rasughan totep, which was a kind of suit made of thick fabrics like wool in white, grey, or black colours and a plain motif. The shape was similar to the traditional clothes from Central Java. A traditional weapon (keris) was also used with two different handle forms of senteki (made from black wood) and lortop (made from palm wood). Root bracelets, rings, and chest watches were the ornaments for a male outfit. The last clothing item was rompi, which was shaped like an undershirt, made of satin, coloured in brown or black, plain, and with no certain design.

Meanwhile, the bottom consisted of several clothing articles as well. The first was ¾ pants made of satin, coloured grey or black with a plain motif. The second was samper embeng, which was a long cloth made from batik. The third was sap-osap, i.e., gloves made of white cotton fabric. The next was setagen, a woven fabric in red, white, and black colours. The fifth was a rather-wide cowhide belt in black or brown colour with plain motifs. The last one was terompah, i.e., cowhide footwear, shaped like sandals or closed slippers.

The traditional clothing for the common people consisted of two parts: top and bottom. The first item on top was odheng santapan, the head covering made from ordinary batik cloth, with a blue lake motif or soga red (storjoan). Another type of head covering was odheng tapoghan, which was made from ordinary batik cloth with floral or flame motifs in soga red colour. The next item was pesa'an, which was a suit jacket, formerly made of cloth from China named lasteng fabric. Nowadays, the fabric used was tetoron with a plain motif and black colour. The last top item was a t-shirt made from satin and coloured in red and white.

Meanwhile, the first bottom clothing item was gomboran, which was trousers characterised by a wide waist size. When stretched, it looked like a sarong with a plain black motif. The next item was sarong palekat, which resembled a sarong in general, made of silk or cotton. The silk one was commonly made in striking colours with golden thread, while the one made of white cotton was decorated with blue or green plaid. The next item was sabbuk katemang raja or sabbuk katemang kalep, which was a wide cowhide belt in black or brown colour, adorned with a pocket in front of it to save money. The weapons used included a dagger, piol (small dagger), clurit (sickle), blade, etc. The last item for the bottom clothing was terompah, i.e., cowhide footwear, shaped like sandals or closed slippers with ropes as clamps. The function of these clamps was to act as binders between the thumb and other fingers. They were worn by ordinary people with a practical function when viewed in its simple, free, and plain form. These items of clothing could not only be used for official events, but also for daily activities at home. Sarongs served not only decoration purposes, but also its main function as an item of clothing for praying.

The Madurese community, in general, and especially in Bangkalan district, were mostly educated in pesantren. They never socially detached themselves from the sarong-wearing tradition. This is not an exaggeration because the influence of the kyai (head of pesantren) was deeply rooted in the society, especially the religious cultural values. During the religious education in pesantren, the kyai plays a very important role in moulding various forms of life behaviour, religious orientation, and the society. Even after the santri leaves pesantren, the relationship between the kyai and his former santri is retained, deep in their hearts and in their social lives (Mansurnoor, 1990: 254). The practice of sarong-wearing by santri was eventually brought to the life outside of the pesantren by sarong-wearing by the public who were not santri initiated through a process of socialising tradition. Socialisation is a process whereby human beings strive to absorb the social culture in their neighbourhood (Sandersoon, 1995: 101). Even social scientists believe that it is through this process that the older generation devotes much of their time to transferring their culture to the younger generation. By relying on tradition and integration, a culture will be preserved (Kleden, 1986: 69–86). Thus, naturalisation of homogeneity of the sarong-wearing tradition in the Bangkalan society would be formed by itself through various transmissions of culture from parents to their children.

The use of sarongs by men in everyday activities was a phenomenon commonly seen in the life of Bangkalan people, especially in the countryside. In this study, the activities generally existing in Bangkalan were divided into three types: (1) religious activities (mauludan, isti-goshah, tingkepan); (2) formal activities (marriage, family gathering, or othok-othok); and (3) daily activities. According to Ismail Bade (58), one of the respondents from Bates Village, from the 1960 to the 1970s, most people in his village wore a sarong for daily, religious, or formal activities. For a rural community, in general, the wearing of trousers for religious or official activities was considered less proper and less polite. He recounted that on his akad (wedding vow) and wedding reception, he wore a sarong as his main outfit. This is in line with Busadin (65), another respondent from Sen-Asen Village. Another respondent from Durjan Village, Karib (52), remembered that before working in Surabaya in the 1990s, he still often wore a sarong for religious, formal, and daily activities. After working in Surabaya, however, he admitted that the tradition of wearing a sarong in everyday life was less suitable for the culture in Surabaya because of the stereotype that sarong-wearing is outdated in the urban community. Gradually, wearing a sarong in daily activities and formal events became less common and was replaced by the use of trousers. Nonetheless, the use of sarongs for religious activities was still a must for him. The change of habit was finally carried back when he returned to Madura, and this habit indirectly affected his family because he considered his status of working in the city to be more modern and more up to date.

The interview then moved on to the respondents aged 30 years and below. One of our respondents from Desa Bungkak, Rosi (29), stated that in his daily life, he had the option to wear a sarong or trousers, although he admitted that wearing a sarong was more dominant because he graduated from a pesantren. In his religious activities, a sarong was mandatory for him. Uniquely, according to his story, at his wedding in 2015, he wore a sarong during the akad event but shifted to trousers during the wedding reception. He admitted that this was a cultural change in the marriage process in the region in modern times that he and his family could accept. A similar opinion was also conveyed by Maulana (25) from Tanah Merah Village. Another case was expressed by Dayat (17) from Bates Village who preferred wearing trousers in his daily and formal activities. According to him, trousers were more appropriate for use by young people for daily or formal activities than wearing sarong, which he thought to be outdated. In addition to being more modern, wearing trousers was considered to be more practical and to allow easier and freer movement in his activities. When the researcher asked whether there was a cultural transition process from his father in the use of sarong in everyday life, he acknowledged that it had happened. His father taught and gave an example in the use of sarongs for religious, formal, and daily activities. Nevertheless, he recognised the influence of his friends and social media on his habits and his views on the use of sarongs for daily and formal activities. Yet, he still wore sarongs for religious activities, because he thought that the use of sarongs was more appropriate and polite for such an activity.

The interview results were in accordance with Pierre Bourdieu's thoughts about distinction and taste, which stated that the opening of a tradition can be seen from the habitus (environment) it originated from. Habitus is a kind of disposition system or a condition related to the existence of a class producing the habitus. Disposition systems could be time-proof and could be inherited; structures could be established and intended to function as forming structures (Bourdieu, 1980: 88–89). As the concept of habitus was successful for uniting the problems of the dichotomy between individual and society, social agents such as Bangkalan society, mostly students, would be more accustomed to dressing by wearing sarongs for every event. These santri's behaviour was an indirect part of the tradition in the community life so that the Bangkalan Madurese society accepted the tradition of sarong-wearing as a commonplace behaviour. However, the habitus provided a shift and distinction between the elderly Madurese people and the young people even within one family. This was affected by some elements such as the influence of the external environment entering the Bangkalan society culture in the use of sarongs as mentioned above.

The shift affected the taste of the people in Bangkalan Madura. As stated by Bourdieu, taste is a reflection of events. It is divided into three groups: the bourgeoisie, the petty bourgeoisie, and the producers (Bourdieu: 1994: 2). In terms of sarong wearing, the application of such a division in Bangkalan Madura society was evident from the brand of sarong used up to the trouser brand worn. Thus, an understanding of the meaning and material of a sarong for the society in Bangkalan could be reflected. The taste of Bangkalan Madura society, especially amongst the older people, tends to keep the tradition with the use of sarongs, which was different from the taste of the young people in Bangkalan who preferred trousers as their formal event clothing as they were more fashionable and more modern.

4 CONCLUSION

Culture is a dynamic entity that changes with time. Similarly, tradition also changes with time. The people of Bangkalan Madura held their religious values so high that they always obeyed it and lived their lives by the religious teaching. On the basis of this premise, sarongs became a characteristic of santri (the Islamic students) who held higher religious knowledge so that they became the role models for the society of Bangkalan Madura and thus sarongs were used in everyday life and events in the area.

In this study, we incorporated the results of interviews in various villages in Bangkalan district and found that the Bangkalan Madurese society accepted the shift from sarongs to trousers amongst old and young people. Nonetheless, young people were more open to accepting and wearing pants in their everyday lives. This was due to sophisticated information and technology so that it could change the way of thinking of young people in Bangkalan as long as it did not corrupt the tradition that had been followed for decades. The shift could also identify that the tastes of young people and old people of Bangkalan Madura society have become different.

REFERENCES

Asrorah, H. (1999). *Sejarah Pendidikan Islam.* Jakarta: Wacana Ilmu.

Bourdieu, P. (1980). *Le sens pratique.* Paris: Minuit.

Bourdieu, P. (1994). *Raisons patiques. Sur la theorie de l'action.* Paris: Seuil.

Kleden, I. (1986). Membangun Tradisi tanpa sikap Tradisional: Dilema Indonesia Antara Kebudayaan dan Kebangsaan. *Prisma, 15*(8), 69–86.

Koentjaraningrat. (2005). *Pengantar Antropologi 1.* Jakarta: Rineka Cipta.

Mansurnoor, L. A. (1990). *Islam Is An Indonesian World: Ulama' of Madura.* Yogyakarta: Gajah Mada University Press.

Munir, M. (1985). Adat Istiadat yang berhubungan dengan upacara dan ritus kematian di Madura. In Koentjaraningrat (Ed.), *Ritus Peralihan di Indonesia.* Jakarta: Balai Pustaka.

Muryadi, D. (2006). ISLAMISASI DI PULAU MADURA: SUATU KAJIAN HISTORIS (Doctoral dissertation, UNIVERSITAS AIRLANGGA).
Pakaian adat tradisi Madura, Kabupaten Bangkalan. Pria Remaja. (n.d.). Retrieved November 17, 2017, from https://jawatimuran.net/2013/06/13/pakaian-adat-tradisi-madura-kabupaten-bangkalan-pria-remaja/.
Pakaian Tradisional Pria Dewasa, Madura Kabupaten Bangkalan. (n.d.). Retrieved November 17, 2017, from https://jawatimuran.net/2013/06/13/pakaian-tradisional-madura-kabupaten-bangkalan/
Rifai, M.A. (2007). *Manusia Madura*. Yogyakarta: Pilar Media.
Sanderson, S.K. (1995). *Sosiologi Macro: Sebuah Pendekatan Terhadap Realitas Sosial*. Jakarta: Raja Grafindo Persada.
Soekanto, S. (2006). *Sosiologi Suatu Pengantar*. Jakarta: PT. Raja Grafindo Persada.
Sugiarti, E. (2009). MARGINALISASI WANITA MADURA. Jurnal Sejarah Lontar, 6(1), 40–57.
Wiyata, A.L. (2013). *Mencari Madura*. Jakarta: Bidik Phronesis Publishing.
Zainalfattah, R. (1951). *Sedjarah Tjaranya Pemerintahan di Dareah-daerah di Kepulauan Madura dengan Hubungannya*. Pamekasan: The Paragon Press.

Urban Studies: Border and Mobility – Kerr et al. (Eds)
© 2019 Taylor & Francis Group, London, ISBN 978-1-138-58034-3

Linguistic landscapes: A study of human mobility and identity change

Ketut Artawa
Udayana University, Denpasar, Indonesia

Ni Wayan Sartini
Universitas Airlangga, Surabaya, Indonesia

ABSTRACT: The focus of this paper is on the Linguistic Landscapes (LLs) of Kuta Village as one of the tourist destinations in Bali. The data for this study are in the forms of photos of outdoor signs taken from the research location, and the data obtained by interviewing a community leader and other informants were analyzed based on the LL theory and then interpreted based on the concept of 'market ideology'. The results of the analysis showed that the languages used in these outdoor signs revealed a diglossic situation. In this context, the Balinese language as a symbol of local ethnic identity is marginalized. The results also showed that Balinese people in the research location tend to choose Indonesian and foreign languages to communicate in everyday life. This can be seen as a sign showing how strong is their desire to have the image, prestige, and power owned by those people who can speak those languages. This phenomenon can have two impacts: positive and negative. One of the positive impacts is that the young Balinese have a strong desire to master foreign languages. In this way, the young Balinese will have a better linguistic capital. With this linguistic capital, it is easier for them to find a job and get material benefit (money) that can be used to build image and power in many arenas of life. One of the negative impacts is that their knowledge of the local wisdom expressed in the local language is decreasing.

Keywords: linguistic landscape, linguistic capital, public signs

1 INTRODUCTION

The fast advancement in communication and technology has turned our present world into a global village. Nowadays, communication has become much easier than it was before. In this context, language plays a vital role in such communication. In any global society the use of foreign words is inevitable. One of the most influential international languages is English. English has a kind of influence on most aspects of everybody's daily life. English language is translated, borrowed, or used as it is in business. Thus, language is a key instrument for human social interaction. It is a means of direct communication between participants of communication either as a written or spoken language. It seems that many of us are not aware of the situation where language has also become a means of communication between human beings and their environment. We communicate with a variety of signs that we encounter in our environment. Signs that we have created serve as an integral part of our lives. The signs may include street names, place names, traffic signs, billboards, or other signs. If we look carefully at our environment, it is clear that our environment is able to speak to us all through the languages that are used in each of these signs. People involved in business use their shop signs, as they constitute a vital area in their business, to display the names of goods and services they deal with. The research location is Kuta district, which is located in Badung Regency Bali Province. Kuta district is divided into three sub-districts, namely NorthKuta sub-district, Kuta sub-district, and SouthKuta sub-district. North Kuta has six villages, which are Kerobokan Kelod, Kerobokan, Kerobokan Kaja, Canggu, Dalung, and Tibubeneng. Kuta sub-district has five villages, namely Kuta, Legian, Seminyak, Tuban, and Kedonganan; and SouthKuta sub-district has six villages, Benoa, Jimbaran, Kutuh, Pecatu, Tanjung Benoa, and Ungasan. The research location is in Kuta Village, known as *Desa Adat* Kuta, which is located in Kuta sub-district, and the focus is to elicit and identify the presence of commercial outdoor signs. Kuta has been chosen to be the research location because Kuta Village is the most famous tourist destination.

There are two types of data used in this study: the data obtained through an interview with a traditional community leader and other informants and other data in the form of photos of outdoor signs taken from the research location. The data in the form of photos of outdoor signs include street names, traffic signs, place names, and billboards. The location of these signs is all the main streets of Kuta, namely *Jalan* Raya Kuta, *Jalan* Kartika Plaza, *Jalan* Pantai Kuta, *Jalan* Bakung Sari, and *Jalan* Legian. The data were collected through direct photo documentation from the streets of Kuta Village. After all outdoor signs were documented, the next process was the classification process. The collected data were divided into two groups: commercial signs and non-commercial signs. Outdoor signs have two main types; they are non-commercial signs and commercial signs. The non-commercial sign is used by charitable organizations and government, which mainly aim to provide information or motivate people to react as a response to the information served without any profit orientation. On the other hand, the commercial sign is used by producers to promote their product to gain profit. The analysis of the data obtained begins with the description of both types of data, and it is followed by an interpretation based on the concepts of linguistic landscapes (LLs) and the ideology of globalization associated with the language as part of culture. In relation to the problems discussed in this research, the following are brief descriptions of the data obtained through interview and the photos data, which showed the process and ideology behind the marginalization of the Balinese language and the implication of this marginalization.

In general, this paper is about the outdoor signs of Kuta as one of the tourist destinations in Bali. However, the focus is on answering these questions: what languages are used in these signs and in which settings or contexts are those languages used, and what can be interpreted from the use of languages in these outdoor signs?

2 DISCUSSION

As a tourist destination, Kuta Village is visited by tourists, either domestic or foreign. The tourists sometimes stay longer in the village. In this situation, many people from that village involved in tourism activities make contact with them over a longer period. This is done, for instance, by those who own accommodations or restaurants, and their employees. The domestic tourists do not speak Balinese; the language used is Indonesian, and for foreign tourists, a relevant foreign language is used like English, French, or Japanese. With regard to the language choice, an informant, who is a security guard in Kuta Market, stated that he used a language relevant to the person he was speaking to. Similar information was also gained from other security informants. In addition, we were informed that beggars used English when they were begging in order to get money from the tourists in Kuta Art Market. This phenomenon showed that the use of relevant language is a way of getting some benefit. This reflects ‘capitalism ideology’. Capitalism ideology served as the background of the marginalization of Balinese people. Communication that occurred among Balinese also showed a language preference. This information was obtained from the interview with another informant. He stated that when Balinese people are talking about business, they use mixed language (Balinese, Indonesian, and English). An observation of Balinese society in general is that the phenomenon of using mixed language is not limited to talking about business matters, but that mixed language is also used by the Balinese for different purposes. It is not only used in informal encounters, but it is sometimes used in a formal situation. For instance, we can also see that a televised program given by 'priests', which mainly talks about Balinese ceremony, uses mixed language. In the tourism area, the use of mixed language in a family domain is quite obvious. An interview with the head of the traditional Village of Kuta revealed this situation. He said that the Balinese language in Kuta is not lost, but it is not thriving because the Balinese language used is *basa Bali pasaran*. He himself admitted to having been using *basa Bali pasaran*. It is also said that many Balinese people find it difficult to speak 'high Balinese'. Problems like these are found not only in everyday conversation but also in the dialogues during the meeting in the traditional village, *desa Pakraman*, especially in the villages that belong to the area of tourism. Thus, this indicates that the marginalization of Balinese is currently and continuously happening. In addition to this, in Denpasar, there are places where 'young parents' can send their children to attend 'play groups'. Favorite places are those that offer English as a means of communication. This also contributes to the marginalization of Balinese. The marginalization of Balinese is also due to the strong influence of the use of the Indonesian language. The collection of photos of all outdoor signs on the main streets of Kuta showed that there are 1132 outdoor signs there; 796 (70.32%) signs are commercial signs and 336 (29.68%) signs are non-commercial signs.

The languages used in these outdoor signs are English and Indonesian. The dominant languages used are a combination of English and Indonesian. Only two outdoor signs are in the Balinese language. From the perspective of LLs, the languages used in outdoor signs can be interpreted as having two functions: the informational and the symbolic. The most basic informational function is that it serves as a distinctive marker of the geographical territory inhabited by a given language community. The LL serves to inform in-group and out-group members of the linguistic characteristics, territorial limits, and language boundaries of the region they have entered. This function is not relevant to the research location. The relevant functional information in outdoor signs found in the Kuta area is to inform the readers that communication and service can be carried out in the languages used in the outdoor signs. The languages used in these outdoor signs showed a diglossic situation, in that the high-status language is used more in public signs compared to a lower-status language. This showed the advantage of a dominant language. Local language is a local identity. The use of the language in public signs of a particular community can give a good feeling to the speakers of that language as a symbol of their identity. In this context, there are only two signs that use the Balinese language.

Studies of the LL are concerned with language in its written form, in the public sphere: language that is visible to all in a specified area. These signs are treated as a text. According to Landry and Bourhis (1997: 25), the language of public signs, advertising boards, street names, place names, commercial shop signs, and public signs on government buildings form the LL of a given territory, region, or urban agglomeration. This study is more on outdoor signs that are relatively fixed signs or fixed texts with regard to their position. Signs might have some degree of stability. For instance we might add the notion of 'mobile' signs, which could be leaflets and flyers being distributed in the street, advertising on vans, buses, and other vehicles that pass through the streets of the area under study, free tourist maps, and other publications available on counters and desks of hotels and tourist information centers. The texts making up the LL may be monolingual, bilingual, or multilingual, reflecting the diversity of the language groups present in a given territory or region. However, language or code preference is never the result of an arbitrary decision. The linguistic code choices in the public sphere serve to index broader societal and governmental attitudes towards different languages and their speakers. In this case, the predominance of one language may reflect the relative power and status of the language. The LL clarifies that the field normally aims to apply its findings in a broader context, making inferences about the motives behind and impact of particular displays of language in public. Landry and Bourhis (1977: 23) explain that LL "serve[s] important informational and symbolic functions as a marker of the relative power and status of the linguistic communities inhabiting the territory"

According to Landry and Bourhis (1997: 25), the informational functions of the LL include serving as a distinctive marker of the geographical territory inhabited by a given language community. This will inform visitors about the linguistic characteristics, territorial limits, and language boundaries of the region they have entered, and it also indicates that the language in question can be used to communicate and to obtain services within public and private establishments within that region. The use of 'foreign' languages, particularly in commercial signs, may also have a symbolic function. In the case of English, for example, shop name signs written in English but located in Kuta might be symbolizing foreign tastes, fashions, or associations between particular products or types of businesses and English-speaking culture. It might also be that English is perceived as being more modern and prestigious than local languages. Thus a prevalence of signs in English often marks a 'tourist space' and can thus be interpreted as serving both informational and symbolic functions.

In this context, English has become a lingua franca. It is a kind of norm to communicate with tourists via a lingua franca as well as to promote the image of the place as tourist-friendly and cosmopolitan. Thus, it can be argued that the use of outdoor signs is part of the social practices that may serve as powerful instruments in the production of the social world in which they form part of the landscape. For instance, the symbolic function of language use in public signs can emerge as a strong marker of social identity. It is clear, then, that the public signs of the LL are potential symbolic markers of status and power. The language use can be seen as a semiotic system that operates as systems of social positioning and power relationship since no choices are neutral in the social world. The LL can provide a way of looking at the power relations in a given community, and it also reflects the relative power and status of different languages in a particular sociolinguistic context. The presence of more than one language in a sign has pragmatic implications of language choice and use in the sense that language preference can be printed in a different font size or given a different color. The outdoor signs in the research location provide visual evidence of the effects of globalization on languages. English has played the role of a linguistic vehicle of globalization. Globalization is a process that is motivated by the dramatic development of transport, communication technologies, and international commerce that is affecting most aspects of life and business including language choice and use. There is a growing body of evidence to show that English is rapidly spreading through the streets of the tourist areas, like in Kuta. In tourist areas, English has become a lingua franca because it has something to do with the symbolic value of English for the local people, as a language of international prestige, or status marker.

Balinese as a local language is marginalized. Human beings are not created alone. They live in the world together with other creatures and other objects. Their lives become more meaningful if they are in harmony with their environment. They cannot be separated from their environment; as the environment may show different characteristics, in order to survive, they developed a socio-cultural system as a way for them to adapt to the environment. The development of their socio-cultural system is tested through their experiences. The elements of the socio-cultural system that are good and beneficial for them continue to be maintained and developed, and then they are inherited by the next generation. Patterns of environmental management that are ideally useful to meet the needs of human life and for the preservation of the natural environment are usually called local knowledge or wisdom. In this context, the local knowledge and wisdom are 'kept' in the languages of the local people. In addition to its function as a means of social interaction, a local language also expresses the cultural values that exist in the society of its native speakers. This is also the case for Balinese. There are many expressions found in the Balinese language that show local wisdom: *kearifan social*, 'social wisdom', and *kearifan lingkungan*, 'environment wisdom'. These expressions need to learned, taught, and reinterpreted. The marginalization of the Balinese language implied that the understanding of local knowledge and wisdom was also decreasing in that the young generation do not understand the meaning of important expressions in the Balinese language, which essentially contain the values of the Balinese culture.

From the results and the description about the marginalization of the Balinese language, it can be concluded that the Balinese language must be maintained or revitalized. Then there is a two-dimensional relevant strategy to be developed and implemented in an effort to revitalize the Balinese language as described below.

2.1 *Ideological dimension*

It is a fact that the high tendency of Balinese people to choose Indonesian and foreign languages to communicate in everyday life can be seen as a sign showing how strong their desire is to have the image, prestige, wealth, and power owned by the people who can speak those languages. With this ability, it is easier for them to find a job and get material benefit (money) that can be used to build their image and power in many arenas of life. Balinese people have shown the characteristics of postmodern people. This means that the people of Bali have been affected by the global culture, a postmodern culture with the ideology of capitalism or market ideology. This happened because globalization united Bali with other countries of global capitalism, and market ideology quickly influenced the socio-cultural system of Bali (Atmadja, 2010: 74). There are different views available on the notion of globalization. One view states that globalization causes "glocalization". Glocalization is seen as a complex interaction between the global and the local that is characterized by cultural borrowing. While becoming global encourages uniformity or homogenization, glocalization encourages heterogenization or diversification of culture that is a mixture between the global and the local (Ritzer, 2006: 104; Steger, 2006: 57).

The occurrence of glocalization cannot be separated from the people who intend to oppose globalization, particularly the concept of becoming global, but they still adopt a global culture that has a strong influence, so the resulting culture is a mixed culture. This can be custom or beliefs. Based on the above description of glocalization and cultural revitalization, then a model of Balinese language revitalization strategy that would be relevant to be developed in the era of globalization is the strategy based on the ideology of glocalization. This is important because it contains the spirit of the ideology of revitalizing local culture while still adopting elements of global culture. By adopting this ideology, Balinese people still maintain the use of the Balinese language and can also use other languages of their choice. But there needs to be an awareness that when Balinese people are talking to their fellow Balinese, the Balinese language must be used.

2.2 *Local dimension*

The marginalization of the Balinese language has implied a lack of understanding of the younger generation of Bali about the values of the Balinese culture. This implication could be so powerful that it could change the cultural orientation of young generations from the orientation of the local culture into a global cultural orientation. This, in turn, could make the Balinese society undergo an 'identity crisis'. Consequently Bali would no longer be admired by the international community. Maintaining the existence of the language of Bali during a time of strong globalization influence is relatively difficult. In general, people prefer foreign language training compared with Balinese language training. Within the scope of the family in the area of tourism, the use of mixed language (Balinese, Indonesian, and foreign language) is evident. Indeed, parents are usually trying to use the Balinese language in communicating with their children, but the use of Indonesian and foreign languages is unavoidable. Therefore, the model of Balinese language revitalization strategy that promotes the understanding of Bali's cultural values as a local culture is very important in order to maintain the Balinese identity and culture. As already known, Bali's cultural values are embodied in various aspects of the Balinese language. Although there seemed to be a group of Balinese people who are concerned with the marginalization of their language, their efforts are weak. One Balinese language program, *tata titi basa Bali*, was broadcast by local TV, and basically showed the lexical equivalence of Balinese, Indonesian, and English. However, the effort of maintaining Balinese is relatively weak; even though there are activities that are supposed to use the Balinese language, a mixed language between Balinese and the Indonesian language is still used. Balinese is taught as a subject from primary to high schools in Bali. Many activities, like speech contests and short story writing, are carried out to help students master Balinese, but the mastery of proper Balinese is still lacking. The teaching of various traditional expressions and traditional songs that contain local knowledge and wisdom can be more intensively done at schools. In Bali there are many traditional activities in the community that still use the Balinese language in the form of 'singing' that give important lessons of local knowledge and wisdom to the community.

3 CONCLUSION

The results of the study described above show that Balinese has been marginalized. It is often noted in the literature that the global capitalist economy could lead to an erosion of local culture in developing countries. Historically, Balinese tourism dates back to the 1920s when Bali was discovered as "the last paradise"; Balinese traditional culture was recreated and even invented for Western audiences (Yamashita, 2013: 49). Yamashita noted that the first five-year development plan began in 1969 under President Suharto's regime. Under this plan, tourism was seen as an important source of foreign currency earnings for Indonesia, and Bali was designated as the most important of Indonesia's international destinations. In this context, the Balinese provincial government adopted the policy of tourism development with special emphasis on culture (2013: 49). In the process of tourism development, Balinese traditional culture has become a cultural resource economically for both the province of Bali and the Indonesian state government. In this way, the local culture has become part of the tourism industry in which touristic culture that is created in the context of tourism has emerged, and through this process, culture has become 'cultural capital'. Language is a part of culture, and then language is also a cultural resource. Communication using a language is then a cultural practice. It is true that human beings are born with the ability to learn a language, but the context in which we learn, and how we use it, could be different. If we want to know the role of language in human life either individually or in groups, the knowledge should come out of the study of grammar, namely by assessing the relations of language with a variety of social activities in society. In this context, it is important to study language as a cultural resource to understand the role of language in one culture and society and the relevance of cultural phenomena in understanding the language. In this regard, the development of tourism in Bali, the influence of the globalization of the ideology of capitalism has strengthened and quickly influenced the socio-cultural system in Bali (Atmadja, 2010: 74). To respond to the fear of losing their identity as Balinese people, the Ajeg Bali movement emerged, namely a movement that aims to preserve the cultural identity of the people of Bali who have been under the influence of globalization (Atmadja, 2010: 3). The intensive use of Indonesian and foreign languages also has a positive effect; namely, through tourism development in Bali, there are more and more people of Bali who can speak Indonesian and foreign languages. The choice of language use in outdoor signs in Kuta is driven by the motivation of choice of language. There are several motivations that encourage the shift of language, such as communicative motivation, economic motivation, social identity motivation, and religious motivation, but the most influential is the economic motivation (Karan, 2011). In short, the LL appears to be a worthy object of investigation since linguistic choices in public spaces index broader social attitudes; linguistic tokens in public spaces may be symbolic markers of status and power. Thus the study of the LL may provide insight into the social identities and ideological orientations of a community.

REFERENCES

Akindele, Dele Olufemi. 2011. *Linguistic Landscapes as Public Communication: A Study of Public Signage in Gabarone Botswana.* Macrothink Institute: International Jurnal of Linguistics 2011, 3(1).

Atmadja, Nengah Bawa. 2010. Ajeg *Bali Gerakan, Identitas, Kultural, dan Globalisasi.* Yogyakarta: LkiS.

Backhaus, Peter. 2006. *Multilingualism in Tokyo: A Look into the Linguistic Landscape*. Clevedon: Multi Lingual Matters Ltd.

Ben-Rafael E. et al. 2006. *Linguistic Landscape as Symbolic Construction of the Public Space: The Case of Israel*. Clevedon: Multi Lingual Matters Ltd.

Cenoz, Jasone. And Gorter, Durk. 2006. *Linguistic Lanscape and Minority Languages.* Clevedon: Multi Lingual Matters Ltd.

Gorter, Durk. 2006. *Further Possibilities for Linguistic Landscape Research.* Clevedon: Multi Lingual Matters Ltd.

Huebner, Thom. 2006. *Bangkok's L..inguistic Lanscapes: Enviromental Print, Codemixing and Language Change*. Clevedon: Multi Lingual Matters Ltd.

Landry, Rodrigue and Bourhis, Richard Y. 1997. *Linguistic Landscape and Ethnolinguistic Vitality: An Empirical Study*. Journal of Language and Social Psychology 16: 23–49.

Ritzer, George. 2006. *The Globalization of Nothing Mengkonsumsi Kehampaan di Era Globalisasi* (Lucinda, penerjemah; Heru Nugroho, penyunting). Yogyakarta: Universitas Atma Jaya.

Urban Studies: Border and Mobility – Kerr et al. (Eds)
© 2019 Taylor & Francis Group, London, ISBN 978-1-138-58034-3

Language transfer in Javanese video clips on YouTube: A sociolinguistic analysis of Cak Ikin's Culoboyo videos

S.D.S. Tungga & T. Suhardijanto
Universitas Indonesia, Jakarta, Indonesia

ABSTRACT: Nowadays, YouTube has become a new media to share videos. People can share their activities, tutorial videos, educational materials, and entertainment. In Indonesia, YouTube is also part of media to keep and share information about cultures such as activities, language, food, art, technology, and other elements of culture. In this paper, we will present our research on the Javanese language in animated videos collected from a YouTube channel named "Cak Ikin". Cak Ikin's animated videos entitled Culoboyo are constantly being produced and uploaded with new or viral social issue themes that occurred in Indonesia. These videos usually get tens to hundreds of thousands of viewers on YouTube. All Culoboyo videos always use Javanese with Surabaya's dialect. However, not all expressions and dialogues are in Javanese but mixed with Bahasa Indonesia. Here, we will focus on linguistic interferences occurring in the Culoboyo animation videos. The objective of this study is to investigate to what extent and for what reasons does the production of Bahasa Indonesia occur in Javanese animated video. The resulting analysis will show the function of Bahasa Indonesia interference in Culoboyo videos. Furthermore, the result can also reveal the vitality of Javanese at present.

Keywords: Bahasa Indonesia, Interference, Javanese, Language Transfer, YouTube

1 INTRODUCTION

Kramer & Bredekamp (2013) state that "culture is no longer a matter of monolithic immobility congealed in works, documents or monuments, but liquefies into our everyday practices with objects, symbols, instruments, and machines". This claim is very appropriate for portraying the conditions of human life nowadays. The increasing prominence of the Internet that has developed into a human need is a cultural phenomenon that calls for studies, particularly in order to find and observe the development of culture. YouTube, as one of the products of the Internet era, has now become a media that is actively used by the public.

Cak Ikin is a YouTube channel that is trying to be part of cultural preservation efforts. This channel produces and uploads animation videos featuring special characters in various topics. With 156.207 subscribers, each uploaded video in the channel has been viewed by tens to hundreds of thousands of viewers. During its earliest time, Cak Ikin published the Grammar Suroboyo series. The first Grammar Suroboyo video was released in 2007 as a film animation, not a YouTube video. Grammar Suroboyo showcases the unique feature of Surabaya dialect called "misuh", which in English, is usually known as the bad, dirty language or swearing. In addition, Cak Ikin also publishes the Culoboyo series in celebration of the National Children's Day in 2010. Culoboyo is different from Grammar Suroboyo in the sense that it does not highlight the "misuh", but presents daily trendy themes and features child characters Culo, Boyo, and Mas Dafa, and an adult character Cak Ikin. To date, Cak Ikin has been publishing Culoboyo videos on YouTube.

Culoboyo videos usually feature new or viral social issues happening in Indonesia. Cak Ikin creatively produces videos in the Javanese language in his attempt to preserve the language, particularly of Surabaya's dialect. These videos have both Javanese and Indonesian subtitles. This is only natural, considering in Indonesia someone will be at least bilingual if they master both the local language and Bahasa Indonesia. As stated by Batran (2015), "linguistics deals with bilingualism as related to the interference between two linguistic systems or to the evolution of a particular language after its exposure to or contact with a foreign language". This tendency is also supported by Krauße (2015) through his survey of 165 informants in Surabaya, Gresik, and Sidoarjo showing that their proficiency in the Bahasa Indonesia (86.1%) is a little higher than their proficiency in the Ngoko Surabayan-Javanese (80.6%). The survey also reveals that in their home, the informants mostly speak Bahasa Indonesia (77.6%), followed by Ngoko Surabayan-Javanese (51.5%) and Krama Surabaya-Javanese (33.9%). These figures show that Bahasa Indonesia has become increasingly prominent in the daily life. Consequently, a linguistic interference occurs from the use of both Bahasa Indonesia and the Javanese language. Such linguistic interference is also reflected in the use of both Bahasa Indonesia and the Javanese language in Cak Ikin's Culoboyo videos. From this perspective, this study aims to investigate to what extent and for what reasons Bahasa Indonesia is produced in Culoboyo videos.

2 METHODOLOGY

Her, we used Simak (observation) method with Simak Bebas Libat Ciakap (SBLC) (observation of uninvolved conversation) technique Sudaryanto (1993) for data collection. This technique was selected because it does not require the researcher to be involved in the observed conversation. Using this technique, the researcher only had to listen to the dialogues and conversations in the source of data.

The source of data is Culoboyo videos accessed from Cak Ikin YouTube channel. This channel contains many videos, most of which use the Javanese language. Only a small number of videos only use Bahasa Indonesia, for example, videos with certain contents such as songs, quizzes, and parodies. However, for the purpose of this study, only videos using the Javanese language were selected as the source of data.

After the data were collected by listing all Bahasa Indonesia words used in the video, they were categorized based on their similarities for analysis. For the analysis, the padan (comparing) method was used. In the padan method, as mentioned by Sudaryanto (2016), the decisive tool of analysis is outside of the language researched.

3 DISCUSSION

The analysis of data shows six patterns of the production of Bahasa Indonesia in *Culoboyo* videos. These patterns show when and why Bahasa Indonesia is used in Javanese dialogues or conversations.

3.1 *Following the title and topic video discussion*

The first pattern of the use of Bahasa Indonesia is that it follows the title and topic discussed in the related video. Here, Bahasa Indonesia is used to describe and discuss a certain topic according to the title or topic of discussion. This is, among others, seen in the videos with the following titles:

a. Kunjungan Raja Salman ke Bali Indonesia
In this video, Boyo speaks a chunk of sentence in Bahasa Indonesia because it follows the title of the video.

1. "Loh awake dhewe iki sedang menyambut kedatangan Raja Arab Cak"
   'We are now welcoming the King of Arab, Cak'
   (Episode: Kunjungan Raja Salman Ke Bali)

Loh awake dhewe iki → Javanese
sedang menyambut kedatangan → Bahasa Indonesia

From the above example, it is known that the sentence has a connection and similarity with the topic of video according to the conversation discussed. The title or topic is "Kunjungan Raja Salman" 'King Salman's Visit to Bali', and the sentence shows that they are welcoming the King of Arab or King Salman ("Loh awake dhewe iki sedang menyambut kedatangan Raja Arab, Cak"). The Bahasa Indonesia part of the sentence is used to show that the sentence given above is following the title and topic of discussion.

3.2 *Effect of conversation partner*

The second is the effect of the conversation partner. It usually makes the response of the language feel different depending on the conversation partner. It is because generally communicators may change their speech sound, syntax, and pronunciation as part of trying to decrease the different-level communication between themselves and the partners as Dragojevic, Gasiorek, and Giles (2015) stated. This is also the case in Culoboyo videos. Here, the effect is the interference of Bahasa Indonesia in the Javanese language, as seen in the following conversation, in which Boyo responds to a police officer who speaks Bahasa Indonesia.

2. Pak Polisi: "Arek cilik mau ikut coblosan?"
   Boyo: "Boleh ya Pak Polisi?"

   'Pak Polisi: "Little kids want to vote?"
   Boyo: "Can we, Sir?"'
   (Episode: Ikutan Mencoblos Pilkada Serentak)

In the above conversation, the police officer asks a question to Boyo in Bahasa Indonesia, and Boyo replies in Bahasa Indonesia, too. In this case, the use of Bahasa Indonesia allows the police officer to give a special treatment to Boyo and allows Boyo to show his respect to the police officer. Although in Surabayan-Javanese—or in the Javanese language in general—people can use Krama or Krama Inggil, some Javanese people, especially those of the younger generation, prefer to use Bahasa Indonesia to show respect. Furthermore, Subroto, Dwiraharjo, and Setiawan (2008) also state:

> *The Javanese language teachers and representatives of the JY stated that in general the JY were not able to use Javanese speech levels correctly and appropriately. They generally were not able to understand well Ng words and their Kr and Kr I correspondences. They even stated that they did not have the courage to use Javanese Kr and Kr I speech levels because they felt incompetent in using the levels and were afraid of making mistakes. Therefore, they chose to use Indonesian when addressing someone older or having a higher status.*

This lack of confidence to use the correct level of the Javanese language can be one of the reasons why Boyo also spoke in Bahasa Indonesia.

In some cases, however, Bahasa Indonesia is just used in the beginning of the sentence. The following conversation between Cak Ikin and KBS Radio shows the use of Bahasa Indonesia in this manner.

3. Cak Ikin: "Halo radio KBS"
   Radio KBS: "Iya, ada apa kewan?"
   Cak Ikin: "mau melaporkan ada dua anak kewan ucul ning tengah embong".

   'Cak Ikin: "Hello KBS Radio"
   KBS Radio: "Yes, what's wrong kewan (animal)?"
   Cak Ikin: "I would like to report that two young animals are loose in the middle of the road".'
   (Episode: Macet Total Karena Bopeng?)

The above conversation shows that Radio KBS and Cak Ikin use two different languages in their sentences. Radio KBS prefers to use kewan "animal" (Javanese) rather than hewan "animal" (Bahasa Indonesia), although it starts its sentence with Bahasa Indonesia, "Iya ada apa". Meanwhile, Cak Ikin also uses Bahasa Indonesia in the beginning of the sentence and then continues the sentence using Javanese, which is shown as follows:

"*Mau melaporkan ada dua anak (Bahasa Indonesia) kewan ucul ning tengah embong* (Javanese Language)"

From the conversation, it can be said that a code switching occurs as a response to the conversation partner (Basir & Ro'ifah, 2015).

3.3 *Information like news or media, proverb, quotes or information from any sources*

Bahasa Indonesia also comes when there is information like news, proverb, or quotes from any source, as seen in the examples below:

4. santai rek, winginane iku Surabaya hujan es, dan menurut Badan Meteorologi dan Geofisika, hari ini akan turun lagi hujan es...
   "Relax, guys, yesterday it hailed in Surabaya, and according to the Meteorological and Geophysical Agency, today it will hail again..."
   (Episode: Bahaya Jajan Bentuk Dot)

5. Santai ae, lek jalene Ibu kita Kartini, "habis gelap terbitlah terang".
   'Relax, Kartini said, "out of dark comes light".'
   (Episode: Habis Gelap Terbitlah Disneyland)

6. Ana pepatah Yo, "tahun depan harus lebih baik", Kon kira-kira wis duwe rencana Gurung ate laopo taun ngarep?
   'There is a saying, "next year must be better", have you had any plan of what you will do next year?'
   (Episode: Tahun Baru Lebih Baik Uang Baru)

The above examples show how Bahasa Indonesia interfaces the Javanese language. It is seen that Bahasa Indonesia is used to quote or retell certain information. According to Hoogervorst (2006), in Surabaya, the use of Bahasa Indonesia to quote in the conversation is common. In terms of embedded language island notion, he considers the Javanese language (Surabaya's dialect) as the matrix language and Bahasa Indonesia as the embedded language. He also mentions that there are four reasons for using the embedded language island: (i) to start a new episode or different scene; (ii) to quote a spoken or written text; (iii) to neutralize comments that expresses an indirect way of thinking, arguments, or hopes; and (iv) to quote Bahasa Indonesia proverbs in the conversation (as seen in Example 6).

The following are the examples of the use of Bahasa Indonesia for neutral comments:

7. Ning kene iku saiki sedang melestarikan watu Cak
   "Here, right now, (people) are preserving the stone, Cak"

3.4 *Emphasizing sentence*

The Indonesian language is also used in *Culoboyo* videos to emphasize sentences previously mentioned in Javanese. This is seen in the following examples:

8. *Ojok mbujuk* Mas Dafa, *jangan memberikan kesaksian palsu!*
   '*Do not be lying* Mas Dafa, *do not give the fake testimony!*'

9. *Maksudmu engko bakal dibangun Taman Hiburan Culoboyoland ngunu ta? Haha tid ak mungkin, tidak mungkin.*
   'Do you mean Taman Hiburan Culoboyoland will be built? Impossible, impossible, impossible.'

Example 8 shows that Bahasa Indonesia is used to emphasize an argument with same expression or linear sentence. In Example 9, on the contrary, Bahasa Indonesia is used to emphasize an argument with negative expression opposing the previous premise.

3.5 *Conjunction*

Culoboyo videos also contain Bahasa Indonesia conjunctions, which have similar equivalences in Javanese, as follows:

– *Tapi,* 'but' (Bahasa Indonesia) = *Nanging,* 'but' (Bahasa Jawa)

This conjunction found in almost every Culoboyo vedeos, for example

10. Cak Ikin: "*Sabar rek sabar*"
    Boyo: "***Tapi** lek mlakune alon ngene ya telat mlebu sekolah Cak*"

    'Cak Ikin: "Be patient guys"
    Boyo: "But, we ride this slow, we will be late for school Cak"'
    (Episode: Operasi Gempa Polisi Mas Dafa)

The use of *tapi* instead of *nanging* is common for Surabayans to express the "but" expression. Therefore, the reason why almost all *Culoboyo* videos use *tapi* instead of *nanging* is that they believe the Javanese equivalence for the word *but* is *tapi.*

Another Bahasa Indonesia conjunction used in *Culoboyo* videos is the subordinate conjunction *dan* "and", which has the Javanese equivalence *lan. Culoboyo* videos use both *dan* and *lan* as the subordinate conjunction.

– *Dan*, "and" (Bahasa Indonesia) = *lan*, "and" (Javanese)

The Bahasa Indonesia conjunction *dan* is used in some of *Culoboyo* videos, although some other videos still use the Javanese conjunction *lan.* The Bahasa Indonesia conjunction is used in the manner explained in Section 3.1, among others in the conversation between Culo and Boyo in the video entitled *Mengapa Surabaya Disebut Kota Pahlawan* "Why Surabaya is Called the City of Heroes". See sentence 11.

3.6 *Other reasons*

This section will focus on the data in which Bahasa Indonesia is used not because of the conditions from the theme or the reasons discussed above, but because of the production of words in the language. The use of Bahasa Indonesia in this manner is seen the following example.

11. ***Makane** Surabaya **akhire** oleh gelar Kota Pahlawan, dan setiap 10 November diperingati sebagi Hari Pahlawan*
    "That's why Surabaya was finally granted the title of the City of Heroes, and 10 November is commemorated as the Heroes Day."
    (Episode: Mengapa Surabaya disebut Kota Pahlawan)

No.	Word in Culoboyo videos	Javanese language	Sentence	Translation
1.	*Soale*	*Sabab/jalaran*	*Iya **soale** sakwise Indonesia meredeka, …*	Yes, because after the Independence of Indonesia, …
2.	*makane*	*Mula/mulane*	***Makane** Surabaya **akhire** oleh gelar Kota Pahlawan, dan setiap 10 November diperingati sebagi Hari Pahlawan*	That's why Surabaya was finally granted the title of the City of Heroes, and 10 November is commemorated as The Heroes Day
3.	*artine*	*tegese*	*Lek abang itu **artine** mandeg*	Red means stop
4.	*Bentuke*	*Wujude/rupane*	*Iya bentuke ya unik kaya dot*	Yes, the shape is unique like pacifier

All the words were listed in Purwodarminto's *Bausastra Jawa* (1939).

In the above example, the word *makane* and *akhire* are Bahasa Indonesia words with Javanese suffix [-e/-ne] (Wedhawati & Arifin, 2006). This suffix manifests as [-e], not [-ne], if it is attached to a word ending with a consonant. In this case, the speaker uses [-ne] for the Bahasa Indonesia word *maka* to produce *makane* and [-e] for the Bahasa Indonesia word *akhir* to produce *akhire*. The Javanese suffix is deliberately attached to Bahasa Indonesia words to make it seems like Javanese words. The Bahasa Indonesia equivalence of suffix [-e/-ne] itself is actually suffix [-nya]. Similar methods of word formation are also found in many *Culoboyo* videos, as follows:

This production of Javanese word has also been discussed by Asmoko (2014) particularly about how the Bahasa Indonesia words transform into Javanese words.

4 CONCLUSION

From the discussions above, it is seen that there are several reasons for the transfer of the Javanese language on Culoboyo videos interfaced by the Bahasa Indonesia occurring in conversations. First, it follows the title and topic discussed in the video. Second, it is because of the effect of the conversation partner. Third, Bahasa Indonesia also interfaces the conversation so as to state certain information like news, proverb, or quotes from any sources. Fourth, Bahasa Indonesia is used because of its emphasizing function. Fifth, Bahasa Indonesia conjunctions are used for two reasons: it has become the common word and it has a special function that correlates to the topic. Additionally, from the data, it is also found that Bahasa Indonesia words are produced as Javanese words.

The use of Bahasa Indonesia that interfaces the Javanese language as seen on Culoboyo videos can reflect the recent condition of the Javanese language, especially of Surabaya's dialect. By observing this tendency, we can reveal the reason for the use of Bahasa Indonesia in a Javanese conversation and the meaning of the conversation itself.

REFERENCES

Asmoko, Y.W. (2014). *Pembentukan Verba Turunan Bahasa Jawa dengan Bahasa Indonesia Berdasarkan Kamus (Analisis Kontrastif)*. Yogyakarta: Universitas Negeri Yogyakarta.

Basir, U.P., & Ro'ifah, A. (2015). Adolescents Java Dialect in Surabaya Bilingual Images and Optional Language Model in Java Community. *International Journal of Humanities and Social Science* (Vol. 5 No. 4), 179–188. Retrieved from http://www.ijhssnet.com/journals/Vol_5_No_4_April_2015/19.pdf.

Batran, M.C. (2015). Language Interference Trigerred by Bilingualism. *PROQUEST SciTech Journals*.

Cambridge University Press. (1999). Retrieved Juni 20, 2017, from Cambridge Dictionary: http://dictionary.cambridge.org/dictionary/english/interjection#translations.

Carter, R., McCarthy, M., Mark, G., & O'Keeffe, A. (2016). *English Grammar Today: The Cambridge A-Z Grammar of English*. Cambridge: Cambridge University Press.

Dragojevic, M., Gasiorek, J., & Giles, H. (2015). Communication accomodation theory. In C.R. Berger, & M.E. Roloff, *The International Encyclopedia of Interpersonal Communication* (pp. 1–21). New York: John Willey & Sons Inc.

Hoogervorst, T.G. (2006). Percampuran Bahasa di Surabaya: Pengaruh Bahasa Indonesia Terhadap Bahasa Jawa Suroboyoan. *PENDETA, Jurnal Bahasa, Sastera, dan Budaya Jilid 2*, 59–74.

Hoogervorst, T.G. (2006). Percampuran Bahasa di Surabaya: Pengaruh Bahasa Indonesia Terhadap Bahasa Jawa Suroboyoan. *PENDETA, Jurnal Bahasa, Sastera, dan Budaya Jilid 2*, 1–10.

Kramer, S., & Bredekamp, H. (2013). Culture, Technology, Cultural Techniques-Moving Beyond Text. *SAGE Journals*, 24.

Krauße, D. (2015). Features of Surabaya Javanese (Suroboyoan). *International Symposium on the Languages of Java (ISLOJ) V.* Bandung.

Poerwadarminta, W. (1939). *Baoesastra Djawa*. Batavia: Groningen.

Subroto, D.E., Dwiraharjo, M., & Setiawan, B. (2008). Endangered Krama and Krama Inggil Varieties of the Javanese Language. *Linguistik Indonesia* (Tahun ke 26), 89–96.

Sudaryanto. (1993). *Metode dan Aneka Tehnik Analisis Bahasa: Pengantar Penelitian Wahana Kebudayaan secara Linguistik*. Yogyakarta: Duta Wacana University Press.

Sudaryanto. (2016). *Metode dan Aneka Teknik Analisis Bahasa*. Yogyakarta: Sanata Dharma University Press.

Wedhawati, & Arifin, S. (2006). *Tata Bahasa Jawa Mutakhir* (Ed. Rev ed.). Yogyakarta: Kanisius.

Urban Studies: Border and Mobility – Kerr et al. (Eds)

Linguistic landscape as a social identity construction of the public space: The case of Batu District

Y. Indarti
Universitas Airlangga, Surabaya, Indonesia

ABSTRACT: In the past decade, Linguistic Landscape Studies (LLS) have been conducted to produce accurate and detailed inventories of urban multilingualism proposed by Bloommaert & Maly in 2014. The aim of this study was to find the social identity construction of the Batu District society through linguistic landscapes. This district was chosen as it is the central governmental and economic area of the tourist city of Batu. A total of 200 linguistic landscapes were collected in eight different areas of the district. The collected data were then analysed using theories proposed by Landry & Bourhis in 1997. The results show that eight languages are shown in the linguistic landscapes, namely Indonesian, Javanese, English, Mandarin, Japanese, Arabic, Korean, and French, of which Indonesian is the most commonly used language, followed by English and Javanese. Furthermore, the Indonesian language still has the highest position compared to the other languages.

Keywords: Linguistic landscapes, urban multilingualism, social identity construction, public space

1 INTRODUCTION

Batu City is known as one of the famous tourist destinations in East Java province as it has extraordinary natural beauty. Until the end of 2015, the Department of Transportation and Tourism recorded a very high number of tourists, reaching 3,580,000 (www.timesindonesia.co.id). As a city with the nickname of *the real tourism city of Indonesia*, tourists visiting the city are not only local people, but also foreigners. Antara news agency (2016) noted that foreign tourists come from China, Malaysia, Singapore, Japan, Australia, Saudi Arabia, European countries, and America.

The rapid growth of the modern tourism sector in Batu City hampers the city's environmental development. It becomes more open to modernity in terms of both population and language. Consequently, the city has developed as a city with a diverse urban multilingual community. Furthermore, having the slogan *Shining Batu*, the city's modern-concept tourism objects are commensurate with international tourism, which can be seen from the naming of the objects in English such as *JawaTimur Park* (Jatimpark 1 and 2), *Batu Night Spectacular*, *Predator Park*, and *Eco Green Park*.

Tourism objects in Batu City are written in mixed codes or one code, which is considered to have high prestige such as English. The use of English in the tourist areas is intended for people with high social status or foreigners, as the former is considered more prestigious than Indonesian or Javanese. As stated by Holmes (2008), a language is considered prestigious if it has a high status, i.e., when it is used in other countries as a means of communication, while a language with low status is only used as a means of local communication. Therefore, the use of English in linguistic landscapes in Batu tourism areas reinforces the image of Batu as an international tourist city. This indicates that language use in Batu also affects language use in the surrounding area.

Administratively, Batu City is divided into three districts, namely Bumiaji District, Junrejo District, and Batu District.[1] Batu District is an area in Batu City that is interesting for further study for several reasons. The first reason is that it is the centre of both governmental economic activities of Batu City. Second, there are many tourist objects, including tourist rides and nature tourism. Thus, it can be assumed that Batu District is the central area of Batu City government policy making. Moreover, Batu District is an icon of Batu City as a national and international tourist destination. In the realm of language use, the district is also a multilingual urban society in which there are a number of mixed codes with their respective functions.

The language situation in Batu District is described as polyglossia, although not all of its people are multilingual speakers. With a total population reaching 211,298 in 2014, the district is multiethnic. In general, the society consists of two groups: natives and the migrant community. The native inhabitants are the Javanese, who are the major residents of East Java province, while the migrant communities are Madurese, Bajau, Bugis, Chinese, and Arabic (batukota.bps.go.id). Furthermore, Indonesian is used to communicate with various ethnicities, as it is the national language and the lingua franca for the people in Batu City, in particular, and for the people of Indonesia, in general. In addition, Javanese is also used as the colloquial language. Indeed, there are also various other languages spoken by various ethnicities as well as foreign languages. Briefly, it can be argued that ethnicities and multilingualism contribute to the polyglossic situation in the district.

2 METHOD

This study applied a qualitative methodology. The data were 200 photos of linguistic landscapes collected randomly from eight sub-districts of Batu District, so that there were 25 pictures taken from each sub-district. The sub-districts were: Oro-Oro Ombo, Pesanggrahan, Sidomulyo, Sumberejo, Ngaglik, Sisir dan, Songgokerto dan, and Temas. The languages that appeared on the landscapes were counted and classified as monolingual, bilingual, and multilingual texts to obtain a general picture, of which languages make up the linguistic landscapes. Having discovered the code preference, the last step was to draw the social identity construction through the use of linguistic landscapes.

3 THE NOTION OF LINGUISTIC LANDSCAPES

The concept of "linguistic landscape" (LL) was put forward by Landry & Bourhis (1997) in their study of ethnolinguistic vitality. The linguistic landscape of a region includes the language of road signs, bulletin boards, advertising, product information, billboards, notes, street names, place names, commercial store signs, façade names, posters, banners on government buildings, and so on. The study shows that language in the public sphere can be considered a major indication of language attitudes.

To examine the linguistic landscape, it is also necessary to know its actor. The actors of the linguistic landscape can be divided into two groups: government (top-down) and private parties (bottom-up). They have different realms in LLs (Ben-Rafael et al., 2006). The first actor is the government (top-down) as the producer of governmental institutions, public bookmarks, public announcements, and street names. Furthermore, the second actor is private parties (bottom-up) as the producers of store markers, private bookmarks, and private entrepreneurs. Furthermore, it is important to know the actors in LLs as this relates to the power or legitimacy of public spaces and policies regarding formal language (Gorter, 2006).

Gorter (2006) states that the texts making up public signs are distinguished into monolingual texts, bilingual texts, and multilingual texts. He further explains that the selection of languages in public signs is based on a number of factors such as the government's or

1. http://www.eastjava.com/tourism/batu/about.html

community's attitude towards a wider range of languages and users. Therefore, a language that dominates the public space represents the relative power and status of its users. Furthermore, an analysis of language in LLs includes the number of public signs, their frequency of occurrence, top-down and bottom-up actors, language choice in bilingual public signs, as well as language choice in multilingual public signs.

4 DISCUSSION

The texts forming the linguistic landscapes may be monolingual, bilingual, or multilingual, reflecting multilingualism (Torkington, 2009). The results of this study show that eight languages are found in 200 public signs located in eight areas of Batu District, namely Indonesian, Javanese, English, Mandarin, Japanese, Arabic, Korean, and French.

Table 1 illustrates the occurrence of the eight languages identified in the linguistic landscapes:

Table 1 illustrates that 35% of public signs are monolingual in Indonesian, Javanese, English, and Arabic, where 28% is in Indonesian, 4.5% is in English, 1.5% is in Arabic, and 1% is in Javanese. This means that Indonesian still dominates the linguistic landscapes in Batu District, followed by English with 4.5% public signs.

In addition, there are bilingual texts and multilingual texts. Indeed, there are nine bilingual texts, namely Indonesian-English, Indonesian-Javanese, Indonesian-Arabic, Indonesian-Japanese, English-Japanese, Javanese-English, Indonesian-Korean, English-French, and English-Korean. It can be seen that the monolingual texts are dominated by Indonesian while English is less than 10%. However, English seems to dominate the bilingual texts in Indonesian-English. The number of bilingual texts with Indonesian-English is 36%. Then, the second highest number of bilingual texts is Indonesian-Javanese, reaching 11%. The number of other bilingual texts is not significant. This indicates that English is the most accepted foreign language compared to the other foreign languages like Arabic, Mandarin, Japanese, Korean, and French.

There are fewer multilingual texts found in the linguistic landscapes of Batu District. Indeed, there are six multilingual texts that are successfully certified, namely 1.5% in Indonesian-Javanese-English, 1% in Indonesian-English-Arabic, 0.5% in Indonesian-English-Mandarin,

Table 1. Monolingual, bilingual, and multilingual texts found in linguistic landscapes.

No.	Codes or languages	Number of public signs	Percentage of public signs
1	Indonesian	56	28%
2	English	9	4.5%
3	Arabic	3	1.5%
4	Javanese	2	1%
5	Indonesian-English	72	36%
6	Indonesian-Javanese	22	11%
7	Indonesian-Arabic	17	8.5%
8	Indonesian-Japanese	3	1.5%
9	English-Japanese	2	1%
10	Javanese-English	2	1%
11	Indonesian-Korean	1	0.5%
12	English-French	1	0.5%
13	English-Korean	1	0.5%
14	Indonesian-Javanese-English	3	1.5%
15	Indonesian-English-Arabic	2	1%
16	Indonesian-English-Chinese	1	0.5%
17	Indonesian-English-Japanese	1	0.5%
18	Indonesian-Javanese-Arabic	1	0.5%
19	Indonesian-Javanese-Mandarin	1	0.5%
Total public signs		200	100%

0.5% in Indonesian-English-Japanese, 0.5% in Indonesian-Javanese-Arabic, and 0.5% in Indonesian-Javanese-Mandarin. The percentage of higher multilingual texts in Indonesian-Javanese-English in public signs than the other two multilingual texts reinforces the linguistic characteristics of Batu District.

The frequency of occurrence of the eight languages found in 200 linguistic landscapes is illustrated as follows:

Table 2 shows the frequency of language occurrence in the linguistic landscapes. This illustrates that Indonesian is the most commonly used language in the linguistic landscapes found in Batu District. Indonesian appears on 176 (53.18%) public signs. This domination of the Indonesian language in the linguistic landscapes indicates the language situation in Batu District, which is described as polyglossic. However, people in the region are not a multilingual community. Most of them still use Indonesian as a means of communication with other ethnicities.

The second highest number for a language in the linguistic landscape is for English, which is seen on 89 public signs or 26.89% of the total public signs. However, the percentage of English used is different in the eight sub-districts. English seems to be dominant in Temas, Sisir, Oro-Oro Ombo Village, Sidomulyo, and Ngaglik. These five sub-districts are well known for their places of interests.

Furthermore, the third highest number for a language is the Javanese language. Javanese is used in 32 (9.67%) total public signs. This suggests that although Javanese is a vernacular language of the native inhabitants, it is only used in regional or local communications.

Finally, the results show that the frequency of occurrence of the other foreign languages in the linguistic landscapes is insignificant. The frequency of Japanese reaches 1.81%, while Chinese and Korean reach the same percentage of 0.6%. Then, French is found in 0.3% of the signs. Thus, foreign languages still have a minor position in the linguistic landscapes of Batu District.

The language choice and language use found in Batu District represent polyglossia, although not all of the people belong to a multilingual community. This is in accordance with the General Provisions of Article 1, the Law of the Republic of Indonesia Number 24 of 2009 on Flag, Languages and Symbol of the Country, and National Anthem which states that the State of the Republic of Indonesia hereinafter referred to as Indonesian is the official national language used throughout the territory of the Unitary State of the Republic of Indonesia; Regional languages are the languages spoken by Indonesian citizens in areas in the territory of the Unitary State of the Republic of Indonesia; and a foreign language is a language other than Indonesian and regional languages. Furthermore, Article 25 states that the Indonesian language as the official language of the country serves a purpose as a means of unifying the various ethnicities, as well as a means of communication between regions and intercultural areas.

The analysis shows that there are eight languages in the linguistic landscapes of Batu District, including Indonesian, Javanese, English, Arabic, Mandarin, Japanese, Korean, and French. Although there are eight languages, Indonesian still dominates the linguistic landscapes. This represents how Indonesian plays an important role as a means of communication in the multilingual community. As people have different vernaculars and speak Indonesian, Indonesian has an informative function for the people in Batu District. Thus, it

Table 2. Frequency of language in linguistic landscape.

No.	Codes or languages	Number of public signs	Percentage of public signs
1	Indonesian	176	53.18%
2	English	89	26.89%
3	Javanese	32	9.67%
4	Arabic	23	6.95%
5	Japanese	6	1.81%
6	Mandarin	2	0.6%
7	Korean	2	0.6%
8	French	1	0.3%

still has the highest position compared to other languages. This is reinforced by the Law of the Republic of Indonesia Number 24 Year 2009 on Flag, Languages, and Symbol of the Country, as well as the Anthem.

The existence of both laws ensures that the Indonesian language will still exist in the future even in the midst of its competition with foreign languages. In other words, although there will be various languages in the future, the government still used the Indonesian language for official and regulation purposes. Indeed, Indonesian will continually be used in the public space as the social identity construction of society in Batu District.

Further discursive construction of language choice and language use in public space illustrates that the use of vernacular Javanese in the linguistic landscapes in Batu District indicates that these areas are occupied by the natives.

5 CONCLUSION

In conclusion, the social ideology construction of Batu District community shows that Indonesian is still in the highest position in comparison to other languages. This is reinforced by the Law of the Republic of Indonesia Number 24 Year 2009 on Flag, Languages, and Symbol of the Country, as well as the Anthem. The Act ensures that Indonesian will be retained as the social identity construction of the community, especially in Batu District and in Indonesia in general.

REFERENCES

A Dictionary of Architecture and Landscape Architecture. 2000.

Ben-Rafael, E., Shohamy, E., Hasan Amara, M. & Trumper-Hecht, N. 2006. Linguistic Landscape as Symbolic Construction of the Public Space: the Case of Israel. In D. Gorter (ed.), *Linguistic Landscape: A New Approach to Multilingualism*. Clevedon: Multilingual Matters: 7–31.

Cenoz, J. & Gorter, D. 2006. Linguistic Landscape and Minority Languages. In D. Gorter (ed.), *Linguistic Landscape: A New Approach to Multilingualism*. Clevedon: Multilingual Matters: 67–81.

Fikriyah S., & Wisnu. 2013. Perkembangan Kawasan Real estate di Surabaya Barat Tahun 1970–2000 (Kontribusi Citraland dalam Perkembangan Kawasan Real estate di Surabaya Barat). *Avatara e-Journal PendidikanSejarah*: 503–518.

Gorter, Durk. 2006a. Further possibilities for linguistic landscape research. In D. Gorter (ed.), *Linguistic landscape: A new approach to multilingualism*. Clevedon [etc.]: Multilingual Matters: 81–89.

Huebner, T. 2006. Bangkok's Linguistic Landscape: Environmental Print, Code-mixing and Language Change. In D. Gorter (ed.), *Linguistic Landscape: A New Approach to Multilingualism*. Clevedon: Multilingual Matters: 31–52.

Landry, R., & Bourhis, R.Y. 1997. Linguistic Landscape and Ethno-Linguistic Vitality: an Empirical Study. *Journal of Language and Social Psychology*: 23–49.

Scollon, R. & Scollon, S. 2003. *Discourses in Place: Language in the Material World*. London: Routledge.

Torkington, Kate. 2009. Exploring the Linguistic Landscape: The Case of the 'Golden Triangle' in the Algarve, Portugal. *Papers from the Lancester University Postgraduate Conference in Linguistics and Language Teaching* 3. Papers from LAEL PG 2008.

Borders and mobility in education and policies

Urban Studies: Border and Mobility – Kerr et al. (Eds)

Policy implementation for fulfilling 30% quota of women's representation in the 2014 legislative member election: A study on the PDI-P and the PPP in Palangka Raya

R.S. Tulis
University of Palangka Raya, Palangka Raya, Indonesia

ABSTRACT: This study evaluates the successful implementation of 30% quota of women's representation in parliament, as well as the mechanisms by which political parties have fulfilled this recruitment quota. Using a conceptual framework of Van Horn and Van Meter, namely policy objectives, implementing agent characteristics, communication, human resources, disposition and the economic, social, and political environment, this study analyzed the recruitment process of women representatives in the Partai Demokrasi Indonesia Perjuangan (PDI-P) and the Partai Persatuan Pembangunan (PPP) in the 2014 legislative elections in Palangka Raya. Moreover, a qualitative descriptive approach was used to analyze the results of this study. First, the implementation for fulfilling 30% quota of women's representation in the 2014 legislative member election in Palangka Raya, plus pre-survey data, can be said to be relatively successful based on the implementation indicators proposed by Van Horn and Van Meter. This success will ensure an effective recruitment function of political parties. Moreover, this will ensure the formal fulfillment of the requirement aspects mandated by the Law 8/2012 on general election for DPR, DPD, and DPRD. Second, this success will allow the correction of the indicators of policy objectives that have not yet been standardized, especially those related to gender-responsive policies as well as the political will of women.

Keywords: policy implementation, women's representation, legislative elections, Palangka Raya

1 INTRODUCTION

Legal norm in Indonesia has a strong foundation in ensuring the rights of women's participation in formal politics. The government attempts to achieve gender equality by developing gender-responsive national policies. This includes equal treatment or treatment that is different but equivalent to the rights, obligations, interests, and opportunities (Nurhaeni, 2009:34). This policy is a consistent and systematic area of concern for the differences between women and men in society in order to remove structural and cultural obstacles to achieve gender equality. Furthermore, this is reinforced with Presidential Instruction 9/2000 on Gender Mainstreaming (PUG) in national development.

The Law 2/2008 on political parties has mandated a minimum of 30% quota of women's representation in politics. According to the Law 8/2012 on General Elections of the House of Representatives, the Regional Representative Council, and the Regional People's Representative Council, each political party participating in the election may nominate candidates for DPR, provincial DPRD and regency/city DPRD for each electoral district by taking into account at least 30% representation of women. It also states that political parties are required to include at least 30% representation of women in the management of the central-level political party. Furthermore, Article 55 requires that "the list of prospective candidates as referred to in Article 53 contain at least 30% of women's representation."

Table 1. Comparison of the number of DPR RI members (Puskapol FISIP UI, 2010).

	Number of DPR RI members	
Election	Women	Men
1999	44 (8.8%)	455 (91.2%)
2004	65 (11%)	485 (89%)
2009	103 (18%)	457 (82%)

Table 2. Data of legislative seats in 2014 election (Puskapol FISIP UI, 2015).

Legislative	Men	(%)	Women	(%)	Total seats
DPR RI	463	83%	97	17%	560
DPD RI	98	74%	34	26%	132
DPRD of province	1776	84%	336	16%	2112
DPRD of region/city	14302	86%	2303	14%	16605
Total	16639	86%	2770	14%	19409

In addition, the Constitutional Court revised Article 56 of 8/2012 to prevent women candidates being placed on the last serial numbers in order to avoid being overlooked by voters. Article 67 clause 2 states that KPU, provincial KPU and KPU of regency/city should announce the percentage of women's representation in the list of candidates of political parties in the daily print media and national media. The minimum quota rules for women is also published in the Commission Regulation 7/2013 on nomination of legislative members, which states that while submitting the list of candidates, parties are required to include at least 30% of women. If this condition is not met, the commission will issue a list of prospective candidates to the political party.

As shown in Table 1, the percentage of women in parliament increased during each election period. The increase observed in 2009 is attributed to the 30% quota of women's representation, but the expected results still have not been achieved. As shown in Table 2, women's representation in the 2014 legislative elections was the same as in the previous election period in which at least 30% threshold has not been met. The highest percentage was found for the members of DPD RI (26%), but their provinces and region quota were below 20%.

Tables 1 and 2 strongly suggest that policies for the fulfillment of the 30% quota of women in politics are ineffective as political parties tend not to meet these requirements. Moreover, several studies have asserted that not all political parties in the region conform to national trends. This paper is a summary of research on a number of facts that occurred in the region, especially the role of political parties in carrying out the mandate of 30% of women's representation in the legislature. This study aims to evaluate the fulfillment of the implementation of the 30% quota of women's representation in the 2014 legislative elections in Palangka Raya, as well as the mechanisms undertaken by the political parties to fulfill this quota, particularly the Partai Demokrasi Indonesia Perjuangan (PDIP) and the Partai Persatuan Pembangunan (PPP) in the city of Palangka Raya.

2 METHODOLOGY

This study is descriptive with a qualitative approach. The data collection methods used were observations, interviews with several key informants and the collection of secondary data related to the focus of research. In fact, the research focused on political parties, the commission and the DPRD of Palangka Raya which are closely related to the implementation of the policy. Determination of key informants was conducted according to the indicators set previously. Data that met the research indicators was analyzed from the beginning until the end of the study period. Data were analyzed using the interactive model of Miles and Huberman (1992: 308), which consisted of data collection, reduction, display, and conclusion testing.

3 DISCUSSION

3.1 *Policy implementation for fulfilling 30% quota of women's representation*

Dye (1987: 3) argues that public policy concerns "whatever government chooses to do or not to do." This understanding equates government policy with government actions, and ensures that any action taken by the government would certainly have goals and targets to be achieved. Van Meter and Van Horn (Winarno, 2014: 14, 9) suggest the implementation of the policy as actions undertaken by individuals or a private government aimed at achieving the objectives set out in previous policy decisions.

In order to evaluate the implementation of the quota rule for representation of women in the 2014 legislative elections in the city of Palangka Raya, Van Meter and Van Horn's policy implementation model (Winarno, 2014: 158–169) was used as the framework of analysis.

The process of choosing cadres as prospective candidates is consistent with the mechanism of political recruitment by political parties. According to Gaffar (1999: 155–156), political recruitment is pursued through two systems, namely open and closed. The open system is the candidate selection process that involves the assessment of cadres' competence. Meanwhile, the closed system is the nomination process that is regulated by political parties and is not known by the public at large (Putra, 2004:209).

Successful implementation can be evaluated by the success of an implementing agency in achieving the policy objectives. According to Van Meter and Van Horn (Winarno, 2014: 159), identification of performance indicators is the most crucial stage, which will make it possible to assess how far the basic measures and policy objectives have been realized. Its success or failure lies in achieving the 30% quota of women's representation in the 2014 legislative elections in Palangka Raya. The overview of existing data shows the achievement of requirements for women's representation. Table 3 points out the fact that the representation of women is above the threshold of 30% provided as the minimum quota for all electoral parties, thus fulfilling the Law 8/2012.

The successful fulfillment of the quota is consistent with the KPU's commitment to always obeying the existing procedure. Through the verification process, the KPU of Palangka Raya firmly imposed a fixed number of publicly nominating candidates of political parties who contested the election in order to meet the requirements of 30% representation of women.

Table 3. List of candidates for members of parliament in the 2014 elections in the city of Palangka Raya (KPU Kota Palangka Raya, 2014).

No.	Party	Men	Women	Total	%
1	NASDEM	18	12	30	40.0
2	PKB	15	11	26	42.3
3	PKS	15	14	29	48.3
4	PDIP	16	12	28	42.9
5	GOLKAR	19	11	30	36.7
6	GERINDRA	19	11	30	36.7
7	DEMOKRAT	17	13	30	43.3
8	PAN	17	10	27	37.0
9	PPP	15	15	30	50.0
10	HANURA	18	11	29	37.9
11	PBB	18	10	28	35.7
12	PKPI	17	10	27	37.0
Total		240	140	344	40.7

An important factor in the implementation of policy is resources, which, according to Van Meter and Van Horn (Winarno, 2014: 160), are the elements that can drive the success of a policy implementation process, including human resources, in addition to financial resources and time. The KPU resources of Palangka Raya can help determine the implementation process, especially the one related to understanding of the electoral rules. Political parties themselves have a big role in fulfilling the 30% quota of women's representation. Several key informants were polled, especially in the process of recruiting candidates. Although the criteria of each party differ, the process of determining women candidates involves a rigorous mechanism in which the quality of cadres is regarded as the main criterion to satisfy the desires of the public. The mechanisms built by PDIP and PPP are further strengthened by the public choice of women candidates proposed by each party.

In the 2014 legislative elections, a total of 344 participants contested the election for legislative seats. Each of the participants competed for 30 seats. The results indicate that of the 30 selected people, 10 were women. The fulfillment of the 30% quota of women's representation in the nomination of legislative members is directly proportional to the election of more than 30% of women candidates who fill the legislative seat in Palangka Raya. This demonstrates the role of the human resources of political parties in the process of recruiting women. Table 4 outlines the success of the PDI-P and the PPP in the recruitment of their cadres for contesting the legislative elections. For the PDIP, this success is primarily due to the strategy of socializing prospective women members who meet both the public expectation and the expected quality of candidates. According to the PPP, the election of women candidates always involves public opinion and the Islamic ideology that considers women who are involved in practical politics as taboo. On account of this opinion, some women disregard their intention to participate despite their capabilities. However, some PPP sources hold that the public opinion is not a barrier in the recruitment of women members.

Van Meter and Van Horn (Winarno, 2014: 166) proposed the criteria for policy implementers to achieve successful implementation. In the context of the fulfillment of the women's quota, the implementing agent is the KPU, with Palangka Raya as the organizer and the political party as the election participant. As a representative of the executive function, the KPU of Palangka Raya city acts as the implementer of the Law 8/2012, which is described in the KPU regulations of procedures. The absolute implementation of the mandate of this law, particularly on women's representation, forces political parties to follow the existing standards and rules.

Its implementation in the realm of political parties is to establish standards and internal rules that allow the management of internal and external party members. The ideology of the party often influences the public opinion in choosing a political party. Some of the objectives of the PDIP in Palangka Raya include empowerment of women and their welfare to encourage women to participate in politics. These objectives are consistent with the PPP's opinion; however, the latter has problems in the quality of candidates. The two parties (PDIP and

Table 4. Women representation from the PDIP and the PPP in the nomination of legislative members in the period 2014–2019 (KPU Kota Palangka Raya, 2014).

No.	Party	Man	Women	Total	Representation (%)
1	PDIP	4	3	7	43
2	PPP	1	1	2	50
3	Gerindra	2	2	4	50
4	Hanura	3	1	4	25
5	Golkar	4	0	4	0
6	PKB	3	0	3	0
7	Demokrat	0	2	2	100
8	Nasdem	1	1	2	50
9	PAN	2	0	0	100
Total		20	10	30	33

PPP) with different ideologies agree on recruitment openness, but differ in method. Some of the basic aspects used to determine candidates are education and political experience. Open political recruitment can be regarded as a characteristic of an agent to achieve the fulfillment of the implementation of the 30% quota of women's representation in legislative positions in Palangka Raya city.

According to the criteria of Van Meter and Van Horn (Winarno, 2014: 158–167), the attitude of the implementing agency is based on the knowledge, understanding, and the deepening of the policy, as well as the direction of implementing the response indicative of acceptance or rejection of a policy. The knowledge and understanding of the implementing agent on the content of the quota fulfillment policy of women's representation in legislative elections will directly determine the success of the implementation process. In the nomination process, the KPU of Palangka Raya has urged each political party to include at least one prospective women candidate among three nominated candidates in the legislative elections.

The knowledge and understanding of political parties is also crucial to the successful implementation of the quota of women's representation. The PDIP of Palangka Raya understands the policy, with no loopholes of non-compliance, because the KPU of Palangka Raya has set rules about the obligation to include the representation of women in proposing candidates. This principle has an impact on the PDIP program on women's empowerment in politics. The PDI-P has an open attitude towards recruitment of prospective candidates. Meanwhile, the PPP not only has the same attitude but also pays due attention to the policy rules. According to the Deputy Regional Leadership Board of PPP in Palangka Raya, the greatest obstacle in the recruitment of women members is the Islamic ideology that shaped the lives of both women and society. Sometimes people tend to think that parties with Islamic ideology are not open to women leaders, which in turn affects the political interests of women. In order to comply with these policy standards as well as the rules of the KPU, the PPP has women's representation in the body of the organization. Similarly, the PDIP conducts women's training and education. The objective of these guidelines is to overcome the exclusive outlook of women in order to strengthen resources to compete in elections.

As Van Meter and Van Horn (Winarno, 2014: 161–162) stipulate, accurate and consistent communication is an indicator of the success of the policy implementation process. As evidenced in Tables 3 and 4, this condition has been met, which is indicated by the absorption of women's representation in nominations as well as public confidence in them. This communication is built from the time of recruitment to the time of election. The KPU of Palangka Raya stated that it had continuously communicated with all parties about the policy of fulfilling the 30% quota of women's representation, in order that the involved parties understand the policy.

The PDIP explained that the party had effectively communicated with the city KPU about equality of perception regarding the rules of electoral implementation, including women's representation. It understands that communication is not limited to the KPU and the involved party, but it can be extended to selected cadres. According to their view, communication is also not limited to candidacy and election but can occur when members are chosen. Although the party's authority is no longer viable when the cadres are already seated as legislative members, the party's ideology and the struggle for policies based on public aspirations must be safeguarded and reflected in the decisions made by the party's legislative fraction. With the same perspective, the PPP argue that the cadres' goal must not be to promote the interests of the party when they become a member of the legislature, but to fight for the public interest which forms the basis of the principle of the PPP.

External indicators presented by Van Meter and Van Horn (Winarno, 2014: 167–169) to determine the success of the implementation process are the stability of the environmental, social, economic, and political aspects. Patriarchal culture in Indonesia is one of the key issues which clearly explains the non-involvement of women in politics. However, Dayak's egalitarian culture has provided the necessary impetus to women to engage in politics. According to the view of DPRD members, the political picture in Palangka Raya is very responsive to gender. This can be seen from the involvement of several women in decision-making. From the viewpoint of the PDI-P and the PPP, the social and family environment is an important factor for women's participation in the political sphere. The policy on the 30% representation of women in government was first considered to be quite difficult to implement. However, considering the growing number of women's participation in the DPRD of Palangka Raya, their presence promises an opportunity to formulate gender-responsive policies such as a legal ban on smoking and domestic violence.

Economic factors also influence women's involvement in nominations, particularly for the success of communal socialization in order to boost their candidacy. From the viewpoint of the PPP's regional leadership, finance must be at the candidate's rather than the party's expense. Given that funds are always needed for campaign activities, both the PDI-P and the PPP agree that the financial contribution of candidates is a factor that supports not only the smoothness of activities in the field, but also the importance of public's acceptance of the quality of candidates. According to the PDIP, acceptability of society is the most important aspect in recruiting women. In general, financial constraints faced by prospective women can be overcome with the support of the wider community and supporters.

3.2 *Future challenges for women's representation in political life*

The factors that determine the successful implementation of a policy, as proposed by Van Meter and Van Horn, have been applied by implementing the policy elements in the nomination and election of members to the 2014 legislative assembly. The basis for high implementation performance is the understanding of political parties and the KPU of Palangka Raya on the contents and purpose of the policy, readiness of resources, open attitude towards women and men without discrimination, interpersonal communication, and wide social support for women's involvement in politics. The necessary measure imposed is the 30% quota increment in the nomination of candidates who have made their way into the list of candidates (Table 3). Evidently, this also includes people's decision to choose the candidates submitted by the PDIP and the PPP based on their quality and ability to meet public expectations (Table 4). In general, the requirement of 30% of women's representation in parliament in the city of Palangka Raya has been met under the Law 8/2012 and other related rules.

Future challenges that may be encountered are public opinion on the importance of women's involvement in politics. From the perspective of the PPP, the main obstacle to recruiting women candidates is the public opinion and Islamic ideology that women need not involve themselves in political life, which is more prevalent in areas far from the information centers. However, this perspective can be overcome by empowering women, especially with knowledge, experience, and education. PDIP and PPP programs based on gender responsiveness consist of progress in fighting for and producing qualified women. These programs can also serve as a model for other areas in Indonesia that are still experiencing difficulties in fulfilling the 30% quota of women's representation.

However, as political coaching for women has been perceived as ineffective by the community, many women show lack of interest in legislative politics. When women candidates are elected to occupy strategic positions in the legislature, there is often a lack of productivity in policy-making related to issues of women who are more sensitive to the rights and interests of women. External factors also influence Indonesia's patriarchal tendency: for example, the people's perception that the obligations and roles of women are only limited to household activities. When women in fact enter into politics or work outside home, they end up with multiple roles which, if unbalanced, can lead to problems. Such problems can certainly be solved by the role of family and society support in the decision-making of women.

A serious future challenge is not the implementation of the 30% quota of women members in nominations by political parties, but it is their representation as legislative members. In fact, political parties are not able to guarantee the selection of their women candidates as legislative members, because ultimately it is the trust of the people who vote for them that decides their destiny. Therefore, the government through the rules and policies is not able to control the decision of the community in selecting women representatives. The 2014 legislative elections in Palangka Raya shows that the implementation of the 30% quota policy is directly proportional to the aspirations of the people of Palangka Raya in that the overall representation of women in the DPRD of Palangka Raya is above 30%, especially above 40% for the PDIP and 50% for the PPP.

4 CONCLUSION

From the detailed analysis, it can be said that the implementation of the 30% quota of women's representation in the nominations has been carried out well by the KPU and all the political parties studied. Some of the aspects that support this success are the understanding and deepening of the policy under the Law 8/2012 and other related rules (both by the KPU and the political parties), open mechanisms provided by the political parties to motivate women candidates in order to fulfill the quota, and the mechanism of openness designed for both men and women to participate in the political arena through the party. It should be noted that "openness and without discrimination" is an important principle to encourage women's representation in the nomination process. Although the nomination of women candidates is important, a future challenge is how to increase the number of women who are actually elected into the DPRD. Increasing the role of women in political decision-making would help eradicate the taboo on women in politics.

In order to encourage the successful implementation and representation of women in the legislature, political parties need to improve their training programs or political education, especially for women, in order to foster their self-confidence. Socialization is also important for the success of women members in politics to eradicate prejudice against women decision-makers. The government also plays an important role in encouraging the participation of women not only as legislative members but also as election organizers and other government agencies.

REFERENCES

Gaffar, Afan. (1999). *Politik Indonesia: Transisi Menuju Demokrasi*. Yogyakarta: Pustaka Pelajar.

Miles, B. Matthew and Huberman, A. Michael (1992). *Analisis Data Kualitatif,* Jakarta: UI-Press.

Nurhaeni, Astuti, Ismi Dwi (2009). *Kebijakan Publik Pro Gender.* Surakarta: UNS Press.

Peraturan KPU Nomor 18 Tahun 2008 tentang Pedoman Teknis Pencalonan Anggota DPR, DPRD Provinsi and DPRD Kabupaten/Kota.

Peraturan KPU Nomor 7 tahun 2013 tentang Pencalonan Anggota Legislatif.

Puskapol UI. Retrieved November 3, 2016, from http://www.puskapol.ui.ac.id/hasil-pemilu#.

Putra, Fadillah. (2004). *Partai Politik dan Kebijakan Publik*. Yogyakarta: Pustaka Pelajar.

Undang-Undang Nomor 2 Tahun 2008 Tentang Partai Politik.

Undang-Undang Nomor 8 Tahun 2012 Tentang Pemimilihan Umum Anggota Dewan Perwakilan Rakyat, Dewan Perwakilan Daerah dan Dewan Perwakilan Rakyat Daerah.

Winarno, Budi. (2014). *Kebijakan Publik: Teori, Proses, dan Studi Kasus.* Yogyakarta: CAPS.

Urban Studies: Border and Mobility – Kerr et al. (Eds)
 ISBN 978-1-138-58034-3

Empowering SMEs and cooperatives: Export capacity building in the era of AEC trade liberalisation

Koesrianti, Dhiana Puspitawati & Nilam Andalia Kurniasari
Universitas Airlangga, Surabaya, Indonesia
Universitas Brawijaya, Indonesia

ABSTRACT: The era of trade liberalisation for the ASEAN Economic Community (AEC) began in December 2015 and brought about economic liberalisation in the Southeast Asian region. This era is competitive and the ASEAN member states (AMSs) almost do not have full power of sovereignty to govern their own economic national matters. In this globalised dependence era, the majority of states in the world have to adjust and adopt as well as adapt their national laws to internationalised rules of law. This trade liberation era also has forced companies and other business entities, including Small and Medium Enterprises (SMEs) and Cooperatives in Indonesia to compete with each other in order to tap the benefits of international trade liberalisation. In this context, SMEs and Cooperatives in Indonesia need some kind of protection from the government that does not oppose international regulations on trade. While the number of SMEs and Cooperatives is 98 per cent, their contribution to Indonesian export is small, at only 19 per cent. They are weak in terms of capacity building and access to capital, information technology, global markets as well as integration with regional and global market chains. This research paper evaluates SMEs and Cooperatives in Indonesia in terms of facing AEC trade liberalisation, i.e., what has done and should be done by the authority is to give proper protection to the SMEs in Indonesia by focusing on the manufacturing SMEs as this sector has the best chance of boosting SMEs' export capacity and building the competitiveness of Indonesian SMEs in order to be equal with other SMEs in the ASEAN region.

Keywords: AEC, SMEs, global value chain, trade liberalisation

1 INTRODUCTION

The ASEAN Economic Community (AEC) as one of the three pillars of the ASEAN Community (AC) was officially established in December 2015. The AEC establishment will create deep economic integration in the region characterised by the free flow of goods, services, skilled labour, capital, and investments. With its slogan, as 'one vision, one identity, one community and its vision' ('ASEAN 2025: Forging Ahead Together'), ASEAN has emerged and presented itself as having great potential as well as promising a strong community on the global stage. The establishment of the AEC will encourage all of ASEAN member States (AMS) as well as their citizens and firms to tap benefits from the economic integration of ASEAN. If Indonesia intends to gain economic benefits from the AEC scheme, Indonesia should have engaged in comprehensive management of its economy. In order to get maximum results from trade liberalisation in ASEAN, Indonesia has to have a strategy to enhance all of its resources, including Small and Medium Enterprises (hereinafter SMEs) and Cooperatives.

In Indonesia, SMEs and Cooperatives were found to have been weathering the monetary crisis in 1998 where larger companies collapsed. In short, Indonesian SMEs and Cooperatives have great economic potential and can play an important role in boosting national economic development. Due to the economic growth in rural areas of Indonesia, the SMEs have grown extensively. There were 57.89 million units (99.9%) in 2013, while large enterprises had around 5.066 units (0.01%). SMEs are able to contribute 97.16% to the job market and to the National GDP amounting to 58.65%. However, in terms of capacity, SMEs are not as strong as large business enterprises, especially in the era of free trade pursued by ASEAN with the establishment of the AEC in 2015.

The development and resilience of SMEs have become an important and integral part in the economic development of a country. Some factors of SMEs' weaknesses that have become important concerns for the government are lack of working capital, marketing difficulties, access to financial resources, lack of technological and managerial skills, human resource capability, productivity, ICT competencies, organisational management, and business performance, as well as business climate. Factors that SMEs should be aware of are access to markets, access to technology, access to financing and access to information, that are critical factors in SME development. In general, the majority of SMEs lack resources and capability compared to their large enterprise counterparts since they are in rural areas. In Indonesia, SMEs lack contribution to exports because the global competitiveness of SMEs is low especially in the global market as well as the domestic market in terms of importing products. SMEs in Indonesia also lack entrepreneurial skill as a result of the absence of training and low-levels of education. SMEs are usually part of a lower tier in the network, but there is a significant threat to their survival.

This article examines the Indonesian SMEs' and Cooperatives' export capacity building in the trade liberation era of AEC 2015. The amount of capital and assets determines the definition of SMEs in Indonesia. SMEs contribute to generate more employment, promote more balanced growth in the country, and exhibit more innovation and dynamism. Single market and production based as the main objective of AEC can be used to give good momentum to Indonesian SMEs, especially in the manufacturing industry to engage in international business through exports.

2 DISCUSSION

2.1 *SMEs' and cooperatives' export capacity building*

The definition of SMEs according to Law No. 20 year 2008 on Micro, Small, and Medium Enterprises, is categorised based on asset and sales revenue. Definitions of SMEs in Indonesia are as follow: 1. Micro Enterprise is a productive enterprise owned by individuals and/or the individual business entities that meet the criteria as defined in the micro enterprise, as provided in this Law. 2. Small Enterprises are productive economic activities that stand alone and get operated by the individual or business entity that is not a subsidiary or their branch companies are not owned, controlled, or grouped as a part either directly or indirectly of a medium or large business 3. Medium Enterprises are productive economic activities that stand alone and get done by the individual or business entity that is not a subsidiary or their branch companies are not owned, controlled, or grouped as a part either directly or indirectly of a medium or large business with total net assets or the annual sales revenue as provided in this Law. Other than based on sales and revenue, there is also a definition based on amount of labour. Thus, the criteria for micro, small and medium enterprises in Indonesia can be summarised as in Table 1 as follows.

Table 1 shows that the total number of SMEs (micro, small and medium) and total employment are very high. However, the contribution of SMEs to export is very small as indicated by Yuri Sato who found that the average SME export share of five ASEAN Member Countries (AMCs), namely, Indonesia, Malaysia, the Philippines, Thailand, and Vietnam was only 23 percent on average, which is lower than that of their counterparts in other East Asian economies, where export share of SMEs is 43 percent in Korea, 40 to 60 percent in China, and 56 percent in Taiwan. In Indonesia for example, large enterprises' contributions jump up to 84 percent while micro enterprises seem to have nothing to do with exports. The SMEs' export shares may go up by taking into account indirect export by subcontracting with particular commodities. Thailand, the Philippines, and Malaysia, which have a high percentage of SME export share, have higher export shares for machinery parts and components in their total exports.

Table 1. SMEs' criteria and total number compared to corporations.

	Assets (excluding land and building)	Sales revenue	Amount of labour	Total number (2013)	Total employment (2013)
Micro	<50 million IDR	<300 million IDR	<5	57.189.393	104.624.466
Small	50–500 million IDR	300 million – 2.5 billion IDR	5–19	654.222	5.570.231
Medium	500–10 billion IDR	2.5 billion – 50 billion IDR	20–99	52.106	3.949.385
Corporates	>10 billion IDR	>50 billion IDR	N/A	5.006	3.537.162

Source: Law No. 20/2008 and Ministry of Cooperatives and SME, 2013.

The big difficulty for SMEs in Indonesia in terms of taking part in regional and/or international markets is the existence of a structural economic gap as a consequence of economic globalisation and trade liberalisation, which in turn creates tight competition. In general, SMEs in Indonesia cannot compete with other enterprises regionally as well as internationally due to missed-strategy in terms of SME development that the government managed long before the establishment of AEC 2015. Indeed, Indonesian SMEs and Cooperatives have not contributed to export activities due to the lack of 'global competitiveness' of SMEs for competing in the global market as well as entrepreneurial skills. SMEs in Indonesia need "globalization of entrepreneurship" as the government policy for SME development. This would come about if the government is able to comprehensively manage SMEs and Cooperatives by providing strategic and structured economic policy for Indonesian SMEs and Cooperatives.

2.2 *Indonesian strategy for SMEs*

By establishing the Southeast Asian region as a single market for goods and services, ASEAN will facilitate the development of production networks in the region and, at the same time, enhance ASEAN's capacity to serve as a global production centre and as a part of the global supply chain. Manufacturing businesses, especially in high-tech industries, require parts and components that come from within and outside ASEAN. Small and Medium Industries (SMIs) in ASEAN play a part in the supply chain, so the distribution of parts and components becomes less costly and less time-consuming. As a result, this will make ASEAN more conducive for a manufacturing base thereby attracting investments in the manufacturing sector.

Accordingly, SMEs are the backbone of ASEAN economies and their development is essential for integrally achieving long-term and sustainable economic growth. Generally, they account for more than 96% of all enterprises and from 50% to 85% of domestic employment in ASEAN Member Countries (AMCs). In addition, the contribution of SMEs to GDP is between 30% and 53%, and the contribution of SMEs to exports is between 19% and 31%. The AMCs have to encourage SMEs by giving greater involvement to SMEs in the production network. Thus, there are huge opportunities for SMEs in this region.

The SMEs have an important role as a key engine of growth for Indonesia's economic development, in areas such as the basis of production, export, and the workforce which also becomes the case for developed countries. As the fourth most populated country in the world, Indonesia is facing numerous challenges, and one of these challenges is a high level of unemployment.

The structure of SMEs in Indonesia is complex since they cover various types of business, such as culinary, garment, etc. The number of unemployed people in Indonesia is 7.0 million (or 8.89%) out of a 127.70 million strong workforce in 2016.

The Ministry of Cooperatives Small and Medium Enterprises (MSME) is the institution that has an authority over the development of and policy on SMEs in Indonesia. SMEs in Indonesia include a very large number of micro enterprises, which are essentially self-employed individuals (or family businesses) that are spread out throughout the rural hinterland. The SMEs are defined as enterprises with less than 100 employees. The majority of them are poorly run with minimal capital, low productivity, and poor products. Compared to the large enterprises, the total number of SMEs is increasing over the years as shown in Table 2 below.

Table 2 shows that the number of SMEs increased from 51.409.612 to 57.198.000 (2008–2013) while the number of LEs only increased slightly, from 4.650 to 4.968 units (2008–2012). SMEs' primary industries are agriculture, forestry, and fisheries. Other sectors are trade, service, transportation, manufacturing and others. It should be borne in mind that the SMEs consist of Micro enterprises and small enterprises. From Table 3, it can be seen that the number of micro enterprises is much higher than small enterprises.

In terms of the total number of employees, SMEs can absorb a much higher number of employees than LEs, and the total number of employees of SMEs increased from 2008 to 2012 (see Table 4).

SMEs can absorb 107.657.509 employees or 97,2% of the nation's workforce. This contributes to 5,8 employee growth. The total number of Small Medium Industries (SMIs) in Indonesia is estimated at 641,143 companies (or 16,1%) of the total number of SMEs, but the contribution to national economic growth is 33% or IDR 489,81 trillion compared to the large industries which contributed IDR 973,53 trillion. The majority of SMEs in Indonesia have only 3–5 workers. However, Law 20/2008 does not state the minimum number of workers in a company, whereas, according to the Law on Workforce, a company should have at least 10 workers and should have company regulations. This issue has been solved by customs that are generally accepted in the community. This may represent obstacles for the development and quality of SMEs in Indonesia in the context of globalisation and trade liberalisation. Therefore, a good strategy to increase competitiveness becomes an important agenda for SMEs in Indonesia, such as by providing some regulations that enhance the role of SMEs in the economic framework of AEC 2015. Indeed, no one denies the significant role of SMEs in providing jobs, reducing poverty, triggering economic growth, ensuring equitable development and ensuring sustainable development.

Table 2. The total number of SMEs and large enterprises in Indonesia (2008–2013).

	Total					
Year	2008	2009	2010	2011	2012	2013
SMEs	51.409.612	52.764.603	53.843.732	55.229.444	56.531.592	57.198.393
LEs	4.650	4.677	4.838	4.952	4.968	N/A

Source: Ministry of Cooperatives and SMEs, 2012.

Table 3. The total number of micro enterprises and small enterprises (2010–2013).

No	Year	Micro	Small
1	2010	53, 210 million	573,060 unit
2	2011	54,650 million	592,087 unit
3	2012	55,860 million	629,042 unit
4	2013	56,510 million	632,056 unit

Source: Ministry of Cooperatives and SMEs 2014.

Table 4. The total number of employees of SMEs and large enterprises in million (2008–2012).

Year	2008	2009	2010	2011	2012
SME	94.024.278	96.211.332	99.401.775	101.772.458	107.657.509
LE	2.756.205	2,674.671	2.839.711	2.891.224	3.150.645

Source: Ministry of Cooperatives and SMEs, 2012.

The basic economic policy for the development of SMEs and Cooperatives is Presidents' Nawa Cita No. 6 and 7, namely, promoting productivity and competitiveness in an international market and promoting economic independence by developing domestic strategic sectors. Furthermore, as the implementation of these policies, the government has a national priority program for year 2018 in which the government has put some important programs for the development of Indonesia's SMEs. National priority programs related to the Ministry of Cooperatives and SMEs are comprised of four main programs: Education, Poverty Alleviation, Development of Business and Tourism, and Area Development. The poverty alleviation programs related to improving access for SMEs and Cooperatives for some programs are as follow:

1. improvement of product quality and access to market;
2. business skill and services development;
3. entrepreneurship development;
4. ease of doing business assurance and protection;
5. increasing access to finance;
6. cooperatives and business partnership.

It should be noted that the government also has a national policy to increase the export of high value-added products and services which include 1). Development of business partnership and networking; 2). Improvement of quality and standard of export products; 3). Improvement of promotion effectiveness and market access.

One of policy frameworks in the SMEs' and Cooperatives' development programs is improving rules and regulations or deregulation in order to promote conducive-business environment for SMEs and Cooperatives. While renewal of Law 20/2008 and Law 25/1992 for the development of SMEs and Cooperatives respectively and their operating regulations will need times, the government has enacted the Negative List under Presidential Regulation No. 44 of 2016 which revoked the previous Negative List 2014. The 2016 Negative List has simplified the categories for the business lines that are open for investment (foreign and domestic) with requirements. Under this Negative List, one category is dedicated to the development of SMEs in Indonesia, business lines that are reserved for or subject to partnership with micro, small and medium enterprises (local SMEs) as well as Cooperatives. Business lines that are reserved for or subjected to partnership with domestic SMEs are: 1). industrial sector, such as cane sugar, which is reserved for partnership with SMEs on the basis of a 'core-plasma' scheme; 2). Public work sector including construction consultancy services and construction implementation services; and 3). Trade sector including areas such as retail trading via mail or internet orders, in particular, for goods under the categories of food, beverages, tobacco, pharmacy, cosmetics, laboratory devices, textiles, apparel and footwear, household and kitchen goods.

In the context of implementation of this regulation, the partnership schemes vary. These include sub-contracting, franchising, trading, distributorship and agency, profit sharing, cooperation, joint venture, outsourcing and other schemes of partnership, but they need further clarification from the Capital Investment Coordinating Board (BKPM). These will be required on a case-by-case basis in terms of the permitted level of foreign ownership in such partnership arrangements. The 2016 Negative List is regarded as a key economic policy for the economic incentive packages introduced by the government to increase foreign direct investment and Indonesia's competitiveness in the ASEAN region by encouraging SMEs to take part in business lines provided by the Negative List.

Other implemented suggestions to boost SME development are by learning best practices from Japan and Singapore's experiences. In Japan, the main driver of SME development is not external assistance, but internal learning and entrepreneurial activities in a competitive environment such as "small but highly competitive" enterprises or "from micro to the world class" enterprises, which could be used as an inspiration for SMEs in Indonesia. SMEs can also have an inclusive pathway, by collective action of small firms in rural and urban communities, clusters, and cooperatives which can help SMEs overcome disadvantages with schemes, such as the one village one product (OVOP) movement, as already implemented in Thailand, in the form of the One Tambon One Product (OTOP). In Singapore, a vast number of SMEs utilise business support programs organised by the governmental enterprise development agency and centres. Later, SMEs, for the competitive and dynamic pathway, are trained mainly through business transactions with customers or foreign suppliers by which their capability is upgraded more than through direct support by the government. In this digitised era, it is also encouraged that SMEs have an on-line presence. If digitally engaged, SMEs are more competitive internationally based on the report which stated that SMEs with basic online capabilities derived 6 percent more of their revenue from international customers than offline SMEs. According to this report, greater use of digital technologies, such as social media, broadband and e-commerce can provide significant benefits for SMEs including higher growth in revenue (up to eighty percent), more likelihood of increasing employment (one and a half times), and more innovation (seventeen times). In this context, the government provides facilities in order to improve the productivity of SMEs and Cooperatives by choosing appropriate technology for SMEs and Cooperatives with guidance and technical know-how to utilise the supportive technology.

3 CONCLUSION

SMEs are the backbone of economic growth in all ASEAN member countries because they account for the majority percentage of national economic growth. While the total number of SMEs in Indonesia is very high, the contribution of SMEs to export is very small. Compared to large enterprises, the SMEs can absorb more employees in the Indonesian workforce. Thus, SMEs are seen to have significant potential to contribute to national as well as regional economic development through participation in the international production network or global value chain. This can be achieved by empowering SMEs' export capacity building which currently is still low. In order to alleviate SMEs' weaknesses, the government provides some policies and programs. While these policies and programs need time to show results, it is better to focus on human capacity building and entrepreneurship as well as giving a grant for start-up businesses and support for R&D to SMEs and Cooperatives.

REFERENCES

ADB Report. 2014. *Asia SME Finance Monitor 2013.* Mandaluyong City, Philippines.

ASEA. 2017. *ASEAN Economic Community Blueprint.* Retrieved November 2, 2017 from http://www.asean.org/archive/5187-10.pdf.

ASEAN. 2017. *ASEAN Economic Community.* Retrieved November 2, 2017 from http://www.asean.org/communities/asean-economic-community.

ASEAN. 2017. *ASEAN Small and Medium Enterprises (SMEs).* Retrieved November 2, 2017 from http://www.asean.org/communities/asean-economic-community/category/small-and-medium-enterprises.

Bhasin, Balbir B. & Sivakumar Venkataramany. 2010. Globalization of Entrepreneurship: Policy Considerations for SME Development in Indonesia. *International Business & Economics Research Journal* 9(4).

Dipta, I Wayan. 2017. Indonesia SME Strategy, ILO/OECD Workshop for Policy Makers on Productivity and Working Conditions in SMEs. Retrieved November 12, 2017 from http://www.ilo.org/wcmsp5/groups/public/---asia/---ro-bangkok/---ilo-jakarta/documents/presentation/wcms_564690.pdf.

Gero, Pieter P. 2014. Optimisme Tinggi Menatap MEA. *Kompas.*

Indonesia Investment. Unemployment in Indonesia. *Indonesia Investment.* Retrieved November 2, 2017 from https://www.indonesia-investments.com/finance/macroeconomic-indicators/unemployment/item255?.

Kimura, Fukunari & Mitsuyo Ando. 2005. Two-Dimensional Fragmentation in East Asia: Conceptual Framework and Empirics. *International Review of Economics and Finance* 14(3): 317–48.

Kimura, Fukunari & Umezaki, So. 2010. Comprehensive Asia Development Plan. *Economic Research Institute for ASEAN and East Asia (ERIA).*

Koesrianti. 2016. Law Reform of Small and Medium Enterprises (SMEs) and Equitable Cooperatives for Competitiveness Improvement in AEC Era. *Dinamika Hukum Journal* 1 (3): 227–340.

Machmud, Z & Huda, A. 2011. SMEs' Access to Finance: An Indonesia Case Study, in Selected East Asian Economies, in Harvie, C., S. Oum and D. Narjoko (eds), Small and Medium Enterprises (SMEs) Access to Finance in Selected East Asian Economies. *ERIA Research Project Report 2010–2014.* Jakarta: ERIA.

Ministry for Communication and Information Technology. 2015. Report: SMEs Powering Indonesia's Success: The Connected Archipelago's.
Ministry of Industry. 2010. *Strategic Plan 2010–2014.*
Narjoko, Dionisius. 2011. Turning Dream into Reality?: Aachieving the Goal of SME Development in AEC, International Conference on ASEAN Vision 2015: Moving toward One Community. *ERI.* Taipei: CIER.
Saptowalyono, C. Anto. 2014. Mendukungsi Kecil di Pasar Besar. *Kompas.*
Sato, Yuri. 2013. Development of Small and Medium Enterprises in the ASEAN Economies, in Beyond 2015: ASEAN – Japan Strategic Partnership for Democracy, Peace and Prosperity in South East Asia, Rizal Sukma and Yoshihide Soeya (Eds). *Japan Centre for International Exchange (JCIE).* Retrieved November 10, 2017 form http://www.jcie.org/japan/j/pdf/pub/publst/1451/9_sato.pdf.
Sembiring, Meliadi. 2008. SMEs Growth and Development in Indonesia. Presented at *ADBI Joint Regional Workshop on SME Development and Regional Economic Integration*: Tokyo.

Urban Studies: Border and Mobility – Kerr et al. (Eds)
© 2019 Taylor & Francis Group, London, ISBN 978-1-138-58034-3

Development of an exercise program to enhance the ability of students in Thai massage classroom: Considerations for promoting traditional medicine education at national and international levels

Kanyarat Peng-ngummuang, Kanocnutya Noiming, Ploythakarn Promsit, Sirinporn Srisanga & Jintana Junlatat
Ubon Ratchathani Rajabhat University, Thailand

ABSTRACT: In this study, we aim to improve the potential of students to empower their finger muscles called Yok-Kra-Dan by an exercise program. A total of 30 2nd-year students of Thai Traditional Medicine (TTM) Program, Ubon Ratchathani Rajabhat University, in the academic year of 2016 were selected as study participants. Data collection was done during February–March 2016 via physical fitness test, Yok-Kra-Dan record form, and general information questionnaire. Data analysis was carried out using frequency, percentage, average, maximum value, minimum value, and paired sample t-tests. The results of this study show that after undergoing the developed exercise program, the participants were more stable and able to do Yok-Kra-Dan longer ($p < 0.05$). The participants had good behavior and attitudes toward finger exercise, as well as good knowledge of exercise and significant better physical fitness in several postures, including pressure of fingers, sit-up, and long jump ($p < 0.05$). These results can be used to develop the finger power of TTM students. Keeping strong fingers is the heart of Thai massage, which leads to effective massage treatment.

Keywords: Exercise program, Finger muscle, Thai massage, Yok-Kra-Dan

1 INTRODUCTION

The benefits of exercise are brought about by several physiological and psychological changes. The physiological effects of exercise include an increase in the endorphin level, body temperature, mitochondrial function, and neurotransmitter production, whereas the psychological effects include a distraction from feelings of depression and anxiety and positive feelings associated with mastery and self-efficacy (Mikkelsen et al., 48). Exercise is important for people of all ages and professions, including Thai traditional medicine practitioners. The body and mind must be healthy in order to provide effective treatment.

Thai traditional medicine (TTM) is a long-standing wisdom, which plays an important role in the healthcare of Thai people and is becoming increasingly popular. TTM consists of four fields, including Thai medicine, Thai pharmacy, Thai midwifery, and Thai massage. Especially, Thai massage is the precious wisdom of Thai people, which is passed and accepted from the past to the present. Thai massage is the treatment of the illness emphasizing pressure, contraction, squeezing, bending, steam, and compression. Thai massage is divided into two types: Rat-Cha-Sam-Nak massage and Cha-Loei-Sak massage. In Thai massage class of Faculty of Thai Traditional and Alternative Medicine, Rat-Cha-Sam-Nak massage is the main type of study. TTM students must both study theory and practice. Massage learners must focus on their hand while pressing on the line or the point. Training for pressing on the point with meditation is learned by a technique called "Yok-Kra-Dan", which is done by sitting, making all fingers like a cup and then lifting the whole body with the fingers (Figure 1). The strength of the finger is the heart of Rat-Cha-Sam-Nak massage. If the massage instructor could not

Figure 1. Yok-Kra-Dan.

force the massage at each point regularly and smoothly, then massage treatment will not work, but can cause injury.

For finger power practice, the strength of the muscles in the body is very important, which is increased by exercising. Exercise impacts both body and mind; it increases flexibility, elasticity, and balance while massaging. Exercise for muscle training increases muscle mass and strengthens the muscles of the body. Each posture is repeated several times, with or without assistive devices. This exercise facilitates the masseur to increase his/her balance. In addition, it can promote good personality and steady and flexible movement.

In the Thai massage class, there are many students who cannot do Yok-Kra-Dan, thereby leading to irregular or shaky massage. This bad massage will be of no value in the future in massage therapy. In this study, we aim to develop the potential of 2nd-year students of Thai Traditional Medicine Program for doing Yok-Kra-Dan, using an exercise program that emphasizes on finger power training and body muscle strength for effective massage therapy in the future.

2 METHODOLOGY

2.1 *Research group*

A total of 30 2nd-year students of Thai Traditional and Alternative Medicine Faculty in the academic year 2016 were examined during December 2016 to March 2017. The inclusion criteria were voluntary participation, healthy physical health, no congestion that impedes exercise, time to exercise, and attendance at every appointment.

2.2 *Collecting data*

The tools used in this study were: (1) questionnaire for collecting general information; (2) physical fitness record before and after the test, including grip dynamometer, standing long jump, weight scale, sixty seconds get up-sit, and sit and reach test; and (3) record of Yok-Kra-Dan duration.

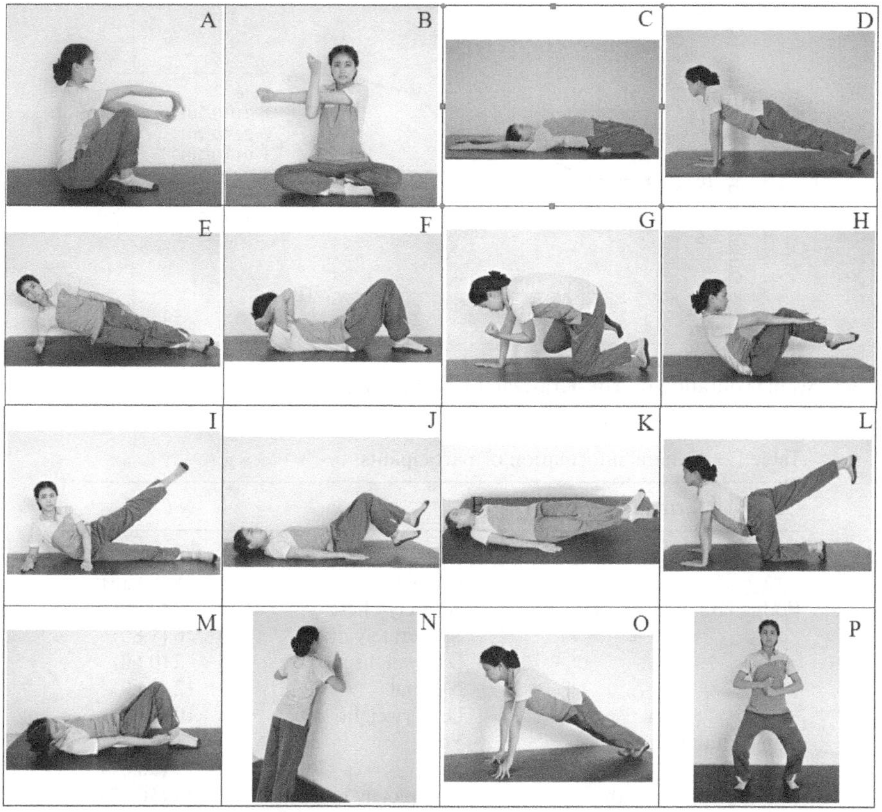

Figure 2. Postures of the exercise program for empowering Yok-Kra-Dan.

The instruments used for data collection were reviewed by three experts and obtained an IOC of more than 0.8. After trying out with a non-experimental group, we found a reliability value of 0.98.

2.3 *Exercise program*

The exercise program to increase the musculature and strengthen the body and fingers consists of 16 postures, including (A) standing and stretching the front arm for 30 s, three sets; (B) stretching shoulder muscles for 30 s each, three sets; (C) abdominal exercise for 30 s, three sets; (D) planking for 20 s, three sets; (E) side planking for 20 s, three sets; (F) sit-up 30 times, three sets; (G) knee lift 30 times, three sets; (H) rowing boat 30 times, three sets; (I) V-shape leg 30 times, three sets; (J) air cycling 30 times, three sets; (K) scissors switch 30 times, three sets; (L) kick back 30 times, three sets; (M) touch the ground 30 times, three sets; (N) wall standing 30 times, three sets; (O) cup-shape finger 30 times, three sets; and (P) giant posture for 30 s (Figure 2).

2.4 *Analysis of data*

The data were analyzed for frequency, percentage, and average to explain the general information and Yok-Kra-Dan behavior of the experimental group. Paired sample t-test was used to test the effects before and after physical fitness test by using SPSS version 16.0.

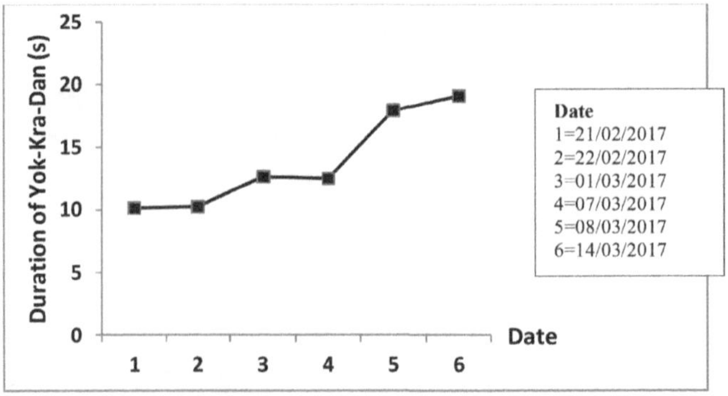

Figure 3. Average duration of Yok-Kra-Dan of 30 participants.

Table 1. General information of participants.

General information		N (%)
Gender	Male	5 (16.67)
	Female	25 (83.33)
Body mass index (BMI)	Severely obese	0 (0.00)
	Moderately obese	6 (1.80)
	Overweight	2 (0.60)
	Normal	12 (3.60)
	Underweight	10 (3.00)
Congenital disease	Yes	1 (3.33)
	No	29 (96.67)
Health (last 6 months)	Completely healthy	17 (56.67)
	Sick (sometimes)	13 (43.33)
	Sick (often)	0 (0.00)
Exercise	Aerobic dance	13 (43.33)
	Swimming	1 (3.33)
	Ride bicycle	1 (3.33)
	Other	15 (50.00)
Duration of Yok-Kra-Dan (seconds)	<10	10 (33.33)
	11–20	7 (23.33)
	21–30	6 (20.00)
	>30	7 (23.33)

Table 2. Physical fitness before and after the exercise program (*significant difference between before and after, $p < 0.05$).

Posture	Before	After	*p*-value
Pressure of the fingers on the scales (kg/body weight)			
Knee down – left	3.62	3.71	0.592
– right	3.73	3.81	0.684
Bottom up – left	4.15	4.95	0.049*
– right	4.18	4.93	0.113
Squeeze power (kg/body weight)			
– left	0.47	0.46	0.410
– right	0.53	0.51	0.290
Sit-up (time)	48.07	54.73	0.024*
Long jump (centimeters)	179.53	181.43	0.000*
Limpness (centimeters)	11.30	11.40	0.889

3 DISCUSSION

Most of the study participants were female. The body mass index values are normal and lower than the standard level. Most of the participants showed a duration time of Yok-Kra-Dan less than 10 s (Table 1). After practicing the exercise program, physical fitness of the participants was found to be improved. In particular, pressure of the fingers on the scales, sitting-up, and long jumping improved significantly ($p < 0.05$), as shown in Table 2. The results from comparison between before and after attending the exercise program on Yok-Kra-Dan duration found that each participant can do Yok-Kra-Dan longer and significantly differ from the beginning ($p < 0.05$) (Figure 3).

4 CONCLUSION

Of the 30 participants in the exercise program, 25 were female. In general, most students who are interested in Thai traditional medicine program are female. This is the case not only in the TTM program in Ubon Ratchathani Rajabhat University, but also in other TTM educational institutions. This may be due to the fact that the TTM profession emphasized on healthcare and promotion of patients closely; however, willingness and high tolerance are important characteristics that make women more suitable for this profession. Most participants had a normal body mass index and some of them showed lower than normal. It can be explained that most of the participants were adolescents with a good metabolic system, and people in this age group generally give great importance to beauty and good shape and hence most of them focus on exercising (93.33%). However, these healthy participants showed less ability on Yok-Kra-Dan. The average duration time on Yok-Kra-Dan is less than 10 s, which indicated that the participants' exercise is not appropriate for increasing finger muscle strength, which is an important part of Thai massage.

In the exercise program consisting of 16 posture exercises, each posture was applied from the Rue-Sri-Dad-Ton, which is a long-standing tradition of Thai massage, and then simple body management was added emphasizing on the muscles of stomach, arms and fingers, which are important areas for Thai massage. The exercise program focuses on increasing the muscle mass by slowly repeating it with the same posture as the set. This method is a very popular way of exercise to increase the muscle mass. It has been proved that exercise can improve the physical performance, muscle strength, and bone mineral density. However, not all types of exercises lead to significant improvement in muscle strength (Chan et al., 2017: 1). A meta-analysis reported that weight-bearing training and progressive resistant training are effective in maintaining the bone mineral density and muscle strength and mass (Nikander et al., 2010: 1; Binder et al., 2005: 1425). Several postures in developmental exercise program were similar to yoga, which also demonstrates improvement in balance and physical performance. It has been showed that yoga exhibits positive effects on mobility, postural control, and gait speed (Tiedemann et al., 2013: 1068). Some postures are similar to traditional Chinese exercises, which are one type of low-level aerobic and gentle. Traditional Chinese exercises combine body posture adjustment, breathing exercise, and mindfulness practice in order to activate muscles and tendons (Ge et al., 2017: 36).

After practicing these 16 exercise postures for 3 weeks, the participants became significantly better in physical fitness, especially in terms of pressure of the fingers, sit-up, or long jumps, compared to beginning of training ($p < 0.05$). Moreover, when doing Yok-Kra-Dan, it was found that its duration was significantly higher than that at the beginning ($p < 0.05$). The findings showed that the developed exercise program is capable of increasing the strength of the body, especially the finger muscles of the participants. This will result in a more effective treatment by Thai massage. Furthermore, this exercise program can be used in Thai massage classes in all academic years. Also, the next experiment can be designed in a way that the exercise posture is emphasized on neurological management. Because muscle strength is exerted through both neural control and muscle quality and quantity, several studies showed that adaptive changes in the nervous system can increase muscle strength in both short and long terms (Gabriel et al., 2006: 133).

REFERENCES

Binder, E.F., Yarasheshi, K.E., Steger-May, K., Sinacore, D.R., Brown, M., Schechtman, K.B. and Holloszy, J.O. 2005. Effects of Progressive Resistance Training on Body Composition in Frail Older Adults: Results of a Randomized, Controlled Trial. *The Journals of Gerontology*, 60(11), 1425–1431.

Chan, D.C., Chang, C.B., Han, D.S., Hong, C.H., Hwang, J.S., Tsai, K.S. and Yang, R.S. 2017. Effects of exercise improves muscle strength and fat mass in patients with high fracture risk: A randomized control trial. *Journal of the Formosan Medical Association*, xx, 1–11.

Gabriel, D., Kamen, G. and Frost, G. 2006. Neural adaptations to resistive exercise: mechanisms and recommendations for training practices. *Sports Medicine*, 36(2), 133–149.

Ge, L., Zheng, Q.X., Liao, Y.T., Tan, J.Y., Xie, Q.L. and Rask, M. 2017. Effects of traditional Chinese exercises on the rehabilitation of limb function among stroke patients: A systematic review and meta-analysis. *Complementary Therapies in Clinical Practice*, 29, 35–47.

Mikkelsen, K., Stojanovska, L., Polenakovic, M., Bosevski, M. and Apostolopoulos, V. 2017. Exercise and mental health. *Maturitas*, 106, 48–56.

Nikander, R., Sievanen, H., Heinonen, A., Daly, R.M., Uusi-Rasi, K. and Kannus, P. 2010. Targeted exercise against osteoporosis: A systematic review and meta-analysis for optimizing bone strength throughout life. *BMC Medicine*, 8, 1–16.

Tiedemann, A., O'Rourke, S., Sesto, R. and Sherrington, C. 2013. A 12-week Iyengar yoga program improved balance and mobility in older community-dwelling people: a pilot randomized controlled trial. *The Journals of Gerontology*, 68(9), 1068–1075.

Urban Studies: Border and Mobility – Kerr et al. (Eds)

Gaming is learning: No more border between children with and without autism spectrum disorder

L.H. Suryawardhani & Y.S. Amalia
Universitas Airlangga, Surabaya, Indonesia

ABSTRACT: Not only in terms of territorial matters, a border also exists in terms of language learning methods between children with and without mental disorders. However, segregation will not be an issue anymore in education when inclusion has become a way of transgression. The present paper is focused on scrutinising the use of chronological mobile game applications as a border-crosser in English language teaching and learning, which can be inclusively applied to language learning activities among children with and without mental disorders, in this case autism spectrum disorders, at the age of five. ChronoloGame is a designed English language learning game consisting of colourful pictures and declarative sentences proposed to be adopted as a learning medium. Overall, the procedures for designing the game follow the theory of psycholinguistics. Therefore, in the light of language learning, children both with and without autism spectrum disorder can play the game together. Themes brought up involve several daily activities. Children are directed to arrange pictures with daily activity themes in chronological order. Since it concerns the learning and teaching of chronological order, the present designed game, for children with mental disorders is suitable, and for children without autism spectrum disorders it is also acceptable.

Keywords: ChronoloGame, Children With Autism Spectrum Disorder, Children Without Autism Spectrum Disorder, Mobile Game Application for Educational Inclusion, Psycholinguistics

1 INTRODUCTION

Recently, in a large number of countries, both developed and developing, educational inclusion has become a centre piece of education policies (Pijl, Meijer, and Hegarty, 1997). The term inclusion is broader in context than the term integration; integration might be interpreted as a means to avoid segregation. Albeit the clear working definition of inclusion remains elusive, the advance of inclusive education needs deeper pedagogical practices (Florian, 2014). Treating all learners differently is, indeed, the appropriate strategy to achieve equity in the inclusive model of the system (Jordan, 2008). Therefore, in educational inclusion there must be flexibility and diversity in the models of teaching methodologies. It is essential to note that inclusion in education is best started in early age education. Educational inclusion from young childhood is expedient in order to familiarise children with the inclusive setting of learning.

Teaching strategies in educational inclusion always have to innovate. Moreover, as the targeted learners of an inclusive setting in education are children, incorporating games into the teaching method can be considered a smart proposed solution, which combines the traditional and the modern approach for the reason of edifying. Children and game are generally known to be inseparable. In addition, beside their function as an entertaining medium, they also serve as a means to educate, as Sudono (2006) has affirmed in stating that playing or gaming is an activity that produces understanding or gives information, giving pleasure and developing imagination. Research that develops games as a learning medium for children abounds (Bryan & Gast, 2000; Carter, 2001; Todd & Tepsuriwong, 2008; Marnik & Sczela, 2008;

Battocci et al., 2009; Fotouhi-Ghazvini, Earnshaw, Robinson, & Excell, 2009; Blum-Dimaya et al., 2010; Rahman, Ferdous, & Ahmed, 2010; Wainer & Ingersol, 2010; Anwar et al., 2011; Hainey, Stansfield, & Connolly, 2011; Chang et al., 2012; Berns, Gonzales-Pardo & Camacho, 2013; Porayska-Pomsta K. et al., 2013; Zakari, Ma, & Simmons, 2014; Bernardini, Porayska-Pomsta, & Smith, 2014). Regardless of the form, whether it is traditional, computerised, or riddle, it has been proven by a number of scholars that the implementation of games in a real-world context for non-gaming purposes in general and the education mission in particular, force motivation and performance (Sailer et al., 2017; Cojocariu & Boghian, 2014; Liu & Chen, 2013; Marnik & Szela, 2008), to improve cognitive ability (Curatelli & Martinengo, 2012), to encourage knowledge (Baytak & Land, 2009), to increase sensibility (Soyluçiçek, 2011), and to minimise distraction (Kayimbasibolgu, Oktekin, & Haci, 2016). The aforementioned goals have proven that the use of games as one of the media for education is beneficial.

In their development, games used for educational purposes have changed following the burgeoning of technology. The rapid advancement of technology has led to the utilisation of computers in developing games. In the past recent years, the employment of computer technology-based games to disseminate literacy to diverse populations of children has gone through dramatic advancements. This technology-mediated model of games is absolutely promising for providing education, specifically in language teaching to non-disabled individuals (Todd & Tepsuriwong, 2008; Fotouhi-Ghazvini, Earnshaw, Robinson, & Excell, 2009; Hainey, Stansfield, & Connolly, 2011; Berns, Gonzales-Pardo & Camacho, 2013 and Porayska-Pomsta K. et al., 2013) and intervention for individuals with special needs (Bryan & Gast, 2000; Carter, 2001; Marnik & Sczela, 2008; Battocci et al., 2009; Blum-Dimaya et al., 2010; Rahman, Ferdous, & Ahmed, 2010; Wainer & Ingersoll, 2011; Anwar et al., 2011; Chang et al., 2012; Zakari, Ma, & Simmons, 2014; Bernardini, Porayska-Pomta, & Smith, 2014). Hence, there is no need for it to be questioned whether the implementation of games results in significant influences in language teaching and learning, especially for children. As an assistive tool, games contribute positively to the betterment of education.

In addition to the advantageous effect of gaming in the process of teaching and learning, Deiner (2013: 310) has suggested that playing is vital to children's social development. However, normal children in general and special children with mental impairment in particular might play differently. The prevalent number of children with neurodevelopmental disorders related to social interaction, verbal and non-verbal communication and behaviour repetition impairment or, to make it simple, children with autism spectrum disorder is counted to be 1 in 150 where formally it was counted to be 1 in 110. The increasing number is due to the increased frequency, and the increased diagnostic ability. The case of autism spectrum disorder occurs in all socioeconomic and ethnic groups caused by either genetic or environmental factors. In addition, it is more likely to occur in boys rather than girls and is mostly observable when children enter school or when children are about five years old. Instead of medication, therapies and interventions are more likely to be seen as helpful for autistic children. Interventions for autism spectrum disorder are varied, ranged from physical to occupational therapy. The use of games can also be one of the intervention tools as well as an educational tool for children with autism spectrum disorder. Many scientists who have an interest in the education of autistic children have conducted research into developing computer-based educational games for children with autism spectrum disorder (Marnik & Szela, 2008; Battochi et al., 2009; Blum-Dimaya et al., 2010; Wainer & Ingersoll, 2011; Rahman, Ferdous, & Ahmed, 2010; Anwar et al., 2011; Chang et al. 2012; Hulusic & Pistoljevic, 2012; Bernardini, Porayska-Pomsta, & Smith, 2013; Saiman et al., 2013; Porayska-Pomsta et al., 2013; Boutsika, 2014; Aziz et al., 2014; Zakari, Ma, & Simon, 2014). Most of the created computer-based games target social interaction, verbal and non-verbal communication and the repetitive behavioural skills of autistic children. The results show the positive effects of using games as an educational as well as an interventional approach for them.

However, due to the integration of autistic children and non-autistic children in the inclusive setting of education, the design of the game to be employed must be applicable for both groups of children. Thus, this situation has boosted motivation to fill the gap by proposing the current contrived technology-based mobile application game to assist the educational inclusive learning process for early elementary aged children, of an average of five years old. The proposed game is named ChronoloGame. It has been developed to be inclusively applied to English language learning activities among five-year-old children with and without autism spectrum disorder.

2 METHOD

The present paper applies a library method of research. The collected articles examine how mobile games' intervention in language education for children both with and without autism spectrum disorder were identified and classified. Articles included in this review (1) focused on children's language learning; (2) utilised interactive multimedia as the primary delivery system for language education and intervention; and (3) targeted language and behaviour abilities. Armed with theories about the implementation of child gaming media in the education of children, the design of the present game took into account the supported findings and studies. ChronoloGame is the designed language learning game consisting of colourful pictures and declarative sentences proposed to be adopted as a learning medium. Themes brought up involve several daily activities. Children are directed to arrange the sequence of several daily activities in chronological order. The current contrived game will play a role as a border crosser, so that, in the light of language learning, children both with and without autism spectrum disorder can play the game together.

3 DISCUSSION

Games are systems composed by components (Sedano, 2012). The creation of games reflects technical and educational evolutions. Thus, the combination between the technical and educational aspect of designing a game should purposively integrate the teaching-learning components in the game design while simultaneously keeping the game engaging and entertaining. Linderoth's dimensions of game design for educational games (see Sedano, 2012) is followed in creating the design of the game. It includes (i) story and fiction, (ii) games and rules, and (iii) technology and platform.

The first to be discussed is the story or the fiction dimension of a game constructed for both non-autistic and autistic children. The game brings up themes related to daily activities. This takes into account that themes related to daily activities are good for non-autistic children and also best for autistic children since children with autism spectrum disorder have problems with repetitive patterns of behaviour such as in doing daily routine activities or rituals (Chang et al., 2012). Hence, seven daily routines are proposed to for adoption in the game design, including rituals before eating, after eating, before sleeping, after waking up, before going out, before entering the house, and tidying toys. The game will consist of these following sentences:

(1) I take my plate. (2) I wash my hands. (3) I pray to God.	Before eating
(1) I wash my hands. (2) I put back my plate. (3) I thank God.	After eating
(1) I wash my hands. (2) I wash my feet. (3) I brush my teeth. (4) I pray to God.	Before sleeping

(1) I thank to God.
(2) I tidy up my bed.
(3) I brush my teeth. } After waking up

(1) I put on my shoes.
(2) I ask my parents. } Before going out

(1) I take off my shoes.
(2) I say salaam. } Before entering the house

(1) I collect my toys.
(2) I put it back. } Tidying toys

The sentences consist of three to five words because the target players of this game are children of around five. The maximum possible words acquired by five-year-old children are limited to around five words. The sentences are also written in a completely declarative manner, explaining the sequences of activities being played in the game; as Mar'at (2011: 67) stated, children of this age are considered to have understood well the syntactic structure of their first language, and thus it is also assumed that they are able to understand complete sentences in English, as their second language. At least by the end of the game they can indicate vocabularies related to daily activities, as well as the structure of the language. Additionally, the font used is Verdana, since children have been shown to read and understand faster in this type of font (Wilkins et al., 2009). The sentences do not stand alone as sentences yet are complemented with colourful pictures. The pictures consist of ten colours: red, yellow, blue, green, orange, cyan, purple, magenta, white and black. This is since those ten colours are assumed to influence the people who see those colours. Beside this, those ten colours have been proven to be favoured by children in research conducted by Kurniawan (2017). In addition, as Deiner (2013: 306) suggested, the pictures used in the game should be single meaning pictures. To add more, the pictures should also represent the sentences and the concepts.

Secondly, the educational game dimension talks about the games and the rules. Beside the English language learning aspect, the present planned game targets the chronological order of daily routines. This is because interventions for children with autism spectrum disorder should have a high degree of structured and planned materials (Deiner, 2013: 304). The structure of the game consists of sequences of colourful pictures and declarative sentences randomly ordered. The players are asked to touch and drag the sentences and the pictures into chronological order. If they are successful in completing the mission, there will be a happy emoji popping up in their screen followed by a clapping sound effect. If not, if they fail to reorder the instructions, there will be a sad emoji popping up on their screen followed by a crying sound effect. By the end of every completed task, there will be a wrap of all steps of the routines in the form of a sound that restates all of the ritual steps which should be repeated after that. The completion of one session for one daily activity will automatically unlock another session of the daily activity.

The present proposed game facilitates all learner with differences in learning strategies through the third of Linderoth's dimensions of game design for educational game technology and the platform used to create the game. It includes three different styles of learning strategies identified by Fachrurrazy (2014: 103): visual, auditory, and kinaesthetic. Visually, the designed game accommodates the visual style of learning strategies by using colourful sequences and pictures of daily routines. The platform of the mobile game application enables the auditory style of learning by allowing sound effects to assist the learning of language. The use of a touch screen gives learners with a kinaesthetic style of learning an opportunity to learn by pointing and dragging the sequences of pictures and sentences into chronological order. The navigation menu and button must be simply and consistently built to avoid children's cognitive overload (Kurniawan, 2017). Overall, the designed game for inclusive education using computer-based technology, especially in the form of mobile game application fulfils children's different needs in learning style.

4 CONCLUSION

In conclusion, educational inclusion, which is broader than just educational integration, is a way of transgression bridging segregation in education in which there is no longer a border between children with and without autism spectrum disorder. A tool presented in this present paper is a current contrived game which hopefully acts as a border crosser in English language teaching and learning in an inclusive educational setting integrating non-autistic and autistic children. The designing of the proposed game named ChronoloGameto could be adopted as an assistive learning medium for educational inclusion of autistic and non-autistic individuals following the theoretical framework of educational computer-based game design and the theory of psycholinguistics; it is hoped it could be a thoughtful and useful educational tool. The choosing of a technology-based mobile application as the platform to create the game incorporates different learning strategies of various types of learners: visual, auditory and kinaesthetic. Thus, since the current proposed game concerns the learning and teaching of language as well as chronological order, the present designed game could be adopted as a learning medium in inclusive educational settings and it is hopefully applicable for all children regardless of their condition, whether they are autistic or non-autistic. However, constructive feedback is very welcome to develop the proposed designed game so that, in real implementation, the designed game for children with autism spectrum disorder is suitable and so that it is also acceptable for children without autism spectrum disorder.

REFERENCES

Amalia, Y. &. -M. (2014). Dialogue in narrative inquiry:Collaboration in Doctoral Study in the USA. In P. Smeyers, D. Bridges, N.C. Burbules, & M. Griffiths, *International Handbook of Interpretation in Educational Research.* London: Springer.

Anwar, A., Rahman, M.M., Ferdous, S.M., Anik, S.A., & Ahmed, S.I. (2011) A Computer Game Based Approach for Increasing Fluency in The Speech of The Autistic Children. *Advanced Learning Technologies (ICALT), 2011 11th IEEE international* conference (pp. 17–18). IEEE.

Aziz, M.Z., Abdullah, S.A., Adnan, S.F., & Mazalan, L. (2014) Educational App for Children with Autism Spectrum Disorders (ASDs). *Procedia Computer Science*, 42, 70–77.

Battocchi, A. et al. (2009) Collaborative Puzzle Game: A Tabletop Interactive Game for Fostering Collaboration in Children with Autism Spectrum Disorders (ASD). *Proceedings of the ACM International Conference on Interactive Tabletops and Surfaces* (pp. 197–204). ACM.

Baytak, A., & Land, S.M. (2010) A Case Study of Educational Game Design by Kids and for Kids. *Procedia-Social and Behavioral Sciences*, 2(2), 5242–5246.

Bernardini, S., Porayska-Pomsta, K., & Smith, T.J. (2014) ECHOES: An Intelligent Serious Game for Fostering Social Communication in Children with Autism. *Information Sciences*, 264, 41–60.

Berns, A., Gonzales-Pardo, A. And Camacho, D. (2013) Game-like Language Learning in 3-D Virtual Environments. *Computers & Education* 60, 1.

Blum-Dimaya, A., Reeve, S.A., Reeve, K.F., & Hoch, H. (2010) Teaching Children with Autism to Play a Video Game Using Activity Schedules and Game-embedded Simultaneous Video Modeling. *Education and Treatment of Children*, 33(3), 351–370.

Boutsika, E. (2014) Kinect in Education: A Proposal for Children with Autism. *Procedia Computer Science*, 27, 123–129.

Bryan, L.C., & Gast, D.L. (2000) Teaching On-task and On-schedule Behaviors to High-functioning Children with Autism via Picture Activity Schedules. *Journal of autism and developmental disorders*, 30(6), 553–567.

Carter, C.M. (2001) Using Choice with Game Play to Increase Language Skills and Interactive Behaviors in Children with Autism. *Journal of Positive Behavior Interventions*, 3(3), 131–151.

Chang, M., Kuo, R., Lyu, C.W., & Heh, J.S. (2012) A Situated Game for Autistic Children Learning Activities of Daily Living. *Digital Game and Intelligent Toy Enhanced Learning (DIGITEL), 2012 IEEE Fourth International Conference* (pp. 217–220). IEEE.

Cojocariu, V.M., & Boghian, I. (2014) Teaching the Relevance of Game-Based Learning to Preschool and Primary Teachers. *Procedia-Social and Behavioral Sciences*, 142, 640–646.

Connolly, T.M., Stansfield, M., & Hainey, T. (2011) An Alternate Reality Game for Language Learning: Arguing for Multilingual Motivation. *Computers & Education*, 57(1), 1389–1415.

Curatelli, F., & Martinengo, C. (2012) Design Criteria for Educational Tools to Overcome Mathematics Learning Difficulties. *Procedia Computer Science*, 15, 92–102.

Deiner, P. (2013) *Inclusive Early Childhood Education: Development, Resources, and Practice*. USA: Cengage Learning.

Faustina, E., & Syukri, M.A. (2014). The Effectiveness of Jolly Phonics in Teaching a 4 Year Old Indonesian Child to Read English Text.

Florian, L. (2014) What Counts as Evidence of Inclusive Education?. *European Journal of Special Needs Education*, 29(3), 286–294.

Fotouhi-Ghazvini, F., Earnshaw, R., Robison, D., & Excell, P. (2009) The MOBO city: A mobile Game Package for Technical Language Learning. *International Journal of Interactive Mobile Technologies*, 3(2).

Hulusic, V., & Pistoljevic, N. (2012) "LeFCA": Learning Framework for Children with Autism. *Procedia Computer Science*, 15, 4–16.

Jordan, R. (2008) The Gulliford Lecture: Autistic Spectrum Disorders: A Challenge and a Model for Inclusion in Education. *British Journal of Special Education*, 35(1), 11–15.

Kayımbaşıoğlu, D., Oktekin, B., & Hacı, H. (2016) Integration of Gamification Technology in Education. *Procedia Computer Science*, 102, 668–676.

Kurniawan, R., Mahtarami, A., & Rakhmawati, R. (2017) GEMPA: Game Edukasi sebagai Media Sosialisasi Mitigasi Bencana Gempa Bumi bagi Anak Autis. *Jurnal Nasional Teknik Elektro dan Teknologi Informasi* (JNTETI), 6(2).

Liu, E.Z.F., & Chen, P.K. (2013) The Effect of Game-Based Learning on Students' Learning Performance in Science Learning–A Case of "Conveyance Go". *Procedia-Social and Behavioral Sciences*, 103, 1044–1051.

Marnik, J., & Szela, M. (2008) Multimedia Program for Teaching Autistic Children. *Information Technologies in Biomedicine*, 505–512.

Mar'at, S. (2011) *Psikolinguistik*. Bandung: PT. Refika Aditma.

Pijl, S.J., Meijer, C.J., & Hegarty, S. (Eds.). (1997) *Inclusive education: A global agenda*. Psychology Press.

Porayska-Pomsta, K., Anderson, K., Bernardini, S., Guldberg, K., Smith, T., Kossivaki, L., & Lowe, I. (2013) Building an Intelligent, Authorable Serious Game for Autistic Children and Their Carers. *Advances in Computer Entertainment* (pp. 456–475). Springer, Cham.

Rahman, M.M., Ferdous, S.M., & Ahmed, S.I. (2010) Increasing Intelligibility in The Speech of The Autistic Children by an Interactive Computer Game. *Multimedia (ISM), 2010 IEEE International Symposium* (pp. 383–387). IEEE.

Sailer, M., Hense, J.U., Mayr, S.K., & Mandl, H. (2017) How Gamification Motivates: An Experimental Study of The Effects of Specific Game Design Elements on Psychological Need Satisfaction. *Computers in Human Behavior*, 69, 371–380.

Saiman, K., Sinnatamby, S., Mustafa, L.M., Alias, N., & Siraj, S. (2013) Impact of Video on Learning in Students with Autism in Malaysia: Future Prospects. *Procedia-Social and Behavioral Sciences*, 103, 459–466.

Sedano, C.I. (2012) Workshop: Designing Games for Specific Contexts. *Procedia Computer Science*, 15, 328–339.

Soelistyarini, T.D., Faiza, E., & Nuswantara, D.A. (2010). *IbM GURU-GURU RINTISAN SEKOLAH BERSTAN DAR INTERNASIONAL: APLIKASI BILINGUALISME PADA PENGEMBANGAN PERANGKAT PEMBElAJARAN.*

Soyluçiçek, S. (2011) Graphical Design Issues on Educational Computer Games for Children. *Procedia-Social and Behavioral Sciences*, 15, 642–645.

Sudono, A. (2006) *Sumber Belajar dan Alat Permainan*. Jakarta: PT. Grasindo.

Todd, R.W., & Tepsuriwong, S. (2008) Mobile Mazes: Investigating a Mobile Phone Game for Language Learning. *CALL-EJ Online*, 10(1), 10–1.

Wainer, A.L., & Ingersoll, B.R. (2011) The Use of Innovative Computer Technology for Teaching Social Communication to Individuals with Autism Spectrum Disorders. *Research in Autism Spectrum Disorders*, 5(1), 96–107.

Wilkins A., Cleave R., Grayson N., and Wilson L. (2009) Typography for Children May be Inappropriately Designed. *J. Res. Read.*, vol. 32, no. 4, pp. 402–412.

Zakari, H.M., Ma, M., & Simmons, D. (2014) A Review of Serious Games for Children with Autism Spectrum Disorders (ASD). *International Conference on Serious Games Development and Applications* (pp. 93–106).

Urban Studies: Border and Mobility – Kerr et al. (Eds)
© 2019 Taylor & Francis Group, London, ISBN 978-1-138-58034-3

Homeschooling as an alternative education system in Surabaya

Wulansary
Universitas Airlangga, Surabaya, Indonesia

ABSTRACT: Independent school, better known as homeschooling, is now more becoming an option for parents in Surabaya. This option might be caused by the discourse that conventional educational institutions fail to fulfill children's needs and see their big potential. This research uses a qualitative approach by interviewing three mothers in Surabaya who prefer to homeschool their children. The purpose of this study is to find out the important considerations of parents who choose homeschooling as an education system for their children and describe the success of this education system as well as comparing the outcomes of this education system with the formal education system in Surabaya. The results show that homeschooling is chosen by families since children have access to their respective course content that supports their family's vision. Other than that, homeschooling is also designed to develop children's uniqueness which parents can directly observe. Homeschooling, in addition, gives best practices for children's character formation and knowledge preference. The emergence of the homeschooling phenomenon in Surabaya, which has also emerged in other large cities, could be an interesting study of the theory of Subculture. Where to be a lifestyle choice, and also can be a benchmark of a particular social class.

Keywords: Homeschooling, Education System, Subculture

1 INTRODUCTION

Homeschooling, as a family-based education, is increasingly prevalent in Indonesia, especially in big cities, like Surabaya. The role of parents is more dominantly needed than in conventional education in educational institutions. A child's learning plan is the result of discussions between both parents and children to determine the way forward and can accommodate the abilities of these children. To support homeschooling activities, extended families are involved, and the environment surrounding the children is conditioned.

Homeschooling is rooted and growing in the United States. Homeschooling is also known as home education, home-based learning or independent schooling. The general definition of homeschooling is an educational model in which a family decides to take responsibility for its own children's education by using the home as its educational base. Choosing to be responsible means that parents are directly involved in determining the process of organising education, the direction and purpose of education, the values to be developed, intelligence and skills, curriculum and materials, and methods and practice of learning (Sumardiono, 2007: 4).

The parental role and commitment are highly demanded. In addition to the selection of materials and home school education standards, they must also conduct exams for their children to obtain certificates in order for them to be able to continue education to the next level. Many Indonesian parents who practice homeschooling take course materials, exam materials and home school certificates from the United States. Certificates from the US are recognised in Indonesia (Ministry of National Education) as documents evidencing a graduate of a foreign school (Kompas, 2005).

According to the Ministry of National Education, the school-house track is categorised as an informal education channel that is the path of a family and environmental education (Article 1 of the National Education System Act – Sisidiknas No. 20/2003). Informal educational activities are undertaken by the family in the environment in the form of independent learning activities. Although the government does not regulate the content standards and processes of informal education services, the results of informal education are similarly recognised to formal education (public schools) and non-formal education once students pass tests in accordance with national education standards (article 27, paragraph 2).

In Law No. 20 of 2003 on the National Education System, education is a conscious and planned effort to create an atmosphere of learning and a learning process so that learners actively develop their potential to have spiritual strength, self-control, personality, intelligence, a noble character and skills needed by the society, nation and state. It also describes the national education system as an integral component of education in an integrated way to achieve national education objectives (art. 1).

The philosophy of homeschooling comes from the idea that "humans are essentially beings of learning and love to learn; we do not need to be shown how to learn. What kills learning pleasure are those who try to divide, organize, or control it" (Holt, 1964). Fueled by the philosophy, in the 1960s, there was a widespread discussion and debate about school education and the school system. As teacher and child observer and educationalist, Holt said that academic failure in students is not determined by a lack of effort on the part of the school system, but is caused by the school system itself.

In Indonesia, especially in big cities like Surabaya, the presence of homeschooling learning systems is in demand and increasing in number. It has been proven by the emergence of homeschooling communities, as well as PKBM (Community Learning Activity Centre) which became the registration centre for homeschooling participants. The Community Learning Activity Centre (PKBM) is a government program for organising informal lines of education. Each PKBM program is divided into Program Package A (for elementary level), B (junior high), and Package C (high school level).

This paper posits that homeschooling may result in a new sub-culture. Thus, this paper questions whether or not the renewal of this alternative education system continues to grow and ultimately be a substitute for the formal education system that is already underway in Indonesia. To answer the issues above, the study employs a qualitative approach by interviewing three respondents who are the parents of homeschooled children.

2 DISCUSSION

2.1 *Home schooling types*

Homeschooling specifically consists of three types; Single Homeschooling, Compound Homeschooling, and Community Homeschooling. Single homeschooling is homeschooling performed by parents in a family without joining others. Usually this type of homeschooling is applied because of a specific purpose or reason that cannot be known or combined with other homeschooling communities. Another reason is due to the location or residence of the homeschooling offender that does not allow them to connect with other homeschooling communities.

In contrast, Compound Homeschooling is homeschooling which is carried out by two or more families for certain activities while the main activities are still carried out by their respective parents. The reason is that there are needs that can be compromised by some families to conduct joint activities. Examples include curricula of consortiums, sporting events (eg tennis athletes), music/art skills, social activities and religious activities.

Meanwhile, a homeschooling community is a composite of multiple homeschooling techniques that put together and define a syllabus, instructional materials, basic activities (sports, music/art, and language), facilities/infrastructure, and learning schedule. The commitment for organising between parents and community is approximately 50:50.

The study chose to investigate single homeschooling or family homeschooling as it applied to three families in Surabaya. They are:

1. Maylia Erna, freelance writer, accounting scholar from Airlangga University, Surabaya. Mother of 3 children: Fairuz (14 years), Asha (10 years), and Dede (5 years).
2. Lyly Freshty, housewife, undergraduate of Planology engineering, ITB. Mother of 2 children: Fair (10 years) and Raya (3 years).
3. Maria Sugiyo Pranoto, housewife, scholar from Universitas Brawijaya, Malang. Mother of 2 children: Grace (8 years) and Louis (5 years).

The informants in this research are domiciled in Surabaya and chose different types of homeschooling. Maylia chose the Cambridge curriculum from the United States with the classic method of homeschooling, Lyly Freshty chose the method of Charlotte Manson, and Maria chose the method of electronics that combines methods with a curriculum according to the condition and potential of the child.

2.2 *Reasons for choosing homeschooling*

All parents want the best education for their children. Similarly, the three informants of this research choose the homeschooling education system for the sake of their children's education in terms of matching their children's basic potential and needs.

According to Maylia Erna, homeschooling gives opportunities to develop values and potentials in her children. The following is an excerpt of her answer:

Maylia Erna:

> "I only see opportunities in this homeschooling education path more accelerate the potential strengthening in children. The focus is not easily distorted or biased by the point of view of a massive institution. My other reason is in the planting of basic values and morals to the child uninterrupted by the bias and morals prevailing in a plural society, ie not in the supposed measure, but more commonly done. Do I exclude children? No, it's just that I ensure mentoring so that children learn to assess, respond to and take decision step by step. After I consider their mature in different environments, I ask them to start learning to decide for themselves."

According to Lyly Freshty, homeschooling is a platform from which to follow her family's vision and character formation. Homeschooling makes her family more free and independent. The following is an excerpt of her answer:

Lyly Freshty:

> "I chose homeschooling system because initially dissatisfied with the existing National education system. But the more here the reason the more personal. By choosing homeschooling system we feel more free and independent as well to define our learning vision for children. We chose the Charlotte Mason method because it is not just talent-based interest, but more to the philosophy of character formation."

Maria Sugiyo Pranoto:

> "I decided to choose homeschooling education for my children because it can accommodate the uniqueness of the child. Also to nurture the love of learning desire that is in each child is not extinguished. In addition, to create a strong bonding between parents and children. We as parents can accompany the process of character building children through exercises, good habits are done consistently."

Children who enroll in homeschooling systems have their own way of learning. The process of learning itself might result in developing children's potential as experienced by Maylia's children. Besides, homeschooling also exercises children's discipline and willingness to study.

Maylia Erna:

"I am satisfied with the potential development of my children with homeschooling methods. Because we are parents who supervise and direct it. Like my first child, she was good with her potential in the writing world. Currently she has several times as a speaker in seminars or workshops, although she was still 13 years old. My second child is not as big as her sister's extrofet, but she is more prominent in the picture world. She expresses herself with drawing activity. If the third child, we have not seen the special potential at this time, because he is still 5 years old."

Lyly Freshty:

"I see Fair quite independent and have the discipline and have her own willing to keep learning. For me that was more than enough. We can discuss the study materials and the rest of her who takes the responsibility of her learning process. It's a development that fits our expectations of her. For talent, nowadays even out many things she can do. Starting from music, writing, as well as speaking in public, she did well. I guess that's because of the pattern of education we choose."

In addition to that, homeschooling as experienced by Maria's children supports the personal needs of her children. Thus, family needs support what their children want and like.

Maria Sugiyo Pranoto:

"I feel fortunate to choose a homeschooling pattern for our child. Because from that we come to know earlier about the potential of her who should be sharpened. Grace loves science, she always does experiments that are sometimes beyond the learning we take. Finally we can buy science books according to her interests. Besides she likes to draw, we only facilitate her needs. But we are relieved, because she knows the interests and talents that we should develop."

In terms of socialising with others, homeschooling does not prevent children from joining non-academic activities apart from the homeschooling activities. More importantly, homeschooling does not hinder children from working together with other peers and exchanging experiences.

Maylia Erna:

"No problem. Because even though the children learn at home with us their parents. They also follow many non-academic activities outside. Meet various people of different ages."

Lyly Freshty:

"No difficulties. I always take the kids out to meet people and work with them. Her peers are many too, because we are members of the community. Often do gathering activities together. Like camping, certain skill workshops, watching movie together, etc."

Maria Sugiyo Pranoto:

"At first I also had concerns about the socialization of children. But it turns out my worries are not proven. My child is easy to socialize because often participate in events in the community. Exchanging experiences with peers. In fact they occasionally create a project together. Like making a book or just a small science experiment."

Concerning conventional school, the three of them said that conventional school would not provide their children with the freedom of potential growth at an early stage. They are afraid that conventional schools do not appreciate the uniqueness of each individual.

Maylia Erna:

"My daughter should be homeschooling. It could be if she going to school, her achievement does not appear early on. I don't want she found her potential talent when grown or even old."

Lyly Fresthy:

> "Yes. In school, all children are not appreciated for their uniqueness. All are considered equal. So I think homeschooling is best for Fair and Raya. If it is a concern, I do protect children from unnecessary intercourse. The story of the abduction, understand the things that are not in time, and other terrible things out there. With homeschooling method I can control everything. Children can still make choices, but more because they know the consequences."

Maria Sugiyo Pranoto:

> "I chose the best for my children's education. And homeschooling is the best. My only worry is that children do not grow optimally, because they are educated with the general in regular school."

From the results of this research, it was found that parents who chose homeschooling patterns for their children's education were parents with well educated characteristics. The three mothers who were informants for this research were all scholars from leading universities in Indonesia, namely Airlangga University, Brawijaya University and Bandung Institute of Technology. The informants were well aware of the importance of an education that is appropriate to their children's potential. In terms of socioeconomic class, they have done with financial affairs, proven by the average as a mother, they are not working to help sustain family economic conditions. Ownership of good cars, houses and outdoor activities is also an indicator. Homeschooling activities outside the home are relatively frequent and nothing is free; there is always money to be spent on the activity.

The main consideration of informants choosing homeschooling patterns is to allow the potential of their children to grow optimally. Also, another consideration is the wish to protect them from an educational system that they feel is not good for a child's growth.

With the advancement of technology, the application of social media is a very effective means for them to be connected. Exchanging information and scheduling activities is conducted through social media. One of the big bonuses is the homeschooling community: Rumah Inspira. Currently the number of members is 57,850 people. Generally, the members of homeschooling families have the same lifestyles, perspectives, and mindsets. On average, they are well educated, visible from the type of activities and style of language they use. The characteristics that can be drawn in terms of similarities include: daring to try new things, being critical, being disciplined, having firm principles in terms of children's education, reading many books, doing the same trend. Given this similarity, it can be assumed that a sub-culture is growing. The realisation that this group grows in a different way from established cultures does not seem to be aware of the self. But as an observer, the growth of the homeschooler community, with an ever-increasing number of members, as well as with its characteristics, is a sub-culture.

According to Cohen (1995), the rise of subcultures represents groups of individuals connected to each other through interactions and interests and similar characteristics, representing locality such as skin colour. The common interests of subculture members also make them identify themselves differently from the culture/culture of the majority (Williams, 2011: 8).

Subculture is born of a spirit of resistance or resistance to the culture that dominates it. It needs a clear identity to distinguish the subculture with the culture that dominates it. Identity is needed to build a commonality and spirit of togetherness, which is shown by the style, behaviour, art system, and even the value system adopted by its members. What they do in their daily life is deliberately set away from the cultural hegemony that dominates it.

Departing from the meaning of the subculture above, the homeschooling community can be categorised as a newly emerging sub-culture even though it is detached from the consciousness or absence of members within the community.

3 CONCLUSION

Homeschooling is an alternative education pattern that does not have to be formal schooling, but rather is based on the independence of learning patterns. Parents play a very important role here. The decision to take the homeschooling education pattern is decided together in the family, with the aim of wanting to provide the best education for children. The failure of the conventional education system results in the assumption that the potential of the child is not considered a uniqueness that must be understood. This is what makes parents choose homeschooling patterns.

The growing number of homeschooling enthusiasts and their interconnected communities, both locally and nationally, creates a new sub-culture. I, as an observer, consider that their existence is a subculture because they all have similar characteristics to each other, both parents and their children, namely: daring to try new things, being disciplined, being critical, having good principles and knowledge in the field of education and patterns of parenting, reading books, being learners, moving in the same trend. It may be the case that members of the community are not knowingly members of a particular subculture, which to this day has been made, and there is no name for this subculture.

REFERENCES

Amalia, Y. &.-M. (2014). Dialogue in narrative inquiry: Collaboration in Doctoral Study in the USA. In P. Smeyers, D. Bridges, N.C. Burbules, & M. Griffiths, *International Handbook of Interpretation in Educational Research.* London: Springer.

Amalia, Y.S. (2010). Educational Reform from Centralizing to Decentralizing Curriculum in Indonesia: What the Teachers Think. *University of Illinois.*

Holt, John. (1964). *How Children Fail.* New York: Pitman Publishing Corp.

Jurianto. (2015). *Espektasi Peserta terhadap Strategi Mengajar dan Materi Ajar di Kelas ELPT Preparation.* Surabaya.

Jurianto. (2015). Strategies for Teaching Writing in EFL Class at a Senior High School in Indonesia. *CELT: a Journal of Culture, English Language Teaching & Literature.*

Mulyadi, Seto. (2007). *Homeschooling Keluarga Kak Seto.* Bandung: PT. Mirzan Pustaka.

Sumardiono. (2007). *Homeschooling Lompatan Cara Belajar.* Jakarta: PT. Elexmedia Kompatindo.

Williams, J. Patrick. (2011). *Subcultural Theory – Traditions and Concepts.* Cambridge: Polity Press Indonesian Constitutions. 1945.

Borders and mobility in gender, identity, and behaviorism

Urban Studies: Border and Mobility – Kerr et al. (Eds)
© 2019 Taylor & Francis Group, London, ISBN 978-1-138-58034-3

Environmentalism and consumerism: The contradiction of globalization in behavior consumption of the urban middle class in Surabaya, Indonesia

D.A. Arimbi, N. Wulan & F. Colombijn
Universitas Airlangga, Surabaya, Indonesia

ABSTRACT: The aim of this research is to study the human–environment interactions in Surabaya, the second largest city in Indonesia. We will focus on middle-class behavior because the middle class is most susceptible to globalizing impacts on the environment. This study uses a qualitative method: observation and purposive sampling (in-depth, open-ended interviews) in Surabaya and the data are taken from 2017 to 2018. The targeted findings of this research are the consumption behaviors of the urban-middle class in Surabaya and how this class negotiates between environmentalism and consumerism. Important findings include models of consumption and environmental awareness of the middle class in Surabaya. One most important finding indicates that the higher the social class the individual is, the less s/he is concerned with environmental issues. The choice of middle-class consumptive lifestyles functions as a strong influence for its members to neglect their environmental consciousness for the sake of prestige and consumption.

Keywords: globalization, environmentalism, consumption, Surabaya

1 INTRODUCTION

The contemporary global environmental crisis is arguably the single most important issue faced by our world. Local issues of human health, threatened by air pollution, unclean water, inadequate collection of solid waste, and a failing sewerage system, are particularly pertinent to the lives of millions of inhabitants of cities in the Global South (Drakakis-Smith 1995).

The North is notorious for its indulgence in consumption, but people in the South aspire to the same 'modern' lifestyle as people in the North. Economic growth will enable them to realize these aspirations. What will happen when the populations of Indonesia, China, India, Brazil, and other emerging economies consume at the same level per capita as the North? This question is not meant to lead attention away from the Northern responsibility for its problematic consumption and production methods, nor to deny emerging economies their fair share of global resources and high consumption levels! But the question does emphasize that the economic development of countries in the South will create problems for the environment.

The aim of this research is to study the human–environment interactions in Surabaya, the second largest city in Indonesia. We will focus on middle-class behavior, because, as will be detailed in the literature review below, the middle class is most susceptible to globalizing impacts on the environment. These global forces produce two contradictory results. On the negative side, global models (from the West, but also for instance from emerging economies such as China) stimulate wasteful consumption patterns. Indonesians follow global fashion trends, aspire to use a car instead of alternative transportation, use air-conditioning (considered 'modern') instead of making use of natural ways of cooling a house, and so on. On the positive side, global models of environmentalism also make an impact on the Indonesian middle class and stimulate its members to behave in an eco-friendly way. Such models of pro-environmental behavior can come from an environmentalist discourse, but also, for instance, from Islam or other religions.

Indonesia is rife with paradoxical examples of this double influence. For instance, as we once noted the car-free day in Semarang was opened by the Governor of Central Java, who came to the event by car. He found it beneath his dignity *not* to arrive by car. Another example is offered by environmental activists who hold a conference in a luxurious hotel on Bali; both the aerial flight and climate conditions in the hotel have no doubt harmed the environment.

The specific purpose of the research is to answer the central question addressed in this proposal, which is how urban middle-class Indonesians negotiate the contradictory global influences of wasteful consumption and environmental care. The research is very important to understand human–environmental interactions at a theoretical level. The results will also add to a global, comparative understanding of the effects of economic growth. And finally the research will produce practical insights that can be used to build environmental awareness and action in Indonesia.

Part of the research needed to be conducted in the Netherlands. Accepting the definition of the Brundtland Commission that sustainable development is development that "meets the need of the present without compromising the ability of future generations to meet their own needs" (WCED 1987), any study of sustainable development must have a time depth of at least one generation. Therefore adding a historical perspective to study contemporary environmental sustainability is imperative. Due to the Indonesian-Dutch postcolonial legacy, there is by far no country in the Global South with equally rich historical sources than the Netherlands, which enable the study of human–environment interactions. Just as important is the fact that the KITLV in Leiden has the best library resources on Indonesia anywhere in the world.

The most innovative aspect of this proposal (the urgency of this research) is the choice to study the dialectics between the contradictory global trends of environmentalism and consumerism of the middle class. Existing studies of middle-class consumption in Indonesia (e.g. Robison 1996; Tanter and Young 1990) do not focus on the environmental consequences of these consumption styles. Existing studies of the environment have too much ignored Indonesia, which is a serious mistake. As the fourth largest country in the world, situated in one of the most dynamic parts of the world, Southeast Asia, Indonesia should be of prime concern for a study on the environment in emerging economies in all circumstances. The study is also innovative in its conscious use of a historical approach, as a necessary prerequisite for an environmental study. To study history is not, as so often, an aim in itself, but is indispensable for testing the sustainability of contemporary institutions. This approach will enrich theoretical debates about both scavenging and middle-class consumptive lifestyles.

2 REVIEW OF RELATED LITERATURE

The literature on environmental problems is enormous. If we focus only on solid waste, the number of scientific publications about waste is as overwhelming as the volume of waste itself. There are also many specialized journals, like *Waste Management*, or *Resources, Conservation and Recycling* about solid waste.

Growing urbanization and economic development have made the issue of urban pollution, and the treatment of waste in particular, more pressing than ever. Economic growth produces more pollution; conversely, growing pollution threatens the sustainability of (economic) development. Urban pollution not only puts economic growth at risk and negatively affects human health but also imperils human well-being in general (Davis 2007).

The influential Brundtland Commission contends that economic and social development can go together with preservation of our resources and does not need to happen at the expense of chances of development for future generations (WCED 1987). This contention, however, is a moot point. David Satterthwaite, for instance, argues that the key to environmental degradation is found in middle-class and elite consumption patterns and the urban-based production and distribution systems that serve them (Satterthwaite 2003: 74). If we accept the working hypothesis that urban poverty is not necessarily the biggest cause of environmental degradation in the Global South (Ghanimé and Jahan 2004; Martínez-Alier 1995), but wealth might create problems, the question arises as to what impact growing prosperity will make.

It is not just growing prosperity that is creating environmental challenges, but the modernization of the world. In the modernized world, or, in Zygmunt Bauman's words, at a time of "liquid modernity", when fixed, traditional frameworks were abandoned, people had the freedom, but were also required, to make individual choices to determine their position in society, still, of course, within societal constraints (Bauman 2000: 18–23). One way to find one's place in modern society is by consumptive choices. While the drive to modernization was perhaps led by the West, societies outside the West (who were often once colonized by the West) adopted modernity as well, but selectively; they picked what suited them best (Colombijn and Coté 2015). In Indonesia, advertisements in newspapers and journals from the mid-twentieth century promoted fashionable clothes, electric lighting, and modern kitchenware. Also less tangible things and certain behavior conveyed the idea of modernity: Western music, American films, smoking cigarettes, and, especially for women, the use of toothpaste and having fresh breath (Basundoro 2015; Husain 2015; Khusyairi and Colombijn 2015; Kusno 2010: 168–81; Maier 1997; Schulte Nordholt 1997: 20).

In a process of objectification, people engage with material objects and as consumers resocialize commodities from which they had become estranged as producers (Miller 1995: 143). Thorstein Veblen used the term conspicuous consumption to describe the deliberately wasteful consumption of the elite (O'Brien 2011: 156–9). The combined result of this stage of liquid modernity, objectification of culture, and conspicuous consumption is an ever-faster level of consumption. Fashion and inbuilt obsolescence compel people to throw away goods that in pre-modern times would have been used much longer. Much more waste is being produced than humans would do if they simply satisfied their material needs.

However, it would be wrong to see economic development of a growing middle class only as a threat to the environment. The current economic development in the Global South may produce several simultaneous effects, especially if the growth reaches larger segments of the population. Growing consumption is likely to produce more waste. Fewer people might be willing to take on the task of waste-picking, which is as a rule looked down upon by the society at large. So, the problem of waste may worsen in two ways when economies develop: more waste is produced and less is recycled.

Nevertheless, these negative environmental effects of development might be offset by growing environmental awareness among the middle class, which takes in global notions of environmentalism. In sum, a better insight into the conditions for environmental awareness in the middle class helps to find tools to reduce the swelling problem of waste in the Global South.

Narratives that explain the status of the environment in general, or waste and consumption in particular, are influenced by political and economic interests (Nunan 2015: 31; Robbins 2004: 12). Therefore the study of consumption, the environment, and waste should be undertaken from a perspective rooted in political ecology.

Paul Robbins (2004: 12) defines political ecology as the "empirical, research-based explorations to explain linkages in the condition and change of social/environmental systems, with explicit consideration of relations of power." These relations of power are, perhaps by definition, uneven, and consequently the study of waste involves a study of the exercise and unequal distribution of power. Seeking social justice is a key aim of a political ecology analysis because there is a normative understanding that there are more sustainable and less exploitative (of both nature and people) forms of human–environmental interactions. The analysis of power is also a matter of scale, and local environmental problems must be studied in the context of national and international influences (Bryant 1992; Jaffe and Dürr 2010; Nunan 2015; Robbins 2004).

3 METHOD OF RESEARCH

The overarching research question we wish to explore is how middle-class people in Indonesian cities produce and process waste, and reflect on their interactions with the environment.

The research took place in Surabaya, the second largest city in Indonesia, with a population of 2.8 million people. Surabaya is a port city, with a comparatively well-developed industry, the largest naval base in the country, and like most cities in the Global South, a large and varied service sector. It is also the capital city of the province of East Java (Dick 2002).

The research covers the particular question of what meaning does the middle class give to the exemplary place of consumerism, the shopping mall. How does the middle class relate to global trends? Why do they sometimes indulge in excessive consumption and other times join global forces to fight environmental damage? How do they negotiate contradictory global messages of consumerism and environmentalism? How do they integrate the use of their car, exemplary object of consumerism, with a green awareness?

The research collected data from 3 malls in Surabaya and interviewed 30 respondents in these malls. The respondents chosen were those who have been categorized as middle-class people that are called *cukupan*, categorized as those who have an income range of 3–10 million rupiah. We also chose some students who are called new-middle-class people because education has helped them be categorized as educated people, and only rich people and middle-class people can apply to be a member of the university. However, the scholarship that the government give means lower-class people can join universities (Gerke, 2005).

4 FINDINGS

As a business and industrial city that absorbs many people, Surabaya also creates problems related to waste and pollution. Traditionally, industries in Surabaya increase their productivity and this enables economic growth and prosperity. The consumption behavior is also one of the effects of the target of economic growth and prosperity. However, this condition can also be a problem for citizens in negotiating the fundamental message toward preserving their environment and the global message of consumerism.

The citizens are just the objects of globalization that leads us to consume more products. As the target of globalization is to make the distant become closer, reachable, and affordable, the citizens tend to choose competitive products that bring them benefits. When the respondents were being asked questions related to their shopping activity and their understanding about eco-awareness, it seems that they know but are not intensively concerned about it.

The data were taken from some middle-class people who were enjoying the malls during the weekend. They comprise several answers from several questions related to renewable products, energy-saving program, and sustainability. The first question was related to how far they understand green awareness or eco-awareness. The results showed that 25 people out of 30 said they understood enough about it, and 5 of them do not understand about it.

Market industry as the actor who provides material products and services has an important role in offering green products or services. Unfortunately, in Surabaya it is not becoming an important focus for the government relating the greening environment program. They still freely spread solid waste, which is mostly plastic waste.

In addition the amount of plastic used has increased drastically coupled with an increase in the intensity of shopping. It is impossible to restrain citizens' habit of shopping. Advertisements invite them to help each other by consuming some products. The more people consume industrial products, the more salary the employee gets; the more salary the employee gets, the more consumptive s/he is. As we know, the word 'prosperity' means that all people become successful and there are equal financial opportunities (Press). So, it is difficult to control their shopping activity since prosperity, especially financial wealth, is the target of life. Some of the respondents said that it is impossible to control their outcome, because the more salary they get, the more money they spend. It means that the more plastics they consume, the more fuel they use for reaching the mall and the more energy they use.

The way they consume actually also gives them their identity. According to Solvay Gerke, the rise of middle-class people in Indonesia is marked by the rise of consumption. By consuming many things people can construct their legitimacy; they can be the arbiters, hawkers, and trend-setters of the 'modernity' (Gerke, 2005). As the researchers analyze through the way the respondents consume, all of them enjoy the shopping mall for getting fashionable things such as clothes and make-up, having coffee at Starbucks, or dinner at McD or at the food court. Those things in Surabaya are costly to imitate; only middle-class society can spend their time and money in consuming them. When the respondents were asked to choose between reducing their consumption habit and reducing the use of plastic waste, they gave answers indicating that whilst they still do some measure of environmentally friendly habits, they cannot reduce what they buy, especially because they buy everyday needs alongside family (N, 2017), reminding us that some Indonesian middle-class people are bold enough to reject the invitation to reduce their consumption lifestyle. On one side, the program that invites them to prevent the environment from being polluted is beneficial, and on the other side it is illogical enough for them because they want to preserve their prestige.

In addition, this rejection will be a new trend for new-middle-class society, students, for example. Even if they do not have income from other institutions, they have a daily budget from their parents. Some of the respondents buy make-up that is an international brand, such as Victoria's Secret. This make-up has helped them to be part of modernity. Those who wear this make-up have been already being members of an international fashion trend. However, being educated in university does not help you to be a member of international modernity (Gerke, 2005). It can be looked at how they prefer to buy make-up, T-shirts, bags, and shoes at the mall rather than the traditional market, as we know that the brands are only located in the high-class malls, such as Galaxy Mall, Tunjungan Plaza Mall, and Grand City Mall.

5 INTRODUCING GOVERNMENT AS THE AGENT OF LOCAL CULTURE

Regarding the green awareness movement, the Surabaya government has already applied this program by inviting its citizens to join in many events. In contrast, this program tends to favor the externalization of ecological costs and blend with some product promotions that let capitalist development increase. From the statements above, we know that on one side the government intensively invites its citizens to keep the environment clean and minimize solid waste, but the business actors, such as stores, give them plastic bags to contain their purchases. Markus Lederer said that this ambivalence actually should be solved by using the term of trade-offs. In this system, the government as the policy maker should be consistent with the effort of greening the land. There will be some sectors who will be traded-off in order to be consistent (Lederer, 2018). The United Nations Environment Program (UNEP) has listed 10 key economic sectors that are "driving the definition trends of the transition to Green Economy including increasing human well-being and social equity and reducing environmental risks and eco-logical scarcities". In addition, according to Urban F, the key characteristic of the Green Economic approach is that social, ecological, and economic issues should be approached in an integrated manner and that they entail some different types of feedback effects. Positive interaction within and across these two issues can be considered as synergies (Urban, 2013). Lederer said that both trade-offs and synergies are the result of interaction that may come about through either deliberate action or as unintended consequences and can themselves become the subjects of bureaucracy management (Lederer, 2018). In contrast, some of the trade-offs and synergies that have been done are adverse for one of two sides; commonly, the government could not control maximally that its side is always get that adverse. It is evident that politics always creates risks for citizens. The government as the agent of local culture forgets to bring its citizens for upholding the local wisdom. In this process, local cultures do not work because of the wave of globalization.

6 CONCLUSION

In Surabaya, the above forms of the symbolic demonstration of social class and group membership exist because of interpretation of status or prestige. With rapid capitalist economic development that has a target of prosperity for all people, prestige and status become negotiable values depending mainly on how they live and their consumption patterns. The international standard of consumption also helps Indonesian people become members of middle-class society on a national or international scale. In the name of increasing the grade of prosperity, globalization brings us to Western culture. In this process, local culture does not work to help people reduce the usage of plastic or solid waste.

REFERENCES

Basundoro, Purnawan, (2015). The two alun-alun of Malang (1930–1960), in: Freek Colombijn and Joost Coté (Eds), *Cars, conduits, and kampongs: The modernization of the Indonesian city, 1920–1960*, (pp. 272–299). Leiden, Boston: Brill.

Bauman, Zygmunt, (2000). *Liquid modernity.* Cambridge and Malden: Polity Press.

Colombijn, Freek and Coté, Joost, (2015). Modernization of the Indonesian city, 1920–1960, in: Colombijn, Freek and Coté, Joost (Eds), *Cars, conduits, and kampongs: The modernization of the Indonesian city, 1920–1960*, 1–26. Leiden, Boston: Brill.

Davis, Mike, (2007). *Planet of slums*. London: Verso.

Dick, Howard. W., (2002). *Surabaya, city of work: A socioeconomic history, 1900–2000*. Athens: Ohio University Press.

Drakakis-Smith, David, (1995). Third World cities: Sustainable urban development 1, *Urban Studies* 32, 659–677.

Gerke, S. (2005). Global Lifestyle under Local Conditions. In Chua, B.H., *Consumption in ASIA: Life style and Identities* (pp. 135–158). Singapore: NUS Press.

Ghanimé, Linda & Selim Jahan, (2004). *The poverty-environment nexus: reinforcing linkages* [Draft October], UNDP.

Husain, Sarkawi, (2015). Chinese cemeteries as a symbol of sacred space: Control, conflict, and negotiations in Surabaya, in Colombijn, Freek and Coté Joost (Eds), *Cars, conduits, and kampongs: The modernization of the Indonesian city, 1920–1960*, (pp. 323–340). Leiden, Boston: Brill.

Jaffe, Rivke and Dürr, Eveline (2010), 'Introduction: Cultural and material forms of urban pollution', in Dürr, Eveline and Jaffe, Rivke (Eds), *Urban pollution: Cultural meanings, social practices*, (pp. 1–29). Oxford and New York: Berghahn.

Khusyairi, John A. and Colombijn, Freek (2015). Moving at a different velocity: The modernization of transportation and social differentiation. In Colombijn, Freek and Coté, Joost (Eds), *Cars, conduits, and kampongs: The modernization of the Indonesian city, 1920–1960*, (pp. 251–271). Leiden, Boston: Brill.

Kusno, Abidin, (2010). *The appearances of memory: Mnemonic practices of architecture and urban form in Indonesia*. Durham and London: Duke University Press.

Lederer, M. (2018). Tracing Sustainability Transformations and Drivers of Green Economy. Approaches in The Global South. *Environment and Development*, 3–25.

Maier, Henk, (1997). Maelstrom and electricity: Modernity in the Indies. In: Nordholt, Henk Schulte (Ed.), *Outward appearances: Dressing state and society in Indonesia*, (pp. 181–197). Leiden: KITLV Press.

Miller, D., (1995). Consumption and commodities, *Annual Review of Anthropology* 24, 141–161.

N, M. (2017). September 9. Eco-Awareness. Galaxy Mall. Niken, Interviewer.

Nunan, Fiona, (2015). *Understanding poverty and the environment*. London and New York: Routledge.

O'Brien, Martin, (2011). *A crisis of waste? Understanding the rubbish society*. New York, London: Routledge.

Press, C.U. (n.d.). *the meaning of prosperity*. Retrieved April 6, 2018, from Cambridge Dictionary: https://dictionary.cambridge.org/dictionary/english/prosperity.

Robbins, Paul (2004). *Political ecology: A critical introduction*. Malden etc: Blackwell.

Robison, Richard (1996). The middle class and the bourgeoisie in Indonesia. In Robison, Richard & Goodman, David S.G. (Eds.), *The new rich in Asia: Mobile phones, McDonald's and middle-class revolution*, (pp. 79–101). London & New York: Routledge.

Satterthwaite, David, (2003). The links between poverty and the environment in urban areas of Africa, Asia and Latin America, *The ANNALS of the American Academy of Political and Social Science* 590: 73–92.

Schulte Nordholt, Henk, (1997). Introduction. In: Nordholt, Henk S. (Ed.), *Outward appearances: Dressing state and society in Indonesia*, (pp. 1–37). Leiden: KITLV Press.

Tanter, Richard & Young, Kenneth (Eds.), *The politics of middle class Indonesia*, Clayton: Centre of Southeast Asian Studies.

Urban, F. &. (2013). Key Issue. In F. &. Urban, *Low Carbon Development.* Milton Park, England: Routledge.

WCED (World Commission on Environment and Development) (1987). *Our common future*. Oxford, New York: Oxford University Press.

Urban Studies: Border and Mobility – Kerr et al. (Eds)

The enforcement of state territoriality and shifting on borderlanders' mobility: The case of Indonesia–Malaysia border in Sebatik Island

L. Puryanti
Universitas Airlangga, Surabaya, Indonesia

ABSTRACT: Indonesia–Malaysia borders have been signified by the nature of porous, permeable, 'soft border' borders in their history and socio-cultural relationships. However, due to their current interests in their own national priorities of sovereignty, defence, and security, the states have made the borders more rigid. Taking the case of the Ambalat Sea Block dispute as a background, I argue that what was in the past was understood as a deterritorialized border gradually changed to be more reterritorialized through the enforcement of state territorialisation from both the states. This paper aims at to investigate the effects of the state territorialization toward the border landers in their everyday-life mobility. The research was conducted in some villages in the Indonesian part of Sebatik Island, a small divided island of the two sovereign states of Indonesia and Malaysia, in the province of North Kalimantan during 2013–2016, with the Bugis community as the research subject. The findings of the research show the state policies on territoriality have affected what used to be an intensified cross-border flow of people and goods into the Sebatik Island and (re)positioned the border population at a crossroads between openness and control of physical borders and socio-political boundaries'.

Keywords: territoriality, Indonesia Malaysia borders, deterritorialization, reterritorialization, everyday life

1 INTRODUCTION

The paper aims to investigate the effects of the enforcement of state territorialization from both Indonesia and Malaysia on the Bugis community's everyday-life mobility on the border of Indonesia in Sebatik Island. This research took place in some villages located on the Indonesian part of Sebatik Island under the administration of Nunukan Regency (*Kabupaten*), the Province of Kalimantan Utara, Indonesia, along with the land border with Malaysia Sebatik and the maritime border with Tawau Sabah Malaysia. The combination of land and maritime borders inevitably form the uniqueness of the site compared with other borders in the world with a single kind of border (Baldacchino 2013).

As a research background, this border island has undergone a gradual process of Indonesian reterritorialization after the case of the Ambalat Sea Block Dispute (2005) and the *Gerakan Reformasi* (Reformation Movement) in 1998. This Indonesian reterritorialization has, through a national policy, encouraged relations between the center and the periphery in Sebatik Island in the areas of society, economy, culture and local politics because the paradigm has shifted from the outer islands being the backyard to the front veranda of the country. In turn, the development of border areas has become a part of the National Development Plan for its strategic roles for the country in the aspects of defense and security, economies, and social cultural dynamics (Nasional 2010). On the other side of the border, Malaysia, which is more developed in its current economic level, also has requirements to protect its border by enactment of the policies for border control. The purpose of the policies, similar to other states in Southeast

Asia, "is not fortifying its borders in response to growing interstate wars, as had traditionally been the case. Rather, particularly since 2001, it has tightened controls to battle illegal migration as well as smuggling and sea piracy." Prior to 2000, the Philippine–Sabah border, including the Indonesia Sebatik–Tawau Sea border in the Celebes Sea, was extremely porous. Fishermen, traders, migrants, and even separatists moved back and forth largely unhindered. This pattern changed in April 2000, when about 20 tourists were kidnapped on the Malaysian resort island of Sipadan, which lies off the northeastern coast of Sabah and was claimed by Indonesia at the time. The Abu Sayyaf group from the southern Philippines, who often mantle themselves in separatist rhetoric, claimed responsibility for the kidnapping (McGahan 2008).

As a result, processes of state (re)bordering that highlight the increasing securitization and impermeability of borders between the two states have heavily impacted the Indonesia–Sebatik island. These recent processes were accentuated by long historical border relations and the 'cooperation' of Indonesia and Malaysia as neighboring countries, including cultural social and kinships relationships that bind the people living at the borderland of these countries. To this date, the intensified cross-border flow of people and goods—legal and illegal—has been accompanied by mutual tightening of sea borders, on which Malaysia has imposed even tighter controls given the flows of migrants from Sebatik that resulted in the (re)positioning of the border population at a crossroads between openness and control of physical borders and socio-political boundaries.

Based on ethnography fieldwork and historical research that I conducted between 2013 and 2016 as part of my dissertation, this paper investigates the changes in the everyday-life mobility of the Bugis community as can be seen in socio-economic aspects of life across the border as a result of the effects of the new state policies on their border. Some stories from the field are very important to explicate my argument in this paper.

2 DISCUSSION

2.1 *The effects of state territorialization on border mobility on Sebatik Island*

The more restricted border between Indonesia and Malaysia has been a new experience for people on Sebatik Island since Tawau was closed to Indonesians from Sebatik Island in 2013, and this creates a hard situation for ordinary people and changes their everyday-life mobilities.

Before the closing, people made the crossing to Sebatik Malaysia by walking across from the districts of Sebatik Tengah and Sebatik Timur to the neighboring Malaysian villages. They did not have any difficulties in getting through. There was no fence or any barriers that clearly showed that it was the borderland zone. A few villages in the districts (Haji Kuning and Sei Pancang Village) also shared a watershed with Malaysian villages. A small simple Indonesian military post was located near the border marker. People just crossed the border without being scrutinized by the guardians since it was part of their everyday activities.

People made the crossing to Tawau in different ways. Before the year 2013, they could go directly to Tawau from the border of Sebatik island to purchase basic needs such as groceries and cooking gas in Tawau. They equipped themselves with a PLB (*Pas Lintas Batas*, which means Border Crossing Pass), which was easily obtained for the locals, or a passport. They could cross the sea border by a passenger speedboat by paying RM10 (around 30,000 Rupiah at that time) or go with some traders with their *jongkong* (wooden boats) that were loaded with cash crops to Tawau. Approaching the Sabindo Tawau Port they would continue with smaller *pete*. This way meant that, as Sebatik Indonesians, they did not have to go through formal Tawau immigration at the Tawau checkpoint, and instead entered Sabah via the backyard of the immigration office. One of the Sebatik Indonesians would take their PLB or passport, along with others, to be stamped at the Tawau immigration. This was a common arrangement prior to the introduction of border control. For this 'facility', everyone including the researcher had to pay RM 4–5 for *uang jasa* (fees for services). Through this 'back door', they were able to officially enter into Malaysian territory. Yet, there were others who entered Sabah illegally, without having passports stamped. This was common among illegal migrants who entered without travel documents by using speedboat services from Sungai

Nyamuk in Pulau Sebatik, which shares a border with Malaysia to Tawau (Dollah 2007). Illegal migrants could take advantage of these daily speedboat services that carried goods and passengers between Pulau Sebatik and the Batu jetty near Tawau port.

However, such relaxed restrictions changed in 2013 when Malaysia introduced tight controls forbidding Indonesians from Sebatik Island from crossing over to Malaysian's domestic port. Since the introduction of new sea and land controls, the Indonesian residents of Sebatik Island have no longer been able to use their previous ways of travel. They now have to go through Nunukan, which has formal immigration facilities. Five ferries with a capacity of 100–158 passengers operate daily to and from Tawau and Nunukan port, namely, KM Mega Express, KM Malindo Express, Mid East Express, KM Tawindo, and MV Labuan Express. A one-way trip from Nunukan, Indonesia, to Tawau, Sabah, takes about two hours with a fee of IDR100,000 (USD11). According to the Consortium of Tawau Ferry Companies, at times, up to nine ferry services operate daily, particularly during busy festive seasons, managed by Tawau Ferry Terminal and Customs, Immigration and Quarantine (CIQ) (Mahadi 2014).

To anticipate this changing situation, the local Sebatik inhabitants who have been accustomed to cross the sea-border for various purposes now enter Tawau officially through Nunukan holding their passport or *PLB*. This international standard of immigration enables these Indonesians to stay in Sabah for one month. Ironically, there is a loophole. When Indonesians from Sebatik Island are officially in Malaysia, they are able to travel easily to the Malaysian side of Sebatik Island and then easily return to their homes by crossing the land border within the island, where there is no border control. They merely return to their own houses on the Indonesian side of Sebatik as the national boundary on this island is, ironically, not controlled. With this strategy, the Indonesia Sebatik Islanders are free to go back and forth between Indonesia Sebatik and Tawau while in the Malaysian official immigration record they are still in Malaysia. While this maneuver is continuously conducted by the Bugis Sebatik community, especially among traders, the tightened border exerts a strong influence on everyday practices on the island; stricter movement across borders has actually lessened dependency on Tawau and motivated Sebatik Islanders to develop new and closer connections with Indonesia. Besides, there were always some people who could not go to Tawau. Many of those were recent immigrants and some of them were deportees and did not have passports or *PLB* and did not share the habit of crossing the border that old residents had. Therefore, people who established themselves on the Indonesian side of Sebatik Island did not automatically or necessarily incorporate Tawau into their mental map. The situation of the Sebatik Bugis border-crossers is similar to Hasting and Donnan's argument, that "local experience of the state and resistance to it cannot be limited to the imaginative representations: attention must be paid to the very material consequences of the actions of states for local population" (1999: 3).

In one interview with a palm oil gardener on the island, the shifting of everyday-life mobility had also affected their perception about crossing border mobility. Haji Kamil, an Indonesian who had been a migrant worker in Malaysia, is a gardener'in Indonesia Sebatik that has not visited Malaysia for years. He said "While we can work here, have our own garden or easily go fishing to the sea, why should we go to Malaysia? Malaysia is only a place for *makan gaji*. Once we have money to buy land in Sebatik, we don't need to return to Malaysia and repeat the bad experience with the Malaysian police. We are happier here in Sebatik Indonesia." A lady, an owner of clothes shop in Indonesia Sebatik, said that Malaysian fashion is out of date compared with Indonesian fashion. Therefore, she bought all the clothes from Jakarta or Surabaya (owner of clothes shop in Sebatik). Many people in Indonesia Sebatik right now are more connected to Indonesian cities because of better transportation facilities.

3 CONCLUSION

Related to the process of reterritorialization, it can be said that people on the border of Sebatik island experience a process of infrastructural and cultural integration with Indonesia, as well as a physical detachment from the bi-national region in which it is immersed. This situation is strongly reflected through their everyday practices, and examining patterns of everyday practices exposes some of the systems of difference and exclusion that nation states have introduced and enforced. The physical barriers that restrict cross-border mobility rearrange identities and boundaries between both Sebatik and Tawau and have reorganized social relations and changed how the Bugis relate to the space where they live. With regard to, the changes on modes of mobilities to immobilities, I suggest that the process of changes the border subjects Indonesian citizens to is fertile ground and open for further research.

REFERENCES

Acciaioli, Greg. 2000. "Kinship and debt; The social organization of Bugis migration and fish marketing at Lake Lindu, Central Sulawesi." In *Authority and Enterprise Among the peoples of South Sulawesi*, edited by and Greg Acciaioli Roger Tol. Kess van Dijk. The Netherlands: KITLV Press.

Acciaioli, Greg. 2004. "From Economic Actor to Moral Agent: Knowledge, Fate and Hierarchy among the Bugis of Sulawesi." *Indonesia* (78):147–179.

Ammarel, Gene. 2002. "Bugis Migration and Modes of Adaptation to Local Situation." *Ethnology* 41 (1):51–67.

Baldacchino, Godfrey. 2013. *The political economy of divided islands: Unified geographies, multiple polities*. Houndsmill, Basingstoke, Hampshire; New York, NY: Palgrave Macmillan.

Bauman, Zigmunt. 1998. *Globalization. The Consequences*. Columbia: Columbia.

Dollah, Ramli & Mohammad, A.M. 2007. "Perdagangan Tukar Barang Malaysia Indonesia: Potensi dan Cabaran." *Jati* 12 (1):83–106.

Held, David, and Anthony G. McGrew. 2000. *The global transformations reader: An introduction to the globalization debate*. Malden, Mass: Polity Press.

Mahadi, Syed Abdul Razak Bin Sayed. 2014. "Indonesian Labour Migration to Sabah: Changes, Trends and Impacts." Doctor of Philosophy Dissertation, School of Social Sciences Discipline of Geography, Environment and Population, The University of Adelaide.

McGahan, Kevin. 2008. "Managing Migration: The Politics of Immigration Enforcement and border Control in Malaysia." Doctor of Philosophy, Graduate School, University of Wisconsin—Madison.

Nasional, Badan Perencanaan Pembangunan. 2010. Rencana Pembangunan Jangka Menengah Nasional (RPJMN) tahun 2010–2014, edited by Kementrian Perencanaan Pembangunan Nasional/BAdan Perencanaan Pembangunan Nasional (BAPPENAS). Jakarta: Kementrian Perencanaan Pembangunan Nasional/BAdan Perencanaan Pembangunan Nasional (BAPPENAS).

Newman, David. 1998. "Geopolitics Renaissant: Territory, Sovereignty and the Wolrd Political Map." *Geopolitics* 3 (Boundaries, Territory and Postmodernity):1–16.

Pelras, Christian. 1996. *The Bugis*. Cambridge, Mass: Blackwell Publishers.

Popescu, Gabriel. 2012. *Bordering and Ordering the Twenty-First Century: Undertanding Borders*. United Kingdom: Rowman & Littlefield Publishers, Inc.

Robinson, Kathryn. 2002. "Inter-ethnic Violence: The Bugis and the Problem of Explanation." In *Beyond Jakarta: Regional Autonomy and Local Societies in Indonesia*, edited by Sakai MInako, 150–154. Adelaide: Crawford House Publishing.

Sack, Robert David. 1986. *Human territoriality: Its theory and history*. Vol. 7. Cambridge [Cambridgeshire]; New York: Cambridge University Press.

Sadowski-Smith, Claudia. 2002. *Globalization on the Line. Culture, Capital and Citizenship at U.S. Borders*: Palgrave.

Sassen, Saskia. 1998. *Globalization and Its Discontents*: New Press.

Tagliacozzo, Eric. 2005. *Secret trades, porous borders: Smuggling and states along a Southeast Asian frontier, 1865–1915*. New Haven: Yale University Press.

Tagliacozzo, Eric. 2009. "Navigating communities: Race, place, and travel in the history of maritime Southeast Asia." *Asian Ethnicity* 10 (2):97–120. doi: 10.1080/14631360902906748.

Tuathail, Gearóid. 2000. "Borderless Worlds? Problematising Discourses of Discourses of Deterritorialization." In *Geopolitics at the End of the twentieth century: The changing world political map*, edited by Nurit and David Newman Kliot. F. Cass.

Warren, James Francis. 2000. *The global economy and the Sulu zone: Connections, commodities, and culture*. Vol. [Philippine]. Quezon City, Philippines: New Day Publishers.

Warren, James Francis. 2002. *Iranun and Balangingi: Globalization, maritime raiding and the birth of ethnicity*. Singapore: Singapore University Press.

Warren, James Francis. 2007. *The Sulu zone 1768–1898: The dynamics of external trade, slavery, and ethnicity in the transformation of a Southeast Asian maritime state*. 2nd ed. Singapore: NUS Press.

Urban Studies: Border and Mobility – Kerr et al. (Eds)
© 2019 Taylor & Francis Group, London, ISBN 978-1-138-58034-3

Civil society and the model of Dayak identity struggle in Central Kalimantan: A framework of neo-Gramscian–Tocquevillian analysis

Anyualatha Haridison & Jhon Retei Alfri Sandi
University of Palangka Raya, Palangka Raya, Indonesia

ABSTRACT: The Dayak communities of Central Kalimantan are one of the ethnic communities in NKRI. The communities have questioned the position of relationships with the state in the political, economic, social, and cultural spheres. The struggle of Dayak identity in the post-reform era indicates the operation of the institutionalization process in the framework of civil society in Central Kalimantan. This study was limited to the post-reform era (2001–2015) with an approach to civil society theory by using a qualitative descriptive method. The results reveal several facts about the Dayak–Madura ethnic conflict tragedy, the formalization of Dayak customary institutions (Perda 16/2008), the application of local content in educational curricula, customs, and customary law enforcement, nature, and prosperity. The analysis shows the model of Dayak identity struggle in two patterns: first, the resistance to the repressive actions of a particular community as a form of resistance to the state; second, a collaborative pattern with the local government to gain legitimacy for the institutional role of the Dayak community in all spheres. The analysis further asserts that the model of the identity struggle by the Dayak community is sporadic. It is not only dominant in one of the models, but also in the combined framework between the neo-Gramscian and neo-Tocquevillian perspectives of civil society.

Keywords: Civil Society, Identity Struggle, Dayak, Central Kalimantan

1 INTRODUCTION

Hobbes's notion (Suhelmi, 2001: 197) about the power of the state as "leviathan" emphasizes that the power of the state becomes an absolute and autonomous political entity of any intervention. This notion will greatly imply the decline of civil society as a place for confronting with the state. According to Chandhoke (1995: 8–13), civil society is a place where people enter into relationships with the state. Civil society therefore becomes a basis or a place for the emergence of rational discourse in order to question the accountability of the state. In addition, Chandhoke (1995) also asserts that the civil society acts as a space for the implementation of relationships between the society and the state in terms of accountability, participation, and autonomy.

The concept of the state as "leviathan" matches the description of the power of the New Order regime in Indonesia. According to Hikam (1996: 243), the politics of depoliticization during the New Order regime caused the splitting of civil society. The communities lost their right to participate in governance and were no longer autonomous neither independent in political life. They became powerless, and access to social, economic, and political environments was obstructed by repressive state agents. These conditions made the existence of an autonomous and independent entity of civil society impossible.

The accumulation of state pressure in all spheres was enhanced by the public distrust of the New Order governance system. Regime and government change had been the persistent demand since the mid-1990s until the outbreak of reformation in 1997. The reform era provided places and positions for people in order to question the power of the state which had large and massive authority. Efforts were made to strengthen its position economically, socially, culturally, and politically. The civil society movement that aimed to pursue political and social change at the national level had an impact on the local level, which occurred in most of the regions in Indonesia.

In this study, civil society is regarded as an "agent", which is related to the realm of associative life, associations and non-association groups, formal and non-formal, non-violent groups, all based on religion, ancestry, ethnicity, and class, operating in non-enforced environments and as a series of relational networks (Alagappa, 2004; Walzer, 1994). Empirically, the constellation of the civil society movement in Central Kalimantan was already perceived before the end of the New Order regime. It was viewed as the Dayak movement to confront with the power of the New Order. According to Kusni (2009), the Dayak movement at that time was considered not as an organized or programmed action, but only as a discourse. However, subsequently, various organizations appeared (e.g. non-governmental organizations and Dayak customary institutions). The discourse of "*putra daerah*" (indigenous people) was one of the facts presented at that time, in order to question the Dayak's access to political leadership and the pattern of resource management by the New Order's power.

This study aims to describe the role of Dayak communities as a civil society agent in the post-reform era without neglecting the sequence of activities that occurred before the New Order era. Towards the end of the New Order Regime until the transition period, the Dayak community was still looking for deep-rooted thinking within the context of identity struggle. The role of the Dayak community as a civil society is elaborated to obtain a picture of the model of its identity struggle in Central Kalimantan, especially in the post-reform era.

This study aims to illustrate the model of Dayak identity struggle in Central Kalimantan. Its main focus is to examine the relationships between the Dayak community and other ethnic communities, corporation, and local government by using a qualitative descriptive method. Data were obtained through the observation and conducting in-depth interviews with several key informants. Secondary data were also used to support the fundamental argumentation and proposition. Data were analyzed according to Moleong (2005), which include recording, classifying, synthesizing, making an overview, data meaning, finding the patterns of data relationships, and making general findings. This study was limited to the post-reform era (2001–2015), based on the events and activities of the Dayak community as an agent or organization in the social and political spheres.

2 DISCUSSION

2.1 *Against hegemony of the state*

In Indonesia, hegemony and control in almost all areas of community life in the New Order regime limited the development of civil society at both the national and local levels. The public distrust was at the point of culmination, and the economic crisis of 1997 was the momentum that triggered Indonesia's democratic transition (Aspinall, 2004). The era of democratic transition or reform implicitly became the beginning of the awakening of civil society in Indonesia. Several studies on civil society in Indonesia revealed that this period of resurrection was marked by the swift demands of society on the democratic agenda, the significant growth of NGOs or civil society organizations, and the emergence of NGOs that are based on or characterized by local culture.

Constellation of democratic transition in various places was finally fulfilled by the central government in the enactment of law 22/1999 on regional government, also known as the law on regional autonomy. This law embraces the principle of decentralization, whose primary purpose is the division of authority in the management of government from the center and secondary purpose is the administration of some function to the region as an autonomous agent (decentralized). The enactment gave new mandates to all local regions to manage their own government. The democracy channel has now been wide open, marked by freedom of expression, open political system through direct regional head elections, and managing their own households by autonomous regions.

The demands of society for democratic life were also institutionalized in Central Kalimantan through the "*putra daerah*" discourse. The people of Central Kalimantan strongly rejected the governor appointed by the New Order government, and preferred that the appointment be made by the region without neglecting the criteria for the development. The ongoing polemic against the "*putra daerah*" discourse was responded by the Central Kalimantan government by issuing Local Regulation 14/1998 (Perda 14/1998) on customary institutions in the province of Central Kalimantan to strengthen the existence of an independent, self-contained institution. During the reform era, many people expected that this customary institution would do much for the empowerment of the Dayak community, but it was still far from reality. For instance, the role of the Dayak community as a civil society was still confronted with the issues of identity and human rights such as civil rights, the right to life and the right to freedom (Budiman, 1990: 34).

The neo-Gramscian analysis is a dominant approach to dissecting the pre-democratic era until the transition to democracy, as Alagappa's proposition implies in several studies of civil society in Asia, including Indonesia. The pressure and format of the New Order government still left various polemics in Central Kalimantan. These issues and demands had long been voiced by Dayaks and Dayak organizations to the government. The main concerning issues were as follows: the economic and livelihood issues of Dayak people who were crushed by the penetration of capitalism (logging companies); the issue of "*putra daerah*" regarding the political leadership such as in the military and government at that time; and the Dayak culture, which was regarded as an underdeveloped, barbaric, and marginal culture. The outbreak of ethnic conflict in Central Kalimantan in 2001 coincided with the readiness to welcome regional autonomy.

This tragedy was discussed many times by Dayak figures, scholars, academics, Western researchers, and even by the Dayak community on various occasions. They concluded that the conflict was generally triggered by a number of interests and rights of the Dayak community which were not accommodated by the government. The ethnic conflict between the Dayak and Madurese was regarded as a purely ethnic or ethnic conflict, which could not be removed under the leadership of the New Order regime. From a political demographic perspective, the transmigration program implemented by the New Order government aimed to cope with population explosions on the islands of Java and Madura, and to extend the management of the vast unproductive lands of the less populated islands. Borneo was one of the targets of this transmigration program in order to make the Maduranese fight in Central Kalimantan. Several analysts believe that the transmigration program was the New Order strategy to extend political support and establishment of the status quo, which was effectively proved. According to Dayak people, the conflict occurred because of the Maduranese culture of ethnic violence that altered the economic sector of the Dayak community. This attitude was not in accordance with the cultural norms and values of Dayak people. In addition, from a social and political perspective, the conflict occurred due to channel obstruction or access to political and economic power in various segments resulting from a pattern of political leadership perceived to marginalize the Dayaks, and then the pattern of economic sector domination that destroyed local wisdom and livelihoods.

Several parties were in agreement that the ethnic conflict occurred due to the failure of security forces (TNI and Polri) to prevent the slaughter of people ("ethnic cleansing"). TNI and Polri are state agents that provide security and protection to conflicting communities, but in reality, they arc powerless. Therefore, it can be said that the state is also powerless. During the ethnic conflict, the Dayak organization played a role in the Institution for Dayak Community and Central Kalimantan Consultation Region/*Lembaga Musyawarah Masyarakat Dayak dan Daerah Kalimantan Tengah* (LMMDD-KT). On the one hand, this organization shouted the slogan "where the earth steps on, is the same place for the heavens need to be upheld", a symbol of resistance against anyone who seeks to destroy and disrespect the Dayak culture. On the other hand, the Dayak ethnic community also had a strong urge to fight.

While constructing the understanding of the Dayak people at that time, the Maduranese were regarded as the representation of the state. Memories of the repressive or "leviathan" nature of Hobbes in the New Order regime still remain, which are easily identified with the ethnic behavior and ethical attitudes of the Maduranese in comparison with other ethnicities. From this perspective, it can be understood that the Dayak community acts as an agent that confronted with the state to gain access to various sectors controlled by other societies. According to Bakker (2002), the principle of voluntarism is a powerful driver of struggle. This principle is reflected in the opposition of the Dayak ethnic community, which is seen as a main action strategy for changing social order. Moreover, Robert Cox (1999) suggests that the resistance is "the agent representing the revolution".

The local government provided the regulation, namely Perda 14/1998, for the customary institution. However, in reality, the implementation of Perda 14/1998 was not effective enough to overcome the hegemony and pressure exerted by other groups on the Dayak community. The exposure to the Dayak–Madura conflict was not to elevate the superiority of Dayak ethnicity, but solely for proving a proposition of civil society. In many aspects, the 2001 experience led to solidarity, a sense of shared destiny, the power to promote non-violent Dayak customs, and the desire to modernize Dayak institutions.

2.2 *Dayak cultural construction as a public culture*

Alagappa (2004) illustrates that the role of civil society is to strengthen autonomy by the institutional and regulatory development of self-governance. It consists of institutionalization; organizational development; finding and solving problems in the wider community by extending to authorities not exercised by the government, preventing the abuse of state power and restricting interference into the private, social, and economic aspects of society; development in terms of articulation, aggregation, and representation of public interest. Furthermore, Alagappa comments on other issues on which local governments cannot work because of their different areas of activity. However, the Dayak cultural sphere is the authority possessed by Dayak Customary Council of Central Kalimantan/*Dewan Adat Dayak Kalimantan Tengah* (DAD-KT) and not by the local government. It is difficult to distinguish between the roles of DAD-KT and the local government because the personnel of DAD-KT are also the elite personnel of the government. Several issues confronted by DAD-KT can be used to construct the Dayak cultural and customary systems in order to gain legitimacy from the wider community and in the national context so that Dayak culture is public. Some of these issues and cases can reinforce this argument in order that the local government and the wider community recognize and accept the various roles of DAD-KT in fighting for cultural identity.

The Dayak customary institution is a community organization which is not only deliberately established but also fairly developed and raised alongside the history of the Dayak customary community with respect to its customary law territory. It is also entitled and authorized to manage various life issues with respect to Dayak customs and customary law. This Dayak customary institution can be measured from the role of DAD-KT in upholding customary law. The enforcement of this customary law related to the attitude and norm of the life of Dayak people and other society should lead to Dayak customary law as an agreed rule. For example, when any party, including the Dayak community itself, violates, it will be dealt with or resolved through customary rules that have been agreed upon. In fact, DAD-KT wanted to revive customary justice for several cases as did their ancestors thousands of years ago. Consider, for example, the case of Thamrin Amal Tomagola, a sociologist from the University of Indonesia, who insulted the Dayak people through his research results in 2011. Another example is the customary court that tried some Indonesian Air Force officers (TNI AU) in West Kotawaringin Regency for violating an order which led to an uproar in 2017.

Dayak titles should be given to Dayaks for their useful services and achievements for the community. For example, the Dayak title was given to the Governor of Central Kalimantan such as Agustin Teras Narang and Sugianto Sabran. However, recently, the Dayak title was also given to those who were non-Dayaks such as the President, the Vice President, and the Governor of Bank Indonesia. As an institution that was authorized to deliver the titles, DAD-KT did this in the interest of Dayak people to increase their popularity at the national level. Several core administrators of DAD-KT know that this award will serve as a reminder to the person who holds it to not neglect the Dayak community, especially in every development policy.

A research study by Sandi (2016) describes the steps taken by DAD-KT to urge the local government to formulate legal regulation and implement protection and learning as well as to introduce Dayak culture, customary law, and knowledge of Dayak indigenous people. Based on the demands, inputs, and encouragement from DAD-KT, the local government set one of the important policies related to local content learning materials in the form of Governor Regulation 22/2011 (Pergub 22/2011). The policy on local content regulates the implementation of 11 types of local wisdom: (1) regional languages, (2) local literature, (3) local arts, (4) customs and customary law, (5) local history, (6) food/cuisine (7) medicine and treatment, (8) design/engraving/sculpture (technology), (9) sports, (10) the environment, and (11) customary rights. Furthermore, this was followed by the implementation of publishing books in order to apply the local curriculum to primary schools in several districts of Central Kalimantan.

According to Alaggapa (2004: 53), in the state sphere, civil society is a territory that influences the functions and policies of the state through the aggregation of interests, forming public opinion and guarding policy changes. Apart from the process and its implementation, this proposition is in fact a manifestation of the role of DAD-KT in the enforcement of Dayak law (customary court), the struggle for the popularity and interests of the Dayak community (title), and the knowledge of Dayak culture to be accepted as a public culture (local content). This is a picture of the autonomy of DAD-KT's self-governance. To some extent, this is also a form of decentralization of power to empower civil society, as Arato (2000) states, or it may be an extension of the authority (Alagappa, 2004) which the local government cannot perform, as well as the resolution of problems developed by society (e.g. the massive action of society to sue the sociologist from the University of Indonesia).

The role of DAD-KT in constructing Dayak culture to be accepted as a public culture can be widely viewed as a normative contribution to the development and empowerment of the Dayak community. This is similar to what Islamic organizations such as NU and Muhammadiyah have done, within this very long institutional range and almost in tandem with the history of Indonesia, who sought to fight for Islam as a public religion, a normative contribution that is freely debated in the public space because there is no longer exclusivity in the agenda. On the one hand, the participation of DAD-KT in the empowerment agenda and as a civil society organization can help maintain consistency and rationality. On the other hand, political leverage, if not strictly protected, can make the Dayak cultural agenda an unlawful public culture and a stigma for society as a whole. For example, DAD-KT became powerless when a Katingan District leader was involved in a case that took place in March 2017. Such incidents can hinder the development of civil society when the interest in struggle is replaced by political interest to leverage the cultural identity of Dayaks.

3 CONCLUSION

The model of Dayak identity struggle in the post-reform era has two phases. First, the struggle of Dayak identity is a purely counter-hegemonic, counter-narrative and even revolutionary tool to confront other communities as a form of representation of the state. Second, the institutionalization of civil society (DAD-KT) as an organization and its accompanying roles in collaborative and accommodative forms is considered as a political struggle of Dayak identity. Although the desire to decentralize governance has been integrated into the civil society space, it has been refracted into the interests of the elite, which indirectly undermines the position of DAD-KT with regard to state policy and struggle for the interests of society at large. From a theoretical implication perspective, Dayak community institutions have, on the one hand, become independent and oppositional and, on the other hand, acted independently and politically in collaboration with the local government. The first phase gives a clear picture of the dominant neo-Gramscian framework. The second phase depicts the institutional phase of DAD-KT using the neo-Tocquevillian framework that describes the role of the local government in making changes to Dayak organizational entities. However, this reflects a weak relationship between DAD-KT organizations. Thus, the logic of the local government to empower the Dayak community in the form of institutionalization of DAD-KT is interpreted as another form of hegemony.

REFERENCES

Alagappa, Muthiah. 2004. Civil Society and Political Change: An Analytical Framework. In Alagappa (Ed.). *Civil Society and Political Change in Asia: Expanding and Contracting Democratic Space*. California: Stanford University Press: 25–54.

Alagappa, Muthiah. 2004. The Nonstate Public Sphere in Asia: Dynamic Growth. Institutionalization Lag. In Alagappa (Ed.). *Civil Society and Political Change in Asia: Expanding and Contracting Democratic Space*. California: Stanford University Press: 455–477.

Arato, Andrew. 2000. *Civil Society, Constitution, and Legitimacy*. Lanham, Md.: Rowman and Littlefield.

Aspinall, Edward. 2004. Indonesia: Transformation of Civil Society and Democratic Breakthrough. In Alagappa (Ed.). *Civil Society and Political Change in Asia: Expanding and Contracting Democratic Space*. California: Stanford University Press: 61–93.

Bakker, Gideon. 2002. *Civil Society and Democratic Theory: Alternative Voices*. London: Routledge.

Budiman, Arief. 1990. *State and Civil Society in Indonesia*. Victoria: Center of South East Asian Studies.

Chandhoke, Neera. 1995. *State and Society: Exploration in Political Theory*. New Delhi: Sage Publication.

Cox, Robert W. 1999. "Civil Society at the TUrn of Millennium: Prospects for an Alternate World Order." *Review of International Studies* 25(1): 3–29.

Haridison, A. 2013. Perjuangan Masyarakat Lokal dan Usaha Memberdayakan Diri: Studi Kasus Pertambangan Emas Rakyat di Merindu, Desa Konut, Kabupaten Murung Raya. *Kritis: Jurnal Studi Pembangunan Interdisiplin* 23(1): 41–61.

Hikam, Muhammad AS. 1996. *Demokrasi dan Civil Society*. Jakarta: LP3ES.

Moleong, Lexy J. 2005. *Metode Penelitian Kualitatif*. Bandung: Remaja Rosdakarya.

Sandi, Jhon Retei A. 2016. *Perilaku Politik Lembaga Adat Dayak dalam Aktualisasi Hak-Hak Adat Masyarakat Dayak Kalimantan Tengah: Kajian di Dewan Adat Dayak Kalimantan Tengah (DAD-KT)*. Malang: Program Pascasarjana Universitas Merdeka Malang.

Siahainenia, Royke Robert. 2017. *Orang Dayak Melawan Tambang: Studi Gerakan Sosial Baru dalam Ruang Publik Virtual*. Salatiga: Satya Wacana University Press.

Walzer, M. 1994. *Toward A Global Civil Society*. Providence RI: Berghahn Books.

Whitehead, Laurence. 2005. Bowling in the Bronx: The Uncivil Interstices between Civil and Political Society. In Burnell and Calvert (Ed.). *Civil Society In Democratization*. London-Portland: Frank Cass.

Widen, Kumpiady. 2017. The Rise of Dayak Identities In Central Kalimantan. In Victor T. King, Zawawi Ibrahim and Noor Hasharina Hassan (Eds.). *Borneo Studies in History, Society and Culture*, Asia in Transition, Singapore: Springer-Institute of Asian Studies (UBD): 273–282.

Urban Studies: Border and Mobility – Kerr et al. (Eds)

Girl marriage and marginalisation of women in the cities of East Java

E. Susanti
Universitas Airlangga, Surabaya, Indonesia

ABSTRACT: This study focuses on how the practices of early girl marriage encourage the process of marginalisation of women. Using a sociological analysis with a gender perspective, this study also aims to examine how the practices of early girl marriage harm the lives of young women and their children. This study was conducted in two regions of East Java Province, Indonesia, with different cultural backgrounds, namely Bangkalan Regency with its Maduranese culture and Blitar Regency with its Javanese culture. The subjects included in this study were young wives, their husbands, parents, parents-in-law, neighbours and local community leaders. Data were collected using a questionnaire administered to 100 respondents and an in-depth interview with 20 informants. The results from this study reveal the role of patriarchal hegemonic culture in the lives of poor communities who still practice early marriage. Women are marginalised for several reasons as follows: a) they are kept in subordinate positions in their families and communities; b) they live in poor conditions; c) they have no protection for their reproductive health. This paper argues that practices of early girl marriage are not a religious but a gender issue.

Keywords: Feminist Perspective, Gender, Girl Marriage, Poverty

1 INTRODUCTION

Early girl marriage is still common in Indonesia although various policies and regulations have been framed to strictly curb these practices. Compared with other regions in Indonesia, the number of early girl marriages in East Java Province is quite large. According to the Indonesia Population Data Survey (IDHS, 2007), the proportion of under-aged girl marriages (i.e. <16 years old) in East Java was 39.4%. Meanwhile, one out of three women (about 40%) who were married for the first time in Java was under the age of 16, especially in West and East Java. A study conducted in Madura has also shown that in order to ease the burden on the family economy, parents encourage girls to marry even though they have not yet attained 18 years of age and still go to school (Susanti H., Emy: 2015).

This study mainly focuses on girls who become victims of early marriage, regardless of the age of their husbands. In an early marriage, women bear more losses because they get pregnant and give birth early. Using a sociological analysis with a gender perspective, this study aims to explain the mechanism of social reproduction of gender-based power relations that support the ongoing practices of girl marriage.

The study not only focuses on the violations of the human rights of women but also aims to reveal the marginalisation of women in the practices of early marriage. It also aims to examine how unequal gender-based power relations in the practice of girl marriage have been reproduced eventually and harm the lives of girls. More specifically, the aim of this study was to reveal how practices of early girl marriage encourage the marginalisation of women in any positions and statuses of women.

2 METHOD

This was a qualitative study supported by quantitative data analysis conducted in two regions in East Java Province, Indonesia. These two places have different cultural backgrounds, for example Bangkalan Regency with its Maduranese culture and Blitar Regency with its Javanese culture. The subjects included in this study were women who were victims of early marriages, their husbands, biological mothers and fathers, parents-in-law, neighbours and community leaders. Data were collected using a questionnaire administered to 100 respondents and an in-depth interview with 20 informants.

3 GIRL MARRIAGE, MARGINALISATION AND POVERTY

The theory of poverty was applied in this study to explain how the reality of poverty continues in the community to influence the practices of early marriage. Poverty is generally associated with deprivation of health, education, food, knowledge, influence over one's environment and so on, which make the differences between truly living and merely surviving. There is another universal aspect of poverty, which makes it particularly painful and difficult to escape. The poor are more vulnerable than any other group to health hazards, economic down-turns and man-made violence (Chambers, 2006).

Furthermore, since women are kept in a subordinate position, vulnerability among poor women is more pronounced. Meanwhile, for poor people, social solidarity is one of the most important assets available to them. To maintain this solidarity and the emotional and physical security it provides, people are willing to make considerable sacrifices to ensure that these social bonds are preserved.

The discussion about cultural bonds and gender relations cannot be separated from power relations analysis. Theories of power relations conceive power as repressive, and suggest that power is viewed as something that can be acquired, like a commodity, and can be exchanged from one person to another through a contractual act. Meanwhile, seeking a non-economist analysis of power, Foucault states that the analysis of power should not be concerned with power in its central location, but rather in its extremities where power overcomes the rules of right. Foucault questions how power is embodied in the institutions at a local and regional level (Foucault, 2002).

4 RESULTS AND DISCUSSION

The results of this study found that marriage of girls under the age of 18 years is common, and girl marriage is even considered as a tradition. When a girl has already begun menstruation or graduated from elementary school, usually she will soon be forced to marry by her parents. The results from this study indicate that the majority of girls in the regencies of Bangkalan and Blitar are married at 11–14 years of age (Bangkalan 64% and Blitar 68%). Some of them are even married at 11 years of age (Bangkalan 16% and Blitar 20%).

Most of the people in these poor communities see the practice of girl marriage as a way to reduce the economic burden of their families. Some of the mothers of girls who were already married before 14 years of age revealed that their families did not have the money to continue their children's education. For these poor communities, marrying off their girls at a young age or under the age of 18 is considered advantageous to help the economic conditions of their families.

Furthermore, the results indicate that the practice of girl marriage in fact exacerbates poverty. Girls are shackled to domestic duties and parenting, making it impossible to engage in productive activities. A woman who got married at a young age said that before being married, she helped her parents in the fields and earned some income. After marriage, she could no longer work in the rice fields, while her husband's income was very low and not always enough to buy milk and supplementary food for their toddler children. This fact shows that girls who are married early become "victims".

The study shows that within her marital household, a young wife only has little bargaining power and ability to make decisions regarding various aspects of her life. A woman's husband and mother-in-law mainly determine her role in the family and control her access to or participation in the outside world. This often remains the case throughout her life and translates to weaker control over resources in her household, tighter constraints on her time, more restricted access to information and health services, and poorer health conditions.

The lack of power and decision-making autonomy has a significant influence on economic decisions. Where women have decision-making power and their priorities are reflected in how household resources are allocated, household expenditures on key areas such as education and health tend to be higher.

Girls subjected to early marriage portray a marginalisation of economic, social, cultural and political conditions. Marginalisation of poor women is more pronounced because they are kept in a subordinate position in gender relations. Unequal gender relations are supported by cultural norms in communities. Therefore, the number of poor women is increasing over time. These symptoms are referred to as symptoms of feminisation of poverty among women who are married at young ages. Married girls suffer due to prolonged poor economic and social conditions. The marriages of these girls often serve to fulfil the gender values of the communities where the status of widows is considered to be much more honourable than unmarried women who are often interpreted as being "not married" or "old spinsters". Gender values treat boys and girls differently.

Because of their low level of education, married girls are often not considered by their husbands and parents-in-law as being capable of making money and/or managing finances or making financial decisions for the household. Married girls who work have less control over their income, face isolation from school, friends and the workplace, and have limited access to social capital and networks.

Young brides often experience overlapping marginalisation because they are still young, often poor and less educated. This affects the resources and assets they can bring into their marital households, thus reducing their decision-making abilities. Child marriage places a girl under the control of her husband and often the in-laws, limiting her ability to voice her opinion and form and pursue her own plans and aspirations.

This study found unequal power relations between a young wife and her husband, parents and in-laws. These unequal gender values are continuously reproduced without considering the "harm" inflicted on the girl's life. Using Foucault's theory (Foucault, 2002), gender-based power relations must be understood by revealing how a triangle of power, rights and knowledge takes place in the lives of poor communities. It is clear that power relates to the rules of the rights that formally delimit power and the effects of truth or knowledge produced and transmitted by power, which, in turn, reproduce this power.

5 CONCLUSION

The results of the study indicate that the impact of girl marriage practices is bad for the lives of girls and their children. The consequences of child marriage have profound effects in the form of lost earnings and intergenerational transmissions of patriarchal and poverty systems.

Based on these results, this study argues that women are marginalised for several reasons as follows: a) they are kept in subordinate positions in their families and communities; b) they live in poor conditions; c) they have no protection for their reproductive health. The practices of early girl marriage are not a religious but a gender issue. The patriarchal system as a social norm and system hegemonises the gender relations in the family and community.

This study also reveals that power relations between young wives and their husbands, parents and in-laws are unequal. The economic impacts and costs of child marriage are likely to be very high for girls who marry early, for their children, families and communities, as well as for society at large. In the light of these facts, unequal gender-based power relations are reproduced in the practices of girl marriage in poor communities.

REFERENCES

Arimbi, D.A. (2017). Politicizing Piety: Women's Rights and Roles in the Tarbiyah Movement in Indonesia. *Religious Studies and Theology, 36*(2), 228–244.

Chambers, R. (2006). *Poverty Unperceived: Traps, Biases and Agende.* Paper University of Sussex.

Field, E., & A, A. (2008). Early Marriage, Age of Menarche, and Female Schooling Attainment in Bangladesh. *Journal of Political Economy*, 881–930.

Foucault, M. (2002). *Power/Knowledge: Wacana Kuasa/Pengetahuan (diterjemahkan dari Power/ Knowledge)*. Yogyakarta: Bentang Budaya.

Gemignani, R., & Q, W. (2015). *Child Marriage and Faith Affiliation in Sub-Saharan Africa: Stylized Facts and Heterogeneity. The Review of Faith and International Affairs.*

Khanna, T., R, V., & E, W. (2013). *Child Marriage in South Asia: Realities, Responses and the Way Forward.* Bangkok: UNFPA Asia Pacific Regional Office.

Nguyen, M.C., & Q, W. (2015). *Impact of Early Marriage on Literacy and Education Attainment in Africa in Q. Wodon (ed) Child Marriage and Education in Sub-Saharan Africa.* Washington: World Bank.

Plan UK. (2011). *Breaking Vows: Early and Forced Marriage and Girl's Education.* London: Plan UK.

Sugiarti, E., Handayani, D., & Wulandari, R.A. (2005). *Marginalisasi Wanita Madura Dalam Taneyan Lanjang*. Surabaya: Universitas Airlangga.

Sukaryanto. (2010). Budaya Nikah Siri di Rembang dalam Perspektif Gender. *Masyarakat, Kebudayaan, dan Politik (MKP), 23*(1).

Susanti, H., & Emy. (2007). *Ketimpangan Gender dan Ketidakberdayaan Perempuan Miskin Perkotaan (Gender Inegulity and Vulnerabillity of Poor Women in the Cities).* Surabaya: Insan Cendekia.

Susanti, H., & Emy. (2009). Perempuan Miskin dan Makna Sosial Kemiskinan. *Jurnal Masyarakat Kebudayaan dan Poliitik*.

Susanti, H., & Emy. (2013). *Kajian Keluarga Miskin dan Perangkap Kemiskinan di Perkotaan (Poor Families and the Trap of Poverty in the CIties).* Surabaya: Insan Cendekia.

Urban Studies: Border and Mobility – Kerr et al. (Eds)
© 2019 Taylor & Francis Group, London, ISBN 978-1-138-58034-3

Ajhemo practice among Madurese women and its correlation with independently healthy life behaviour

Sri Ratnawati
Universitas Airlangga, Surabaya, Indonesia

ABSTRACT: *Ajhemo* practised by Madurese women is closely related to the habit of planting trees, flowers and foliage in backyards, long before the government had encouraged its citizens to plant medicinal plants. Through their own initiative, these women began to plant with the aim of using as *tambha* (for medicinal purposes). The practice of *ajhemo* raises the question of why Madurese women maintain their habit of consuming medicinal plants while they are living in the modern era. This study applies Spradley's methods of ethnography specific to the cognitive anthropological school, assuming that each society has its own unique knowledge system that can be learned by understanding the perception and mind organisation related to the material phenomena that occur around the people. Data were collected by means of an interview with some Madurese women living in Branta. This study finds that *ajhemo* is a daily practice carried out by Madurese women based on the concept of *ajhemo ekariyan*. The concept has encouraged Madurese women to carry out the *ajhemo* practice through generations up until today since the concept of health and beauty resulting from *ajhemo* has been associated with women. Therefore, Madurese women are mainly concerned about their health and beauty in order to fulfil their bigger responsibilities in their families.

Keywords: ajhemo, health, beauty, Madurese women

1 INTRODUCTION

Ajhemo means drinking *jamu/jhemo* (traditional medicinal drinks), which is practised by Madurese people throughout generations for both curative and preventive purposes. It is a traditional Madurese medicinal system that has existed long before the development of the modern medicinal system. In her article, Jane Beer (2012: 32) explains that although modern medicine and beauty experts seem to have just discovered this idea, the Javanese have practised it for centuries. Foreign experts have admitted that *jhemo*, developed by ancestors in the 18th century, has existed long before the birth of modern medication. This fact shows that Indonesia has developed its own unique medicinal system based on local wisdom.

In Indonesia, *jhemo* was first introduced by the Javanese royal family and was then disseminated to the Sumenep royal family. When the power of Sumenep royalty began to fade, the knowledge about how to make *jhemo* has been spreading in the community until today. Although the Madurese learned about *jhemo* from the Javanese, they developed their own *jhemo* based on Madurese cultural values, emphasising physical strength which is perhaps inspired by their admiration for *Joko Tole*, the character in the legend of Joko Tole. This character has inspired the Madurese to integrate the image of physical strength and might into its *jhemo* products.

The *ajhemo* practice among the Madurese is based on their internal needs. The implementation of *ajhemo* is closely related to the way Madurese people promote a healthy life based on their local wisdom, which has been practised by Madurese for centuries. While explaining the importance of *ajhemo* to their children, Madurese parents say *ajhemo ekabares* (drink *jamu* in order to be healthy or to be cured). From a normative perspective, the standards of *bares* and *sake'*[1] of the people living in the Branta Pesisir region are determined by *ajhemo*. *Ajhemo* is considered to be the manifestation of *bares*[1] and *sake'*[2] practised through generations. The habit of consuming *jhemo* is always perceived as a process to make someone healthier, in order to reach an almost universal agreement among *jhemo* users that *ajhemo ekabares* (meaning drinking *jamu*) will make you healthier.

The concept of *bares* and *sake'* incorporates the ethical dimension composed of values and norms which are believed and practised by the local community. In order to define this concept, we need to understand the culture of an ethnic group using an emic approach, for example, by understanding how the people in the Branta Pesisir region perceive the concept of *bares* and *sake'* based on their own culture (from a native point of view). Since the concept of health has been learned, taught and practised by the Madurese, several factors may affect health beyond clinical facts. These factors are socio-cultural factors, economic factors and education. Capra (1988:395) defines health as a multidimensional phenomenon involving interrelated physical, psychological and socio-cultural aspects. Each individual within the community may define *bares* and *sake'* differently based on his/her personal experiences; therefore, the definitions may be very relative in meaning. These definitions may represent not only a medical perspective but also a cultural and religious perspective. This means that there is no single definition for health from a medical perspective. Cultural values and religious values also contribute to the meaning of health. Based on a religious perspective, health is affected by the religiosity and spirituality of an individual. The correlation between *ajhemo* and *bares* is based on individual experience of consuming *jhemo* for years or even for a lifetime.

Before discussing the meaning of *bares*, we need to look at the definition of health proposed by the World Health Organization which states that health is "a state of complete physical, mental, and social well-being and not merely the absence of disease or infirmity". Meanwhile, Law Number 23 Year 1992 defines health as a state of physical, mental and economic well-being. Based on these definitions, health must be perceived as a holistic unity incorporating physical, mental and economic well-being.

The degree of health known as psycho-socio-somatic health well-being is the result of four different factors, namely environment, behaviour, heredity and healthcare service (i.e. preventive, proactive, curative and rehabilitative health program). Among these four factors, environment and behaviour are considered to be the most dominant factors affecting the health behaviour of the traditional Madurese community. For example, if they experience a toothache, they prefer to drink *jhemo selokarang* rather than going to a dentist.

Hence, understanding why people consume *jhemo* is not as simple as understanding why people use medications. People commonly use medications because they are sick or have illnesses. Meanwhile, people drink *jhemo* not necessarily because they have illnesses. They may consume *jhemo* because they believe the existing myth about *jhemo*. In the context of tradition, *ajhemo* provides mystical experiences, namely restoring life energy, strength, beauty and health. It heals not only physically but also mentally.

This study aims to find the underlying reasons for the motivations of Madurese women to consume *ajhemo* as part of their daily life by interviewing some Madurese women from Branta Pesisir, Madura. This study aims to prove the causes of healthy habits and beauty as a result of the *ajhemo* practice. Data were analysed using Spradley's methods of ethnography that are associated with cognitive anthropology (2006). These methods of ethnography assume that every community has its own unique system of knowledge. They are applied to reveal conceptual tools used by *ajhemo* consumers to classify, organise and interpret their social and natural universe (Kaplan, 2002).

---

1. *Bares* means *to be healthy* or *to be cured*. In the social context, these definitions are interchangeable depending on the situation. In order to avoid the ambiguity of definition, in this study, the word *bares* is defined as health as an important concept defining the ideas describing its meaning. The people of Branta Pesisir have a different understanding of the meaning of *bares* which depends on their personal experiences.
2. *Sake'* means illnesses.

2 DISCUSSION

2.1 *Ajhemo practice for Madurese women*

Etymologically, the word *ajhemo* is derived from the base form *jhemo*, meaning *jamu*. The word *jamu* is derived from the Old Javanese word *jamuju*. In Webster's Dictionary, *jhemo* is defined as the types of specific plant with medicinal functions, ingredients or concoctions of medicinal recipes. Indonesia has become the second largest home of medicinal plants after Brazil. There are about 3,500 to 40,000 species of medicinal plants in Indonesia (Wiwaha, 2012).

In general, the concoction of *jhemo* refers to traditional recipes passed on by ancestors (Handayani, 2004: 46; Limananti, 2003: 12; Siswanto, 2012: 204, Beers, 2001: 15). Every region in Indonesia has its own unique medicinal plant species that can be processed into different forms of *jamu*. This unique process is also known as folk knowledge, traditional knowledge, Western science or traditional ecological knowledge (Batiste, 2005; Duit, 2007). As mentioned in the Regulation of Ministry of Health Number 003/Menkes/Per/I/2010, *jamu* includes traditional ingredients from plants, animal ingredients, minerals, and extracts or a mixture/concoction of these ingredients, which are traditionally used to cure illnesses in accordance with social norms.

2.2 *Ajhemo: The healthy lifestyle*

Ajhemo has become *"cara hidup"* (a way of life) for people in the Branta Pesisir region (Cheney, 1996: 157). In the KBBI (Indonesian General Dictionary), the word *cara* refers to a method or attitude performed by an individual (or a group) to reach specific goals. In the context of *ajhemo*, the *ajhemo* practice serves not only as the concept of health but also as myths such as *ajhemo ekabares, ekauwad, ekaserra'ngoda* and *ekaradin*. For the people of the Branta Pesisir region, *ajhemo* is a construct that is completely different from the construct of health. These people perceive the concept of *ajhemo* as a mixture of rationality and myths.

Healthy lifestyle practised by the people of Branta Pesisir throughout generations has become a culture that has existed long before the development of the medical concept and perspective which compete with each other to develop the construct of health. The discourses discussed above have illustrated how *ajhemo* has become a folk practice followed by the people of Branta Pesisir. On the other hand, the role of *ajhemo* becomes stronger if the habit has become a family tradition. For example, in Elis' family, when a *jhemo* peddler comes to their house, all the family members gather and voluntarily drink *jhemo* together, including their children. From the onset of puberty, the *ajhemo* habit follows the menstrual cycle of women. After delivering a baby and during the breastfeeding period, a woman compulsorily drinks *jhemo* so that both the mother and the child remain healthy. In this regard, *ajhemo* has become a daily beverage, which the Madurese call *ekariyan*.

Ajhemo has become a cultural expression of people in the Branta Pesisir region. The health quality of the community depends on their understanding of the function of *jhemo*. Since ancient times, *ajhemo* has become a habit practised from generations to generations. As this habit has been deeply rooted and developed on a daily basis according to their needs, it is hard to withdraw from their daily activities. Bordieu calls this habit a cultural disposition (Bordieu,…). It is very difficult to separate cultural disposition from the memory of the users of certain cultures. Several cases of the *ajhemo* phenomenon among fishermen's families indicate that their children are apparently accustomed to seeing their parents drink *jhemo* not only when they are sick but also in good health conditions. These events influence the minds of the children as they grow. Consequently, their perception about *ajhemo* also develops into comprehensive knowledge that *ajhemo* is strongly related to curing illnesses while the truth he/she perceives is not so.

The habit of consuming *jhemo* is always perceived as a process to make someone healthy. This agreement applies almost universally among *jhemo* users through the saying *ajhemo ekabares* (drinking *jhemo* makes you healthy). Although *jhemo* users drink it occasionally, the collective agreement on the benefits of *jhemo* remains unchanged, which is greater than its failure.

The belief about the benefits of *jhemo* can be verified by the experiences of *jhemo* users. During their lifetimes, *jhemo* users stay healthy and never complain about their health. Some of them even say *tada' orang mate polana ajhemo* (no one dies because of drinking *jhemo*). This belief is merely a dogma taken for granted without any critical logic. In Bourdieu's terminology, this type of acceptance is known as *doxa. Doxa* is defined as beliefs and unconscious values which are deeply rooted, fundamental and considered universal.

Here we can see how close the *ajhemo* practice is to the health of a woman. Women must *ajhemo* because they experience menstrual periods every month. The *Ajhemo* practice follows their biological structure (i.e. menstrual cycle and childbirth cycle). The biological factors become the main reason for *ajhemo* which, in turn, develops into a cultural principle. The distinctive biological structure of a woman makes them drink *jhemo* more frequently than a man. A girl tends to drink more than one type of *jhemo* from her childhood when compared with a boy who only consumes two types.

In the absence of a clear-cut time frame for the administration of *ajhemo* to children, most parents assume their own right time to introduce it. In general, parents practice *ajhemo* along with their children. They give their children *jhemo* with an extremely sweet taste, such as *sinom*. Madurese children (both boys and girls) drink the same types of *jhemo*, as listed in the table below.

Ajhemo in childhood		*Ajhemo* for teenagers	
Girl	Boy	Girl	Boy
Konyi' accem	*Konyi' accem*	*Konyi' accem*	*Konyi' accem*
Bherres kencor	*Bherres kencor*	*Bherres kencor*	*Bherres kencor*
		Sere konce	–
		Beluntas	*Beluntas*

During childhood, Madurese boys and girls consume the same types of *jhemo.* The first types of *jhemo* introduced to them are *konyi' accem* and *bherres kencor*, which are very sweet. This is a wise strategy performed by Madurese parents to introduce *jhemo* to their children. Although there is no specific guidelines on at what age a parent should introduce *jhemo* to his/her children, they often introduce *ajhemo* from early ages (5–6 years old) by inviting their children to drink *jhemo* together.

When they enter their teenage years, boys and girls begin to undergo changes, indicating their transition to adulthood. For example, girls experience their first menstrual cycle as a sign of adulthood. From these ages onwards, a mother will consider the menu of *jhemo* for her daughters by adding *jhemo sere konce*. This type of *jhemo*, known as *araksa aba'* (body treatment), is specially prepared for girls, but not for boys.

From these descriptions, we can see that the *ajhemo* practice is closely related to women. Women undergo menstrual cycles and the childbirth process because of their biological structure that makes them different from men. On account of this unique biological structure, women consume more types of *jhemo* than men. From their childhood, Madurese girls tend to consume more types of *jhemo* than boys who drink only two types. This is because women experience childbirth and undergo the breastfeeding process. They are expected to consume special *jhemo* on these special occasions. In the Madurese tradition, women are expected to drink *jhemo* for 40 days after giving birth (*malang ore'*).

Whether consciously or not, these biological conditions develop a different attitude, which, in turn, generate distinctive behaviour patterns between men and women. Therefore, it is normal for women to be more concerned about the health of themselves and their families. For example, a mother is concerned about the health of herself and her family. She will notice any symptoms of illnesses experienced by her husband and children. For example, if children have a fever, then the mother makes *jhemo konye'*, or if the husband has a fever, she will make *jhemo kunir* with egg yolks.

Unlike the people living in big cities who, when they get sick, will immediately go to a doctor, a mother living in rural areas is expected to be independent and have a sufficient understanding of traditional medicines taught by her ancestors throughout generations. As stated by one of our respondents, she must always be ready to cure herself when she gets sick. She said:

> *Atiya mon sake' aduuuhhhh bengko ta' ron karon. Poko'na mon bebine' se sake' lalakon ajheletakan,na' –kana' ta karamot, neser asalakola mole tade' nase', neser.... kalaparan. Sambi sake' ejakaje alalakon, ajhelen dhibi' nyare jhemo* (If I get sick, my household chores will be neglected. There will be no one to take care of my children and there will be no food for my children to eat when they come home from school and they may be starving. Poor them! Therefore, I must keep working at my household chores although I am sick. I prepare my own *jhemo* (in order to be healthy).)

The responsibility as a mother has become internalised in Madurese women. Neglected children and unfinished household chores when they get sick become their burdens. Madurese women are fully concerned about the noble role of a mother. Therefore, a mother is expected to be able to prepare her own *jhemo* to cure her illnesses so that she gets better to continue her household chores.

As a result, a mother is very concerned about her own health in order to be able to fulfil her major responsibilities. As a source of righteousness, a mother symbolises the characteristic of *sepi ing pamrih* (sincerity), and thus the moral responsibilities towards her family, that is, doing everything for her husband and children, become her priority. It seems therefore logical that a woman's life is closely related to *jhemo*.

Unlike a modern lifestyle that only emphasises the material aspect and tends to be personal, a traditional lifestyle may be carried out individually or in a group. When a modern woman wants to look beautiful, she will use cosmetics and go to a beauty salon for facial treatments. This information is often transmitted via social media and advertised on television. In comparison, when the women of the Branta Pesisir region want to look beautiful, they drink *jhemo* as told by their ancestors *ajhemo ekaradin ekaserra' ngoda*.[3]

For Madurese women, consuming *jhemo* is a necessity that cannot be ignored. In fact, *ajhemo* becomes even more important after they get married. In order to please their husbands, Madurese women undergo a series of body treatments (*araksa aba'*), including *ajhemo Sari Rapet* for mothers and *jhemo Harum Pengantin* for girls. We must admit that the existence of a modern medical system is inseparable from the traditional medicinal system that originated from our ancestors who used several medicinal plants and concocted them into *jhemo*.

Doxa tends to legitimate the existing social order through the *ajhemo* practice of the people of the Branta Pesisir region. Furthermore, the *Ajhemo* practice is imposed on the youths of the Branta Pesisir region by their parents. As reported by a girl studying in a boarding school who was going to marry soon, one of her marriage preparations was "*ajhemo harum pengantin*". She learned this *jhemo* preparation from her mother who had taught her to consume it twice a week in order to make her body smell good when she gets married. A girl is expected to smell good not only during the wedding ceremony but also every day. Therefore, she must undergo certain body treatments to make her body not only smell good but free from body odours every day. Hence, a woman is obliged to drink *jhemo* regularly.

The spirit of *ajhemo* among the women of the Branta Pesisir region is inseparable from their *araksa aba'* practices. These practices include beautifying themselves and maintaining their beauty, slimming their bodies, birth control and treating their vaginas. Women are expected to be concerned about themselves and have the ability to maintain their beauty and health. The efforts to maintain health begins with themselves, namely by *ajhemo* regularly. The habit of *ajhemo* makes one healthy and eventually dignified because by regular *ajhemo* an individual can perform his/her daily activity and gain the meaning of life for himself/herself and other people.

---

3. *Ajhemo ekaradin* means drinking *jhemo* makes you beautiful. *Ajhemo ekaserra'ngoda* means drinking *jhemo* makes you stay young. These two discourses function as a stereotype rather than a truth.

3 CONCLUSION

Ajhemo ekariyan is a habit that describes the attitude of a woman who consumes *jhemo* periodically. This discourse originates from ancestral teaching that advocates women to follow the *ajhemo* practice regularly. *Ajhemo ekariyan* is a social construct based on a female perspective. Among the Madurese, *ajhemo* is more socialised among women than among men. *Ajhemo ekariyan* is strongly correlated with *araksa aba'*. *Araksa aba'* can be considered as the way of life practised by the people of the Branta Pesisir region for a long time, which keeps developing in accordance with individual interests. It encourages people to *ajhemo ekariyan* and make*jhemo* as the object of *nom-enoman*. All of these concepts describe the empirical structure of thought. Therefore, the dynamics of the people of the Branta Pesisir region can be perceived by understanding their historical dynamicity and several perceptions which have been strongly rooted and have become part of the healing process practised even today. Within this framework, the *ajhemo* process is often mixed with myths which are intentionally (or unintentionally) created to maintain a healthy lifestyle and create inner beauty in Madurese women.

REFERENCES

Capra F. 1988. *The Turning, Science, Society, and the Rising Culture.* Toronto: Bantaum Books Education, UK.

Handayani, L. et al. 1977. *Inventarisasi Jamu Madura yang dimanfaatkan untuk Perawatan Kesehatan dan Pengobatan Gejala Penyakit yang Berkaiatan dengan Fungsi Reproduksi pada Wanita*. Surabaya: Pusat Penelitian dan Pengembangan Pelayanan Kesehatan.

Helton, T., Pesek & H. Nairm. 2006. *Healing Across Culture: Learing FromTraditions Eco Health*. Retrieved October 10, 2017 from www. Springerlink.com.

Jordaan, E. Roy. 1985. *Folk Medicine In Madura*. Netherland: Lieden

Kaplan, Elisabeth. 2002. *'Many Paths to Partial Truths': Archives, Anthropology, and the* Kier, Sarah Elizabeth. 2000. *Our Roots, Our Strength: The Jamu Industry, Women's Health and Islam In Contemporary Indonesia.* Doctoral Dissertation. United States: University of Pittsburgh.

Liesbeth Hesselink, 2011. *Healers On The Colonial Market: Native doctors Indies Midwives in The Ducth East Indies.* Lieden: KITLV Press.

Managestutu, Subehan, Widyawaruyati, Aty, Zaidi, Syed Faizal Haeder, Suresh, Awale & Shagetoshi, Kadota, Traditional Medicine of Madura island in Indonesia. *Journal of Traditional Medicine*. 24(3): 90–103.

Mutmainnah, 2009. Pemanfaatan Jamu Madura oleh Perempuan di Kabupaten Bangkalan. Journal *Masyarakat Kebudayaan dan Politik*. 22 (2).

Niehoof, Anke. 1985. *Women and Fertility in Madura.* Netherland: Leiden *Power of Representation*. Archival Science. Kluwer Academic Publisher.

Romaidi, G. 2011. Tingkat Preferensi Masyarakat Madura Terhadap Obat Tradisional (jamu) Madura. *El-Hayah*. 1 (1).

Spradley, James. 2006. *Conformity and Conflict: Reading in Cultural Anthropology*. Pearson U.S.A.

Webster-Merriam. Merriam Webster's Collate Dictionary 11th Edition. Springfield: Massachusetts, USA.

Urban Studies: Border and Mobility – Kerr et al. (Eds)
© 2019 Taylor & Francis Group, London, ISBN 978-1-138-58034-3

Muslimah identity on the Wardah Muslim-segmented cosmetic products

N.C. Fajri
Universitas Airlangga, Surabaya, Indonesia

ABSTRACT: The spreading of products targeting Muslims is currently a trend in Indonesia, for example, *Wardah* cosmetic products. The appearance of these products cannot be denied as the fact is that Indonesia is the largest Muslim populated country, but these products have become widespread in recent years. The emergence of Muslim-segmented products brings about the idea that the beauty of Muslimah is different from that of non-Muslims. The aim of this article was to find the Muslimah identity created by Muslim-segmented products in Indonesia. The focus of this article is the Muslimah identity and its meaning as portrayed in the Muslim-segmented products *Wardah*. In this article, we examine the advertising of the Muslim-segmented products *Wardah* using *Stuart Hall*'s identity theory and *Roland Barthe*'s semiotics using a descriptive qualitative research method.

The emergence of Muslim-segmented products strengthens the borderline that Muslimah must be differentiated from non-Muslim women. The religious border, which was blurred once, has been re-constructed by the appearance of Muslim-segmented products. This shows the phenomenon in Indonesia of re-creating the gap, by using religion. As a result, this gap blocks one from creating his or her identity freely.

Keywords: Identity, Muslimah, Muslim-segmented products, Wardah

1 INTRODUCTION

Muslim-segmented products are beginning to emerge. Not only food, but also other products such as fragrance, detergent, shampoo, and cosmetics have been widely used. The emergence of these products is correlated to the fact that Indonesian people are mostly Muslims. According to the census taken by the Central Bureau of Statistics (BPS) in 2010, 207,176,162 (87.18% of the total population) Indonesians were converted to Islam. In addition, the spreading of issues related to religion in Indonesia makes the producers try to "commodify religion". This commodification is successful, since Muslim-segmented products attract consumers and compete in the market.

One of the Muslim-segmented products surviving the trade competition and being well-known today is the *Wardah*. In fact, the *Wardah* cosmetic product is not a new brand; it is in the market for more than 20 years in Indonesia. Its well-known appearance attracts significant attention of Indonesians. With the tagline "*inspiring beauty with halal product*", *Wardah* is the first cosmetic product in Indonesia claiming to be *halal*, meaning "permitted" or "lawful", in contrast to other cosmetic products which have taglines such as natural and safe for skin. This Islam-related tagline attracts Muslimah who are doubtful about using cosmetics at first. Consequently, this terminology attracts Muslimah to be more open and to apply makeup.

In Islam, there is prohibition called *tabarruj*, the concept of displaying oneself in terms of attitude and clothing. This idea also prohibits the use of perfumes and overapplication of makeup to show off. Besides, some people might think that makeup will block *air wudhu* (ablution water) in the skin's pores. Taking these problems into account, *Wardah* targets the Muslimah who worry about their own cosmetics.

In fact, *Wardah* products are not very different from other products. At present, *Wardah* not only provides makeup products but also skin care products. The Muslim-segmented shampoo product for veiled women promises freshness after use, which is very different from that offered by *Wardah*. It has no significant difference from other products. The only striking point of *Wardah* products is their tagline: *halal*. From this point, it can be implied that *Wardah* tries to depict another identity based on the product. This product, unconsciously, places a barrier between Muslimah and non-Muslim women. It leads to the urgency of portrayal of being a good Muslimah by using a halal product in order to strengthen the Muslimah identity.

The *halal* labelling informs the consumers that there are no forbidden substances in the products. The absence of the *halal* label can imply that the product has no permission from the authorised institution to be classified on the list of *halal* products. Lukmanul Hakim (2014), Managing Director of the Assessment Institute for Foods, Drugs, and Cosmetics Indonesia Council of Ulama (LPPOM MUI), stated that *halal* cosmetics fulfil the requirements of *halal* criteria. All the substances must be *halal*, pure, and produced in an area free from contamination. In fact, *Wardah* is not the only cosmetics brand in Indonesia using the *halal* label. Other cosmetics brands, such as Viva and Purbasari, include the *halal* label in their sachets, but *Wardah* uses "halal" as the main tagline.

In this study, a Muslim woman's identity in *Wardah* advertising also shows a form of opposition to the assumption about a veiled woman who is often considered inferior, but this study does not use advertising media obtained from television or YouTube, but through the social media of Instagram. The selection of Instagram is based on the fact that in this media, all the talent, as well as a brand ambassador can come from various circles, ranging from actresses, models, designers, and social influencers to even businessmen. All these talents have the same promotional space—which is different from the television that would only be controlled by talents derived from the background of actresses or YouTube that is dominated by beauty vlogger. In Instagram, this same space can show the identity of Muslim women who want to display *Wardah* thoroughly because it reaches all the talents that indirectly represent the Muslim woman according to *Wardah*'s idea.

Photostructural building is difficult to stand on its own. Photos, at least, mingle with other structural buildings, such as text (title, explanation, or comment). Thus, the totality of information is mediated and presented by two different structures (Barthes, 2010: 2). A unique feature appears on the official Instagram account of Wardah @wardahbeauty compared to the Instagram accounts for other cosmetic products. *Wardah* highlights the ambassador brand dominated by a hijab woman. It is not surprising, indeed, because this brand targets Muslim women. It is interesting that the *Wardah* brand ambassadors are not always shown with thick makeup. The brand ambassador who portrays *Wardah* cosmetics is not someone who applies daily makeup. They are often presented with make up as it is, so this is the opposite of most cosmetics advertisements.

In addition, on Instagram, *Wardah* also highlights the figure of influencers. These figures are talented and skilled in their respective fields, but it is not highlighted that they have an expertise in makeup. These figures include social media influencers, lifestyle bloggers, designers, MCs, and entrepreneurs. This is certainly very different from other brands such as Maybelline or Purbasari, which do not use the *halal* tagline. Maybelline and Purbasari feature a beauty vlogger who is currently increasing in popularity. From this distinction, it appears that *Wardah* uses different symbols from other cosmetics brands. These symbols can be interpreted through the semiotics put forward by Roland Barthes, which can then show the Muslim identity in *Wardah* products. The brand *Wardah* then indirectly gives the suggestion that Muslim women have different needs from those of other faiths.

2 METHOD

In this study, we use a descriptive qualitative research method, which can show about community life, history, behaviour, as well as organisational behaviour (Chony, 1997: 11). Therefore, to show the world and the life of the human object to be investigated, qualitative research is used. Through qualitative research methods, the identity of Muslim women in *Wardah* cosmetics advertising can be seen, described, and understood.

Primary data used in this study are uploaded onto the Instagram official account of cosmetic wardah @wardahbeauty. Focused data are uploads by 2017, which can show the current Muslim identity that *Wardah* displays. Secondary data are obtained from books related to advertising, hijab, and women. After data has been collected in the form of images and captions, identity analysis is done using Roland Barthes' semiotics method.

3 MUSLIM IDENTITY IN *WARDAH* MUSLIM-SEGMENTED COSMETICS PRODUCTS

In general, women want to look beautiful. One of the most prominent elements of a woman's quest for beauty is cosmetics. There is a verse in the holy Qur'an, which states that a woman should not be dressed except in front of her husband. This leads to a view that a woman should not apply makeup because it is forbidden. The *Wardah* brand is aimed at Muslim women, which has been upset about the use of makeup because it was feared that makeup on her face would block the wudhu water (ablution water) getting to the skin. In response to these problems, *Wardah* emerged with its slogan of inspirational beauty with *halal* products. The emphasis of the slogan suggests that *halal* can be interpreted as being permissible.

The slogan of inspirational beauty with *halal* products is promoted through talent as well as a brand ambassador for *Wardah*. Unlike other cosmetics brands that mostly use the talent of an actress or beauty vlogger, through Instagram, *Wardah* brings talent that promotes talent. Talent *Wardah* includes Tatjana Saphira (actress), Raline Shah (actress), Dewi Sandra (actress, singer), Inneke Koesherawati (actress), Natasha Rizky (actress), Zaskia Sungkar (designer), and Miranda (designer). In addition, *Wardah* also has a social media influencer who not only sells physically but also talents and works in various fields, such as Chiki Fawzi (animator), Lulu el Hasbu (vlogger, designer), Dwi Handayani Syah Putri (entrepreneur), Ghaida Tsurayya (entrepreneur), and Sarah Sofyan (designer).

The selection of this talent indicates that *Wardah* wants to release the negative stigma of society about the veiled woman. The contribution of a woman is often ignored, and among the contributions, there is a gap that needs to be filled. The above talents are expected to give a new perspective in society that *hijabers* can also be creative and active. Hijab is not a barrier to work. Talent social influencers are expected to influence the community and show the success of a woman in hijab. The talent can be a track record that reinvents women's experiences. The veil is an existing practice from the past until now with the variables that each part is attached to the cultural system. This meaning of this practice is not necessarily the same if it is placed in different cultural contexts ideologically, amongst different people, and at different times (Guindi, 2003: 39). Therefore, the re-identification of Muslim women's identity breaking through boundaries that curb their creativity is demonstrated. *Wardah*'s brand ambassadors are not all identical with makeup in the photos on Instagram, not the least of which do not seem to display makeup on her face. It can be seen from the following picture.

Photo on instagram @wardahbeauty's account.

Photo on instagram @wardahbeauty's account.

Photo on instagram @purbasari_indonesia's account.

Photo on instagram @revlon.id's account.

From the pictures above, we can see the difference between *Wardah* talent and other brands' talent. *Wardah* certainly shows Muslim identity through its talents. Through the picture, the Muslimah in *Wardah* advertisements (later be called Islam cosmetics) are representations of women who display makeup as natural as it is and do not accentuate themselves with thick makeup. In this context, Islam cosmetics is a makeup that makes the Muslimah look fresher without showing off her skills in applying the makeup. This portrayal is shown in order to keep the concept of Islam and *halal*, as the core of the *Wardah* brand.

In the first picture, there are two talents for *Wardah* cosmetics, namely Chiki Fawzy and Dewi Sandra. Both seem happy to sing in a modest house, which is a home for the deaf. In the photo, it casts two *Wardah* brand ambassadors using natural makeup on their faces. The makeup is not glamorous, and hence shows a fresh impression and not conspicuous. Chiki Fawzy in the photo plays ukelele. The use of ukelele—not a guitar—can give the impression of closeness and more cheerful music. Clothes used are of pastel colours, namely light blue,

Picture on account of @wardahbeauty's instagram.

light brown, and pink. In the photo, Chiki Fawzy appears wearing a veil in a turban style, showing that she is trendy. In addition, she uses a watch that adds to the trendy and young impression.

Dewi Sandra, in the photo, also wears pastel-coloured clothes. The hijab used by Dewi Sandra is simple and describes her as having a simple and motherly image. This photo is taken from the perspective of an eye-angle to see the interactions of the people in the photo. This reveals *Wardah*'s desire to show the activities of Muslim women in the photo. It is a candid photo that can reflect the natural act, not artificial action, of Muslim women in *Wardah* ads.

The caption of the photo is as follows:

> *"Inspired by the heartfelt beauty of friends @ dcfingertalk give love to everything she does, @chikifawzi and @dewisandra share happiness by making a song that is translated in sign language for @dcfingertalk friends. #BeautyFromHeart"*

The above caption shows that *Wardah* wants to show Muslim women who share with others. Through a brand ambassador with talent in arts, Muslim women are portrayed creatively as being able to create a song that is translated into a sign language. This is a breakthrough because the deaf cannot listen to Dewi Sandra's voice and the ukulele-playing of Chiki Fawzy, but they are still able to see the sincerity of Sandra and Chiki Fawzy through the song in sign language as *Wardah* Muslimah brand ambassadors. At the end of the caption, the tag #CantikDariHati further reinforces that *Wardah* emphasises morality and behaviour to describe the identity of Muslim women, not the beauty of the outside that is makeup.

The meaning of the connotation that can be seen is in the photo, even though the *Wardah* brand ambassadors do not use glamorous makeup, but both still become more conspicuous than others because of makeup. The colours of their clothes are similar, that is, pastel colours, which give a colourful impression because it uses a variety of colours at the same time, but not excessive. This is in accordance with the advice of a Muslim who should not appear conspicuous. The image to be conveyed in this photo is the figure of a Muslim woman who is cheerful, creative, sincere, and sharing.

4 CONCLUSION

The above explanation shows that *Wardah* advertisements on the Instagram account @ wardahbeauty portray a unique Muslimah identity. Instagram can be considered a social media, which can represent the portrayal because there are more Instagram influencers who get highlighted than YouTube or television influencers. Islam cosmetics, based on the above explanation, is the idea that shifts the view that using makeup should not be so simple as makeup is part of art to freely express oneself. In this explanation, Islam cosmetics show the simple use of makeup in order to make the user fresh. The focus of Muslimah identity on the *Wardah* advertisement is that the Muslimah's activities and movements should not be restricted by the condition. A good Muslimah is the one who is beautiful inside and brave enough to show her talents publicly. The identity of a Muslimah is not only in her beauty, but also in her activeness and creativity. The choice of an Instagram social influencer – not a YouTube beauty vlogger – as the brand ambassador is the manifestation of the idea that Muslimah need makeup to prevent certain negative perspectives about Muslimah such as that they are incapable, passive, and monotonous. Moreover, there is not an effort to show Muslimah's skill in applying makeup. This shows that *Wardah* advertisements have created an identity construction portraying that Muslimah need makeup, which is different from non-Muslim women.

REFERENCES

B, R.S. (2015). *Analisis Wacana Citra Perempuan dalam Iklan Televisi Produk Kosmetik Wardah "Kisah Dibalik Cantikmu" dan Maybelline "Clear Smooth All in One".* Yogyakarta: Universitas Gajah Mada.

Barker, C. (2014). *Kamus Kajian Budaya.* Depok: Penerbit PT Kanisius.

Barthes, R. (2004). *Mitologi.* Yogyakarta: Kreasi Wacana.

Barthes, R. (2010). *Imaji, Musik, Teks.* Yogyakarta: Jalasutra.

Barthes, R. (2012). *Elemen-elemen Semiologi.* Yogyakarta: IRCiSoD.

Bramantio. (2010). *Barbie dalam Cerpen Indonesia: Konstruksi Kecantikan dan Perempuan Artifisial/Fisikal/Ideal.* Surabaya: FIB.

Candrawati, Y. (2015). *Analisis Wacana Kritis Identitas Perempuan Muslim Indonesia dalam Iklan "Wardah Exclusive Series: Dewi Sandra Goes to Paris.* Surabaya: Universitas Katolik Widya Mandala.

Elsy, P. (2012). *Dinamika & Perubahan Sosial Budaya Kota-kota di Jawa Timur—Bab VI: Dinamika Budaya Masyarakat Kota Surabaya.* Surabaya: Arr-Ruzz Media.

Guindi, F.E. (2003). *Jilbab: Antara Kesalehan, Kesopanan, dan Perlawanan.* Jakarta: Serambi Ilmu Semesta.

Hapsari, N.F. (2012). Critical Discourse Analysis of Identity Performativity of Moslem Women on Personal Accounts of Salafi Women in Facebook. *Urban Mobility International Conference.* Surabaya.

Ida, R. (2016). *Metode Penelitian Studi Media dan Kajian Budaya.* Jakarta: Prenada Media Group.

Ramadhani, D.A. (2015). *Pengaruh Label Halal Terhadap Brand Switching Produk Kosmetik dan Faktor-Faktor yang mempengaruhi Konsumen Mengonsumsi Kosmetik Berlabel Halal.* Bogor: Institut Pertanian Bogor.

Saraswati, L.A. (2017). *Putih: Warna Kulit, Ras, dan Kecantikan di Indonesia Transnasional.* Tangerang: Marjin Kiri.

Wibowo, W. (2003). *Sihir Iklan: Format Komunikasi Mondial dalam Kehidupan Urban Kosmopolit.* Jakarta: Gramedia Pustaka Utama.

Wuryanto, A.G. (2015). *Representasi Perempuan Berhijab dalam Iklan Wardah Versi "Girls Day out", "Exclusivee Series", dan "Perfect Match".* Surabaya: Universitas Airlangga.

Borders and mobility in maritime, spatial movement and locality

Urban Studies: Border and Mobility – Kerr et al. (Eds)
© 2019 Taylor & Francis Group, London, ISBN 978-1-138-58034-3

Movements around island and waterfront reclamation projects

T. Kerr
Curtin University, Perth, Australia

ABSTRACT: This paper describes an agenda that seeks to mobilize researchers across borders in investigating how community frontiers emerge in controversies over environmental change and spatial reorganization. That is, an agenda directed at research collaboration that transcends nation states in furthering understandings of the impacts of environmental change and mobility in the discursive production of borders. This paper describes the development of this agenda in investigating controversies around island and waterfront reclamation projects in Western Australia and Bali, and its possible trajectory in studying discursive nodes referencing island and waterfront reclamation at other sites in Indonesia as well as in Singapore and Malaysia.

Keywords: reclamation, environmental mobility, community frontier

1 INTRODUCTION

This investigation of controversies emerging around the development of new islands and waterfronts in the eastern rim of the Indian Ocean is aimed at providing fresh insights into how collective identities, places and spaces of governance are negotiated (Kusno, 2011). New knowledge from the project will enable the negotiation of shifting expectations and tensions around environmental change in the context of emerging spaces of collective action, rapidly-developing media technologies (Lim, 2014) and heightened geopolitical anxieties (Pan, 2014) particularly around the reclamation of land from sea in Southeast Asia (Casey & Sussex, 2012; Forbes, 2015) and Australia (Kerr, 2015c). As such, this investigation addresses the question of how land-reclamation projects impact on the representation of peoples, their places, spaces and policies. The investigation focuses on how claims around land reclamation projects become public, popular and consequential.

Building from Kusno's (2011) multimodal reading of texts around reclamation and subsequent development works in Jakarta Bay, and extending the forms of collaboration developing from research on the social negotiation of meanings around new island and waterfront sites in Perth and Broome (Kerr & Cox, 2016; Kerr & Cox 2013), a transnational group of researchers could work together in studying discursive nodes referring to reclamation projects currently planned or under construction in Indonesian, Western Australian, Singaporean and Malaysian waters. A group of researchers learning from each other as they identify nuances in the discursive dynamics of each node would arguably be in a position to extend important insights on the creation of terra nullius for investment projects (Tsing, 2012), the politics of proximity in ecological crisis (Kerr, 2015a), antagonism in popular movements (Laclau, 2006) and on the cultural studies' contribution to area knowledge (Heryanto 2013).

2 DISCUSSION

2.1 *Research mobility in advancing knowledge on the impacts of environmental change*

Multi-site investigations have been shown to produce seminal knowledge on the impacts of environmental change across varying cultural contexts and at varying spatial scales. For example, Hajer's (1995) multi-country study of acid rain discourses provided important insights into the dynamics of environmental change representation in policy production, and Tsing's (2005) ethnography of multiple sites and communities furthered knowledge on the production and relation of frontiers of environmental change and universalizing discourses. Their studies have shown that analysis and comparison of the cultural specificity of multiple nodes across a study area can advance knowledge on the cultural and political impacts of environmental change at the various spatial and temporal scales in which environmental changes and impacts are represented and contested. In other words, contrasting and comparing within a multi-nodal investigation can shed light on the complex relations between impacts of specific instances of environmental change and broader discourses of capital, indigeneity, environmental rights, religion, modernization, and so on.

This multi-nodal investigation sets out to address important questions on how environmental change influences the discursive complexities of nationalism, transnationalism (Tsing, 2005) and populism (Alvares & Dahlgren, 2016; Heryanto, 2013). Its multi-nodal approach to analysing discourses around land reclamation schemes enables nuances in articulations of localized movements responding to environmental change to be studied, and new understandings realized. These new understandings may help advance knowledge on articulations of the 'frontier' as theorized not just by Laclau (2005) but also by Tsing (2005) and Kusno (2011) in thinking through the transnational relations of otherwise, apparently, localized environmental impacts. In doing so, the investigation may provide insights for addressing urgent issues on spatial rights and people's sovereignty (McCarthy & Robinson, 2016), particularly on the emerging use of the United Nations Declaration on the Rights of Indigenous Peoples in contests over land and water rights in Australia and Southeast Asia (Kerr & Cox, 2016; Fay & Denduangrudee, 2016). Also, the multilingual sharing of knowledge in analysing multiple context-specific nodes will help build understandings between researchers and engaged communities in the study area, particularly helping to overcome the paradox of Australians and Indonesians being nearest neighbours who, mostly, know little of each other (Ang, et al 2015). The trial stages of this multi-nodal investigation has led to the building of relationships between a variety of people and peoples and already begun to provide teaching materials (Kerr & Cox, 2013) for furthering understandings of one another.

2.2 *Approaching community borders through antagonism and environmental change*

Utopian island and waterfront spaces cannot be created without radically challenging existing social practices and their place relations. Spatial utopias, going back five centuries to Thomas More's description of the artificially created island of Utopia, were intended to stabilize and control social processes but, as such, they cannot not be materialized without radically challenging and disrupting the social practices of existing places (Harvey, 2000). Disruption to a system of meaning—particularly the construction of land from a marine environment (Kerr, 2015c) – stimulates antagonism as people construct a community border between them and the enemy that they hold responsible for preventing the realization of their lived-out identities (Howarth 2000). Antagonism may contingently unify a popular movement if people's heterogeneous demands against the enemy are incorporated by an empty signifier linking their demands in a chain of equivalence (Laclau 2005). People's sensual engagement with a familiar environment can stimulate the emotional investment in an empty signifier—the shared desire—that motivates them to act as a unified community frontier against the outsider perceived to be threatening their environment (Kerr, 2015a). People will seek to conserve a familiar coastal environment when they sense it being threatened particularly by the radical change of dredging and land reclamation. They are driven and unified by desire to continue experiencing objects of a familiar environment, and this shared desire for a continuation of past experience can enable localized popular movements to overcome the imposition of land reclamation projects by even the most well-resourced property or petroleum consortium (Kerr, 2015a; Kerr & Cox, 2016). The resistant communities' emotional investment is stimulated by affect, which has been shown to play a substantial role in localized political action (Thrift, 2008; Kerr 2015a). Psychoanalytical theory can explain how this political dynamic operates through subjects' emotional response to an object (Laclau, 2006). Lacan's theory of excess towards an object and Freud's association of sensual experience with drive for the maternal object provides a way of understanding how affective investment in objects enables antagonism, as well as the dissatisfaction through which community laws and ideals are produced (Copjec 2004).

These theoretical approaches help explain the production of localized popular movements responding to land reclamation projects in Western Australia and, upon initial survey, at Benoa Bay in Bali (Kerr, 2015b). However, there are nuances in popular formations around land reclamation sites and in the geographies of media assemblages through which claims about reclamation projects are contested (Kerr, 2015a; Primayanti, Nuraeni & Fitriawan, 2016). For example, the plan to infill Benoa Bay, Bali, for construction of the Tirta Wahana Bali Internasional resort, led by the Jakarta-based Artha Graha Network, is resisted passionately by local Balinese communities, citing damage to sacred sites, coastal environment and local industries. Their resistance is backed by local tourism and fishing businesses in an alliance with Balinese environmentalists, rock musicians and other artists supported by environmentalists and musicians from Australia and other countries (Kerr, 2015b). Meanwhile, in Jakarta Bay, Artha Graha is just one of many large property developers and government interests driving island reclamation projects forward despite resistance from local residents, fishing and informal businesses whose concerns have been articulated by environmental and social organizations and have been incorporated into sizeable demonstrations led by religious organizations. Preliminary research shows that the nodes in this study contain a variety of local, national and transnational concerns, which require other ways of knowing (Rose, 2004) in their analysis. Therefore, this project's methodological approach engages first with the particularity of each discursive node—requiring a particular theoretical understanding of its unique context (Padawangi, 2014) and systems of knowledge production (Muecke, 2016)—before engaging in an analysis of the relations between and across nodes.

Despite differences between the nodes, evidence suggests that a large reclamation project conceived by the private sector is more likely to meet significant local community resistance than one conceived by the state (Kerr, 2015c), particularly where waterfronts have been prominent symbols within nationalist discourse (Kerr, 2016; Kusno, 2011; Perera 2009). In Singapore, the state has introduced reclamation projects to increase its land area from 580 km2 to about 700 km2 since independence in 1965 (Casey & Sussex, 2012). Public opposition to Singapore's reclamation projects has tended to come from outside its own citizenry: from Malaysia, which has accused Singapore of narrowing the Johor Strait and from other Southeast Asian communities regarding the mining of coasts and islands for the reclamation infill. However, because of its internal political and spatial discourses, the Singapore Government is less likely than its neighbouring governments to face the crisis of land reclamation stimulating an antagonism that consolidates populist action by citizens who feel betrayed by their political elites (Dahlgren, 2009). In other words, the particularity of a movement responding to a land reclamation scheme is influenced not just by people's sensual relation with the local environment, but by the discourses, institutions and practices around them. Hence, this project will advance knowledge about the particular processes through which antagonism around environmental change stimulates, disrupts, reshapes and consolidates the representation of popular identity.

2.3 *Multimodal analysis of multiple modes by engaging multiple communities*

The methodological approach to this investigation is a necessary innovation in triangulating on the complexities of environmental change through 1) multimodal discourse analysis 2) across multiple discursive nodes 3) involving multiple research communities. The methodological approach builds on my individual and collaborative investigations into the social negotiation of island and waterfront construction in Perth, Broome and Benoa Bay (Kerr & Cox, 2013; Kerr 2015b; Kerr, 2015c; Kerr & Cox, 2016). The approach also draws and builds on decades of research into the politics of environmental change in the study area: From studies on the pioneering acts of consumer associations to anti-dam movements in Indonesia and Malaysia (Aditjondro, 1998; Zainal, 2016), to the cultural complexity of environmental contests around resort development in Bali (Warren, 2012) and waterfront development in Jakarta (Kusno, 2011), and to the confluence of indigenous identity and environmentalism in Australia (Birch, 2016). The approach is informed by the experience of creating space for intervention into the enduring political impacts of land reclamation (Creagh, et al., 2016). It is also informed by scholarship on the politics of representing centres, peripheries and frontiers in architecture, digital assemblages, popular media and government (Kusno, 2011; Srinivasan & Fish, 2017; Tsing, 2005; Volkmer, 2014).

A wide variety of public and media texts will be collected in the multimodal discourse analysis (O'Halloran, 2013) of the articulation of localized movements around the construction of new islands and waterfronts. The texts include references not just to the new built environments but to their relation with pre-existing communities, buildings, landscapes and seascapes. Texts for analysis, therefore, can include project plans, renderings and video simulations, investment proposals, meeting minutes, master plans, urban planning regulations, television, print, radio and internet news, social media posts, advertisements, websites, signs, and recordings of public forums, demonstrations, musical compositions, landscapes, seascapes, buildings and heritage objects. Several thousand of these texts would be collected and assembled in a digital archive from which corpora would be selected for transcription and analysis by close-reading, reflexive coding and query by qualitative data analysis software. This methodology builds on the techniques of multimodal discourse analysis developed in the investigation of a popular movement responding to a major reclamation project off the coast of Perth (see Node 1 below). This multimodal discourse analysis approach enables a thorough examination of the articulation of popular movements through complex processes involving affect, intertextuality and interdiscursivity (Kerr, 2015a). That is, the approach can identify and explain the particular processes through which claims around a radical environmental change become public and have a consequential impact on government policy.

This investigation will gather and analyse data from six discursive nodes around the construction of particular islands and waterfronts (Figure 1). These cosmopolitan nodes trace the Eastern Rim of the Indian Ocean where Indonesian, Malay and English languages are dominant public languages but other languages are spoken by Indigenous and migrant people. In Broome, for instance, a mix of Indonesian/Malay was once the dominant street language spoken not only by pearl industry workers recruited via Singapore and Kupang but also adopted by Aboriginal people and Europeans (Demin, 2007). These cosmopolitan nodes, therefore, offer an opportunity to compare, contrast and reflect on sedimentations of culture and their relationships with contemporary capitalism and colonial legacies. They also offer an interesting chance to reflect on the development of assemblages around electronic communication technologies given that the nodes correspond approximately to the sites of stations connecting the submarine telegraph cables laid in the late 19th century that linked Penang, Singapore, Jakarta, Banyuwangi (a proposed sand mining site in Node 3), Broome

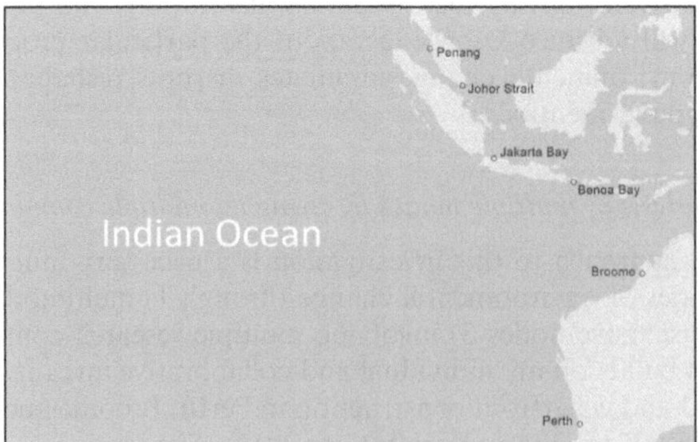

Figure 1. Map of nodes, referencing reclamation sites, selected for this study.

and Perth with many other places in the world. Other nodes could also be included, such as the ongoing and planned reclamation works in Makassar and Malacca, but only time, funding and the availability of collaborating researchers will determine whether they can become part of this investigation. For now, the six discursive nodes in the investigation will be:

1. **Perth**, Australia: the (proposed) 345-hectare, six-island, North Port Quay residential resort;
2. **Broome**, Australia: the (proposed) Liquid Natural Gas refinery and port at James Price Point as well as harbour dredging and waterfront reclamation;
3. **Benoa Bay**, Indonesia: the (proposed) 700-hectare, multiple-island, TWBI residential resort;
4. Jakarta Bay, Indonesia: 5,100-hectares of land reclamation involving multiple developers constructing seventeen islands (eight are under construction);
5. **Johor Strait**, Singapore and Malaysia: Phase 1 of Singapore's Tuas port development and, nearby, the 1,370-hectare Forest City residential resort development in Malaysian waters (both projects are under construction);
6. **Penang**, Malaysia: the (proposed) 1,698-hectare, three-island, South Reclamation Scheme.

The feasibility of this multi-nodal discourse analysis approach has been tested by comparing and contrasting the significant objects and findings of the multimodal discourse analysis of Node 1 (Kerr, 2015a) against those identified in pilot studies for nodes 2 and 3 (Kerr, 2015b; Kerr & Cox, 2016). Communications with researchers on the approaches to nodes 4, 5 and 6 are underway.

As analysis of the discursive nodes gets underway, community research workshops will be organized through local universities to help identify significant discursive objects for further research and to test interpretations in the corpus for each respective node. These workshops are expected to be funded through grant applications for this project, and organized in conjunction with collaborators working in local university campuses: Curtin University's Australia-Asia-Pacific Institute and its International Centre for the Indian Ocean Region and Universitas Gadjah Mada's Department of Communication. Other partners contacted for inclusion in the study include the National University of Singapore's School of Design and Environment and Universitas Udayana's Faculty of Cultural Knowledge. The community workshops, trialled at the Matagarup/Heirisson Island reclamation site (Kerr & Cox, 2013), involve activities that enable the identification of significant objects, texts and discursive mechanisms as well as intertextuality and interdiscursivity in the respective discursive node. This workshop process not only improves interpretation and the manual identification of discursive effects, it also helps to overcome the limitations of using techniques of corpus-assisted discourse analysis on multilingual corpora (Freake, et al., 2011). Through these workshops as well as site visits, archival searches, research meetings and place readings, the project will be open to subject-making encounters so that each discursive node is analysed critically in its cultural and linguistic specificity. With this ethics of openness (Zylinska, 2005), the methodology enables identification of the institutions and practices through which people negotiate the radical uncertainty of environmental change. Preliminary findings will be verified with members of research communities during the development of emerging themes and theoretical understandings across the nodes through processes of writing, presentation, discussion and critical reflection. If grant applications are successful, collaborating researchers will attend international workshops to discuss research approaches as well as preliminary and key findings from the project and share the dissemination strategies being developed in participation with their local research communities. This will enable the multimodal, multi-nodal discourse analysis with multiple research communities to build collaboration and understanding while producing rich theoretical insights into the cultural and political impacts of environmental change.

3 CONCLUSION

This paper has presented an argument for an ambitious project which can only feasibly be implemented if a sufficient cohort of engaged researchers can be mobilized across the bureaucratic borders of nation-states and academic institutions. If the mobilization is successful, this project will open up an intercultural space for furthering understandings of relations between identity, antagonism, representation and environmental change. This paper argues that a collaborative methodological approach that includes local ways of knowing in the analysis of discursive nodes within a multi-nodal transnational project will offer a breakthrough in understanding the cultural and political impacts of radical environmental change.

REFERENCES

Aditjondro, G.J. (1998). Large Dam Victims and their Defenders: The Emergence of an Anti-dam Movement in Indonesia. In Hirsch, P. & C. Warren (Eds.), *The Politics of Environment in Southeast Asia: Resources and Resistance*, 29–54. London, Routledge.

Alvares, C. & Dahlgren, P. (2016). Populism, Extremism and Media: Mapping an Uncertain Terrain. *European Journal of Communication*, 31(1), 46–57. DOI: 10.1177/0267323115614485.

Ang, I., Tambiah, Y. & Mar, P. (2015). *Smart Engagement with Asia: Leveraging Language, Research and Culture*. Report for the Australian Council of Learned Academies. Retrieved from www.acola.org.au.

Birch, T. (2016). Climate Change, Recognition and Social Place-Making, In E. Vincent & T. Neale (Eds.), *Unstable Relations: Indigenous People and Environmentalism in Contemporary Australia* (pp. 252–272). Crawley, WA: UWA Publishing.

Casey, A. & Sussex, M. (2012). Energy Transit States and Maritime Security in the Malacca Strait: The Case of Singapore. *Australian Journal of Maritime and Ocean Affairs*, 4(1), 25–36, DOI: 10.1080/18366503.2012.10815698.

Copjec, J. (2004). *Imagine there's No Woman: Ethics and Sublimation*. Cambridge, Massachusetts: MIT Press.

Creagh, R., Kerr, T., Cox, S. & Ryder, R. (2016). Securing Space for Hospitality in a Settler-colonial City. In T. Kerr & J. Stephens (Eds.), *Indian Ocean Futures: Community, Sustainability and Security*, 229–246. Newcastle upon Tyne: Cambridge Scholars Publishing.

Demin, S.B. (2007). *Once in Broome*. Broome: Magabala Books.

Dahlgren, P. (2009). *Media and Political Engagement: Citizens, Communication, and Democracy*. New York: Cambridge University Press.

Fay, C & Denduangrudee, H.S. (2016). Emerging Options for the Recognition and Protection of Indigenous Community Rights in Indonesia. In J. McCarthy & K. Robinson (Eds.), *Land and Development in Indonesia: Searching for the People's Sovereignty*, (91–112). Singapore: ISEAS Publishing.

Forbes, V.L. (2015). ASEAN, China and Malaysia: Cautious Diplomacy, Trade, and a Complex Sea. *International Journal of China Studies*, 6(2), 129–148.

Freake, R., Gentil, G. & Sheyholislami, J. (2011). A Bilingual Corpus-assisted Discourse Study of the Construction of Nationhood and Belonging in Quebec. *Discourse & Society*, 22(1), 21–47, DOI: 10.1177/0957926510382842.

Hajer, M.A. (1995). *The Politics of Environmental Discourse: Ecological Modernization and the Policy Process*. Oxford: Oxford University Press.

Harvey, D. (2000). *Spaces of Hope*. Edinburgh: Edinburgh University Press.

Heryanto, A. (2013). The Intimacies of Cultural studies and Area Studies: The Case of Southeast Asia. *International Journal of Cultural Studies*, 16(3), 303–316, DOI: 10.1177/1367877912474541.

Howarth, D. (2000). *Discourse*. Buckinghamshire: Open University Press.

Kerr, T. & Cox, S. (2013). R. Briggs, N. Lucy & S. Mickler (Eds.). *Setting up the Nyoongar Tent Embassy: A report on Perth media*. Perth: Ctrl-Z Press.

Kerr, T. (2015a). Negotiating Green Space between Ecological Threats and Beloved Objects. *Continuum: Journal of Media & Cultural Studies*, 29(3), 402–418, DOI: 10.1080/10304312.2015.1025368.

Kerr, T. (2015b, December 11). *To the Beach: A Critical Examination of Creation and Control in Environmental Communication*. Keynote address at the Indonesia International Conference on Communication (IndoICC) 2015 at Universitas Indonesia, Jakarta, Indonesia.

Kerr, T. (2015c). *To the Beach: Community Conservation and its Role in 'Sustainable Development'*. Crawley, WA: UWA Publishing.

Kerr, T. (2016). Australian Border Works: Representation of Coastal Places in Anzac Centenary Books. In T. Kerr & J. Stephens (Eds) *Indian Ocean Futures: Community, Sustainability and Security*, 49–66. Newcastle upon Tyne: Cambridge Scholars Publishing.

Kerr, T. & Cox, S. (2016). Media, Machines and Might: Reproducing Western Australia's Violent State of Aboriginal Protection. *Somatechnics*, 6(1), 89–105, DOI: 10.3366/soma.2016.0176.
Kusno, A. (2011). Runaway city: Jakarta Bay, the Pioneer and the Last Frontier. *Inter-Asia Cultural Studies*, 12(4), 513–531, DOI: 10.1080/14649373.2011.603916.
Laclau, E. (2005). *On Populist Reason*. London: Verso.
Laclau, E. (2006). Why Constructing a People Is the Main Task of Radical Politics. *Critical Inquiry*, 32(4), 646–680.
Lim, M. (2014). Seeing Spatially: People, Networks and Movements in Digital and Urban Spaces. *International Development Planning Review*, 36(1), 51–72.
McCarthy, J. & Robinson, K. (2016). Land, Economic Development, Social Justice and Environmental Management in Indonesia: The search for the people's sovereignty. In J. McCarthy & K. Robinson (Eds.), *Land and Development in Indonesia: Searching for the People's Sovereignty*, (91–112). Singapore: ISEAS Publishing.
Muecke, S. (2016). Indigenous-Green Knowledge Collaborations and the James Price Point Dispute. In E. Vincent & T. Neale (Eds.), *Unstable Relations: Indigenous people and environmentalism in contemporary Australia* (pp. 252–272). Crawley, WA: UWA Publishing.
O'Halloran, K. (2013). Multimodal Discourse Analysis. In K. Hyland & B. Paltridge (Eds.), *The Bloomsbury Companion to Discourse Analysis* (pp. 120–137). London: Bloomsbury Academic.
Padawangi, R. (2014). Counter-Hegemonic Spaces of Hope? Constructing the Public City in Jakarta and Singapore. *ARI Working Paper*, 219, April 2014. Retrieved from https://ari.nus.edu.sg/Assets/Uploads/docs/wps/wps14_219.pdf.
Pan, C. (2014). The 'Indo-Pacific' and Geopolitical Anxieties about China's Rise in the Asian Regional Order. *Australian Journal of International Affairs*, 68(4), 453–469, DOI: 10.1080/10357718.2014.884054.
Perera, S. (2009). *Australia and the Insular Imagination: Beaches, Borders, Boats, and Bodies*. New York: Palgrave Macmillan.
Primayanti, N.W., Nuraeni, R. & Fitriawan, R.A. (2016). Analisis Framing Berita Reklamasi Teluk Benoa pada Harian Kompas and Bali Post Edisi Juni 2013 – Desember 2014. *Jurnal Sosioteknologi*, 15(1), 68–85.
Rose, D.B. (2004). *Reports from a Wild Country: Ethics for Decolonisation*. Sydney: University of New South Wales Press.
Srinivasan, R. & Fish, A. (2017). *After the Internet*, Cambridge: Polity Press.
Thrift, N. 2008. *Non-representational Theory: Space, Politics, Affect*. Milton Park, Abingdon: Routledge.
Tsing, A.L. (2005). *Friction: An Ethnography of Global Connection*, Princeton: Princeton University Press.
Tsing, A.L. (2012). On Nonscalability: The Living World Is Not Amenable to Precision-Nested Scales. *Common Knowledge*, 18(3), 505–524, DOI: 10.1215/0961754X-1630424.
Volkmer, I. (2014). *The Global Public Sphere: Public Communication in the Age of Reflective Interdependence*. Cambridge: Polity Press.
Warren, C. (2012). Risk and the Sacred: Environment, Media and Public Opinion in Bali. *Oceania*, 82(3), 294–307, DOI: 10.1002/j.1834-4461.2012.tb00135.x.
Young, D. (2006). *Broome to Java Submarine Telegraph Cable*. Draft Nomination for Historic Engineering Marker Plaques and Ceremony Report published by Engineers Australia, Barton: ACT.
Zainal, Z.I. (2016). *Environmental Attitudes in Selected Contemporary Malaysian Novels in English: An Eco-Marxist Perspective*. Doctor of Philosophy. University of Malaysia, Kuala Lumpur.
Zylinska, J. (2005). *The Ethics of Cultural Studies*. London: Continuum.

Urban Studies: Border and Mobility – Kerr et al. (Eds)

Maintaining expressions of prohibition (*pamali*) as signaling the existence of tengger community's culture

D. Handayani & M. Lutfi
Universitas Airlangga, Surabaya, Indonesia

ABSTRACT: The expression of prohibition (*pamali*) is a way of parents in the past advising or forbidding their children from doing something that is considered taboo or dangerous. Each region has expressions of prohibition observable in everyday life as this is a form of oral tradition passed down from generation to generation. Oral tradition always experiences development as a result of the intersection of a tradition with modernisation so that adjustments are necessary for a region to preserve the customs as an ancestral legacy. This study attempts to explore the potential of oral tradition by identifying expressions of prohibition existing in the Tengger community. Implementation of the expressions of prohibition covers the sexes, both male and female, different age groups (elderly, adolescent, children), marital status (unmarried or married), and certain situations that occur in the community. One of the targeted areas of research is the Tengger community located in East Java. The people of Tengger are known to have local wisdoms that are positive, harmonious, adaptive, and religious. The study shows that the process of internalisation and socialisation of cultural values of the community has run very well to maintain the value of tradition as a religious system in the practice of everyday life. The religious system refers not only to the belief in religion but also the belief in the strength of the ancestor's advice, including expressions of prohibition as both a living memory and a living tradition of its generation. For the people of Tengger, belief in magical and mystical things still exists, while the assumption of expressions of prohibition can be identified given the potential of the oral tradition to be almost neglected and marginalised. There is also an assumption that something derived from orality is only a memory.

Keywords: maintenance, expressions of prohibition, oral tradition, Tengger community

1 INTRODUCTION

Basically, language is a code or system that serves social interaction and development of a particular culture in society. Serving social interaction, the role of language is very important as a means to cooperate, exchange ideas, and think critically. Serving the development of culture, language is necessary for shaping the character of society. Character building is related to the philosophy of life and the desire to congregate with other communities. Thus, it can be said that language has two main functions, as a means of communication and a cultural instrument that unites community groups with other groups (Nasution & Warjiyati, 2007: 7). As a living medium in society, language can be presented in writing and speaking to others to convey certain intentions. As a means of communication, language has quite complex functions, namely: (1) instrumental, as a means to meet material needs, (2) regulatory, to regulate and control inter-individual behaviour in social relations, (3) interactional, to create social networking between groups, (4) personal, as a medium of self-identification and expression, (5) heuristics, to explore, learn, and understand the world around, (6) imaginative, to express one's imagination, and (7) informative, as a medium of conveying messages in communication (Halliday, 1978: 21).

In every speech, the comprehension of the intention depends on the understanding of the context used by the speakers of the language. The use of language is closely related to what is said, what is captured, what is understood and what is interpreted in one's mind, which is derived from an observable context. Furthermore, Wijana (1996: 10–12) states that speech involves several aspects, such as speakers and hearers, the context of speech, purpose, action or activity of a person as a verbal product. As a verbal product, language can be presented orally to other speakers in societal life, which is further known as an oral tradition.

Oral tradition can serve as a cultural force which refers to the formation of identity to build human civilisation. Furthermore, oral tradition is used not only as a narrative, but a concept of cultural heritage that should be maintained and nurtured. In a cultural talk show broadcast by TVRI on April 17, 2015, Dr. Pudentia stated that research into oral tradition is not merely a narrative, but a concept of inheriting a culture and part of us as social beings. In addition, the oral tradition includes community and context and can be utilised to resolve conflicts in the community. The discourse of oral tradition is not just a fairy tale, mythology, expressions, mantras, and legends with various messages within, but also contains a cognitive system of society based on local wisdom that should be re-identified. This is because the transformation of values and norms of life change from time to time.

One example of the forms of oral tradition to be analysed in this study is the expression of prohibition (*pamali*). In this case, a prohibition is usually in the form of a prohibition or abstinence sentence against something mystical, magical or known as "superstition". People in the past, however, stressed that the form of *pamali* should be avoided because it would become a reality if it was violated and not executed. As a result, things previously considered mystical changed to be realistic. Each region must have a view of life and belief in the form of *pamali* to maintain safety and the environment. However, in this modern time with its sophisticated technology, presumably *pamali* has begun to be abandoned by the society and it seems to have become "dormant". By looking at the phenomenon, we see that many disasters have occurred as a result of *pamali* violation that can sometimes give a warning to the perpetrators. The tsunami and the sinking of large ships in a tragic way may still come to mind. After the incident, there was a rumour that the beach and the ships must have been used by young people to commit inappropriate conduct, which then resulted in disaster. This is a warning to the public that belief in the expressions of prohibition (*pamali*) should be maintained properly to ensure survival.

An area that became the target of research on *pamali* is Tengger located in East Java Province. Tengger is an indigenous community that has, until now, been resilient in holding various traditions of ancestral heritage, thus resulting in positive, conducive and harmonious local wisdoms (Santoso, 2012: 7). In addition, the Tengger community has always upheld the teachings of the ancestors and has faithfully inherited customs in the era of modern culture. In this case, Sutarto (1997: 54) states that Tengger society is also open; they are not a primitive or isolated tribe different from the Javanese. Based on the explanation, the researcher has become interested in the issue of *pamali* and attempted to do this research because *pamali* can have positive effects on the wider community. Even though we now live in a modern era with sophisticated technology, transformations of cultural tradition or oral tradition should be maintained from generation to generation.

The research method used in this study is a qualitative method, which is a research procedure that produces descriptive data derived from written or oral data that exist in a language community (Djajasudarma, 2006: 11). Thus, the data were based on facts, obtained using strategic and systematic steps in order to facilitate the description, classification and analysis related to metaphorical forms that use animal names in the Javanese and Indonesian languages used by the community, especially in East Java.

On the basis of the method above, this study uses three strategic stages, which are (1) data collection method, (2) data analysis method, and (3) method of presentation of data analysis results (Sudaryanto, 1988: 57). Twenty people in Tengger as informants were purposively involved in collecting the data. The method used for collecting data was watching and listening, which included listening to oral communication that was directly conducted by people in everyday communication in informal situations. The data collection was done with no interference in the interaction, so that the data was natural. The data were then recorded and orthographically written on data cards. In addition to using oral data sources, the data collection was also done by searching for written data sources. To collect the necessary data, the researchers also used conversation, so that there was direct communication between the researchers and informants.

2 DICUSSION

The development of oral tradition in the past is actually difficult to exist at present because it is marginalised by time. The related community seems to be "dormant" because of the intersection of the old tradition with "modernisation". Expressions of prohibition (*pamali*) are part of an oral tradition that should be preserved and cared for by the community, starting from the family domain. In reality, the related values and norms are still used as a collective memory in the present time, which can be used to educate the present generation in strengthening their identities. One of the targeted areas of research is Tengger community located in East Java. Tengger people are known to have local wisdoms that are positive, harmonious, adaptive, and religious to maintain the values of tradition as a religious system in the practice of everyday life. In addition, the expressions of prohibition serve not only as a living memory but also as a living tradition for the next generation. Implementation of the expressions of prohibition includes the sexes, both male and female, different age groups (elderly, adolescent, children), marital status (unmarried or married), and certain situations that occur in the community.

Therefore, this research was conducted in Tengger as a crucial location to explore the potential of oral tradition, especially in terms of expressions of prohibition. For example, the community still believes in such expressions of prohibition as (1) "Ojo kawin karo tonggo sing omahe adep-adepan" (Do not marry a neighbuor living across the street), (2) "Wong perawan ojo mangan nang lawang" (Young unmarried woman cannot eat in front of the door), (3) "Nek ketemu nang ndalan, ojo ngobrol karo bojone" (If you meet your husband/wife on the street, do not talk there). The data were obtained from reliable sources who believe that Tengger people still firmly maintain traditions, beliefs and customs. The expression of prohibition in number (1) advises against marrying a neighbour whose house faces yours because it will lead to prolonged conflict, number (2) advises a young unmarried woman not to eat in front of the door because it makes it difficult for her to get married, and number (3) suggests that a husband and a wife should not have a talk in the street because this may make any conflict in their household known to others. In addition, other expressions of prohibition maintained by the Tengger community include (4) prohibiting pregnant women from soaking the laundry, (5) prohibiting people from speaking dirty words while they are in the area of Mount Bromo, (6) prohibiting women from killing animals during pregnancy, and 7) prohibiting women from getting pregnant out of wedlock. The expressions of the prohibition are still believed by the Tengger community as an identity to maintain their survival in the region. They are used as an instrument to safeguard the safety of every citizen and villager. For instance, prohibiting a young woman from getting pregnant before marriage is strongly upheld. If this happens, this is truly a great disaster, both for the girl's family and the safety of the village. Therefore, when there is a girl getting pregnant out of wedlock, it is necessary to hold "village cleaning", and the girl and the man are paraded through the village to avoid any misfortune.

3 CONCLUSION

Expressions of prohibition in the Tengger community are still maintained as signaling the existence of local culture, which is implemented in everyday life. In the era of increasingly advanced globalisation, the culture of Tengger's community still exists as a potential of ancestral heritage or customs passed down from generation to generation. Therefore, it can be said that the preservation of the expressions of prohibition (*pamali*) has constituted a tradition until now for the next generation.

REFERENCES

Alwasilah, Chaedar. 1985. *Sosiologi Bahasa*. Bandung: Angkasa.

Aminuddin. 1988. *Semantik: Pengantar Sutudi tentang Makna*. Bandung: Sinar Baru.

Bloomfield, Leonard. 1933. *Language*. Holt, Rinehart and Winston, Inc.

Chaer, Abdul. 1994. *Semantik Bahasa Indonesia*. Jakarta: Rineka Cipta.

__________. 1994. *Linguistik Umum*. Jakarta: Rineka Cipta.

__________ & Leonie Agustina. 2004. *Sosiolinguistik:Perkenalan Awal*. Jakarta: Rineka Cipta.

Cruse, D.A. 1986. *Lexical Semantics*. Cambridge: Cambridge University Press.

Departemen Pendidikan Nasional. 2000. *Kamus Baru Bahasa Indonesia*. Jakarta: Balai Pustaka.

Djajasudarma, Fatimah. 1999. *Semantik I: Pengantar ke Arah Ilmu*. Makna. Bandung: Refika.

__________. 1999. *Semantik II: Pemahaman Ilmu Makna*. Bandung: Refika.

__________ 2006. *Metoda Linguistik: Ancangan Metoda Penelitian dan Kajian*. Bandung: Refika Aditama.

Halliday, MAK. 1978. *Languge as Social Semiotics*. London: Edward Arnold.

George, Yul & Gilian Brown. 1996. *Analisis Wacana*. Jakarta: Gramedia Pustaka Utama.

Goddard, Cliff. 1998. *Semantic Analysis*. Oxford: Oxford University Press.

Keraf Gorys. 1996. *Diksi dan Gaya Bahasa*. Jakarta: Gramedia Pustaka Utama.

Leech, Geoffrey. 1981. *Semantics: Study of Meaning*. New York: Pelican Book.

Lyons, John 1968. *Linguistics Semantics*. Cambridge: Cambrridge University Press.

Pudentia. MPSS. 2015. *Metodologi Tradisi Lisan*. Jakarta: Buku Obor.

Sudaryanto. 1988. *Metode Linguistik*. Yogyakarta: Gajah Mada University Press.

Sudikan, Setya Yuwono. 2000. *Metode Penelitian Kebudayaan*. Jakarta: Citra Wacana.

Sumarsono & Paina Partana. 2004. *Sosiolinguistik*. Yogyakarta: SABDA.

Ullman, Stephen. 2007. *Pengantar Semantik*. Adapted by Sumarsono. Yogyakarta: Pustaka Pelajar.

Wijana, I. Dewa Putu & Mohammad Rohmadi. 2006. *Sosiolinguistik: Kajian Teori dan Analisis*. Yogyakarta: Pustaka Pelajar.

Wardaugh, R. 1986. *Introduction to Linguistics*. New York: Mc Graw Hill Books Company.

Urban Studies: Border and Mobility – Kerr et al. (Eds)
© 2019 Taylor & Francis Group, London, ISBN 978-1-138-58034-3

Multiculturalism and local wisdom in the Gilimanuk-Bali community

I.B.P. Manuaba
Universitas Airlangga, Surabaya, Indonesia

ABSTRACT: This study aims to investigate a multicultural community and local wisdom. In this study, the writers investigated how to preserve local wisdom and the multicultural community in Gilimanuk-Bali. This ethnographical research applied a qualitative method. Data used in this research consist of field data and documents which were collected through observations, in-depth and open interviews, and certain documents. This research was supported by related studies, such as literature, cultural studies, philosophy and history. The findings of this study indicated that local wisdom, which has been inherited by Gilimanuk residents, affects their social harmony and integration. Well-preserved local wisdom has formed Gilimanuk residents into a harmonious multicultural community. This research also indicates that a harmonious community which is not segmented into exclusive groups and which lives together in diversity regardless of different languages, social activities, and domicile may be an effective solution to creating social harmony and national integrity.

1 INTRODUCTION

Indonesia is facing a series of multidimensional crises. One of these crises is the national disintegration threat caused by multicultural communities. Conflict, disharmony, and disintegration potential successively threaten Indonesia as a result of certain groups' provocation with the intention of disintegrating the nation.

The residents of Gilimanuk-Bali are a multicultural community with various ethnicities, religions, and cultures. However, this diversity does not represent a threat for them. It helps them to build a harmonious, peaceful, and prosperous community. It is one of the Indonesian multicultural communities which is able to treat its diversity not as source of conflict, but as an "asset" in creating a harmonious and peaceful community. Managing the diversity by maintaining local wisdom is the core of this study.

The study on the local wisdom of multicultural Gilimanuk residents is important because it is expected to provide useful findings to preserve and develop a multicultural nation with national harmony and a national integration framework. This study is expected to develop a social harmony model to save Indonesia from disintegration. This study focuses on the role of local wisdom to develop and maintain social harmony and national integrity. Studies on multicultural community local wisdom have scarcely been conducted while national disintegration and disharmony still threaten this nation.

This study is important to remind people who cannot accept the diversity of Indonesia. An intense study on investigating the values of local wisdom shared by a multicultural community is important to prompt a person or a community to re-learn long-neglected local wisdom owned by each region. The writers realise that there is much local wisdom which has not been utilised well. Therefore, the findings of this study are important for developing social harmony and national integration based upon local wisdom.

Implicitly, the writers have stated that "*state of the art*" in this study is the role of local wisdom shared by multicultural Gilimanuk-Bali residents in maintaining social harmony and national integration. Johnson (as cited by Sunaryo and Joshi) defines local wisdom as indigenous, a set of knowledge invented by a group of people and passed through generations which is used to help them live harmoniously within their environment. The knowledge develops within the environment, adjusting their needs. It is the result of their creativity, innovation, and continuous effort to adapt to the new conditions of their surroundings involving internal inputs and external effects. Therefore, the knowledge is not outdated, primitive, and static (Noor and Jumberi, 2010).

There are several studies related to this study. However, almost all of these studies had their own focuses. In one of his scientific articles entitled, Budiono (2010) focused his study on verbal intelligence. He found that verbal intelligence based on an established cultural foundation will help children in facing their future. Noor and Jumberi (2010) conducted intensive research which focused more on the role of local wisdom in wildlife conservation by the Dayak tribe. In one of her scientific articles, Wahjuni (2010) explained the results of her research regarding the importance of cultural wisdom in preserving the environment. Research conducted by Noorinayuwati et al. (2010) described the role of local wisdom on peat land utilisation in Kalimantan.

This is a study on culture which focuses on local wisdom development in maintaining multicultural society. The ethnographical approach utilises cultural materials. Tilley (2001:258) defined materials of culture as humanity issues produced by a society. By citing the theories proposed by Levi-Strauss, Tilley defined cultural study as "*...study non-verbal aspects of human culture provides an essential foundation for material culture studies*" (Tilley, 2001:258). Ethnographical study concentrates on ideas, meaning, and languages within the scope of cultural studies. In cultural studies, a theory proposed by Parsons is commonly applied. He systematically proposed a theory which described the relationship among culture, personalities, and social structure by developing an abstract but universally applicable model (Putranto, 2005:51).

This study is related to culture as an actual environment of practices, representations, languages, and customs of a particular community. According to Hall (as cited in Barker, 2009: 8), culture relates to contradictive common senses which are deeply rooted and form the life of a community.

Understanding culture means exploring how the meaning is produced symbolically in a language through a significance system. In this study on local wisdom, the writers will not only describe what the local wisdom is but also describe how the community – as the owners of the local wisdom – utilise them to maintain their social condition as a multicultural community.

Therefore, the aims of this study are to examine the local wisdom shared by multicultural Gilimanuk-Bali residents and to understand the multicultural community in Gilimanuk-Bali. These aims are focused on the development of local wisdom's potential to maintain multiculturalism.

2 MATERIALS AND METHODS

This is a qualitative study applying an ethnographical approach (Spradly, 1987). Muhadjir (1998:94) stated ethnography is a research model which is closely related to anthropology. It examines cultural events and presenting the ideas of subjects as the object of the study.

Data used in this study was collected through several strategies. They were participative observation, open and in-depth interviews, and document summaries. The unit of this study was local wisdom shared by multicultural Gilimanuk-Bali residents.

Local wisdom of Gilimanuk-Bali residents was chosen as the unit of the study because it plays the role of "adhesive tape" which unites the residents to maintain social harmony and national integration. Despite their multiculturalism, multi-ethnicity, and different religions, peace and harmony are continuously maintained.

This study relied more on emic data obtained from the subjects of the study, consisting of public figures and indigenous elders from among Gilimanuk-Bali residents. They are considered to understand their local wisdom and are able to explain the role of local wisdom in maintaining social harmony and national integrity. These data were combined with ethical data taken from document summaries and previous research.

3 RESULTS AND DISCUSSION

3.1 *Condition of society of Gilimanuk-Bali*

Gilimanuk is located on the westernmost part of Bali, connecting two islands. The residents come from different ethnicities, religions, and cultural backgrounds. All of them live together as a multicultural community which is united in diversity (*Bhinneka Tunggal Ika*).

In the 1930s, Dutch colonial authorities transported 100 prisoners – mostly from Lombok – from Candiloka to Gilimanuk. Since that time, many people have come to live in Gilimanuk, a place which used to be deep jungle and where no one wanted to live. These settlers came from Java, Madura, Bali, Bugis, and bird hunters seeking birds in that area.

Since 1975, the status of Gilimanuk was improved from *Banjar Dinas* into a village headed by Gusti Made Berata. In 1981, Gilimanuk was improved into a *kelurahan*. Until now, there have been 14 *Lurah* successions in Gilimanuk.[1]

Gilimanuk residents come from different religions. All of them are migrants. Based on data obtained from Kelurahan Gilimanuk, there are 3 Buddhists, 1,318 Hindus, 2,634 Muslims, and 95 Christians. Gilimanuk residents also come from different ethnicities. Although there is no official record owned by Kelurahan Gilimanuk, based on information obtained from local residents, the residents are Javanese, Madurese, Balinese, Lombok, Bugis, and Makassar.

3.2 *Local wisdom of the multicultural Gilimanuk community*

Multiculturalism in Gilimanuk is initiated by appreciating their local wisdom. This local wisdom plays a big role in maintaining a multicultural community. One of the important local pieces of wisdom which supports interaction among residents is communication flexibility. A unique code-mixing is used by the unsegmented community in their daily communications. This code is accepted by the residents without questioning its origin.

> "... local wisdom and values which have been shared and utilised to maintain good communication among residents where the flexibility of communication as the most important aspect. Language (a code) we use in our daily communication does help us develop rapport, intimacy, and unity among the residents". (Diatmika, 2017)
>
> "We use a unique code-mixing. Some words were adopted from Javanese, Balinese, and Indonesian. Our accent is also unique. Therefore, it would be difficult to recognise who we are, whether Javanese or Balinese, whether we are Muslims, Hindus, Buddhists, or Christians. The language has made us a big family." (Diatmika, 2017)

Besides the language, there are other local wisdom potentials which are utilised to maintain multiculturalism. They are community awareness, mutual cooperation, assimilated domicile, and faith in the religions. Community awareness and high understanding as a multicultural society endorse the residents to appreciate and respect each other. Therefore, all Gilimanuk residents feel that all of the residents live in this village as a family despite their different ethnicities and religions.

> "It feels like this awareness, this understanding has been taught to us since the first settler generation; ... our residents have lived together as a family, without segmentation." (Diatmika, 2017)

Mutual cooperation is one local wisdom which is still maintained by Gilimanuk residents. Social activities have been on a monthly basis for them. All of them participate regardless of their ethnicity and religion. Another interesting activity practiced routinely by the community is sermon.

---

1. All of the information was obtained from documents owned by Kelurahan Gilimanuk especially documents relating to a short history of Gilimanuk. These documents were copied under the permission of BASOSPOL Jembrana Regency.

"… the sermon is delivered by a local figure (usually by indigenous leaders) to remind the residents about the importance of maintaining the unity, social rapport, and mutual cooperation." (Diatmika, 2017)

The sermons are continuously implemented and internalised in the society.

Another local wisdom is resident assimilated neighbourhoods. The residents live in a neighbourhood which assimilates their different ethnicities and religions. Sacred places of the multicultural Gilimanuk community (temples, monasteries, mosques, *Mushola*, Catholic churches, and Protestant churches) are maintained and preserved by an inter-religious community.

Based on historical records, there have been no conflicts among this multiethnic and multireligion community. On the contrary, they preserve these sacred places as a responsibility.

The implementation of inter-religion tolerance and harmony is indicated from the religious celebrations:

"In our daily activities – even in religious celebrations—we help each other. …when most of Christians celebrate Christmas and New Year, Muslims and Hindus help. During *Galungan* celebration, we (Hindus) install *penjor* among Muslim houses. There is no single *penjor* destroyed. During Eid (al-fitri), we (Muslims) are supported by Christians, and Hindus. … our (Buddhist) monastery were cleaned and preserved by Hindus and Muslims. There are Muslims and Hindus working in the monastery. Everyone is warmly welcomed to come to the monastery". (Michael, 2017)

These local wisdoms are passed through generations by multicultural Gilimanuk residents. Social harmony and peaceful situations are well maintained. Public facilities (such as sports and art facilities), which enable them to conduct social activities regardless of their ethnic and religion differences, are also preserved.

"Our local wisdom will be passed through generation. The importance of maintaining and preserving these values is getting higher. Our next generations will be more assimilated through marriage, social interaction, and other processes. Friendships and familial bonds are developed by inter-ethnic and inter-religion individuals since their childhood. Our youngsters do social activities, such as exercising and watching movies together. If there is an event or performance (such as *joged bumbung*) we usually dance together". (Punia, 2017)

Because of local wisdom preservation, Gilimanuk residents feel united as a family despite any differences. This society has achieved a level of harmony at which they can appreciate and sincerely feel other residents' feelings. These attitudes are inherited and implemented in their social activities.

Harmonious social interaction has been established among Gilimanuk residents. The interaction stimulates a sense of community, intimacy, tolerance, and mutual cooperation. This condition is supported by public facilities which enable them to conduct social interaction regardless of their diversity. Among these facilities are: language, arts, social activities, food stalls, sport courts (volley ball and soccer) and other public spheres such as the village hall and *pos siskamling.*

Gatherings conducted by Gilimanuk residents function to conduct activities which are intended to maintain harmony, intimacy, and unity among the residents. Gilimanuk also adopts *desa pekraman* and *desa kedinasan.* Although *desa pekraman* is intended to maintain social practices based on Hindu tradition, it does not mean that the Hindus are exclusively separate from their Muslim, Buddhist, and Christian neighbours. It is merely because the role of *desa pakraman* is to preserve traditional Balinese Hindu practices. Meanwhile, *desa dinas*, due to its function which is related to more general and official affairs, involves all residents without exception.

The harmonious interaction is strengthened by flexible use of inter-religious greetings. These greetings have strengthened the intimacy among the residents. Besides greetings, unique code-mixing has also strengthened social rapport among the residents. This "code-mixing" is a mixture of Javanese, Balinese, and Indonesian. In their daily communication, Gilimanuk residents also mix Balinese, Javanese, and Madurese accents with Indonesian. This "mixture" increases the connectedness among Gilimanuk residents.

Gilimanuk local leaders are neutral and moderate towards their beliefs. They always avoid particular thoughts which potentially cause conflict in the community. The main point is each member of Gilimanuk community has to respect and appreciate each other. Each community member realises that they are a big family, regardless of their religion, ethnic, and cultural backgrounds.

This local wisdom significantly affects each member of Gilimanuk community's character building. The most concrete effect can be identified from their attitudes relating to tolerance, togetherness, appreciation, harmony and a sense of community.

In Gilimanuk community, there is no clear hegemony from the majority group

> "Although Gilimanuk Muslims is the largest population, they do not dominate the community. When Hindus celebrate Galungan and establish *penjor*, there is no single *janur*[2]/ *ambu penjor* damaged. If a member of community conducts religious events or ceremonies, all of community will help regardless of their religions. We are unity in our diversity; we are diverse in our unison as an assimilated Gilimanuk society. *Adzan*[3] has to be delivered beautifully and easy listening because its voice is broadcasted by using loudspeaker. Therefore it won't disturb other community members". (Punia, 2017)

The tolerance among multicultural community members assures every conflict can be neutralised and solved using a unanimous solution. The inherited tolerant characteristic enables members of Gilimanuk society to live together harmoniously in sincerity and intimacy.

3.3 *Opinions regarding multicultural Gilimanuk*

The Gilimanuk community does realise that if their diversity is well-preserved, it will have good results. They consider their diversity an asset rather than a problem. They are very thankful they live altogether in their diversity, and are diverse in their unity.

> "As the part of community, indeed, we are the actors. We did realise that our community is multi-ethnic, multi-religion, and multicultural since long time ago. We have to be able to accept this condition. Therefore, our multiculturalism is our asset. We are freely interacting with other people and learning their culture and religion. … this is the reality which we have to accept and maintain well. We don't want to be like people outside (Gilimanuk) who are conflicting and battling caused by different ethnicities, religions, and culture." (Parto, 2017)

Besides their appreciation of their diversity, Gilimanuk residents have accepted that multiculturalism has been integrated as their characteristic. Therefore, they preserve and develop this characteristic. Gilimanuk residents do not want to be like other communities (outside Gilimanuk) who accentuate their differences, causing conflicts, unsolvable problems, and segmented communities where there is majority and minority rule.

Based on the results of interviews conducted on September 1st–4th 2017, Gilimanuk residents admitted that their positive attitude towards their diversity was mainly supported by interpersonal fluent communication. In Gilimanuk society, there was no suspicion. This positive attitude is closely related to their genealogy. They inherited social harmony and peaceful lives. The generation of Gilimanuk residents who are the subjects of this research have inherited positive values from previous generations. Gilimanuk society prioritised an egalitarianism, intimacy, and unity. These values provide harmony in their lives. The harmony has been maintained well in Gilimanuk society.

This positive attitude promoting inexplicable multiculturalism is supported by Gilimanuk' local leaders. These local leaders keep inspiring their people to appreciate and respect each other, and to preserve social harmony which has been developed. Gilimanuk residents and local leaders prevent external influence which may destroy the established social harmony. Gilimanuk's community optimistically believes that they are able to pass on the social harmony and peaceful life to the next generations and prevent external influences.

---

2. Coconut leaf.
3. Muslims call to prayer.

"...Optimistically we say: 'yes, we can.' We have a positive attitude on our multicultural condition. Optimistically we assure that established social harmony will never change in the future. We will be a harmonious peaceful society in the future as far as there is no extreme external influences which may change our understanding." (Diatmika, 2017)

Through a good understanding of this multicultural condition, violence, disturbance, and negative attitudes are successfully prevented. Gilimanuk residents never engage in internal conflict because they uphold the meaning of multiculturalism and diversity.

Although Gilimanuk's society is multi-ethnic and multi-religious, there is no ethnic segregation that is usually the source of conflict in most areas in Indonesia. The absence of segregation stimulates ethnic and cultural assimilation and reduces—even eliminate—internal conflicts and clashes.

4 CONCLUSION

Based on the analysis above, it can be concluded that the multicultural Gilimanuk community carries on its social life based on local wisdom which has been passed down through generations. The local wisdom consists of values which served as guidelines for maintaining social harmony. These values were: a) *communication value*, implemented by their "unique" language, indicated by mixing several ethnic dialects; b) *togetherness* which is developed from their common feeling as immigrants to the area; c) *assimilation*, indicated by their assimilated neighbourhoods where multi-ethnic and multi-religious residents live together; d) *social harmony*, developed from their understanding and consciousness about the importance of respecting each other and appreciating their diversity; and e) *mutual relationships*, indicated by intimate relationships and mutual cooperation which are preserved by the society.

The important model proposed in this study shows that the more segmented a community is into exclusive groups and neighbourhoods the more vulnerable it is to social disharmony. The writers propose a multicultural society preservation model which organises the residents into assimilated societies with a shared common code (language), social and cultural activities, and domicile rather than separating the residents into "exclusive" neighbourhoods. The writers believe that this model will help to solve inter-ethnic and inter-religious conflicts happening in multicultural societies.

REFERENCES

Barker, C. (2009). *Cultural Studies: Teori dan Praktek (translated by Nurhadi from original book Cultural Studies, Theory and Practice).* Yogyakarta: Kreasi Wacana.

Budiono, R. (2010). Pengenalan Kearifan Budaya Lokal sebagai Salah Satu Langkah Awal Pembentukan Kecerdasan Verbal-Kult. Kalimantan Tengah.

Diatmika, I.M. (2017, September 1–4). Kearifan Budaya Lokal pada Masyarakat Multikultural Gilimanuk-Bali. (I.B. Manuaba, Interviewer).

Michael. (2017, September 1–4). Kearifan Budaya Lokal pada Masyarakat Multikultural Gilimanuk-Bali. (I.B. Manuaba, Interviewer).

Muhadjir, N. (1994). *Metodologi Penelitian Kualitatif, edisi III.* Yogyakarta: Rake Sarasin.

Noor, M., & Jumberi, A. (2010, August 27). *Kearifan Budaya Lokal dalam Perspektif Pengembangan Pertanian di Lahan Rawa.* Retrieved from Balitra Litbang: http://www.deptan.go.id./lokal/kearsifan.

Norginayuwati, A.R., Noor, M., & Achmadi. (2010). *Kearifan Lokal dalam Pemanfaatan Lahan Gambut untuk Pertanian di Kalimantan.* Balai Penelitian Pertanian Lahan Rawa.

Parto. (2017, September 1–4). Kearifan Budaya Lokal pada Masyarakat Multikultural Gilimanuk-Bali. (I.B. Manuaba, Interviewer).

Punia, K. (2017, September 1–4). Kearifan Budaya Lokal pada Masyarakat Multikultural Gilimanuk-Bali. (I.B. Manuaba, Interviewer).

Putranto, H. (2005). Budaya dan Integrasi Sosial: Menelusuri Jejak Karya Talcott Parsons. In M. Sutrisno, & H. Putranto, *Teori-teori Kebudayaan* (p. 51). Yogyakarta: Kanisius.

Tilley, C. (2001). Ethnography and Material Culture. In P. Atkinson, A. Coffey, S. Delamont, J. Lofland, & L. Lofland, *Handbook of Ethnography* (p. 258). London: SAGE Publications.

Wahjuni, S. (2010, August 27). *Pemulihan Lingkungan dengan Kearifan Lokal: Kasus Studi Gunung Barangrang.* Retrieved from http://pengasuh bumi.

Urban Studies: Border and Mobility – Kerr et al. (Eds)
© 2019 Taylor & Francis Group, London, ISBN 978-1-138-58034-3

Cultural capital of traders on Pahing Sunggingan market in Boyolali

Jatmiko Suryo Gumilang, Mahendra Wijaya, Bagus Haryono & M. Si
Universitas Sebelas Maret, Indonesia

ABSTRACT: One of the attractions of Pasar Sunggingan is the market day of Pahing, where traders from all directions sell various merchandise every day. However, as the era advances, various changes are also anticipated, for example the existence of market conditions when Pahing day disappears gradually. The purpose of this study was to determine the cultural capital of traders and socio-cultural impacts of post-revitalization of Pasar Sunggingan in Boyolali. This is a qualitative study conducted using the phenomenology method. Based on the research result, it can be concluded that the impacts of revitalization of Pasar Sunggingan include (1) a change in cultural elements such as Javanese language, value and norm, market organization, merchandise, trade, place of worship, music, art, and dance, (2) Pahingan tradition as the custom of Sunggingan residents from generation to generation and the source of economic activity for the Sunggingan community, and (3) the community's strategy in preserving Pahingan tradition by facilitating continuous socialization so that the market still proceeds on Pahing day.

Keywords: Pahingan tradition, Sunggingan Market, Elements of Culture

1 INTRODUCTION

Market is a meeting place between the seller and the buyer. According to Kotler (1997), "The market is made up of all potential customers who possess similar specific needs and may be willing and able to execute an exchange to satisfy their need and desire." Market is a meeting place between sellers and buyers who have social interaction.

Indonesia cannot be separated from various forms of diverse cultures. The existence of its cultural pluralism includes religion, language, clothing, customs, norms, tribes, and livelihoods which overshadow a region. As a result, a particular region may form a particular community with a tradition that is firmly embedded in the community. Indonesia has a wide range of markets, ranging from traditional markets to modern markets. However, a market that has certain uniqueness is a traditional market. The latest data from the Regional Revenue Service in 2013 revealed that the number of markets located in Central Java was 2,729 markets, which consisted of both modern and conventional markets. In 2013, Boyolali had 109 markets, with 39 conventional markets, 6 departmental stores and 64 supermarkets (Regional Revenue Service, 2013).

One of the traditional markets located in Boyolali is Pasar Sunggingan. It is located at the heart of Boyolali City, which makes it a strategic region. Pasar Sunggingan attracts people to come on its market day, i.e. on Pahing day, where traders from all directions hawk various merchandise every day. This also becomes an allure of its own where the prospective buyers are willing to travel long distances to find the desired goods and ultimately purchase them with the traditional bargaining process and the traditional market style.

This study also applies four pillar elements, which are the essence of Pancasila. The fourth precept states "Democracy led by wisdom of wisdom in representation/deliberation," which means the government policy affects the system that has been running in a place. One of the policies of the government is to revitalize traditional markets. However, the government policy on market revitalization brings about changes in these market conditions, which can create a considerable impact. As stated by Bagus Haryono, the advent of information technology applications (IT) at home or office, personal computers (PCs), Tablets (Tabs), and the Internet in everyday life has influenced community information. One of the impacts of market renovation is the change of traditions, which also causes a huge impact, ranging from changing the structure of market buildings as well as patterns and market systems, and declining traders profit to loss of livelihood (Good: 2017). In fact, market renovations undertaken by the government aims to improve the look of market buildings by transforming them into more modern forms. However, the impact of this renovation has led to several gaps in the market itself. This impact has been felt by some agents such as traders, local community leaders, consumers, DPRD and even the regent. However, there is a gap between one agent and another. Some of the agents want to maintain Pahingan tradition as a form of local wisdom of the Boyolali society; however, some agents want to change the traditional market into a more modern market. This study aims to explore the cultural capital of traders on Pahing day in Boyolali Sunggingan Market.

2 DISCUSSION

The natural and geographical conditions of Sunggingan, located at the heart of Boyolali City, make it a strategic region. As a result, the region has the potential to develop various businesses. Therefore, the majority of Sunggingan residents work in the business sector, as one can see along the main road that crosses Sunggingan. There are many banks, shophouses and kiosks which can support the needs of Sunggingan residents. The Sunggingan region is famous for trading various goods and services. Many homes are renovated and built into shop (house and shop) for business purposes. It can be concluded that the majority of the people work as traders. Thus, Sunggingan people financially depend on the existence of Sunggingan Market. Therefore, they cherish Pahingan tradition on Pahing day based on the Java calendar.

From these data, it can be concluded that although the Sunggingan community is described as a modern society, they follow the traditions of their ancestors, i.e. Pahingan tradition, in the Sunggingan region even today.

Pahingan tradition had continued in the past. Sunggingan market itself comes from the routine activities of the community and migrants who trade in the region and bring their merchandise in an *In-sunggi* manner. *In-sunggi* is one way of carrying heavy items by raising goods on the head of a person and holding them with both hands so that the goods do not fall, hence the name Sunggingan. It is also the existence of a market that makes it interesting inside the Sunggingan region.

Furthermore, the viscosity of traditions found in Sunggingan Market makes Sunggingan Market a place for traditional customary preservation in Boyolali City. Several changes have taken place in Sunggingan Market due to the government's revitalization program. This revitalization program is one of the government's programs that are aimed at updating market buildings and expected to provide comfort for the community. This is related to the fourth precept of Pancasila, which states: "People are led by wisdom of wisdom in representation/deliberation." Indonesia is a democratic country, but because of its populated society, only some people determine the prevailing policies and regulations. These representatives are called premises of the councils. Any policy established by popular representatives often reaps the pros and cons because there is a personal interest in it. Similarly, the market revitalization policy that reaps the pros and cons provides various kinds of impact. As stated by Bagus Haryono, the advent of information technology applications (IT) at home or office, personal computers (PCs), Tablets (Tabs), and the Internet in everyday life has influenced community information (Good: 2017). The impact of the market revitalization program includes the change of the building and the change of the kiosk in Pasar Sunggingan. However, the changes lead to some difficulties in the Sunggingan Market since the tradition is meaningful and hereditary to the community around Pasar Sunggingan.

During Pahingan tradition held on Pahing day, researchers observed the activities of the event, with the aim of providing an overview of this tradition. During the observation, researchers observed the course of buying and selling activities in the Pahingan. Traders and consumers from both local communities and communities outside the Sunggingan region participated in the event. Based on these observations, researchers asserted that Pahingan tradition is an interesting momentum that has an implicit meaning. Similarly, the activity has its own significance.

Pahingan tradition has two meanings. First, it is a sign of custom made by Sunggingan community members from generation to generation. Second, the existence of Pahingan tradition is very important for the Sunggingan community as a means of seeking fortune, and it becomes the main livelihood for them. In other words, the traditions of Sunggingan Market are a source of economic activity for the Sunggingan community itself.

This tradition is considered as a cultural heritage of the local community in Sunggingan, which is preserved until now. Although there are some obstacles that exist in Pahingan tradition, this tradition is still implemented and preserved.

Several strategies and efforts have been undertaken to preserve Pahingan tradition. First, the Sunggingan community socializes the use of Pahingan tradition to their generation. This continuous socialization on Pahing day will gradually accustom them to the tradition.

3 CONCLUSION

Tradition cannot be separated from the elements of space and time. This can be seen from Pahingan tradition which is held on Pahing day in Sunggingan Market. Spatial elements in Pahingan tradition are related to the process of socialization, where the community within and outside Sunggingan can meet in one place with the same goal of buying and selling activities. This tradition has been kept since ancient times. The placement of space and time emerges repeatedly; therefore, this routine results in a habit or folkways. On account of this habit, Sunggingan people as well as people outside Sunggingan gather in one space and time without an invitation. The function of the elements of space and time in Pahingan tradition as a strategy for Pahingan tradition exists even today. Culture is abstract, but can be felt with the elements in it. The elements of culture include language, knowledge system, social organization, living equipment system and technology, livelihood and religious system, and art, which can be seen in Pahingan tradition. Although there are some changes in the elements of culture, they do not alter the form of elements that have been inherited in Pahingan tradition itself. As the era is growing, the government launched a new regulation by revitalizing traditional markets. This aims to give comfort to society through buying and selling. However, several negative impacts are faced by some agents. One of the agents affected by this policy is the traders in the traditional market. The traders in Pasar Sunggingan achieve profitability and experience the impact of changing cultural elements in Sunggingan Market. For example, this includes the change from the use of the Javanese language to a mixture of Javanese and Indonesian languages, the knowledge system of Pahingan tradition, changes in values and norms, changes in the organization and market, changes in merchandise sold by traders in Sunggingan Market, changes in the merchant's livelihood system, the renovation of places of worship located in Sunggingan Market, and the change from traditional to modern music and dance. Therefore, the existence of Sunggingan Market as a traditional market is an important economic source of the Sunggingan community.

REFERENCES

Al Masitoh, Eis. 2013. *Upaya Menjaga Eksistensi Pasar Tradisional: Studi Revitalisasi Pasar Piyungan Bantul.*

Badan Pusat Statistik. 2013. Dinas Pendapatan Daerah (Dipenda) Kabupaten/Kota. Retrieved April 19, 2017 from http://jateng.bps.go.id/index.php/linkTabelStatis/997.

Elizabeth. Nottingham. K. 1997. *Agama dan Masyarakat (Suatu Pengantar Sosiologi Agama)*, Jakarta: Gravindo Persada.

Haryono, Bagus. 2017. The Impact of Information Technology on Family Values Transformation. *Knepublishing*. Retrieved October 20, 2017 from https://www.knepublishing.com/index.php/Kne-Social/article/view/881/2288.

Heung-Ryel, Kim. 2015. *Culture and Tourism-Oriented Local Traditional Market Strategies in Korea.*

Ikbal, Yanuar. 2012. *Metode Penelitian Sosial Kualitatif.* Bandung: Refika Aditama.

Ismawati, Esti, 2012. *Ilmu Sosial Budaya Dasar*. Yogyakarta: Ombak.

Janette Webb from David Hawkey. 2016. *On (not) assembling a market for sustainable energy: heat network infrastructure and British cities.*

Koentjaraningrat. 2009. *Pengantar Ilmu Antropologi.* Jakarta: Asdi Mahasatya.

Kuncahyawati, Helina. 2016. *Pemberdayaan Pasar Tradisional dan Pedagang Pasar Menurut Peraturan Daerah Kabupaten Purworejo Nomor 6 Tahun 2014 (Studi Kasus: Pasar Krendetan).*

Lexy J. Moleong. 2006. *Metode Penelitian Kualitatif.* Bandung: Remaja Rosda Karya.

Pahk, Sang-hyoun. 2016. *Misappropriation as market making: Butler, Callon, and street food in San Francisco, California.*

Poerwanto, Hari. 2006. *Kebudayaan dan Lingkungan Dalam Prespektif Antropologi.* Yogyakarta: Pustaka Pelajar.

Ritzer & Goodman. 2012. *Teori Sosiologi Klasik—Post Modern Edisi Terbaru (Trans: Nurha di).* Yogyakarta: Kreasi Wacana.

Sachari, A., & Sunarya, Y. Y. 2001. *Desain dan Dunia Kesenirupaan Indonesia dalam Wacana Transformasi Budaya.* Bandung: Penerbit ITB.

Satori, Dja'man & Komariah, Aan. 2012. *Metodologi Penelitian Kualitatif.* Bandung: Alfabeta.

Scott, Jhon. 2012. *Teori Sosial.* Yogyakarta, Pustaka Pelajar.

Sholihah, Ummu. 2016. *Strategi Pengembangan Pasar Tradisional Dalam Meningkatkan Ke. puasan Pedagang (studi kasus di pasar kliwon, karangwelas, banyumas jawa tengah). Jawa Tengah*: Institut Agama Islam Negeri (IAIN) Purwokerto.

Soemardjan, Selo. 1981. *Perubahan Sosial di Yogyakarta.* Yogyakarta: Gadjah Mada. University Press.

Somantri, Gumilar Rusliwa, 2005, *Memahami Metode Kualitatif.* Depok: Universitas Indonesia.

Soo, Hyun Taek. 1998. *Budaya dan Kekuatan: Pengertian Ilmu Sosial Bourdieu*, Seoul: Penerbit Nanam.

Susilo, Dwi. 2011. *Tentang Dampak Operasi Pasar Modern Terhadap Pendapatan Pedagang Tradisional Di Kota Pekalongan.* Pekalongan: Universitas Pekalongan

Sutopo, H.B. 2002. *Metodologi Penelitian Kualitatif.* Surakarta: Sebelas Maret University Press.

Wicaksono, Lulud N & Harsasto, Priyatno & Puji Astuti. 2013. *Persepsi Pedagang Pasar Terhadap Program Perlindungan Pasar Tradisional Oleh Pemerintah Kota Semarang Studi Kasus Pedagang Pasar Peterongan Semarang Selatan.* Semarang: Universitas Diponegoro.

Widjaja, H.A.W. 2000. *Komunikasi: Komunikasi dan Hubungan Masyarakat.* Jakarta: Bina Aksara.

Za, W.A. Amir. 2016: *Modal budaya dan kesannya ke atas kebergantungan kerja komuniti nelayan di Malaysia: Kajian kes nelayan Kuala Terengganu.*

Urban Studies: Border and Mobility – Kerr et al. (Eds)

"The Legend of Nusantara": Disguising the boundary between locality and globality in *Indoeskrim Nusantara*'s commercial advertisement

Milawaty
Universitas Airlangga, Surabaya, Indonesia

ABSTRACT: The objective of this study was to reveal the potential motive behind the mixture between local and global items in an ice cream commercial advertisement, *Indoeskrim Nusantara*, through any system of signs. For this purpose, semiotics of Roland Barthes was chosen along with Arjun Appadurai's ideas of globalisation flow. The result of this study indicates that the advertisement has attempted to localise global products by adding Javanese local values to combine elements into a single entity. Globalisation flow (i.e. technoscapes) plays a major role in combining a mixture of opposite elements. As a result, globalisation leads to cultural heterogenisation in which the society tends to localise the global material. Apparently, this disguises the boundary between locality and globality.

Keywords: commercial advertisement, global, *Indoeskrim Nusantara*, local, signs

1 INTRODUCTION

Indonesia is a multicultural country with hundreds of tribes spread over its regions; therefore, it is known as a country with abundance of local values. The history of Indonesia reveals the existence of various kingdoms in the past, which show their cultures and beliefs transmitted through many local narratives in the form of myths, fables, fairy tales or legends. One of the most notable examples is the story of *Brama Kumbara*, a king of *Madangkara* Kingdom which directly borders *Majapahit* Kingdom. This is a famous story in Indonesia which was broadcast on radio in the 1980s, as well as appeared in the movies and in a recent television drama. The latest one was a commercial advertisement of an ice cream entitled "*Kisah Legenda Nusantara*" or "The Legend of Nusantara", which was made by an Indonesian food company, *Indofood*. It picked the story of *Brama Kumbara* to promote their ice cream product, *Indoeskrim Nusantara*.

This has gained massive attention of people since its broadcast via the online media. The trigger mechanism is that along with a historical setting, it also involves some electronic tools that can only be found in the globalisation era, which include modern communication technology, such as walkie-talkies and smartphones with GPS technology, and electronic equipment, such as a dispenser and a freezer, as well as the use of English. Therefore, this study aims to explore the potential motives behind the mixed elements of locality and globality in one "bowl" by examining the representative local and global signs contained in it. The theory of semiotics proposed by Roland Barthes is a reliable tool that can be used to deliver the messages to the people. As the advertisement deals with globalisation issues, Arjun Appadurai's theory of globalisation provide a fine explanation of the complexity found within it.

1.1 *Barthes' semiotics*

As Ferdinand de Saussure [in Barthes, 1967: 9] states, any system of signs including images, gestures, musical sounds, objects, and the complex associations of them that constitute systems of signification is called "semiology". A sign consists of what Saussure called a *signifier* and *signified.* As the descendant of Saussure, Barthes borrows these two terms in his theory of semiotics [Barthes, 1967]. In his book entitled *Elements de Semiologie* (1964), Barthes gives exact definitions of the elements of sign. A *signifier* is defined as the mediator, the material and the "expression" of the "thing" [Barthes, 1967:47], while a *signified* is the substance, or the mental representation of the "thing", which Saussure called a "concept" [Barthes, 1967:42]. Barthes offers an example, an "ox". The word "ox" functions as a *signifier*, while the *signified* is its concept, a big four-legged herbivore animal with a pair of horns. This is the first level of signification which Barthes called *denotation*, carrying a *denotative sign* [Barthes, 1967:89]. While Saussure's theory ends at this level, Barthes adds a new level called *connotation*, the second level of signification carrying a *connotative sign*. In connotation, the meaning is related to the culture, knowledge and history; in other words, it refers to the environment in which the system stands [Barthes, 1967:91–92]. This level is usually called *myth* [Budiman, 2001, in Sobur, 2006:71].

Returning to the previous example, the ox in India is related to its religion and beliefs; therefore, it is referred to as a sacred animal, which is not to be eaten, but to be honoured. Hence, as Barthes touches this cultural level, his semiotic analysis becomes deeper and more reliable in terms of understanding the messages of any cultural products, such as movies, dramas, television programs and advertisements.

1.2 *Appadurai's ideas of globalisation flow*

Through his essay *Disjuncture and Difference in Global Cultural Economy* (1990), Arjun Appadurai promoted his idea of globalisation flow by rationalising the complexity of the globalisation era. As Appadurai [1990:296] states, "the complexity of the current global economy has to do with certain fundamental disjunctures between economy, culture, and politics...". To explore this disjuncture, he proposes five interrelated dimensions of global cultural flow: *ethnoscapes*, *finanscapes*, *technoscapes*, *mediascapes* and *ideoscapes.* Ethnoscapes is the movement of people including tourists, immigrants, refugees and entrepreneurs or businessmen across nations. Finanscapes is the flow of money across nations due to currency markets, national stock exchanges and commodity speculations. The movement also influences the spread of technology to every corner of the world—which is called technoscapes. Techno-mobility causes the distribution of media and its cultural content throughout the world—which is called mediascapes—whether in the form of newspapers, magazines, television stations including its programs and advertisements, film production studies, etc. Mediascapes may lead to ideoscapes, which is the movement of political notions and ideology around the world. He points out that the current global flow occurs in and through the growing disjunctures of these five dimensions (Appadurai, 1990:296–301).

2 METHODOLOGY

This study uses a qualitative approach to fulfil the objectives of analysing the commercial advertisement *Indoeskrim Nusantara*. Signs are the main data source that carries the required messages. As signs in an advertisement are apparent in scenes, they were collected through the scenes presented in 2 minutes 58 seconds. The signs are in the form of visual (codes) and linguistic messages (language) as Barthes [1967:10] claims for every meaning in any visual messages of a movie, advertisement, press photograph, etc. This idea will be confirmed through the linguistic messages that follow. It is analysed in two steps: the identification of messages using two levels of Barthes' semiotics, which are denotative signs consisting of the *signifier* and the *signified* and connotative signs consisting of the *myth*; and the interpretation using Appadurai's theory of globalisation flow.

3 DISCUSSION

3.1 *The signs: Locality vs globality*

According to the Indonesian dictionary, the word "local" is defined as (something) that occurs in a particular place or is limited to a specific area. Locality itself means a cultural space that holds social and ideological portraits represented through the interaction of its figures and the cultural dynamics that express and keep the values of humans in cultured life [KBBI, 2017]. Thus, locality is commonly associated with originality, indigenous culture or a separate characteristic owned by a particular region obtained through inheritance, in other words tradition. Consequently, locality is often represented by traditional entities. The opposite of local is global, which means (something) that is related to or about the whole world [KBBI, 2017]. Globality is a global phenomenon that has no recognition of boundaries, whether they are local or regional. It covers multinational-level cultures owned by the world societies. In other words, it leaves locality and is freed from traditional styles and values. Thus, it is often associated with modernity and represented by modern entities. Both locality and globality are represented in the commercial advertisement *Indoeskrim Nusantara*, as evident from the co-existence of traditional and modern entities expressed through "signs" operating in the scenes.

Scene	Denotative sign	Connotative sign
	*Udeng* (the signifier) is a piece of Batik cloth which is shaped and tied around the forehead (the signified). Sleeveless cloth (the signifier) is a fabric outfit that covers the upper part of the body with no sleeves (the signified). *Jarik* (the signifier) is a Batik cloth with no stitching which is used to cover parts of the body (the signified). Talisman necklet (the signifier) is a black ornamental string with a square pendant that is worn around the neck (the signified). Hair bun (the signifier) is a hairstyle in which the hair is drawn back into a tight coil at the top or the back of the head (the signified). *Walkie-talkie* (the signifier) is a handheld communication device that connects two people at a particular distance (the signified).	*Udeng* is the name for a Javanese headband. Together with the sleeveless cloth and *Jarik*, these are typical of an ancient Javanese outfit, especially for men. Talisman neck let is a traditional necklet, which is commonly used by the lower classes on Java in the past. Hair bun is the traditional Javanese hairstyle. *Walkie-talkie* (the signifier) is a technology used in the modern era globally (the signified).
	Double-pendant gold necklace (the signifier) is an ornamental string with two or more pendants made from gold which is worn around the neck (the signified). Upper-arm gold bracelet (the signifier) is a circular ornament made from gold which is worn on the upper arm (the signified).	Accessories including the double-pendant gold necklace, the upper-arm gold bracelet and the golden belt, which represented both local and historical values of the Javanese, are of ancient style that were only used in the Javanese royal period.

(*Continued*)

(*Continued*)

	Golden belt (the signifier) is a band made of gold which is worn around the waist to fasten the body outfit (the signified). *Sanggul* or hair bun (the signifier) is a hairstyle in which the hair is drawn back into a tight coil at the top or back of the head (the signified). *Gada* (the signifier) is a batter with an enlarged round tip made of wood, iron and so on (the signified). A giant hawk (the signifier) is a big bird of prey with broad rounded wings and a sharp beak (the signified). Smartphone (the signifier) is a mobile phone that functions as a personal computer, usually with additional features such as a touch screen and wireless Internet access (the signified) [KBBI, 2017]. GPS (the signifier) is the Global Positioning System, which has an accurate global navigational and surveying facility based on the reception of signals from an array of orbiting satellites (the signified). Internet (the signifier) is an electronic communication network that connects computer networks and computer facilities, organised around the world by telephones or satellites (the signified) [KBBI, 2017]. A golden carved phone case is a type of phone protector carved in gold, which resembles a local figure or animal (the signified).	*Sanggul* or hair bun is the traditional Javanese hairstyle for women, especially in ancient times. *Gada* is a weapon owned by a royal knight, which often appears in folklore. It is believed to keep the power of a deity in eastern cultures. A giant hawk is an animal of myth which is related to a historical figure as a rider, especially Brama Kumbara in Javanese culture. Smartphone, GPS and Internet are the products of technology in the modern era, which are used globally. A golden carved phone case is a product resulting from the mixture of globality (phone case) and locality (a local figure or animal).
 	A golden carved freezer (the signifier) is a food and beverage cooler which has golden carving on the case (the signified). Ice cream (the signifier) is a soft frozen food made with sweetened and flavoured milk fat (the signified).	A golden carved freezer is a product resulting from the mixture of globality (freezer) and locality (Javanese carving). Ice cream is a product enjoyed globally.

(*Continued*)

(*Continued*)

 	Gallon (the signifier) is a tube made from plastic which functions as a water container (the signified). Dispenser (the signifier) is an automatic electronic device for heating and cooling water for drinking (the signified) [KBBI, 2017]. Photograph (the signifier) is an image taken by a camera (the signified). Rice cooker (the signifier) is an electronic device to cook rice (the signified). Microwave (the signifier) is an electronic device to heat any kind of food (the signified). Laptop (the signifier) is a small, portable personal computer that could be placed on the user's lap, consisting of a single device including a keyboard, a display screen and a microprocessor, which is usually equipped with a rechargeable battery (the signified) [KBBI, 2017].	Gallon, dispenser, photograph, rice cooker, microwave and laptop are the products of technology in the modern era, which are used globally.

The use of English emphasises the global elements of the advertisement. It should be noted that the main character, Brama Kumbara, operates the GPS using the English language such as "Head west", "You are on the fastest route despite the usual traject". The wife of Brama Kumbara also uses an English word, i.e. "hubby", as the contact name for her husband.

Denotative sign	English (the signifier) is the language of British people (the signified).
Connotative sign	*English* is the international language as well as the official language of multinational communities.

3.2 *"The imagined world": Disguising the boundary between locality and globality*

Globalisation is a process in which the world societies interact or are interconnected with each other in all aspects of their lives (Lodge, 1991:56 in Anastasya, 2011:1). It is a term commonly related to the centre-periphery model or cultural homogenisation, i.e. Americanisation. However, in recent decades, this term has been contested by its opposite, i.e. cultural heterogenisation. Arjun Appadurai, one of the pioneers of the term, argues that cultural homogenisation has failed to consider the fact that "when forces from various metropolises are brought into new societies, they tend to become indigenized..." [Appadurai, 1990:295]. Not all world societies will receive the "raw material" of globalisation that will end in homogenisation. Some, affected by the ideology owned, accept it by attaching some modification based on their perspectives, creating a new cultural dimension different from the original form, i.e. to remain localised. This cultural creativity, by the mixing of global and local elements, ends up with abundance and complexity of cultural space. It blurs the boundary between the local and the global.

The findings of this study reveal that the mixture between local and global elements is Indonesian, especially Javanese local values with products of modern technology. The combination of local and global aspects presented in the advertisement denote the way the food agent for *Indoeskrim Nusantara, Indofood,* perceives the quick and extensive influence of globalisation in Indonesia. Indofood, the mother company of *Indoeskrim Nusantara,* is a major Indonesian food company based in Jakarta, Indonesia, exporting many of its products across countries.

While constructing the images of its product in this global era, Indofood involves some aspects of globality in its framework. Technology becomes the choice by which the global aspects are attached as globalisation enters the world along with the development of technology. Technology makes everything possible and interconnected regardless of regional distances. This is due to the spread of technology which Appadurai calls technoscapes, i.e. the movement of technology at high speed across various types of impervious boundaries [1990:297]. As a core point in globalisation, technology becomes the appealing points in the commercial advertisement *Indoeskrim Nusantara*.

Landscapes are deeply perspectival constructs. They are affected by the historical, linguistic and political situatedness of different sorts of actors, such as nation-states, multinationals, diasporic communities and even sub-national groupings and movements [Appadurai,1990:296]. While constructing the advertisement's concept, Indofood practically takes the Javanese local values as it is based on Java Island with a notable Javanese culture. The involvement of Javanese values in the combination of local and global aspects results in the existence of modified entities such as a golden carved phone case and a freezer, a rice cooker and a microwave. These entities lead to a new cultural space where there are no clear boundaries between the local and the global. This cultural space indicates the existence of what Anderson called "imagined worlds", i.e. the multiple worlds constituted by the historically situated imaginations of persons and groups spread around the globe (Appadurai 1989 in Appadurai 1990). The lines between the "realistic" and "fictional" landscapes are blurred, while the construction of an "imagined world" tends to involve chimerical, aesthetic or even fantastic objects [Appadurai, 1990:299].

4 CONCLUSION

The sign analysis indicates that the commercial advertisement *Indoeskrim Nusantara* involves both local and global aspects, i.e. Javanese local values and products of modern technology, respectively. It is the effect of globalisation flow, especially technoscapes, that has caused this movement. The localised global products presented in the advertisement are motivated by the way the agent perceives globalisation and takes it into their own framework that is culturally influenced. The commercial advertisement *Indoeskrim Nusantara* is the product of global cultural flow with the aim of receiving the advance movement, especially global technology, that causes a disguised boundary between the elements involved (i.e. locality and globality) and brings together a seemingly imagined world.

REFERENCES

Anastasya, M. (2011). *Globalisasi dan Kawai i Guzzu: Analisis Teori Globalisasi Appadurai dalam Studi Kasus Karakter Hello Kitty*. Jakarta: Universitas Indonesia.

Appadurai, A. (1990). *Disjuncture and Difference in the Global Cultural Economy: Theory, Culture, and Society.*

Barthes, R. (1967). *The Elements of Semiology translated.* New York: Hill and Wang.

Barthes, R. (2012). *Elemen-Elemen Semiologi Terjemahan.* Yogyakarta: Jalasutra.

Bronner, S.J. (2007). *The Meaning of Folklore: The Analytical Essays of Alan Dundes.* Logan: Utah State University Press.

Kamus Besar Bahasa Indonesia Edisi V. (2017). Retrieved November 18, 2017, from Kemendikbud Web site: http://kbbi.kemendikbud.go.id.

Karyanto, P. (2011). Bahasa, Kekerasan dan Media: Analisis Teks Media Terhadap Kasus Penyerangan Jemaat Ahmadiyah Indonesia dalam Pemberitaan Media Massa Nasional.

Khusyoiri, J.A. (2006). From Concordia to Sarinah: The Dynamics of Changing Names on a Malang Building Between 1930s and 1960s. *International Seminar on Indonesia Urban Cultural Symbolism in Early Independent.* Yogyakarta: Universitas Gadjah Mada and Netherland Institute for War Documentation.

Santoso, E.B. (2003). Manfaat Prasasti dalam Penulisan Sejarah Lokal: Kajian atas Sejarah Lokal Kediri. *Lemlit Unair*.

Sobur, A. (2006). *Semiotika Komunikasi.* Bandung: PT Remaja Rosdakarya.

Sobur, A. (2012). *Analisis Teks Media.* Bandung: PT Remaja Rosdakarya.

Tim Nasional Penulisan Sejarah Indonesia. (2010). *Sejarah Nasional Indonesia: Zaman Kuno.* Jakarta: Balai Pustaka.

Urban Studies: Border and Mobility – Kerr et al. (Eds)

Urban environmental quality and human well-being assessment: Towards walkable neighborhood (A case study of Dr. Soetomo Hospital, Surabaya)

E.T. Sunarti, A.B. Tribhuwaneswari, O.E. Rachmalisa & R.P. Kurniasanti
Institut Teknologi Sepuluh Nopember, Surabaya, Indonesia

ABSTRACT: Health and mortality risks are indirectly associated with factors such as physical inactivity and sedentary lifestyles with the car-oriented urban approach. Therefore, walkable communities should be encouraged, with a focus on cultural and environmental aspects that are necessary to Indonesian people. However, social and economic borders can be observed in a typical street where only the lower-middle class can be seen walking, whereas the higher-income groups mainly use cars as a primary transportation mode. In order to study this movement gap, a provincial health center (Dr. Soetomo Hospital) was selected as a relevant case study. This would aim to equally provide a sense of security, comfortability, and convenience to all kinds of people. This study uses a walkability assessment tool to assess people's tendencies to choose their mobility methods and links with the surrounding conditions. The results of this study is used to gain insight into Indonesian attitudes towards walking, in order to foster the physical and psychological well-being by stimulating lively and diverse activities in public space.

Keywords: Environment, Human Well-being, Lifestyles, Public Space, Walkability

1 INTRODUCTION

There is now overwhelming evidence that regular physical activities have important and wide-ranging health benefits. Health and mortality risks are indirectly associated with factors such as physical inactivity and sedentary lifestyles with the car-oriented urban approach. This evidence supports the view that physical inactivity is one of the most important public health problems of the 21st century, and may even be the most important (Blair, 2009).

Walking is now gaining attention as a key factor in the promotion of healthier, environmental-friendly, and socially active communities (Jackson, 2013). It can also be considered as the basis for a sustainable city, providing social, environmental, and economic benefits, often being the only way many people can access daily activities. It also brings life to streets, and livable streets in turn contribute to safer urban environments (Moura, Cambra, & Goncalves, 2017). However, social and economic borders can be observed in a typical street where only the lower-middle class can be seen walking, while higher-income groups mainly use cars as a primary transportation mode. In order to study this movement gap, a provincial health center (Dr. Soetomo Hospital) was selected as a relevant case study. This would make it possible to equally provide a sense of security, comfortability, and convenience to all kinds of people.

According to a recent study conducted by Stanford University, Indonesia is ranked as the lowest in terms of managing walking on a daily basis. Nevertheless, research shows that people tend to walk or bike in a convenient and safe place. Hence, promoting physical activity through walkable and bikeable neighborhoods becomes necessary. The aim of this paper was to encourage walkable communities, in order to focus on cultural and environmental aspects (Matthews, 2017). This study presents the correlation of people's perception measures to understand the impact of increasing environmental quality on walking behavior. The results of this study will be used to gain insight into Indonesian attitudes towards walking, in order to foster the physical and psychological well-being by stimulating lively and diverse activities in public space. Therefore, this study aims to support urban design for more walkable environments that can minimize the movement gap.

As one of the basic physical activities, walking makes a city sustainable through improvement in a built environment from social, environmental, and economic perspectives (Abbey, S & NZ Transport Agency, 2009). The purpose of walking itself can be seen as a leisure, an exercise or recreation, access, or traveling to workplace (Leslie, Butterworth, & Edwards, 2006). Thus, it is important to consider the main aim and quality of sidewalks, paths, and other various pedestrian infrastructures.

From a public health perspective to increase regular moderate-intensity physical activity, walking is the behavior that is most likely to be amenable to influence (Moudon, et al., 2006). Physical attributes of local walking environments may be related to walking factors. Furthermore, there is a generalized lack of consensus on the meaning of walkability. As Lo (Lo, 2009) points out, although many researchers have evaluated the relationships between urban environment and pedestrian behavior, all have a different definition on how to measure walkability.

There is also a new perspective on the social setting within the environment (Brown, 2007). Physical, social, psychological, and cultural aspects in the form of sensory stimulations largely contribute to the behavior of users, especially the walking behavior. As a result, many urban designers contemplate within these two main factors, namely the physical and non-physical aspects of walking behavior. This leads to many theories about walking environment itself.

To measure walkability, we must first understand various factors such as whether a city is considered as walkable. Some researchers have pointed out that several factors such as pedestrian characteristics, walking purpose, urban context, culture and environment can affect how we perceive walkability (TRB, 2000). Pedestrian characteristics are classified based on gender, age groups, and health conditions. Conversely, walking purposes are categorized based on the purpose of walking itself in two ways: walking for leisure and walking for utilitarian needs.

The general ideas of walkability are mainly associated with perceiving the 5 C's and the extent of 7 C's, which are all coherently recognized. ***Connectivity*** indicates the extent to which the users and their environment are linked with each other within the pedestrian network (Tal & Handy, 2012). ***Convenient*** refers to the efficiency, feasibility, and the practical use of the pedestrian way (Cervero, Sarmiento, L, & Gomez, 2009). ***Comfortable*** comprehends the ease and pleasantness of users in their walking experiences (Saelens & Handy, 2008). Furthermore, ***convivial*** are factors that identify the liveliness and leisure aspect links between the environment and users (Carvero & Duncan, 2003). ***Conspicuous*** determines whether the neighborhood is clear enough in terms of direction, visibility, and route or building identification (Ewing, et al., 2013). Meanwhile, ***coexistence*** refers to the ability of pedestrian space and modes of transport to coexist in the same limit of time and place (Koh & Wong, 2013). The last C is ***commitment***, which indicates the ideas of engagement, accountability, and responsibility for its walking environment by the government and other stakeholders.

Although theories about walkability continue to grow, they extend no further than pooling parallel factors such as the scale of the analysis (Lee & Talen, 2014), purpose of the walking trip (Handy, 2005), and pedestrian types (COST, 2010). Measurements are examined over the walking purpose and pedestrian group, in order to tailor the local conditions as a means to adopt the results and tools in the urban context (Guo & Loo, 2013). Using the layout of 6 C's on walkability, this study considers the engagement of the pedestrian in Dr. Soetomo Hospital to explore the relationship between the users and their environment. The purpose of the study was to understand the city-hospital complex from a public viewpoint, as well as the complexity and multidimensional characteristics of humans in relation to the surrounding settings.

2 METHOD

This study on the public space of Dr. Soetomo Hospital was conducted in 2017, which included three streets: Prof. Dr. Mustopo, Dharmawangsa, and Airlangga streets. As the primary and direct relationships with the hospital, these areas reflected the area zoning for the health facility. Based on previous similar research, different methods were used to understand the pedestrian's perception towards walkability. To measure the user's tendencies towards walkability factors, the research method included interviews and a questionnaire.

Street surveys were conducted on a random sample of 120 adults living in the neighborhood. Street users were asked about their walking experience along certain streets, and to rank the importance factor that supported walking behavior. The participants were encouraged to raise issues that were not included in the questionnaire. Walkability assessments were based on C factors: connectivity, convenient, comfort, comfortable, convivial, conspicuous, and coexistence (Grant & Herbes, 2009; Moura, Cambra, & Goncalves, 2017). These factors have a higher influence on Indonesian people's behavior.

3 DISCUSSION

The exact number of population who visited RSUD Dr. Soetomo was not known because it was not fixed. As Fraenkel and Norman E. Wallen (1993) described, a sample of 100 subjects was considered as essential. Therefore, the required number of samples was rounded up to 120 subjects. As mentioned previously, a variety of walkability assessment methods have been developed globally, but only a few have tested the validity of the obtained results.

A total of 120 valid questionnaires were obtained. The characteristics of the respondents are provided in Table 1. Low reliability can be related to a high degree of subjectivity.

Table 1. Characteristics of the respondents.

Characteristics	n = 120	Percentage (%)	Characteristics	n = 120	Percentage (%)
Gender			**Avg. duration**		
Female	67	55.83	<5 min	16	13.33
Male	53	44.17	5–9 min	48	40.00
Age			10–14 min	39	32.50
<18 years	1	0.83	>15 min	17	14.17
18–24 years	43	35.83	**Avg. distance**		
25–44 years	42	35.00	300–399 m	29	24.17
45–64 years	27	22.50	400–799 m	61	50.83
65+ years	7	5.83	800 m–1.6 km	26	21.67
Disability			>1.6 km	4	3.33
Disabled	1	0.83	**Usual time**		
Occupation			Morning	34	28.33
Retired	4	3.33	Noon	30	25.00
Government employee	19	15.83	Afternoon	21	17.50
Unskilled worker	2	1.67	Night	1	0.83
Skilled worker	6	5.00	**Purpose**		
Business person	13	10.83	Access	74	61.67
Student	46	38.33	Recreation	9	7.50
Unemployed	5	4.17	Working/studying	23	19.17
Others	24	20.00	Others	13	10.83
Wages			**Transportation modes**		
Less than Rp 1,000,000	29	24.17	By foot	67	55.83
Rp 1,000,001–2,500,000	24	20.00	Bike	21	17.50
Rp 2,500,001–5,000,000	6	5.00	Motor taxi	2	1.67
Above Rp 5,000,000	24	20.00	Taxi	2	1.67
Education			Angkot	18	15.00
None	0	0.00	Bus	1	0.83
Elementary school	11	9.17	Private vehicle	9	7.50
Junior high school	6	5.00			
Senior high school	44	36.67			
University	52	43.33			
Others	5	4.17			

To determine the extent of bias introduced by the sample, we weighted the sample data by gender, age, disability, occupation, income, education, and other characteristics that support walkability data. Among the 120 subjects, one respondent had hearing impairment.

The people in the neighborhood with different needs and characteristics represented a wide range of walkability factors and various frameworks. The data collected from the first questionnaire were classified using the 6 C's key concerns. This was followed by the calculation of the percentage of pedestrian consciousness in each walkability category. Table 2 presents the generated weight for each sidewalk factor. The results indicated that convenient, comfortability, conviviality, and conspicuous had similar attributes for each factor for all streets. The degree of obstruction, pavement quality, meeting place and seating, and pedestrian maps were the most important factors used to grade the quality of walkability in the area. However, connectivity and coexistence indicators show a different grading on the key point. For example, the assessment results for connectivity indicate linkage for Airlangga, path directness for Dharmawangsa, and continuity for Mustopo. Meanwhile, for coexistence, the results indicate the highest key concerns, that is, driver's behavior for Mustopo and Dharmawangsa and crossing location and potential conflict for Mustopo. Considering the physical attributes of the street, somehow, the results were relevant to the pedestrian impression about walking.

RSUD Dr. Soetomo consisted of three types of corridors with a hospital-like environment. These corridors were evaluated using a participatory approach based on the 6 C's factors as the indicators for the walkability level. This analysis provides a set of examples that indicate the users' viewpoint on their built environment associated with their walking experience, as given in Table 2.

Table 3 indicates that the most concerning aspects for this neighborhood were convenient, convivial, and comfortability of sidewalk facilities as only a few improvements were required, according to Aghaabbasi & Shah (2017). However, Dharmawangsa corridors felt the need for improvement, as indicated by a score of 4 on the LRI scale. Interestingly, all of the corridors expressed a stable opinion about the user's connectivity aspects. Conversely, the convenient

Table 2. 6 C's key concerns and corresponding weights defined by the characteristic of each street.

		Value (%)					
Indicator	Corridor	Awful	Bad	Fair	Good	Excellent	LRI
Connectivity	A	0	0	22.5	37.5	40	1
	D	0	1.25	20	37.5	41.25	2
	M	0	0	22.5	30	47.5	1
Convenient	A	0	10	22.5	40	27.5	3
	D	0	6.25	23.75	46.25	23.75	4
	M	0	0	7.5	45	47.5	2
Comfortability	A	0	2.64	30.7	51.28	15.38	3
	D	0	6.34	29.11	44.3	20.25	2
	M	0	2.5	12.5	57.5	27.5	2
Convivial	A	2.5	2.5	15	62.5	17.5	3
	D	1.25	2.5	15	67.5	13.75	3
	M	0	0	20	67.5	12.5	2
Conspicuous	A	0	5.14	12.82	48.71	33.33	2
	D	0	3.8	12.66	46.83	36.71	2
	M	0	0	22.5	20	57.5	1
Coexistence	A	0	12.5	30	40	17.5	2
	D	0	10	32.5	35	22.5	2
	M	0	12.5	32.5	32.5	22.5	3

Airlangga Corridor: A; Dharmawangsa Corridor: D; Prof. Dr. Moestopo Corridor: M. LRI (Level of Required Improvements): 1 = very limited improvements are required; 2 = limited improvements are required; 3 = some improvements are required; 4 = many improvements are required; 5 = too many improvements are required.

Table 3. Level of required improvements based on pedestrian values and walkability factors.

Indicator	Corridor	Value (%)					
		Not at all	5 min	5–9 min	10–14 min	>15 min	LTW
Connectivity	A	8	40	32	16	4	2
	D	12.24	32.65	32.65	20.4	2.06	3
	M	19.04	42.86	23.81	14.29	0	2
Convenient	A	6.9	13.78	44.83	27.59	6.9	3
	D	3.23	14.51	43.55	35.48	3.23	4
	M	14.29	33.33	33.33	19.05	0	3
Comfortability	A	17.65	2.94	41.18	29.41	8.82	4
	D	12.31	3.07	44.62	30.77	9.23	4
	M	6.9	20.69	48.28	20.69	3.44	4
Convivial	A	36.36	18.18	24.24	9.1	12.12	3
	D	34.78	20.29	23.19	14.49	7.25	3
	M	45.71	31.43	17.14	0	5.72	2
Conspicuous	A	3.85	30.77	38.46	23.07	3.85	4
	D	2	32	44	18	4	4
	M	0	52.94	17.65	23.53	5.88	3
Coexistence	A	0	16.67	23.33	40	20	5
	D	3.23	17.74	17.74	45.16	16.13	4
	M	3.12	25	31.25	25	15.63	3

Airlangga Corridor: A; Dharmawangsa Corridor: D; Prof. Dr. Moestopo Corridor: M. Initial average duration: <5 min = 13.33%; 5–9 min = 40%; 10–14 min = 32.50%; >15 min = 14.7%. LTW (Level of Tendency to Walk): 1 = very limited tendency to walk; 2 = limited tendency to walk; 3 = some tendency to walk; 4 = more tendency to walk; 5 = lots of tendency to walk.

aspect, as shown in Table 3, indicated 23.75% (fair) and a cumulative of 16.25% based on the LRI scale, which was considered to be unfavorable for its walkability factors. The respondents mainly reported the difficulty of access due to the obstacles and the unease of mobility-assisted users in using public facilities.

The results indicate that meeting and seating places and pedestrian maps and wayfinding were the most important factors that supported hospital neighborhoods. Conviviality was related to social spaces, and conspicuous was related to legibility and wayfinding. It was interesting to note how convenience (obstructions on pedestrian walkways) and comfortability (pavement quality, which was related to the pleasure level when walking along the streets) highly affected the walking motivation of the people. Indeed, each street has its own characteristic concern for walking improvements. The figure below shows which street urgently requires an improvement and which factor has the highest walkability level.

Table 3 categorizes the people's tendencies towards improvements in walking duration based on the walkability assessment. As mentioned previously, each factor showed different levels of reliability in the redevelopment of sidewalk design. The tool considered the people's desire for the connectivity factor, which showed excellent conditions but a small effect on the walking motivation of people. In contrast, conspicuous, coexistence, and comfortability showed high levels of walking behavior when these elements were improved.

Linking the scale of assessment with the improvements in walking duration was proved important in this study. The scale presented in Table 3 was a measure of an element to facilitate improvements for street users. Meanwhile, the increase in walking motivation (Table 3) was analyzed to determine the extent of the impact that the improvement of the 6 C variables towards the user's desire had on walking behavior. This inverse relationship indicates that the worst quality is the most important priority for future design development, but which will have the greatest effect on the increase in walking duration from a community perspective.

4 CONCLUSION

Previous studies on urban design suggested that creating walkable environments will lead to a healthy, efficient, safe, and dynamic public space. This study was conducted to evaluate the sidewalks in health facility neighborhoods using the 6 C's assessment tool. The lessons learned from this study are considered important for transforming a typical Indonesian street to attract people to walk and create more exciting environments.

The current interest lies in making physical improvements in the area, such as pavement quality, trees, seating features, and crossing lines. However, more importantly, the streets need to support and encourage a wide variety of commerce and preserve community gathering places to strengthen a pedestrian-friendly neighborhood. Further work from this study will aim to address the distinctive needs of pedestrians to create a desirable place for walking.

REFERENCES

Abbey, S & NZ Transport Agency. (2009). *Predicting walkability.* Wellington: NZ Transport Agency.

Aghaabbasi, M., Moeinaddini, M., & Shah, M. (2017). A new assessment model to evaluate the microscale sidewalk design factors at the neighbourhood level. *Journal of Transport & Health*, 97–112.

Blair, S.N. (2009, January). Physical Innactivity: the biggest public health problem of the 21st century. *BMJ Journals*, 1–2.

Brown, B. (2007). Walkable route and perceptions and physical features: converging evidence for en route walking experience. *39*.

Carvero, R., & Duncan, M. (2003). Walking, Bicycling, and Urban Landscapes: Evidence From the San Francisco Bay Area. *93*(9).

Cervero, R., & Kockelman, K. (1997). Travel demands and the 3D's: density, diversity, and design. *2*(3).

COST. (2010). *Pedestrians' Quality Needs Final Report part B1 – Functional needs.* Hector Monterde i Bort (coord).

Fraenkel, J., & Wallen, N. (1993). *How to Design and Evaluate Research in Education.* New York: McGraw-Hill.

Guo, Z., & Loo, B.P. (2013). Pedestrian environment and route choice: evidence from New York City and Hong Kong. *28*.

Jackson, L. (2013). The relationship of urban design to human health and condition. *Research Gate*, 191–200.

Koh, P., & Wong, Y.D. (2013). Comparing pedestrians' needs and behaviours indifferent land use environments. *26*.

Lee, S., & Talen, E. (2014). Measuring walkability: a note on auditing methods. *19*(3).

Leslie, E., Butterworth, I., & Edwards, M. (2006). Measuring the walkability of local communities using Geographic Information Systems data. *7th international conference on walking and livable communities* (p. VII). Walk 21.

Lo, R.I. (2009). Walkability: What is it? *2*(2).

Matthews, S. (2017, July 12). *Mail Online*. Retrieved August 28, 2017, from dailymail.co.uk: http://www.dailymail.co.uk/health/article-4688148/Do-live-world-s-laziest-country.html

Moudon, A.V., Lee, C., Cheadle, A.D., Garvin, C., Johnson, D., & Schmid, T.L. (2006). Operational definition of walkable neighborhood: theoretical and empirical insight. *3*(1).

Moura, F., Cambra, P., & Goncalves, A.B. (2017). Measuring walkability for distinct pedestrian groups with a participatory assessment method: A case study in Lisbon. *Elsevier Landscape and Urban Planning*, 282–296.

Sunarti, E., Tribhuwaneswari, A., & Rachmalisa, O. (2017). Understanding the linkages of behavioural observation. *Global Journal of Engineering Science Research Management*, 44–58.

Tal, G., & Handy, S. (2012). Measuring nonmotorized accessibility and connectivity in a robust pedestrian network.

TRB. (2000). *Highway capacity manual 2010.* Transportation Research Board

Urban Studies: Border and Mobility – Kerr et al. (Eds)

Trialectic city space based on an immigrant view through urbanisation: A study on settlement migrants in Surabaya city

I.Y.A. Rohmah
Universitas Airlangga, Surabaya, Indonesia

ABSTRACT: Surabaya is the second largest city in Indonesia. The interesting thing about Surabaya is that it includes the migrant community from outside the region, such as those who have moved from villages to the city or from outside the city to Surabaya. The presence of immigrant communities in Surabaya city signifies the urban "socio-historical" phenomenon. This provides an interesting idea for researchers to conduct research on urban space "trialektika" based on the views of immigrants who become urbanised by moving to the city of Surabaya. This study was conducted by a qualitative method of an intrinsic case study type. The results are based on the views of migrants from villages to the city (urbanisation) or from outside the city to the city of Surabaya. Immigrants assume that the city of Surabaya can provide convenience for them in meeting the needs of life because it is considered easier to get a job in the city of Surabaya than in their area of origin. The city of Surabaya is considered a mental and ideational field, and the assumption is that the city space is the most ideal residence. Their assumptions about the city of Surabaya as a migration destination city are driven by their own enthusiasm and positive assumptions and then the influence of assumptions from those around them about the Surabaya city.

Keywords: Trialectic, City space, Urbanisation

1 INTRODUCTION

Indonesia's total population was more than 262 million people in 2016. The community partially lives in villages and some live in cities. Villagers are mostly farmers, as well as landowners and labourers, while the people living in cities are mostly factory employees and office employees.

Population growth is experiencing a very significant increase every year. The number of live births is continuously increasing in both urban and rural contexts. In addition, there is a larger population of productive age in Indonesia, which may provide demographic bonuses to the country if the population of productive age is able to improve its economy.

In relation to other population problems, the increase of the population in Indonesia has both positive and negative impacts on the social and economic sectors of the community. In villages, with an increasing number of people, the amount of agricultural land is reduced. Among the causes of reduced agricultural land are the facts that much agricultural land is used for houses, the villagers have inherited agricultural land as offspring, whose numbers are increasing, resulting in lower agricultural income, and much agricultural land has been converted for other businesses.

The existence of problems in terms of other agricultural factors such as agricultural revenues makes it less encouraging for villagers to innovate in other business areas. However, often new efforts by rural communities do not always provide substantial benefits, so villagers are encouraged to migrate to other places where it is considered easier to do business than in agriculture. In this case, villagers are most likely to move from the village to a city, which is known as urbanisation.

Urban space, in the assumption of people who migrate to the city, is a space that has several positive attributes that can provide a livelihood. City space is a phenomenon as a "space-social-historical", in which case city space is a trialectic. In the neo-Marxian space analysis, Edward Soja studied it using three perspectives with interpretive and explanatory purposes.

Indonesia has several large cities, one of which is Surabaya, the second largest city after Jakarta. The city of Surabaya is considered an industrial city because the industry sector in Surabaya is quite well developed. The city of Surabaya has a number of divided spaces, including space for industrial operations, residential spaces, other business spaces such as trade centres, and others (Nanang Martono, 2011: 119). According to Anthony Giddens, this encourages rapid changes that sometimes cannot be controlled. It has pushed the city of Surabaya towards increasingly modern change.

People of Surabaya come from various regions, with immigrants comprising the majority, who came to Surabaya to finding a job. The community considers the city of Surabaya as an industrial city and a trading centre that will provide facilities for the community to meet the needs of life, including in terms of finding a job.

Surabaya's city space is always alive and full of the activity of people who are busy with work, leading to assumptions that the city space is a space to get material advantage or economic advantage, but only for those who live in the city. The pattern of community thinking about Surabaya city is full of materialistic social practices that encourage the keiinginan people from outside the region, such as people from the villages to come to the city of Surabaya.

City space is often a gathering place for migrants from outside the city of Surabaya as a trading business space. The trading business space is in the form of a modern market and traditional market. There are many traditional market spaces in Surabaya; one of them is the traditional market room Gubeng. Traditional market Gubeng is a legal market; the market has official permission to stand, and people get permission to conduct trading activities.

The objective of this study is to explore the traditional market space of the Gubeng sub-district of Surabaya. The research is very interesting because the researcher has interpreted and explored the space of Surabaya city, as a destination of urbanisation by using the perspective of the analysis of the trial or the urban space as the "space-social-historical" phenomenon, as proposed by Edward Soja.

2 METHOD

In this study we use a qualitative research method of an intrinsic case study type (Ambo Upe & Damsid, 2010: 108). Research using qualitative methods focuses on humans as the main subject in various social phenomena. In this type of case study, the intrinsic case study aims to understand the researcher in a case that cannot represent other cases (Denzin & Lincoln, 2009: 301). The case referred to here is the one with certain characteristics.

The data collection in this field is achieved by two techniques: interview and observation techniques. Interview techniques used by researchers are of the tersetruktur type (Sugiyono, 2013: 145). When researchers use structured interview techniques, they have planned data—what data will be sought? While the observation technique used is a non-participant observation technique, that is, in the process of observation, the researchers did not participate in the agenda to be studied.

In the analysis of the research data, researchers used data analysis techniques (Milles and Huberman, cited in Sugiyono, 2013: 246). Data analysis techniques, according to Milles and Huberman, include several stages, namely data collection, data reduction, data display, and conclusion.

3 DISCUSSION

The city of Surabaya has had a long history of offering better city life since the time of the kingdom and colonial Dutch colonisation (Jo Santoso, 2006: 116). The city of Surabaya has experienced setbacks and progress in the course of its history. As time passes, many of Surabaya's basic elements have changed; almost all the elements have changed, making Surabaya a social phenomenon that has different characteristics from other cities. The change of many basic elements in Surabaya city means many people in the city or outside the city want to know more about the city of Surabaya.

The city of Surabaya is an open city that always changes; global modernisation has a significant influence on changes in the urban spatial layout and social order (Piotr Sztompka, 2010: 82). As Aguste Comte explains, changes in modernisation have distinct characteristics to their social order, namely (1) centralised labour, (2) the classification of the division of labour based on the profit gained, (3) & (4) the existence of special treatment, both antagonism and protagonism, between the owners of capital and workers and between classes of workers, (5) the emergence of social imbalance, and (6) the existence of a more open system of economy (J.W. Schoorl, 1991: 20). The existence of modernisation in urban development, especially in the city of Surabaya can also happen as part of Westernisation; it is possible that the culture and lifestyle of the people of Surabaya will move towards the Western culture.

The Surabaya city space is divided into several sections, including industrial space consisting of various factories, residential space, and trading room, traditional and modern. One of the traditional market spaces in Surabaya is the Gubeng traditional market in the Gubeng sub-district of Surabaya city, which is in the middle of town. Its location is very strategic; the market is located near the highway and the new Gubeng city of Surabaya. The location of Gubeng traditional market is also close to the residential area. Gubeng traditional market's operations have legal status from the government of Surabaya city; this gives a sense of calm to traders who occupy the market to trade daily.

Traditional market conditions in Gubeng at this time still require special attention from the city government. Attention is necessary to restructure the use of traditional market facilities by traders. It is necessary to maintain the quality and quantity of the market. So far, the existence of traditional markets is still growing in the midst of modern market trade competition. However, as the large number of modern markets in Surabaya city offer many excellent products and high-quality facilities that can be used by consumers or buyers, it has an impact on traditional markets with a decrease in the number of consumers or buyers.

Traders in traditional markets in Gubeng are mostly migrants from outside the city of Surabaya. The traders who are the subject of this research are from outside the city of Surabaya, coming from Lumajang, Jombang, Malang, Bangkalan (Madura), and Purworejo Regency. Traders in the traditional market in Gubeng have mostly undergone urbanisation from villages to town and from outside the city to the city of Surabaya. The average subjects of the study are part of the permanent immigrant population. They urbanise from outside the area to the city of Surabaya and have been long-settled in the city of Surabaya so that they have the status of a resident of Surabaya based on identity cards (KTP).

The main objective of the residents who urbanise to the city of Surabaya is to find a decent job. The immigrants carry out a positive assessment of the city of Surabaya (Sugiono Soetomo, 2009: 21). Robert Potter explains that the existence of the city and the urge to urbanise are always interrelated with the influence of the elite community groups in the city. The elite community groups in the city of Surabaya include a group of migrants from outside the town and also native citizens of Surabaya. Elite groups in urban areas are usually characterised as having a large amount of capital in setting up an urban economy. In addition, the elite group is also a group of urban leaders who have the authority to organise the elements of the city.

The informants in this study assumed that by living in Surabaya, immigrants would be able to easily get a job and do new business such as trading. As the first informant explained when interviewed, the purpose of him moving from Jombang to Surabaya was to trade. Informants feel more positive selling vegetables in the traditional market of Gubeng Surabaya than in the traditional market of Jombang Regency. This is because vegetables in the traditional market of Gubeng Kota Surabaya are more saleable than in the traditional market of Jombang Regency.

The first informant is a woman of about 50 years of age. The woman had come to Surabaya with her husband. However, the woman's husband did not work, and the only one working was the woman. The couple had lived in Surabaya city for 20 years. The informant's job remains as a trader of vegetables whose daily income is not more than Rp. 50,000. The informant had come to Surabaya with her husband and two children. A long time ago, her two sons got married and left the woman and her husband. Her two children live in a house separately from their parents, in a different sub-district in Surabaya. It encourages the informants who have to work on their own as vegetable traders in traditional markets.

The second informant is an old woman (50 years of age) who is a trader selling fish in the traditional market in Gubeng. She had lived in Surabaya city for 35 years. She is a permanent immigrant who had moved to become a resident of Surabaya city. She trades in the traditional market of Surabaya city to make a far greater profit than by selling in the old area of Madura. However, in reality, she does not often get a greater profit from the trade. The fish sold by the second informant are marine fish including pindang fish, milkfish, and fish tenggiri. Sometimes, fish older than 3 days are not sold, and have to be thrown away. The result is that she does not get any profit, but makes a loss. This often happens, but the trader retains her status as a marine fish seller in Gubeng traditional market.

The proceeds from the sale of fish in traditional markets are just enough to meet the needs of eating each day. Therefore, in order to meet the needs of life other than to eat, the second informant often asks for help with costs from other family members. Surabaya city space seems to hypnotise the second informant's mindset. The merchant still assesses the city of Surabaya to be the best city for meeting all the necessities of life. Traders in fish still assess the area of origin of their birth to be no better than the city of Surabaya. Besides, the traders think Surabaya is more profitable for them; the informants also consider the city of Surabaya to be more qualified and have more facilities, making it easier to take advantage of public facilities of the city.

The second informant moved to the city of her own volition. In addition, prior to urbanisation, these informants have been influenced by positive spirits in the surrounding community about the city of Surabaya. She and her family members, based on this, did move from the village in Madura to Surabaya. She changed her status to be a resident of Surabaya city because, from the beginning, she had planned to settle for a long time and open a trading business in Surabaya. She and her family members visit her native place in Madura only during Islamic religious holidays.

Next is the third informant's argument, who is a migrant from Madura to Surabaya. Her age is 45 years. The woman had lived in Surabaya for 1 year. She sells some staple foods. She has complete public facilities and does many types of work that are easy to find. She claimed to prefer to be a trader of basic necessities in the traditional market in Gubeng despite the fact that the number of buyers in this traditional market is restricted because of competition with the modern market. However, informants remain optimistic about surviving trading in the traditional market in Gubeng Kota Surabaya, although the income earned is not much.

On the basis of the explanation from several informants in this study, it can be explained that Surabaya city has since developed into an industrial city, and the number of changes in the city in terms of public facilities and inventory of the city evidences that Surabaya city attracts people from outside the region, making Surabaya a full social historical space (Bagong Suyanto, 2017: 1–2). Edward Soja highlights intrinsic space from three perspectives. First, the urban space is a materialised space that encourages the transfer of individuals from outside the region into the city, i.e., urbanisation. Second, looking at every individual's assumption about urban space, every individual has an idealistic assumption about urban space, as if every individual has the ideal image of city space. Third, the process of migration from villages to urban spaces is influenced by individual and group experiences in urban residences; the urbanisation affecting urban space is fully occupied by migrants.

From the first perspective, according to Edward Soja's terms trialectika, the migrants from outside Surabaya who came to Surabaya city understood that the city of Surabaya has the majority of population busy with their respective occupations, assuming that the people work to generate profits of high magnitude. In that case, it is as if the city of Surabaya is a materialistic city. In addition, the assumption is also encouraged with the realisation that life in the city of Surabaya needs a lot of capital to survive rather than living in the village or area of birth.

From the second perspective, the migrants in Surabaya city, based on interviews with a number of informants, have the assumption that the city of Surabaya is a city space that has an idealistic value. Idealist values are intended as the assumption of the immigrants about the availability of public facilities, and the spatial city of Surabaya is considered more beautiful and comfortable; the assumption is that the space of Surabaya is a reflection of a city that is developed (Gilbert and Gugier, 2007: 55). It is, in fact, the case that large cities are the spaces where powers are exercised and privileges exist.

The informants' assessment is far from being pessimistic or frustrated when they live in Surabaya. The guarantee of well-being in life is always present in the imagination of the immigrants. Although, in reality, the newcomers who came to Surabaya city a long time ago have still not all got the welfare that is intended for them, that is, life welfare from family that can fulfil family necessities, in terms of living harmoniously with family without any deficiencies and allowing a decent life for families left behind in their home regions.

Surabaya city spaces look beautiful and are well assessed by the informants; however, in reality, not all of them can be enjoyed by them. Beautiful spaces are just false hopes, just as a dream of life to be able to enjoy the beauty and facilities that exist is a false hope. However, the immigrants who do not have enough capital, both financial and non-financial, and the migrants who live in the city of Surabaya can only try to work hard with the work they have; sometimes, the results of work do just enough to meet the needs of food to survive in the city (Damsar and Indrayani, 2013: 198). Basically, social practices in modern society cannot take away the skills of each individual. Therefore, if the migrants only have little financial and nonfinancial capital and income earned from the work, then the needs of recreation, health, education, and others sometimes cannot be fulfilled.

From the third perspective, the informants interviewed gave a response about Surabaya as a city open to newcomers. The newcomers who come to Surabaya are basically self-willed and some are influenced by other people. As the fourth informant who came from Lumajang district, the migrants came to Surabaya based on self-impetus; immigrants understand that Surabaya is an open city to newcomers. In addition, migrants also understand that the city of Surabaya provides certain urban spaces that can be occupied by immigrants; they can even legally buy land in the city space.

Another informant came to Surabaya city through, in addition to self-will, influence from other family members and the community. The influence given to informants by family and society is the result of experiences that they had already felt living in the city of Surabaya. There are also some informants who moved to Surabaya to follow their husband and family members who have lived in the city of Surabaya. Even after the death of husbands or other family members, surviving migrants persist and have family back in Surabaya.

4 CONCLUSION

The city of Surabaya represents a "space–social–historical" phenomenon. Here, we used three perspectives to highlight urban intrinsic space so that it can perform interpretive and explanatory analysis. From the first perspective, immigrants coming from outside the region to the city of Surabaya have the main intention to work. Immigrants believe that working in the city of Surabaya will make it easier to earn more than working in their home areas.

From the second perspective, this study shows that urbanisers have the assumption that the space of Surabaya city is a city space that has many social practices from the people who are busy working to meet the needs of life. Immigrants who came to Surabaya also consider that the economy of Surabaya city is better than their home economy.

From the third perspective, the space of Surabaya city provides living space for immigrants, in which case, the city of Surabaya as an open city provides an opportunity for migrants from outside the region to live in Surabaya as either permanent or non-permanent residents. The existence of the openness of the city of Surabaya makes people who are out of town migrate to the city of Surabaya; there are people moving from villages to the city of Surabaya and from out of town to the city of Surabaya. One of the city spaces in Surabaya, which is a full space inhabited by the immigrant community, is the traditional market space. Immigrants take advantage of the traditional market space in Surabaya city to earn household income. However, in reality, immigrants working in Surabaya and especially in traditional markets do not have a high and fixed income.

REFERENCES

Basundoro, P. (2010). Dari Kampung Desa ke Kampung Kota: Perubahan Ekologi Kota Surabaya dalam Perspektif Permukiman pada Masa Kolonial. *Jantra, 4*(5).

Basundoro, P. (2012). Rakyat Miskin dan Perebutan Ruang Kota di Surabaya Tahun 1900–1960-an. *Masyarakat Indonesia, 38*(2), 427–457. Retrieved from http://ejournal.lipi.go.id/index.php/index/index

Damsar & Indrayani. 2013. *Pengantar Sosiologi Ekonomi*. Jakarta: Kencana Prenadamedia Group.

Denzin, Norman K. & S. Lincoln, Yvonna. 2009. *Handbook of Qualitative Research*. Yogyakarta: Pustaka Pelajar.

Gilbert, Alan & Josef Gugier. 2007. *Urbanisasi & Kemiskinan*. Yogyakarta: PT. Tiara Wacana Yogya.

Itafarida, S. (2015). Undercurrents: Unearthing Hidden Social and Discursive Practices Space, Mobility, Memory: The (Re) construction of 'Green Living' of Citra Garden Structured-Settlement House Residents in Sidoarjo, East Java, Indonesia. *Inter-Asia Cultural Studies Conference*. Surabaya: Inter-Asia Cultural Studies Conference.

Itafarida, S. (2016). Kaum Muda Urban Dan Ekoliterasi: Kontestasi Nilai Dan Praktik Green Living Di Pemukiman Modern Di Surabaya Jawa Timur. *Art and Urban Culture International Conference*. Surabaya: Art and Urban Culture International Conference.

Martono, Nanang. 2011. *Sosiologi Perubahan Sosial: Perspektif Klasik, Modern, Posmodern, dan Poskolonial*. Jakarta: Rajawali Pers.

Santoso, Jo. 2006. *(Minyiasati) Kota TanpaWarga*. Jakarta: KPG dan Centropolis.

Schoorl, J.W. *Modernisasi (PengantarSosiologi Pembangunan Negara – negara Sedang Ber kembang)*. Jakarta: PT. Gramedia Pustaka Utama.

Soetomo, Sugiono. 2009. *Urbanisasi dan Morfologi*. Yogyakarta: Graha Ilmu.

Sugiyono. 2013. *Metode Penelitian Kuantitatif, Kualitatif, dan R & D*. Bandung: Alfabeta.

Suyanto Bagong. 2017. *Analisis Ruang Neo-Marxian*. Materials in Postgraduate Sociology of Airlangga University.

Sztompka, Piotr. 2010. *Sosiologi Perubahan Sosial*. Jakarta: Prenada.

Upe, Ambo & Damsid. 2010. *Asas – asas Multiple Researches*. Yogyakarta: Tiara Wacana.

Urban Studies: Border and Mobility – Kerr et al. (Eds)
© 2019 Taylor & Francis Group, London, ISBN 978-1-138-58034-3

The influence of social mobility on cultural values: A case study on Chinese-Indonesians in Surabaya, Indonesia – preliminary research

R.A. Saputra
Universitas Airlangga, Surabaya, Indonesia

ABSTRACT: Surabaya is one of the important cities in Indonesia with a high prevalence of multiculturalism. According to the National Centre of Statistics, after Javanese ethnic people, Tionghoas or Chinese-Indonesians account for 7% of Surabaya's total population. Multiculturalism creates a limitation in social mobility within the scope of communications: for example, the blasphemy case against a former governor of Jakarta, Basuki Tjahya Purnama (Ahok), which became the object of much debate. Some people may argue that it is just a slip of the tongue, which may not be considered rude in Chinese-Indonesian communities; however, according to the majority of the Indonesian society, the governor's speech was unethical. Therefore, this study aims to discover the influence of social mobility on communication and cultural values that Chinese-Indonesians hold with the long-term purpose of bridging the disparity in cultural understanding. This study was conducted by using a quantitative method with two variables, namely social mobility as an independent variable and cultural values as dependent variables. Cultural value indicators are language and social norms. The hypotheses of this study are twofold: 1) social mobility does not influence cultural values and 2) social mobility significantly influences cultural values. The study included a random sampling of 20 Chinese-Indonesians, and was analysed using a linear regression analysis technique. The results of this study proved that social mobility influences cultural values by 5% and social mobility does not significantly influence cultural values. As this study reports a preliminary piece of research, there is much room for improvement, which, in the long run, can hopefully bridge the disparity between cultures in the Indonesian society.

Keywords: Social Mobility, Cultural Values, Ethic, Chinese-Indonesia

1 INTRODUCTION

A long history of the existence of Chinese ethnic people in Indonesia, that is, Chinese-Indonesians, has left some polemics in its social and political life. Since the Dutch colonial government of the East Indies (which later become known as Indonesia), policies against the Chinese have been enacted. At that time, Dutch and Chinese merchants were allies, with Chinese merchants acting as tax collectors on behalf of the Dutch government. However, within a short span of time, the Chinese stopped working with the Dutch, and the sentiments and stereotypes of Chinese descendants have remained until today. Consequently, the presence and the existence of Chinese-Indonesians are often taken seriously and indifferently by the local community as well as the government. Sometimes sentiments are even expressed in extremes, either with full resentment or through expressions of love where there is only black and white opinion. In short, prevalent attitudes towards this community have already existed in government institutions, reflecting implicitly on government policies. The limited ability to socialise, communicate and to jump into political positions has sparked the curiosity of the author to find out how social mobility influences current cultural values among Chinese-Indonesians.

One of the most recent cases is the blasphemy case against Basuki Tjahya Purnama (Ahok), which has become the topic of much debate among many Indonesians. Whether it was a "slip of the tongue", or ethically wrong, or completely wrong, depends on how Indonesian citizens interpret what happened. As a result, it has created a tension between races and religions. Therefore, with the rising tension between racial, religious and cultural positions, this study aims to act as a bridge to address the disparity between cultural values so that cases such as "Ahok" can be avoided in future.

This study is a preliminary piece of research that aims to explore how social mobility influences the cultural value of Chinese-Indonesians in Surabaya, Indonesia, by finding similarities in social values across Indonesia and where the gap is located. In the long run, this research is expected to truly help in understanding how social mobility influences cultural values. Surabaya is chosen for this study because this city has a large number of Chinese-Indonesian communities, of which approximately 7% of Surabaya's total population is classified into these ethnic groups, according to the Indonesian National Statistics (Badan Pusat Statistics).

This research used a quantitative methodology that involved a questionnaire. This analysis used a linear regression technique to understand the influence of the variable X (social mobility) on the variable Y (cultural values). Using a random sampling technique, a sample of young adult Chinese-Indonesians aged 20–30 years was randomly selected for the study. This age group was chosen because young people tend to have more exposure to social media and their perception of cultural values is quite different from older people. According to Uma Sekaran (1985), an appropriate sample size should represent at least 10% of the total population. However, as this was a preliminary piece of research, only 20 individuals were used for the study to find the influence between social mobility and cultural values.

2 DISCUSSION

2.1 *Validity and reliability test*

2.1.1 *Validity test*

A validity test was carried out to examine the extent to which measurement tools measured what they were supposed to measure on a questionnaire. Validity or invalidity of the instruments was confirmed by comparing the correlation index of product moment with a significance level of 5% along with the table calculations (0.482, n– = 15). If the r value is greater than the r-table value of 0.482, then the item is considered valid and vice versa.

According to Table 1, several items were found to be valid, notably the question numbers P75, P76, P77, P78, P88, P89 and P90. This is reflected in the correlation value (r) for valid items, which is smaller than the r-table value of 0.482, or has a significance value of more than 0.05. Therefore, this study identified seven valid items for the variable X (social mobility).

According to Table 2, several items were found to be valid for the variable Y (cultural values), which include the question numbers P3, P4, P7, P11, P18, P28, P29, P30, P31, P38, P39,

Table 1. Validity test of the instrument questions for the variable X (social mobility).

Items	r	Significance	Results
P75	0.723	0.002	Valid
P76	0.697	0.004	Valid
P77	0.771	0.001	Valid
P78	0.598	0.019	Valid
P88	0.638	0.011	Valid
P89	0.638	0.011	Valid
P90	0.688	0.005	Valid

Source: Author's data analysis, 2017.

Table 2. Validity test of the instruments questions for the variable Y (cultural value).

Items	r	Significance	Results
P3	0.741	0.002	Valid
P4	0.753	0.001	Valid
P7	0.568	0.027	Valid
P11	0.523	0.046	Valid
P18	0.516	0.049	Valid
P28	0.652	0.008	Valid
P29	0.801	0.000	Valid
P30	0.839	0.000	Valid
P31	0.680	0.005	Valid
P38	0.602	0.018	Valid
P39	0.637	0.011	Valid
P42	0.560	0.030	Valid
P44	0.525	0.045	Valid
P45	0.631	0.012	Valid
P48	0.640	0.010	Valid
P56	0.584	0.022	Valid
P59	0.598	0.018	Valid
P63	0.731	0.002	Valid
P64	0.630	0.012	Valid
P65	0.790	0.000	Valid
P66	0.671	0.006	Valid
P67	0.796	0.000	Valid

Source: Author's data analysis, 2017.

Table 3. Reliability test of the variables.

Variables	Items	Cronbach's alpha coefficient	Results
X (Social mobility)	7	0.825	Reliable
Y (cultural values)	22	0.962	Reliable

Source: Author's data analysis, 2017.

P42, P44, P45, P48, P56, P59, P63, P64, P65, P66 and P67. This is reflected in the correlation value (r) for valid items, which is smaller than the r-table value of 0.482, or has a significance value of more than 0.05. Therefore, this study identified 22 valid items for the variable Y (cultural values).

2.1.2 *Reliability test*

Reliability is an index that represents the accuracy and stability of the instruments whether they are reliable or not. Instruments are considered reliable if they have a reliability coefficient of 0.6 or above. In this study, the reliability test used was Cronbach's alpha. If the alpha coefficient is less than 0.6, then the instrument is not reliable and vice versa. The results of the reliability test are summarized in Table 3.

According to Table 3, the variables in this study were valid and reliable. As a result, the research can be conducted accordingly. Subsequently, the data were analysed by using a simple linear regression analysis and testing several classical assumptions such as residual normality, multicollinearity and heteroskedasticity.

2.2 *Simple linear regression analysis*

To find the influence of the independent variable X (social mobility) on the dependent variable Y (cultural values), statistical data analysis was carried out at several stages

Table 4. Results of the simple linear regression analysis.

Variables	*B*	$t_{calculated}$	Significance	Result
Constant	52.143			
Social mobility (X)	0.568	0.867	0.402	Not significant
A		= 0.050		
Determination coefficient (R^2)		= 0.055		
t-tabel ($t_{15,0.05}$)		= 2.131		

Source: Author's data analysis, 2017.

using the SPSS software program. Based on the calculation, the results are summarized in Table 4.

From the analysis, the result of the model was found to be **Y = 52.143 + 0.568 X**.

The model interpretation can be analysed as follows:

a. $\beta_0 = 52.143$. This regression coefficient value indicates that the total score of the variable X increases without any influence of the dependent variable Y. In other words, an increase in social mobility occurs even before cultural values is factored in, indicating that social mobility can increase before any influence exerted by cultural values.
b. $\beta_2 = 0.568$. This positive regression coefficient value indicates a linear influence between social mobility (X) and cultural values (Y). In other words, if social mobility increases, then cultural values will also increase and vice versa.

2.2.1 *Determination coefficient (R^2)*

The determination coefficient measured the accuracy of the model or matched the linear regression obtained from the parameter assumptions based on social mobility and cultural values. It was also used to measure the proportion of variance obtained by the model. According to Table 4, the size of the contribution from the independent variable to the dependent variable was 0.055 (5.5%). This result explains that the contribution from social mobility in the regression equation to cultural values was 5.5%, whereas the remaining 94.5% was contributed by other factors that were not included in this research. It can be concluded that social mobility is not the sole influence on cultural values. Thus, cultural values in the society are derived not only from external factors but also from internal factors, as discussed by Koentjaraningrat (1987) in his theory of cultural values.

2.2.2 *Hypothesis testing*

The regression model also includes a hypothesis to determine whether an independent variable has a significant influence on a dependent variable. The result of the hypothesis is as follows:

H_0: there is no significant influence on the variables X and Y.
H_1: there is a significant influence between the variables X and Y.

Decision-making:
H_0 is rejected if $|t_{calculated}| > t_{table}$ or the significant value is $< \alpha$.
H_0 is accepted if $|t_{calculated}| < t_{table}$ or the significant value is $> \alpha$.

According to Table 4, the variable X (social mobility) has a statistical test (t) of 0.867 with a significance of 0.402. Therefore, the calculated t value is less than the t-table value, which means that $\mathbf{H_0}$ is accepted. This result indicates that the variable X does not significantly influence the variable Y. Although social mobility has a positive influence on cultural values, it does not significantly influence cultural values. As mentioned by Steward (cited in Tucker 2013), a researcher can understand cultural values by first examining the cultural core that represents an important component in determining the ability of the culture to survive. The cultural core consists of the technology, knowledge, labour and family organisation, which can be used to collect resources from the environment (Tucker 2013). Therefore, it is not necessary that only social mobility should affect cultural values.

3 CONCLUSION

From the results of this study, it can be inferred that social mobility does not significantly influence cultural values. Nevertheless, there is a 5% influence of social mobility on cultural values among Chinese-Indonesians. Moreover, the first hypothesis in this study proves that social mobility influences cultural values. As this was a preliminary piece of research, further work will target specific cultural values, not just general ones. For example, we will choose tolerance values instead of combining all the indicators into one general category for cultural values.

This study aimed to discover the extent of the influence between the two variables X and Y. In the cultural context, certain things cannot be only measured by quantitative means as many other factors also influence cultural values in a positive direction. Therefore, for future research, a new set of questionnaires will be designed to specifically target certain cultural values in order to bridge the disparity in cultural values that Chinese-Indonesians hold to avoid a repeat of the "Ahok" case.

REFERENCES

Buzney, Catherine & Jon Marcoux. Cultural Materialism. University of Alabama Department of Anthropology Anthropological Theories: A Guide Prepared by Students for Students. *Anthropology*. Retrieved November 5, 2017 from http://anthropology.ua.edu/cultures/cultures.php?culture=Cultural%20 Materialism.

Estetika Indonesia. ____. Pengertian Kebudayaan. *Estetika Indonesia*. Retrieved October 29, 2017 from https://estetika-indonesia.blogspot.co.id/2015/12/pengertian-kebudayaan-menurut-menurut.html.

Harris, Marvin & Orna Johnson. 2007. *Cultural Anthropology,* 7th edition. Boston: Pearson.

Koentjaraningrat, 1987. *The Javanese Culture*. Oxford: Oxford University Press.

Lumen. ____. Cultural Ecology. *Cultural Anthropology*. Retrieved October 29, 2017 from https://courses.lumenlearning.com/culturalanthropology/chapter/cultural-ecology/.

Nikaolaus, Tsiopinis. 2005. *Definitions of Truth as Cultural Paradigms and a Rubric to Foster Communication within The Science and Religions Dialogue*. Retrieved October 29, 2017 from www.metanexus.net.

Puspitaviani, S., & Rahayu, S.D.I.S. (2015). *Aktivitas Ekonomi Etnis Tionghoa Di Tuban Tahun 1945–1959* (Doctoral dissertation, Universitas Airlangga).

Rahayu, S.D.I.S., & Purwanto, B. (2010). *Pendidikan etnis Tionghoa di Surabaya pada pertengahan abad ke-19 hingga tahun 1942* (Doctoral dissertation, [Yogyakarta]: Universitas Gadjah Mada).

Uma Sekaran. 2016. *Research Methods for Business.* United States: Wiley Books.

Urban Studies: Border and Mobility – Kerr et al. (Eds)
 ISBN 978-1-138-58034-3

Local government capacity in managing fishery conflict in the Indonesia–Malaysia maritime border zone

M.A.P. Sari
National Institution of Public Administration, Samarinda, Indonesia

M.R.K. Muluk & Sujarwoto
Universitas Brawijaya, Malang, Indonesia

ABSTRACT: Effective decentralization is determined mainly by local government capacity in managing local development. In the case of archipelago countries like Indonesia, local government capacity in managing national border issues such as fishery conflict between countries is vital to prevent and solve any disputes among actors between nations. In this study, we aim to explore issues of fishery conflicts between Indonesia and Malaysia in Sebatik Island as well as the capacity of Nunukan local government to solve the conflict. A single case study was used to explore and to identify the main issues of the fishery conflict and key local government capacities to manage the conflict. Observation and interviews with key informants were conducted in August 2017. The findings show three issues of fishery conflict between Indonesia and Malaysia at Sebatik: illegal fishing, fisherman robbery, and the dependence of Sebatik fisherman on Tawau (Malaysia) market. All issues are related to the low capacity of local government in terms of resources and structure capacity in managing the fishery conflicts. The findings suggest that despite the minimum authority that they have in managing fishery conflicts, local governments still need to enhance their capacity in managing fishery conflict, due to the geographical condition of the border closer to the local government in the regency, so they are more aware of the cause of the problem at the conflict site.

Keywords: local government capacity, fishery conflict, maritime border zone

1 INTRODUCTION

Indonesia shares maritime region with neighboring countries in Southeast Asia. The Nunukan Regency in North Kalimantan has a district named Sebatik Island, which is directly adjacent to Sabah, Malaysia. As seen from its economic structure, fishery is the most significant contributing sector in Sebatik Island and is the primary livelihood for the community (Nunukan in Number, 2016). Like maritime border area, in general, Sebatik Island faces various problems in the management of fisheries, including conflicts of fishery (Solihin, 2012).

According to Solihin (2012), one of the conflicts involving two fishery actors in Sebatik and Tawau (Malaysia) that has long been happening is the dependence of Sebatik fishermen on Tawau. Another fishery conflict that occurs in Sebatik Island is illegal fishing.

Management of fishery conflicts is a challenge that needs to be faced by countries that have maritime border zones (Charles, 1992). This issue is widely discussed by scholars, and requires in-depth understanding for resolving. It is possible to be achieved through the community linkages of all dimensions, including law, culture, and social organization (Bavinck, 2005: 806).

Because of decentralization, fishery conflicts have become the responsibility of the local government as well as the closest point of contact (Dirhamsyah, 2005), due to which the local government needs to equip the capacity and ability to withstand and adapt the changes in solving cross-border fishery conflicts (Pomeroy et al., 2007). Unfortunately, the majority of the local governments in areas in Indonesia do not have the capacity to solve problems and conflicts occurring in the maritime border zones due to the lack of resources (Dirhamsyah, 2016).

Many scholars focus on studying the ways of organizational capabilities in enhancing its responsiveness to new problems and how these abilities could be best institutionalized within governments (Andrew et al., 2013). According to a study conducted by Christensen and Beth (2008), in four decades, there has been a growing sentiment in the multi-discipline literature about the influence of organizational capacity in reaching the optimum performance, which is also related to the settlement of fishery conflicts (Bennett et al., 2001; p. 2). Fishery conflict becomes the main concern as one of the issues that is difficult to be managed by the government due to the vast area of scope, the complexity of the problems, and the interest of many parties on the area (Carbonetti, 2014; p. 6). In addition, studies by Dang et al. (2017) on the Vietnamese policy and by Biziouras (2013) on piracy in the Gulf of Guinea and Somalia were carried out, both citing the need for better measures taken by the government.

Due to the few of systematic and empirical investigations on fishery conflict and the capacity of local government to handle the issue (Bennet, 2001; p. 2), it is deemed necessary to make the mentioned issue as the focus of the study. On the basis of the above discussions, in this study, we address the following two questions: (1) what are the key issues of fishery conflict between Indonesia and Malaysia at *Sebatik* Island? (2) To what extent *Nunukan* local government capacity can solve fishery conflict between Indonesia and Malaysia at *Sebatik* Island?

In this study, we develop a conceptual model of local government capacities in managing fishery conflicts by integrating Dang et al.'s (2017) government capacity, Cigler's (2007) local government capacity in routine emergencies, Murshed-e-Jahan et al.'s (2014) communication strategies, and Kusumasari et al.'s (2010) local government capacities in managing disaster to structurally explain the local government's capacity. From this model, we can understand two things: (1) Whether the local government capacity is currently capable of handling fishery conflict and (2) What capacity building should be done by the local government to overcome the existing fishery conflict?

2 METHODOLOGY

The method of case study is best applied for this study. There are two conditions that can be appropriately applied to the case study: (1) when a study conducted addresses descriptive or explanatory question and (2) if we want to explain an event (Bromley, 1986). According to Yin (2009), there are four types of case study, of which we used single-case study here to explore and identify the main issues of the fishery conflict and key local government capacities to manage the conflict so that it can be the representative for another case of fishery conflict.

Primary data were collected through observation and in-depth interviews with key informants conducted in August 2017. Key informants include the fishermen, the manager of the fishermen, and the officials or managers in the government agencies of Nunukan Regency who have a direct link or relation in the management of fishery conflict. Despite interviews, this study also spread the list of questions to structural and extension officers in the Fisheries Agency.

Secondary data were collected through literature studies, including the past scientific studies and documents concerned with the study. The profile of local government agencies was also considered as an important source of secondary data. The other source of the secondary data were the activities of program planning documents stored in the strategic plan (Renstra) and also the annual report of agency performance (LAKIP). The collected primary and secondary data were analyzed by the descriptive analysis technique.

3 RESULTS

3.1 *Geographical location of Sebatik Island*

Sebatik island is one of the farthest islands located in the Nunukan Regency, North Kalimantan, with four of its five sub-districts directly adjacent to the Malaysian border. As for the maritime area, Sebatik island shares borders with Malaysia and the Philippines. It is also considered as the territorial sea zone, continental shelf, and exclusive economic zone in the Sulawesi Sea. In this case, the territory of Sebatik island is currently experiencing changes after the shift in ownership of Sipadan and Ligitan islands from Indonesia to Malaysia.

The strategic geographical location of Sebatik island causes fishery to become the most significant contributor to their economy, even in 2015, 78% or 3.30815 tons of fish capture in Nunukan Region originated from Sebatik island.

3.2 *Key issues of fishery conflict at Sebatik Island*

The fishery conflicts occurring in the border of Indonesia and Malaysia have become one of the most interesting dimensions to be analyzed (Song, 2017) due to the complexity of fishery conflict in the border that involved parties from different countries. Thus, any situation occurred could not be solved at the fishermen level, but to be solved at the inter-governmental level (Solihin, 2012). According to interview results, several problems exist in the island.

The first problem is illegal fishing. Almost every year, there are some cases of fishing violation on Malaysian or Filipino fishers in the Indonesia–Malaysia maritime border zone. The majority of illegal fishing practice undertaken by foreign fishermen in Sebatik island are the ones that deliberately perform illegal fishing in the Exclusive Economic Zone (ZEE) of Indonesia without proper permit and documents. They utilize advanced fleet and equipment, which includes Lengkong boats and Purse Seine equipment, and generally, the ship has been specifically designed to serve that purpose, according to an informant. Another violation that is commonly committed by foreign fishermen is not storing the fishing gear on hold. This violates Constitution Number 21/2004 regarding fishery in Indonesia.

Another conflict of fishery handled by the fishermen of Sebatik island is robbery on fishermen. On the basis of the information from one of the key informants, it was known that most foreign robbers in Indonesia commonly act as fishermen. Because of the similar posture with Indonesians, they are hard to detect. Robbery patterns also come in periods of activity and inactivity. These acts implant a sense of fear and worry among the local fishermen. Moreover, majority of them are scared to report the case of robbery due to safety reasons, preferring to report the incident to the port officers. This concern had once disappeared at least for 3 years, but has become infamous in the past year. In 2016, there were four noted cases in the database of the port officials, despite the fact that there are still plenty of cases left unreported.

Finally, Sebatik fishers have problems in the marketing of their catch. As mentioned earlier, most of the fishermen's catch were sold to Tawau, Malaysia. This is due to the geographical location of Sebatik island that is closer to Tawau than to Nunukan or other areas in North Kalimantan Province. There are also several aspects that make Tawau market more promising than domestic markets: the higher selling price due to the transactions using Malaysian currency (Ringgit) and the higher purchasing power of the people of Tawau.

The dependency of Sebatik fishers on Tawau market is not only due to the geographical location but also due to the bond between the fishermen of Sebatik and the capital owners of Tawau, which they usually call Tauke. This long-lasting connection goes as far as the Tauke lending capital without guarantee to the Sebatik fishers. From the interviews, it is also known that all catch were sold to Tauke through middlemen. Sebatik fishermen cannot sell their products directly to Tauke, because of the government regulations related to fish sales in Malaysia. The majority of the middlemen are Indonesians who somehow found a way to obtain Malaysian identification card. The scheme of sale and distribution of the fishing results of Sebatik fishers is reinforced by previous research conducted by Solihin (2012), which is depicted in Figure 1 by Solihin (2012; p. 58) himself.

Figure 1 shows that fish catching of Sebatik fisherman has become the input industry in Tawau, and even the processed products of the industry became consumption in Indonesia. This condition is obviously very detrimental to the fishermen of Sebatik, in particular, and Indonesia, in general, as the economic activity mostly occurred in Malaysia. In fact, from the result of the catch purchased by Tauke, it then becomes Tauke's source of export commodities.

The magnitude of Tawau market demand for the fish catch in Sebatik island should be a lucrative opportunity for the fishermen of Sebatik. However, the dependence of Sebatik fishermen to

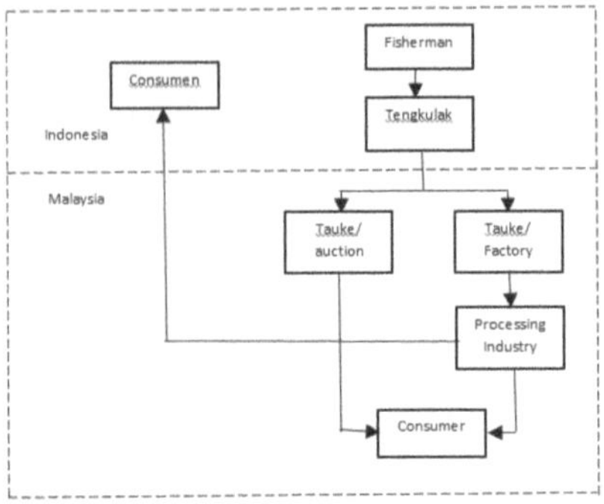

Figure 1. Distribution pattern of fish capture in Sebatik.

Tauke affected the fishermen on the low bargaining capabilities of Sebatik fishermen regarding the selling price of their catch. The fishermen state that the price of the capture fisheries that are being sold in Tawau market is flat, the Indonesian fisherman are only benefited through the fluctuation of ringgit currency; thus, when the currency weakens, they experience loss.

3.3 *Local government capacity in managing fishery conflict in Sebatik Island*

Since the local government's policy on fishery is executed by the Fisheries Agency, in order to understand the local government's capacity in the management of fishery conflict, this study focuses on the Fisheries Agency capacity that represents local government capacity.

The Fisheries Agency employed 108 people in 2016, of whom 36 (33%) are civil servants. The Fisheries Agency still does not have an adequate number of employees compared to the scope of responsibility. According to the Fisheries Agency's LAKIP, there are only three employees in the resources surveillance. This number is lacking considering the amount of work burden that is in the field of supervision, such as handling licensing and handling robbery. The existence of additional employees is not of much help because of the implementation of supervisory work, which requires employees with formal legality.

However, 63% of employees have bachelor's and master's degrees, while the rest have diploma, high school, junior high school, and even elementary school education, meaning that most employees have an adequate level of education. In addition, the employees of the Fisheries Agency also have educational background suitable for their current employment, which is related to the fishery sector. This is indirectly in line with the answers of the fishery service officials, who mostly stated that they have skills related to conflict management.

Besides the lack of personnel, the lack of financial support owned by the local government has caused an impact on the low achievement of the performance of the Fisheries Agency in the management of fisheries conflicts. The condition can be seen from the Fisheries Agency's LAKIP, whereby of the 250 surveillance operations targeted in 2011–2016, only 50 were conducted by 2013. The activity is then continued with law counseling. Unfortunately, only three counseling could be conducted by 2013 of the total target of 28.

The lack of budget owned by the Fisheries Agency affects the condition of facilities and equipment they have. Currently, they only have five units of vessels commonly used for surveillance operations. The number of ships can be felt to be less since the Fisheries Agency did not only conduct operation solely on Sebatik island but also on certain waters of Nunukan Regency, which are the entrance of the fishermen. To combat the shortage, the Fisheries Agency usually cooperates with the navy (LANAL) and the Department of Transportation in Nunukan Regency in the implementation of the surveillance operation.

Furthermore, the other infrastructure owned by the Nunukan Regency local government is the Fisheries Port of Indonesia (PPI), Sebatik. PPI itself was built in 2000, but it has been a while that the condition of PPI has become damaged and unable to be utilized up to the maximum level. PPI has wide functions, such as a place for fisherman landing their catchment and also refueling their vessels, but in fact, the PPI functions are not maximized due to the inadequate supporting facilities and infrastructure such as electrical power, water facilities, bridge, building of fish auction place, storage building, employee mess, and ice factory (Renstra Dinas Perikanan 2014–2021). Therefore, during this time, PPI Sebatik post was only used as the surveillance station for the maritime and fisheries resources.

In terms of policy, the passing of Act Number 23 in 2016 regarding the local government carries significant changes toward the division of affairs between the central government, provincial level, and city districts for the maritime and fisheries matter. Along with the shift of certain Nunukan Regency Fisheries Agency authority, there are certain changes that realization fails to do. The said authorities are the ones regarding the maritime, shore, and small islands, along with the surveillance of the utilization of fishery resources. With the shift in authorities since 2016, the local government has no authority to conduct the supervision and security as well as empowerment of the surveillance community (POKMASWAS), whereas this authority is directly related to the effort of managing fishery conflict.

Policies were made between 2011 and 2016 regarding the fishery conflict in the area. The first policy is the distribution of help to the group of fishermen in the form of fishing gear, including gill nets, boats, boat engines, and cooler boxes, intended to help them and reduce the dependency of capital from Tawau. The second policy is the surveillance operation of maritime and fishery resources. This is conducted by the Fisheries Agency with the involvement of all the security elements (army and police) to prevent and overcome the illegal fishing and fisherman robbery. The third act is the legal counseling in the usage of maritime and fisheries resources. Because of the lack of resources, until 2016, the counseling activity is only done for at least three times from the total 28 targeted activities. The fourth activity is the formation of surveillance community (POKMASWAS). The target of POKMASWAS is to create the mechanism of surveillance of the maritime resources, which is community-based. The achievement of the POKMASWAS target itself was not reached, whereby until 2016, there were only three groups among the total targeted 80.

4 DISCUSSION

The condition of each capacity of the Nunukan Regency government has been mentioned in the past, which illustrates the weakness of owned capacity in the management the issues of illegal fishing, the robbery of fisherman, and the dependence of Sebatik fisherman on Tawau market. The inadequate budget, number of employees, as well as number of ships cause inabilities of the Fisheries Agency to conduct safety and security operations optimally. Of the total of 250 operations targeted in 2011–2016, only 57 operations were successfully carried out. This caused an increase in the number of cases of illegal fishing and fishermen robbery.

It is known that the incapability of the local government in the management of fishery conflict is also due to the suboptimal utilization of Sebatik PPI. Essentially, the port should be the place to conduct fishery economic activities. However, due to the lack of supporting facilities, it can only function as the place for recording as well as retribution withdrawal of catch that is marketed to Tawau. The function of Sebatik PPI that is not optimum has made Tawau the only capture fisheries market of Sebatik fishermen. However, if the function of Sebatik PPI is optimized, it can provide the added value to the Sebatik fisheries and Indonesia, in general. Therefore, the fishermen of Sebatik can break through the foreign market rather than only dependent on Tawau market.

In reality, the local government of Nunukan Regency has high-quality employees as well as a policy that supports the management of fishery conflict, but there are limitations such as inadequate financial support, reduction of local government authority, inadequate number of employees, as well as lack of equipment, leading to the ineffectiveness of the policy implementation by the local government. From the above data, this study found that the managing of fishery conflicts in the Indonesia–Malaysia maritime border zone by local government was not effective, as we can see in Figure 2, which shows that local government faces a shortage of resources and structure. These limitations are the reasons for the failure of managing fishery conflicts

Considering that the local government has limited authority in marine management, it is necessary to establish patterns of technical work between the regency government, provinces

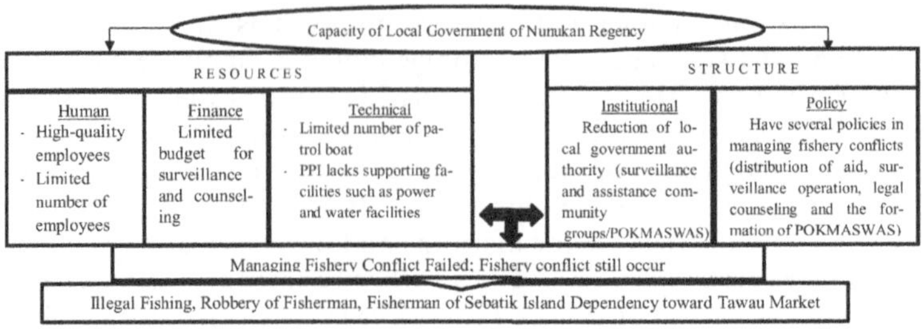

Figure 2. Final conceptual framework of local government's capacity of Nunukan Regency in managing fishery conflicts.

government, and central government to achieve the effectiveness in managing fishery conflicts in the border of Indonesia and Malaysia. Programs such as providing assistance to fishermen regarding sanitation and standard operation in port management also need to be implemented. This policy is related to the national program in 2018 that will establish the international port in Sebatik.

Despite the minimum authority in managing fishery conflicts, the local governments still need to enhance their capacity in managing it. This is due to the geographical condition of the border closer to the local government in the regency, so they are more aware of the cause of the problem at the conflict site and also because the local government has skilled personnel who can still be involved in managing fishery conflicts.

REFERENCES

Bavinck, M. (2005) Understanding Fisheries Conflicts in the South: A Legal Pluralist Perspective. *Society and Natural Resources*, 18(9), 805–820.

Bennett, E., Neiland, A., Anang, E., Bannerman, P., Rahman, A., Huq, S., Bhuiya, S., Day, M & Clerveaux, W. (2001). Towards a Better Understanding of Conflict Management in Tropical Fisheries: Evidence from Ghana, Bangladesh and the Caribbean. *Marine Policy*, 25(5), 365–376.

Biziouras, N. (2013). Piracy, State Capacity and Root Causes: Lessons From the Somali Experience and Policy Choice in the Gulf of Guinea. *African Security Review*, 22(3), 111–122.

Bromley, D. B. (1986). *The case-study method in psychology and related disciplines*. John Wiley & Sons.

Carbonetti, B., Pomeroy, R., & Richards, D. (2014). Overcoming the Lack of Political Will in Small-Scale Fisheries. *Marine policy*, (44), 295–301.

Charles, A. (1992). Fishery Conflicts: A Unified Framework. *Marine Policy*, 16(5), 379–393.

Christensen, R & Beth, G. (2008). Capacity for Public Administration: Analysis of Meaning and Measurement. *Public Administration and Development* 28, 265–279.

Cigler, B. (2007). The "Big Questions" of Katrina and the 2005 Great Flood of New Orleans. *Public Administration Review*, 67(s1), 64–76.

Dang, N, B., Momtaz, S., Zimmerman, K., & Nhung, P, T, H. (2017). The effectiveness of Formal Institutions in Managing Marine Fisheries for Sustainable Fisheries Development: A Case Study of a Coastal Commune in Vietnam. *Ocean & Coastal Management*, 137, 175–184.

Dirhamsyah. (2005). Maritime Law Enforcement and Compliance in Indonesia: Problem and Recommendations. *Maritime Studies*, (September-October).

Kusumasari, B., Alam, Q., & Siddiqui, K. (2010). Resource Capability for Local Government in Managing Disaster. *Disaster Prevention and Management: An International Journal*, 19(4), 438–451.

Murshed-e-Jahan, K., Belton, B., & Viswanathan, K. (2014). Communication Strategies for Managing Coastal Fisheries Conflicts in Bangladesh. *Ocean & Coastal Management*, 92, 65–73.

Solihin, I. (2012). Strategi Pengembangan Perikanan Tangkap di Wilayah Kabupaten Nunukan Kalimantan Timur, Perbatasan Indonesia-Malaysia (Strategy Development of Fishing Fish at Regency of Nunukan in East Kalimantan, Indonesia-Malaysia Border) *Doctoral Dissertation*. Institut Pertanian Bogor.

Song, A., Scholtens, J., Stephen, J., Bavinck, M., & Chuenpagdee, R. (2017). Transboundary Research in Fisheries. *Marine Policy*, 76, 8–18.

Yin, R. (2009) *Case Study Research, Design and Methods*. USA: Sage Publication.

Urban Studies: Border and Mobility – Kerr et al. (Eds)
© 2019 Taylor & Francis Group, London, ISBN 978-1-138-58034-3

Coffee stall: Politics identity of Cangkrukan

L. Santoso & M.G.R. Pandin
Universitas Airlangga, Surbaya, Indonesia

ABSTRACT: *Cangkrukan* (hanging out) has become a phenomenon in metropolitan cities. This indicates the presence of these cangkrukan communities in *warung kopi* (roadside coffee stalls) in the city. In many ways, *cangkrukan* has been used as a pattern to identify identity construction, i.e., formed through the habit of *cangkrukan* at *warung kopi*, and to understand social functions which take place when the communities gather at roadside coffee stalls. This article aims to obtain a profound understanding of the meaning to be conveyed to the public about the reality of the presence of *cangkrukan* communities in both Surabaya and Sdoarjo using phenomenology. Not only does *cangkrukan* in coffee stalls constitute an economic function, it also has sociocultural and sociopsychological functions. Through *cangkrukan*, every citizen identifies themselves as neutral individuals without any fear of discriminatory treatment and social injustice. In these coffee stalls, every individual participates in every mutual social relationship and this reality creates a dynamic social cohesion within the urban community. The formation of organic solidarity in a coffee stall is a proof that, humanely, people require the presence of others without being dependent on one or two strong figures but rather on each other by being connected to the shared hopes and beliefs.

Keywords: cangkrukan, coffee stalls, social cohesion

1 INTRODUCTION

Surabaya and Sidoarjo can be said to be prototypes of cities where industrialisation is one of their main supporting pillars. Industrialisation has become the vein of economic growth in both cities where all development is highly dependent on the industrial sector. Therefore, it is not surprising that the industrial sector has always been the 'golden goose' in terms of development orientation. At every city corner, the development of factories and warehouses, for the sake of industrialisation and other related factors, has become an inevitable view.

Despite their dynamic development as cities, Surabaya and Sidoarjo have marked their achievements as industrial cities. Their development as industrial cities has lured people outside Surabaya and Sidoarjo to make both cities their destinations to seek for jobs. The abundance in the industrial sector marked by the establishment of factories on every corner has clearly shown that both cities are 'havens' for job seekers. Urban citizens overflow on every city corner while hoping to get a better job from the one they previously had.

The presence of this urban community in Surabaya and Sidoarjo has impacted on the cities' population rates as well as the availability of residences. The demand for residences increases while the availability of residential space is extremely limited, forcing regional governments to generate policies that would convert agriculture areas into residential complexes. The massive development of residential industries in Surabaya and Sidoarjo is an inevitable phenomenon to fulfil the needs of this urban community.

The presence of this urban community in both cities has provided an interesting side to the living habits of the communities in Surabaya and Sidoarjo. In another context, the development of Surabaya and Sidoarjo into industrial cities has also contributed to a new identity formation pattern in terms of the community's custom which is distinctively different from the previous one. One of the interesting practices in Surabaya and Sidoarjo is the phenomenon of *cangkrukan* or hanging out to chat at roadside coffee stalls (*warung kopi*) or street cafés. This phenomenon is interesting as it is a part of a distinct and unique practice of the urban community in the city filled with developing industrialisation.

In almost all corners of Surabaya and Sidoarjo, spots to have *cangkrukan* can be found. What is interesting is that on every one of these city corners, coffee stalls are located both by the roadside and in places that are open spaces, including spaces for street pedestrians. The occurrence of these coffee stalls has become a new phenomenon, pushing cafés or similar businesses aside. At certain hours, these stalls will be crowded by groups of people having *cangkrukan* while enjoying coffee or other condiments or refreshments. Every day, this phenomenon can be easily found and has become a common view embellishing the scenery of Surabaya and Sidoarjo.

This activity is unique as, besides being rarely found in other cities, it also forms a habitual pattern which eventually manifests into an identity construction which, in the end, turns into a city trademark. The practice of *cangkrukan* has become a common phenomenon in industrialised societies. This is possibly due to certain factors. They can be identified as: (1) the condition of industrial society has been overtly busy with routine work hours, making the people constantly attempt to find a different 'atmosphere' that is detached from their routine, and (2) the lack of public space availability within the city has resulted in people always keenly taking advantage of every open space as a public space to meet and exchange ideas or information.

Cangkrukan has seemingly become a trademark of Surabaya and Sidoarjo. It has seemingly marked the meaning of the subjects' presence amidst the so-called metropolitan cities. The endless work routine, along with the nuance of 'busy-ness' constantly flowing, has made everyone constantly perform economic activities and compete with one another. The city ticks like a working clock from the dawn of the day to the dawn of the next day. Both cities have similarities, thus it is not an exaggeration to say that the people's life rhythms in both cities are of similar pace.

Borrowing Marxian terminology, social change happens and always starts from changes in the material or production sector. Such a perspective is interesting to investigate in terms of what it reveals. The reality of *cangkrukan* at coffee stalls has become a habitual activity for almost the majority of the public in both cities. *Cangkrukan* is a representation of industrial cities. Amidst the hustle and bustle of the working rhythm, the people of Surabaya and Sidoarjo always try to find themselves some space to meet and interact with others. The meeting points are no longer formal nor 'luxurious' like those found in malls or cafés. Instead, we can find such meeting points at roadside coffee stalls. There is a different atmosphere at these roadside coffee stalls.

The meaning of the subjects' or the 'agents' presence at a coffee stall has become significant. In addition, the identity construction clinging to the habit of *cangkrukan* as a marker of the 'subjects' presence is worthy of the city's attention. A metropolitan city does not always have to be in line with glamorous life styles, shopping in malls, etc. The presence of *cangkrukan* communities at coffee stalls is an essential part marking that there are 'the others' in a city worthy of attention. It signifies that there is a community which also contributes to the life pulse of the hectic activities in metropolitan cities. Presumably, *cangkrukan* can be 'people's house' born from the community itself, which can serve as a catalyst to minimise conflicts that may arise in an urban community.

The reality of the presence of *cangkrukan* communities in both cities is the focus in this research. In order to obtain a profound understanding of the meaning to be conveyed to the public, a phenomenology approach is chosen to be the main instrument. This approach is chosen so that the researchers can reveal the hidden 'chamber of mind' through the activity of *cangkrukan* at coffee stalls. Phenomenology takes every researcher to be directly involved so that every analysis is not merely based on theoretical assumptions but rather on the disclosure of direct life experience (*labenswelt*). In addition, such an approach can also be utilised to discover the hidden realities in a life experience.

2 CANGKRUKAN AND CITY IDENTITY

Cangkrukan at coffee stalls is actually not a new phenomenon in Surabaya and Sidoarjo. Such a phenomenon has long taken place, especially since both cities rapidly developed as metropolitan cities with industrialisation and commerce. It is not an exaggeration to state that *cangkrukan* attracts the interest of many who want to conduct a research on such a practice although it should be admitted that research on *cangkrukan* is still highly limited.

Muchammad Hamka Mudhowillah (2014:2) stated that *cangkrukan* became phenomenal as it is being massively done by many people of Surabaya and also because such an activity has become an opportunity for communication, socialisation, information sharing, and also an entertainment hub. *Cangkrukan* is considered as an inevitable hub for communicating and socialising because with cangkrukan, everyone can talk about anything with any theme. In addition, *cangkrukan* serves as an information hub where all the news and the latest or trending information can be shared in an event. Cangkrukan can also serve as an entertainment hub since in cangkrukan, people can temporarily rest their minds from all the fatigue. Later on, cangkrukan develops into an identity shared by the people of a city.

Identity is a feature which is inherent in an entity, both in individuals or social groups, be it in the form of ethnicity or race. As a unique feature, identity is an instrument used to distinguish one another. Identity is obtained in two ways: (1) identity which by default is given to someone at birth, such as being a male or female, and (2) the identity obtained through a quest (Castel, 2001:42). Identity acquired since birth is actually a biological identity and irreplaceable, while the identity perceived through a quest process is obtained through a long process of incessant life experience. Identity can – possibly – alter and be replaced with other identities.

Identity formed through a quest process is usually formed when it is continuously done which eventually becomes a habit. In the context of culture as an activity pattern, identity is inherent in every custom, constantly practiced by a group of people or an ethnic group in an environment. The same pattern of activity, though not necessarily identical, is performed by many people. When practiced in such a pattern in a similar way, a social group identity will gradually be formed and become inherent (Erikson, 1988: 24).

This means that identity construction is attached to the form of patterned activity as a cultural expression. Routine pattern similarities usually take place as a manifestation of an adaptation process to the environment. Different environments will result in a different adaptation process and vice versa. This notion becomes clear when it is placed in the context of reviewing a community, including *cangkrukan* communities at coffee stalls. The similarity of living activity patterns and rhythms will gradually form a unique feature that distinguishes the *cangkrukan* community from others. The identity comes from strong similarities, so strongly attached to the mutual life experience that it becomes a social identity. In addition, a *cangkrukan* community can also be seen as a marker of its subjects' presence with different identities in the city. Furthermore, *cangkrukan* will manifest itself in a city's identity. *Cangkrukan* and cities have become an integral part.

The same goes for *cangkrukan*. Its presence does not merely constitute a hang out activity in a certain place, which, in many cases, is at roadside coffee stalls. This habit of hanging out by a group of people which is continuously done will eventually form a distinguishing feature in other groups. Not only does it manifest in a certain action, it also manifests itself in unique and distinguishing statements. Various terms are often born from *cangkrukan* and become important evidence that *cangkrukan* has constructed itself as an awareness of the presence of a marker in a community living in a city.

3 CANGKRUKAN AND WARUNG KOPI: A CULTURAL PHENOMENON

The existence of *warung kopi* as the venue of *cangkrukan* should be perceived as a public space without domination. *Cangkrukan* is a space with no social hierarchy. Nobody feels in a different class from each other. *Cangkrukan* is a space where everyone can 'strip' themselves. Everyone is open to mutually identify themselves without anything to hide. This is contrary to formal city spaces like offices, institutions, malls and so on where people hide their self-image.

It is due to this very openness that *cangkrukan* is required by the urban citizens who start to lose the meaning of togetherness. *Cangkrukan* is an important cultural space for people's cultural resistance against the increasingly inhumane movement of a city. Borrowing a term from Mudji Sutrisno (2009), a cultural space is a space referring to the value in daily living. Through *cangkrukan*, urbanites try to define their presence within the city.

For the city, coffee stalls, with the phenomenon of *cangkrukan* in them, have always been considered untidy and filthy, the arena where the poor and lazy people assemble. Cities have always been uncomfortable with the activities of sub-urban communities, as if everything done by the poor is always wrong and dirty. In fact, they only want to buy (a situation) what they cannot buy in cafés or karaoke lounges.

The habits of the people in these two cities as urban villages have made them more easily mix with their colleagues, classmates, and people of the same profession or perhaps other identities that unite them. The terminology of being migrants – mostly – implies that one may not be familiar with his/her surrounding environment or neighbours. They live in a housing complex or flats near their working places or campuses. They are not close to the neighbours, because they are not childhood 'buddies' where closeness is more directed to the people of common identity, or of the same profession in the same work field or daily routine activities. Neighbourhood seems to be a mere residential space where people relieve their fatigue and tiredness. Even if people meet up, it is usually only at certain times like during the working days off or in the evening.

Cangkrukan becomes a marker of how a city provides space for its residents to manifest themselves. Each citizen from various social layers, ethnicities, both the formally educated and those who are not, will have the same position to engage in public activity so long as it does not cause any outcry in public spaces. Nevertheless, it must be acknowledged that the phenomenon of *cangkrukan* could also potentially bring social loss, such as conflicts of interest, outcry, traffic jams, and a number of other social issues such as alcohol consumption, drugs, and so on. The possibility of such social problems emerging in *cangkrukan* does not make it identical to negative impressions or social prejudice among other communities. Furthermore, *cangkrukan* can be considered as the gathering of lazy people to waste their time where it could actually be used for useful activities. Other than just smoking and drinking coffee, people who perform *cangkrukan* are deemed to be present in a mere social event without any clear orientation.

Such negative views of *cangkrukan* cannot be ignored as, in reality, this activity can actually bring a number of problems. Rather, what is required here is to put any event undertaken by other communities in the right context. From a certain point of view, the phenomenon of *cangkrukan* should be seen as a cultural phenomenon, requiring it to be placed in a distinct context. Putting things into a different perspective will certainly cause a different perspective as well. Normality and abnormality in a cultural event are merely a habit and consensus where what is abnormal at a certain period, can become the new normal. This means any place can present its own normality and abnormality, depending on how the reality is defined.

All these times, the city has always got the characteristics of regularity and orderliness. The city has the character of workers and bureaucrats; therefore, there is always a normative boundary in interpreting *cangkrukan* as the event of an unemployed and anti-social group. *Cangkrukan* is seen to be a representation of the loss of hard work and regularity in a social order known as the 'city'. That is why *warung kopi* is always stereotyped with particular classes losing in a socio-economic competition. *Cangkrukan* is – seemingly – always synonymous with the working class who 'loses' or students who cannot stand 'heavy workloads' in their college with all of its complexity, or even the urban poor people who can only buy a cup of coffee and a cigarette to then stay for hours at a coffee stall.

This perspective blatantly debases the presence of *cangkrukan* at coffee stalls. In addition to such stigmatisation, the presence of the subjects at coffee stalls is also considered futile. Nevertheless, in other contexts, we need to realise that social problems could arise anywhere. Even in a space which is considered to bring 'piety' and 'bureaucrat', social problems can appear and manifest themselves. Social problems can arise from a formal activity. Aren't humans paradoxical beings who can present two opposite potentials of piety and depravity, order and wildness, as well as love and hatred? In well-behaved formal-bureaucratic spaces, corrupt behaviours can be born as well as the elements of hatred, unfair policies, and so on. In a formal space, many people can conceal their self-images by presenting formal and good characters, although the opposite can also happen. This means that every place – actually – has the potential to ignite various problems that cannot be necessarily stigmatised with other places.

In almost all corners of Surabaya and Sidoarjo, *cangkrukan* can be found. As a cultural event, *cangkrukan* cannot be detached from the standpoint of the non-participants. This event should be put in the context of the needs of 'subjects' who are present and participate in it. Through their participation (having *cangkrukan* at coffee stalls), there will always be something new which occurs beyond expectation.

Every individual in a coffee stall is a free being who may speak and express anything, as long as no one suffers. It is then not a surprise to witness various expletive mockeries and swearing in a *warung kopi* which often appear in every conversation, where otherwise it would be grounds for indecency in formal places. It is then confirmed that *cangkrukan* at *warung kopi* is an activity allowing people to 'strip' themselves of all forms of fake imaging. While in a *warung kopi*, all inherent identities and built images can blend into equality. That is why a *warung kopi* can be a marker of the melting of formality where people identify one another in equality. The self-images presented are the same as demonstrated by the language used and the food appetite.

Such equality has made coffee stalls the hub of public spaces without domination. All kinds of speeches are presented without any barriers and limitations. Even when there are other facilities in a *warung kopi* such as Wi-Fi, chess boards or playing cards, they are just parts of an effort to get people close to each other. Someone who is very respected in a certain hierarchical formal space, can be verbally ridiculed upon losing a chess game or card game. Such a situation has made *warung kopi* egalitarian. Someone who has always 'lost' in his everyday formal social relations can have the chance to be a 'winner', although only in just a game.

Such a reality is sometimes present and sometimes goes unnoticed. In *warung kopi*, people who (may) identify themselves as marginalised or the 'other class' in a city's hierarchy can actually get their spot to internalise themselves as significant subjects who are present in an interaction between humans in a city. Only by involving themselves (in a coffee stall) that every momentum can simply take place. Therefore, to construct people's habits by participating in the same space and the same discourse is a form of wisdom. Without all of this, what emerges is only cultural prejudice towards every *cangkrukan* happening at a *warung kopi* with all the swearing words accompanied by domino cards, playing cards and chess boards.

In a different context, the presence of coffee stalls in every corner of Surabaya and Sidoarjo actually confirms that these two cities have become fully urbanised. As industrial cities, Surabaya and Sidoarjo necessitate the presence of the urbanites to take part in their activities, either to work or to build a family. The presence of this urban community in both cities has provided a lively atmosphere with their working activities, whose lives are under the pressure of working hours, doing the same routines every day from morning to evening, being stuck in traffic jams on the way home from work and finally arriving home, tired and exhausted to meet the next day only to undertake the same routine without any differences. Therefore, the concept of the weekend can only be constructed in the metropolitan community, after five days from Monday through to Friday living under the pressure of working hours.

Coffee stalls are citizens' destination as they open up every time and are also relatively affordable by all social layers. People use their working break time to go to coffee stalls even only for one or two hours, or when they come home from work and get stuck in a traffic jam, people can just stop by the nearest roadside to find a coffee stall. Usually, based on some informants, *cangkrukan* has always been planned together in a group or community. *Cangkrukan* marks the presence of two people or more at the same coffee stall with the same purpose or usually the same need. Almost never is *cangkrukan* done individually. Through *cangkrukan*, the urban citizens are actually building a cultural phenomenon, marking the presence of an urban community identity amidst the metropolitan city.

As a cultural event, enjoying a cup of coffee in a coffee stall will always be present in any level of social relationship. In rural communities, enjoying coffee is a hereditary tradition done before going to the rice field to work accompanied by some snack or while visiting a neighbour's house. The same thing goes for industrialised societies in the metropolis, where workers commonly enjoy a cup of coffee in the morning before starting their working routines in the industrial sector. Elsewhere, coffee is inevitable on menu and with dishes served while hanging out in cafés in the afternoon.

4 CONCLUSION

For the last few years, *warung kopi* (coffee stall) has become a significant part of the two cities' dynamics – Surabaya and Sidoarjo – as they transform into metropolitan cities. As metropolitan cities, Surabaya and the nearby cities such as Sidoarjo have grown rapidly, not only in infrastructure construction, but also in terms of their superstructures in the form of life habits, lifestyle, occupancy rates, and so on.

Warung kopi manifests into an alternative space to have *cangkrukan* other than in big cafés, pubs, malls or karaoke lounges. Besides its relatively cheap price, *cangkrukan* in *warung kopi* does not require various formal rules like other formal places. People can spend hours having *cangkrukan* in warung kopi without having to feel uncomfortable even if they only buy a cup of coffee and a cigarette. For them, meeting with other friends is hard to do and constitutes priceless value. Chat topics in a coffee stall vary from everyday talk to more complex and complicated topics.

In a particular context, *cangkrukan* in coffee stalls constitutes a counter-culture against the establishment of certain social classes. It may even be a critique of the city's populist development for the benefit of certain social classes only. *Cangkrukan* is clearly a cultural resistance for people who have experienced loss in the formal life competition in big cities. In addition, *warung kopi* does not only function as a *cangkrukan* venue, but also serves as a social space where every citizen can socialise and interact with each other as well as releasing all the aggravation of the tough life competition in a big city. Not only that, *warung kopi* has also become a new promising career opportunity for the urban community who do not have access to the formal sector or have been laid off from their old jobs. Through *warung kopi*, the economic movement of the lower-middle class people is actually helped and, at the same time, people can have a cheap 'meeting' club amidst the invasion of malls and cafés.

Cangkrukan has the function of a public space without domination because in *warung kopi*, no social barriers segregate one citizen from others. *Warung kopi* is the most neutral room for drinking coffee and have some cigarettes while, at the same time, being a safe place to 'curse' about issues which are getting people's attention. It is a place without domination because in *warung kopi* everyone can have their own arguments and even propose issues to be discussed. This function is so difficult to get in formal spaces like offices, factories, and malls. It is through *warung kopi* that the city residents, especially the urban citizens who have long been marginalised by the capital-bearing mainstream groups, can define the meaning of their existence as city subjects.

REFERENCES

Booree, G. (2004). *Personality Theories trans.* Yogyakarta: primasophie.

Castel, M. (2001). *Power of Identity.* London: Blackwell.

Cavallaro, D. (2004). *Critical and Cultural Theory translated by Gunawan Admiranto.* Yogyakarta: Jalasutra.

Erikson, E. H. (1989). *Identitas dan Siklus Hidup Manusia translated by Agus Cremers.* Yogyakarta: Niagara.

Heidegger, M. (1969). *Identity and Difference*. New York: Harper and Row Publisher.

Heryanto, A. (2015). *Identitas dan Kenikmatan, Politik Budaya Layar Indonesia, KPG.* Jakarta.

Koentjaraningrat. (2004). *Kebudayaan Mentalitas dan Pembangunan 20th edition.* Jakarta: Gramedia.

Nugroho, I. P. (2013). *Fenomenologi Politik, Membongkar Politik Menyelami Manusia.* Magelang: Sanggar Pembasisan Pancasila.

Silas, J. (1996). *Kampung Surabaya, Menuju Metropolitan.* Surabaya: Yayasan Bhakti Surabaya and Surabaya Post.

Triwikromo, A. Y. (1999). *Pemulung Jalanan, Konstruksi Marginalitas dan Perjuangan Hidup dalam Bayang-Bayang Budaya Dominan.* Yogyakarta: Media Pressindo.

Borders and mobility in media, technology and global research

Urban Studies: Border and Mobility – Kerr et al. (Eds)
© 2019 Taylor & Francis Group, London, ISBN 978-1-138-58034-3

Awareness and preparation for cross-border future careers under the one belt, one road initiative proposed by the People's Republic of China (PRC): A case study of students majoring in Chinese at UBRU, Thailand and HUFL, Vietnam

S. Songsukrojiroad & L.L. Chuyen
Ubon Ratchathani Rajabhat University, Thailand

ABSTRACT: The objectives of this study were to 1) investigate the awareness and understanding of students majoring in Chinese at UBRU and HUFL about the One Belt, One Road Initiative (BRI) proposed by the People's Republic of China (PRC) in relation to their future career opportunities in ASEAN countries and other regions, and 2) explore the ways for developing Chinese teaching to suit 21st-century learning and skills for the two targeted groups of students. To collect data, questionnaires consisting of 268 items (140 in Thai and 128 in Vietnamese) were distributed to the two groups. The findings revealed significant differences between the students at UBRU, Thailand and HUFL, Vietnam in terms of four areas of the first objective pertaining to the students' awareness and understanding of the BRI in relation to their future career opportunities. Conversely, the findings of the second objective pertaining to exploring the ways for developing Chinese teaching to suit 21st-century learning and skills for the students at both universities shared common expectations of the two groups in the aspects of (1) fulfilling other updated and important contents such as economic and financial literacy when teaching in the classroom, apart from teaching only Chinese linguistics and language *per se*, (2) integrating teaching techniques by reducing passive learning and increasing active learning in the classroom, and (3) giving more outside-class assignments to emphasize on self-learning from media and Internet literacy.

Keywords: 21st Century Learning and Skills, Awareness, Chinese major students, Cross-Border of Future Careers, HUFL; One Belt, One Road Initiative, UBRU

1 INTRODUCTION

Since the official opening of the ASEAN Community, all ASEAN countries have attempted to develop themselves in various fields such as education, science and technology, economy, trade and investment, quality of life, and so on by seeking cooperation from other ASEAN countries and from countries in other continents through various collaborative dimensions.

To build the cooperation between the People's Republic of China (PRC) and ASEAN countries, including those in other continents, the megaproject One Belt, One Road Initiative, also known as the Belt and Road Initiative (BRI) or the Silk Road Economic Belt and the 21st-Century Maritime Silk Road, was proposed by China's most powerful leader Xi Jinping. It is a development strategy that focuses on connectivity and cooperation between Eurasian countries, primarily the PRC, the land-based Silk Road Economic Belt (SREB), and the oceangoing Maritime Silk Road (MSR). The strategy underlines China's push to take a larger role in global affairs, and its desire to coordinate manufacturing capacity in various areas with other relevant countries. The BRI is believed to bridge the 'infrastructure gap' and thus accelerate economic growth across the Asia Pacific region and Central and Eastern Europe. Certainly, Vietnam and Thailand are also expected to coordinate with the

PRC. Thus, preparation of human resource development through effective education management for future career opportunities of the young generation of Thailand and Vietnam is considerably prioritized by these two governments. To the authors' knowledge, the BRI has created increasing and diversified needs for languages in the world to promote a better understanding and cooperation among people from different languages and cultures. Therefore, it is believed that the BRI is also one of their opportunities to look for good and sustainable jobs after their graduation.

As Chinese is one of the important foreign languages of ASEAN people for their current and future careers, it is believed that learning this non-native language (Mandarin Chinese) will help students majoring in Chinese at UBRU and HUFL to gain more opportunities for their future careers under the BRI. Meanwhile, Chinese teaching techniques adopted by professors at UBRU and HUFL should also be developed to suit the 21st-century learning and skills necessary for their students.

The objectives of this study were to (1) investigate the awareness and understanding of students about the One Belt, One Road Initiative (BRI) proposed by the PRC in relation to their future career opportunities, and (2) explore the ways for developing Chinese teaching to suit 21st-century learning and skills for the two targeted groups of students. The scope of this study focused mainly on the results obtained from the survey using questionnaires distributed to undergraduate students majoring in Chinese (year 1 to year 4) at UBRU, Thailand and HUFL, Vietnam in 2017. The research questions are as follows: 1) what are the awareness and preparation for future careers in relation to the BRI of UBRU and HUFL students majoring in Chinese? 2) How should Chinese teaching be developed to suit 21st-century learning and skills for the two targeted groups of students? This study hypothesized that the awareness and preparation for future careers of students majoring in Chinese at UBRU, Thailand were different from that of the students at HUFL, Vietnam.

2 LITERATURE REVIEW

Based on the P21's Framework for 21st Century Learning (www.21org. 2017: online) that was developed to strengthen the skills and knowledge of students to succeed in their work, life and citizenship skills, as well as to strengthen the support systems necessary for 21st-century learning outcomes, it has been observed that thousands of educators and schools worldwide have gradually been working on this framework in order to develop their students to suit the 21st-century skills such as critical thinking, problem solving, reasoning, analysis, creativity, perseverance, oral and written communication, leadership, teamwork, collaboration, ICT literacy, media and Internet literacy, civic, ethical, economic and financial literacy, and global awareness.

For the 21st-century education, Jenkins et al. (2009) stated that "*the new media literacies*: a set of cultural competencies and social skills" that young people need in the new media landscape primarily involve social skills developed through collaboration and networking. These new skills that young people in this digital era need to comprehend and be able to conduct are as follows: (1) *Play*—the ability to experiment with the surroundings as a form of problem solving, (2) *Performance*—the ability to adopt alternative identities for the purpose of improvisation and discovery, (3) *Simulation*—the ability to interpret and construct dynamic models of real-world processes, (4) *Appropriation*—the ability to meaningfully sample and remix media content, (5) *Multi-tasking*—the ability to scan the environment and shift the focus on salient details, (6) *Distributed cognition*—the ability to meaningfully interact with tools that expand mental capacities, (7) *Collective intelligence*—the ability to pool knowledge and compare notes with others toward common goals, (8) *Judgment*—the ability to evaluate the reliability and credibility of different information sources, (9) *Transmedia navigation*—the ability to follow the flow of stories and information across multiple modalities, (10) *Networking*—the ability to search, synthesize, and disseminate information, and (11) *Negotiation*—the ability to travel across diverse communities, discern and respect multiple perspectives, and grasp and follow alternative norms.

Table 1. Criteria of rating scale for data analysis.

Scoring measurement (percentage of responses)	Rating scale
85–100%	Highest
75–84%	High
65–74%	Medium
45–64%	Low
0–44%	Lowest

3 RESEARCH METHODOLOGY

A total of 140 UBRU and 128 HUFL undergraduate students majoring in Chinese (year 1 to year 4) in 2017 were selected for this study. To collect data, questionnaires (in Thai and Vietnamese) were distributed to the two groups of students. Data are presented as percentages. The data analysis comprised (1) using percentage for analyzing general data of the respondents, and (2) analyzing the relationship between the general data of the respondents and responses in each aspect and open-ended questions answered by the respondents. For data analysis on four main aspects of the first research question, criteria of rating scale were computed from Likert scales (cited in Boonchom, 2010). A five-point Likert scale and scoring measurement (percentage of responses) are presented in Table 1.

4 RESEARCH OUTCOMES

The results of the investigation were divided into four areas: (1) understanding the overall impact of the BRI proposed by the PRC, (2) awareness of the BRI and people mobility among ASEAN countries and other continents resulting from the BRI, (3) future career expectations of the respondents after their graduation, and (4) targeted working place in the first five years after their graduation.

Figure 1 shows the proportion of the two groups of the respondents: (1) 140 UBRU Chinese-major students (male 11% and female 89%), and (2) 128 HUFL Chinese-major students (male 15% and female 85%).

Table 2 reveals significant differences between the responses from the UBRU students and the HUFL students. On the Thai side, the highest rating scale for understanding the overall impact of the BRI was rated for the Environmental impacts and Connectivity and Cooperation aspects. The high rating scale was shown for the Socio-cultural aspect, while the Economic aspect and the Political–security aspect were rated at the medium and low levels, respectively. In terms of the Vietnamese side, the highest rating scale was not found. The Economic and the Political–security aspects were rated at a high level, while the Connectivity and cooperation and the Socio-cultural aspects were rated at a medium level. Surprisingly, the Environmental impacts were shown at a low level.

Table 3 reveals the similarities between the responses from the UBRU students and the HUFL students. On the Thai side, the highest rating scale was found in the aspects of awareness of the important role of China as the leading country in extending BRI cooperation, awareness of the people mobility phenomenon along the routes of the BRI, and awareness of paying attention to the BRI by other countries around the world. Moreover, the aspect of awareness of the importance of using Chinese for future career in relation to the BRI was rated at a high level, while the aspect of awareness of paying attention to the BRI by the home country of the respondents was rated at a medium level. In terms of the Vietnamese, the highest rating scale was found for the aspect of awareness of the important role of China as the leading country in extending BRI cooperation and awareness of the importance of using Chinese for future career in relation to the BRI. Furthermore, the high rating scale was found for the aspects of awareness of paying attention to the BRI by both other countries around the world and the home country of the respondents. However, the aspect of

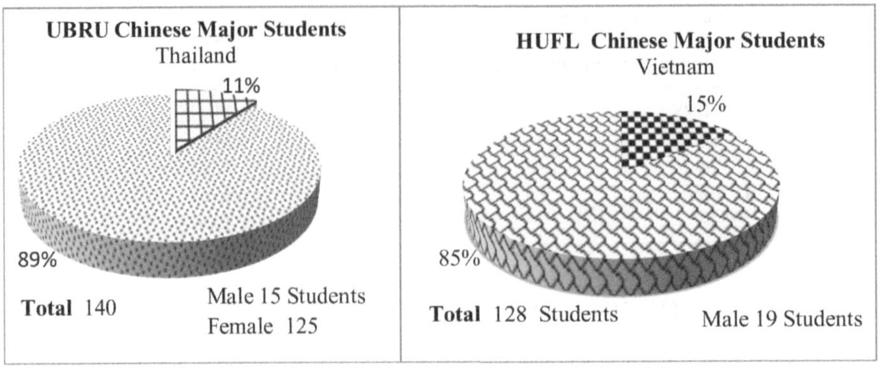

Figure 1. General data of the respondents.

Table 2. Comparison of understanding the overall impact of the BRI.

	UBRU students, Thailand		HUFL students, Vietnam	
Aspect	Percentage	Rating scale	Percentage	Rating scale
1) Economic aspect	68%	Medium	80%	High
2) Socio-cultural aspect	83%	High	65%	Medium
3) Political–security aspect	55%	Low	76%	High
4) Environmental impacts	90%	Highest	62%	Low
5) Connectivity and cooperation	88%	Highest	72%	Medium

Table 3. Comparison of awareness of the BRI and people mobility.

	UBRU students, Thailand		HUFL students, Vietnam	
Aspect	Percentage	Rating scale	Percentage	Rating scale
1) Awareness of the importance of using Chinese for future career in relation to the BRI	82%	High	86%	Highest
2) Awareness of the people mobility phenomenon along the routes of the BRI	92%	Highest	70%	Medium
3) Awareness of the important role of China as the leading country in extending BRI cooperation	95%	Highest	88%	Highest
4) Awareness of paying attention to the BRI by other countries around the world	85%	Highest	80%	High
5) Awareness of paying attention to the BRI by the home country of the respondents	72%	Medium	78%	High

awareness of the people mobility phenomenon along the routes of the BRI was only rated at a medium level.

Table 4 presents the expectation of the UBRU respondents concerning their future careers. Lecturer of Chinese and Interpreter/Translator were rated at the highest level, while Import & Export, Office worker of Chinese company, and Tour guide were rated at a medium level. The other careers were rated at the lowest level.

Table 5 presents the expectation of HUFL respondents concerning their future careers. Bank employee and Import & Export were rated at the highest level, while Interpreter/Translator and Diplomat were rated at a high level. Surprisingly, Tour guide was rated at a low level.

Table 6 shows significant differences between the responses from the UBRU students and the HUFL students. On the Thai side, the highest rating scale was found in the aspect of

Table 4. Comparison of the expectation of UBRU students (Thailand) for their future careers.

	UBRU students, Thailand	
Aspect	Percentage	Rating scale
1) Lecturer of Chinese	88%	Highest
2) Interpreter/Translator	85%	Highest
3) Import & Export	72%	Medium
4) Office worker of Chinese company	68%	Medium
5) Tour guide	65%	Medium
6) Others	28%	Lowest

Table 5. Comparison of the expectation of HUFL students (Vietnam) about their future careers.

	HUFL students, Vietnam	
Aspect	Percentage	Rating scale
1) Bank employee	88%	Highest
2) Import & Export	86%	Highest
3) Interpreter/Translator	78%	High
4) Diplomat	72%	High
5) Tour guide	56%	Low
6) Others	24%	Lowest

Table 6. Comparison of targeted working place after graduation.

	UBRU students, Thailand		HUFL students, Vietnam	
Aspect	Percentage	Rating scale	Percentage	Rating scale
1) Work in hometown	66%	Medium	70%	Medium
2) Work in other places in home country	88%	Highest	86%	Highest
3) Work in China	68%	Medium	54%	Low
4) Work in ASEAN countries	65%	Medium	92%	Highest
5) Work in other continents	40%	Lowest	80%	High

working in other places in home country, while choosing to work in China, work in hometown, and work in ASEAN countries were rated at a medium level. The lowest level was found for the aspect of working in other continents. On the Vietnamese side, working in ASEAN countries and working in other places in home country were rated at the highest level, while choosing to work in other continents was rated at a high level. Interestingly, working in hometown and working in China were rated at a medium and a low level, respectively.

5 CONCLUSION AND DISCUSSION

The findings from this study have revealed interesting phenomena regarding studying Chinese for future careers by the UBRU Chinese-major students (Thailand) and the HUFL students (Vietnam). The highest level rated by the UBRU students was for the Environmental impacts and Connectivity and cooperation aspects, while the Political–security aspect was rated at the lowest level. Perhaps the perception of the risk of investment and infrastructure construction of China and partner countries would have led to destroy all the aspects of natural environments. However, connectivity and cooperation under the BRI are still considered to be two main key factors that can help Thailand and other ASEAN countries develop their economic growth more rapidly. In addition, the Socio-cultural aspect rated at a high level reflected the awareness of these Thai students about encountering the phenomena of people mobility and cross-cultural diversities after the BRI becomes a more powerful megaproject among ASEAN countries which will boost the economic growth of ASEAN countries even further. On the other hand, the UBRU students majoring in Chinese did not understand much about the political and security aspect. This might be because they mainly paid attention to studying only the Chinese language without awareness of the impact of international politics.

The important findings from the HUFL students revealed that these Vietnamese students placed importance on learning more about the economic aspect besides studying only the Chinese language. In addition, the of Political–security aspect rated at a high level also revealed their desire to work in the fields related to finance, banking, trade or investment after their graduation. However, the Environmental impacts aspect rated at the lowest level indicated that they overlooked the risk of environmental disaster resulting from the BRI.

The important findings from Table 2 revealed the similarities and differences between the responses from the two groups of students. Both Thai and Vietnamese respondents rated the highest scale for the aspect of awareness of the important role of China as the leading country in extending BRI cooperation. However, Thai students also rated the highest level for the aspect of awareness of paying attention to the BRI by other countries around the world and the aspect of awareness of the people mobility phenomenon along the routes of the BRI, while Vietnamese rated the lowest level for awareness of the people mobility phenomenon along the routes of the BRI. These results indicated that both Thai and Vietnamese students were aware of the important role of China as a leading country in helping Thailand, Vietnam, and other countries to improve their economic growth and other developments. However, the lowest rating by Vietnamese respondents for the awareness of the people mobility phenomenon along the routes of the BRI was surprising. This was perhaps because the Vietnamese respondents only emphasized on how to obtain good careers related to using Chinese for communication in their future works without awareness of people mobility from other places or countries, in their own country and among other ASEAN countries. Another interesting point shared by the Thai and Vietnamese students was a common expectation of using Chinese for future careers in relation to the BRI. Thus, the expectation of the BRI that people from other countries will be able to use Chinese, besides English, for communication and work along the routes of the BRI seems possible.

The findings from the data analysis of future career expectation of the respondents after their graduation revealed some interesting points in terms of the highest level rated by both groups. The Thai respondents rated the highest level for Lecturer of Chinese and Interpreter/Translator, while the Vietnamese students rated the highest level for Bank employee and Import & Export. This future career expectation of the Vietnamese respondents that combined their responses in item 5.1 revealed that they placed importance on learning more about economic aspects besides studying only Chinese language. In comparison, most of the Thai respondents preferred working as teachers of Chinese and as translators to working in the fields of banking or finance and trade. This was perhaps because the Thai respondents might have thought that most of their family background was not from commerce or businesses. Furthermore, being a teacher in a Thai setting, especially in the rural area, seems to be more respected by Thai people rather than working in other careers. Another surprising finding from this study was that both respondents rated for being a tour guide at a medium and a low level, respectively. One of the reasons might be that being a tour guide for Chinese tourists does not relatively pay well.

The findings from the data analysis in this item revealed significant differences between the responses from the UBRU and HUFL Chinese-major students. The Thai students rated the highest level for the aspect of working in other places in their home country, while the Vietnamese students rated for working in ASEAN countries and working in other places in their home country. Moreover, the Thai students rated the lowest level for the aspect of working in other continents, while the Vietnamese students rated the high level for working in other continents. These findings indicated that the Vietnamese students were ready to walk out from their comfort zone (hometown and home country) to look for new opportunities in cross-border regions for their future careers, while the Thai students perhaps felt more comfortably to work by using Chinese for their work in Thailand. However, the authors believe that the impacts of the BRI and the people mobility phenomenon among ASEAN countries in the near future will make both Thai and Vietnamese students studying Chinese have more awareness of working not only in their home countries but also in other countries.

Furthermore, the findings of the second objective, namely exploring the ways for developing Chinese teaching to suit 21st-century learning and skills for students at both universities, shared a common expectation of both groups of students in the aspects of (1) fulfilling other updated and important contents such as economic and financial literacy when teaching in the classroom (besides teaching only Chinese linguistics and language), (2) integrating of teaching techniques by reducing passive learning and increasing active learning in the classroom, and (3) giving more outside-class assignments to emphasize on self-learning from media and Internet literacy. These findings also reflected the readiness of the new generation to learn new things from various sources.

To successfully transform both groups of students, including those from other countries, into more competent individuals in the 21st century, there is a need to transform the learning ecosystem into purposeful, generative, mindful, and result-based learning. Active learning both in and outside class should be well designed by both lecturers and students. Furthermore, both perception and awareness of the challenge of changes in this globalization era should be adopted in teaching contents. Finally, there is also a need to develop new skills through an effective training program or system for students before their graduation in order that they will be able to serve the demands of social, industrial, and economic sectors. The authors believe that the One Belt, One Road Initiative proposed by China will help encourage the two universities to design appropriate Chinese instructional programs to help graduates to be well-equipped with knowledge and skills required for the 21st century so that they will be able to compete with graduates from other universities and have ample opportunities to be successful in their future careers.

REFERENCES

Jenkins, H. et al. (2009). *Confronting the Challenges of Participatory Culture: Media Education for 21st Education Century*. Cambridge: The MIT Press.

The Belt and Road Economics from initial plan. Accessed 18 October 2017 from: https://en.wikipedia.org/wiki/Belt_and_Road_Initiative.

The P21's Framework for 21st Century Learning. Accessed on 19 October 2017 from: http://www.p21.org/our-work/p21-framework.

Urban Studies: Border and Mobility – Kerr et al. (Eds)
© 2019 Taylor & Francis Group, London, ISBN 978-1-138-58034-3

Articulating Indonesian migrant domestic workers' activism in Hong Kong and the use of communication technology

I. Wahyudi
Universitas Airlangga, Surabaya, Indonesia

ABSTRACT: Indonesian migrant workers have been sent to several countries to fill the need for labour, especially in the domestic sector. Hong Kong is one of the countries to have received Indonesian Migrant Domestic Workers (IMDWs) since the early 1990s. In this paper, I will discuss recent conditions of IMDWs in Hong Kong along with their activities in advocating for their rights by the utilisation of communication technology. Hong Kong is a major receiving country for migrant workers, particularly women from South East Asia who work as domestic assistants or caregivers. Hong Kong's economy, with its better working conditions and attractive salary, are magnets for migrant workers. It offers an opportunity for an improved financial future and for paid work that many women may not be able to obtain in their home country. Many migrant workers use communication technologies to contact their friends and families back home. These tools have succeeded in mediating migrant workers' interactions across national boundaries and time. Scholars agree that utilisation of communication technologies by migrant workers has significantly increased their capability to access and share information with fellow migrants and their relatives back home. Not only that, IMDWs also utilise communication technologies for the purpose of migrant rights activism. I identify IMDWs' activities in Hong Kong in relation to migrant rights activism and the utilisation of media and communication technology. The review provides a basis for further research into the discourse of IMDWs, their social conditions in Hong Kong, and their involvement in media activism.

Keywords: Indonesian migrant domestic workers, communication technology, migrant rights

1 INTRODUCTION

Indonesian migrant workers unite, we cannot be beaten! We are workers! We are not slaves! Wage increase now! Now, now, now! (Personal communication, 16 June 2013).

Such slogans are commonly voiced by Indonesian migrant domestic workers (IMDWs) in Hong Kong when they gather and rally in public places, such as Victoria Park and the front of the Indonesian Consulate office at Causeway Bay. The slogans encapsulate the angst of their daily life in the host land. Furthermore, the slogans portray their struggles in resisting the unfortunate conditions of their working environment, such as long hours, working in isolation and vulnerability to workplace abuse. In this paper, I will discuss recent conditions of IMDWs in Hong Kong along with their activities in advocating for their rights by the utilisation of communication technology.

IMWs have been sent to several countries to fill the need for labour, especially in the domestic sector. Hong Kong is one of the countries to have received IMDWs since the early 1990s (Constable, 2007, pp. 30–31). Hong Kong is a major receiving country for migrant workers, particularly women from South East Asia who work as domestic assistants or caregivers. Hong Kong's economy, with its better working conditions and attractive salary, are magnets for migrant workers. It offers an opportunity for an improved financial future and for paid work that many women may not be able to obtain in their home country.

I identify IMDWs' activities in Hong Kong in relation to labour rights activism and the utilisation of media. The context of Hong Kong as a popular destination for migrant workers is also discussed. The review provides a basis for further research into the discourse of IMDWs, their social conditions in Hong Kong, and their involvement in media activism.

2 THE DISCOURSE AROUND INDONESIAN MIGRANT WORKERS

It is projected that, in 2020, the Indonesian population will reach 271 million people (Rastika, 2014). Indonesia's population growth has created significant issues, especially in employment. Unstable national economic growth has made gaining employment very difficult. Opportunities for Indonesians to work abroad with better incomes are viewed as one option for stabilising the economy.

The migrant labour sector has always been important in the Indonesian economy. The Indonesian Government's *Third Five-Year Development Plan* (1979–1984) constructs migrant workers as exports and as one of the major non-oil commodities (p. 39). During the period of the plan, the Indonesian Government came close to meeting its goal to transport 100,000 migrant workers (Hugo, 2005, p. 57). The following years saw an increase in the number of migrant workers sent abroad. During the *Fourth Five-Year Development Plan* (1984–1989), the target increased from 100,000, to 250,000 migrant workers. However, the actual number surpassed this target and reached 292, 262 (Hugo, 2005, p. 57). In 2013 alone, the government sent 512,168 migrant workers (BNP2TKI, 2016). Many IMWs operate in low and semi-skilled areas such as agriculture, manufacturing, construction and domestic work (Hall, 2011, p. 38). Though, financially IMDWs earn more than they would in their home country, several scholars highlight unfavourable experiences such as abuse, extortion, and bondage practice in the area of labour migration in Indonesia (Blackburn, 2004; Hall, 2011; Killias, 2009; Morgan & Nolan, 2011).

IMDWs working in Hong Kong are mostly female. Women and labour migration are intensely discussed by scholars. Women's migrant labour, according to Annette Lansink (2009, p. 129), is the result of growing economic and male employment insecurities, which has made households and individuals rely on women for their survival. In line with Lansink, Danah Boyd (2006) considers that migration is exceedingly gendered, and promotes gender inequality. As stated by Susan Martin (2007, p. 1), "gender inequality can be a powerful factor in precipitating migration, particularly when women have economic, political and social expectations that actual opportunities at home do not meet". Martin further explains that, traditionally, women have migrated to join their husbands or fathers. However, women now migrate without their husbands or family for work. Work performed by women is also gendered, as it involves jobs that are associated with female occupations, which are low paid and are in the domestic sphere, such as cleaning or caring for children (p. 3). Martin notes that "domestic service is a common occupation for migrant women … in almost all parts of the globe" (p. 3). The same applies in Indonesia.

In the context of Indonesia, Olivia Killias investigates the background history of female IMW mobility. She identifies Javanese Indonesians as among those early migrants sent abroad during the Dutch administration.

Female labour export continues to the present day under a system controlled by the Indonesian Government, and recruitment processes are regulated under strict rules. Nevertheless, Killias identifies problems with the migrant worker recruitment process performed by agencies under the Indonesian authorities' control. She considers the contemporary system of migrant labour in Indonesia, especially in the field of domestic service, as similar to bondage practices (2009). She investigates recruitment processes and the moment of pre-departure, which lasts for months and involves different actors, such as local brokers, local authorities and recruitment agencies. According to Killias, the pre-departure process affects migrant workers' situations both before and after departure to the destination country, and this process has "eventually given rise to contemporary forms of bondage in transnational domestic service" (2009, p. 152).

Female IMDWs are facing several challenges in their struggle to earn an income from abroad. As critically discussed by Killias, the problems faced by Indonesian female domestic workers are complex from the early stage, which is the recruitment process. Moreover, Guy Morgan and Chris Nolan explain, "(i)n theory, the Indonesian migrant worker recruitment is robust ... In practice, however, there is significant variability ... that can contribute to less-than-ideal outcomes for workers" (2011, p. 3). The migrant worker recruitment process in Indonesia is based on a formal system that enlists young adults based on job orders from employers in destination countries (p. 3). The legal process of gaining employment begins at the village level. The candidate must have permission from her family and written consent from the village chief, to work abroad (p. 4). The regional authority processes the contract, which is signed by the candidate before undertaking pre-departure training on local customs, language, and the expectations of the host country (p. 3).

Morgan and Nolan note that, during this process, inappropriate practices occur that place migrant worker candidates in a difficult position. They (2011) note several distressing issues: 1) recruitment fees of variable amounts that candidates must pay; 2) contracts that are only signed with the recruitment agency, not the end employer, which lead to contract substitution, pay discrepancies and even fraud at the expense of the worker; and 3) problems with the employer or recruitment agency while performing work (Morgan & Nolan, 2011, pp. 3–4). These inappropriate practices occurred in almost all destination countries.

Moreover, poor practices occur not only in the recruitment process. Sakdapolrak (in Hall, p. 50) explains that IMWs' poor treatment occurs at all stages of the migration process. According to Andy Hall (2011), IMWs are still "at high risk of maltreatment, human rights abuse, and exploitation with inadequate social protection from either origin or destination countries" (p. 50). Low-skill jobs handled by migrant women, especially in the Asian context, are not limited to domestic services. They are also connected with sex-related industries. Nicola Piper (2003, p. 724) argues that these conditions are "reflecting gendered sociocultural conditions of varying nature in the sending and receiving countries". Human Rights Watch has repeatedly documented how IMWs have accepted bad treatment.

Besides experiencing minimum-standard working conditions, IMWs also have limited access to medical treatment and are at risk of being trafficked. Ford, Lyons and Schendel have given attention to the issue of labour migration and trafficking (2012). They discuss the definition of trafficking in relation to the concept of labour migration by comparing different views from scholars and NGOs. The ILO utilises a "forced labour continuum", which identifies three categories of labour: 1) trafficked victims of forced labour, 2) non-trafficked victims of forced labour, and 3) successful migrants (p. 10). Many IMWs, as categorised by the ILO, belong to the first category because they have "least freedom of movement and are the most vulnerable" (p. 10).

The Indonesian Government is aware of these working conditions. In 2012, the government contributed to efforts to provide new and more sustainable protection for IMDWs (Human Rights Watch, 2012). The *1990 International Convention on the Protection of the Rights of All Migrant Workers and Members of Their Families* (the 'Migrant Workers Convention') was adopted by the Indonesian parliament to create more reliable regulations for IMWs. This international treaty guarantees migrants' rights and obliges government protection against employer, agent and public official abuse. The Indonesian Government signed the Migrant Workers Convention in September 2004. However, in practice, there has been no significant action taken by the Indonesian government to implement the treaty (Human Rights Watch, 2012).

3 THE USE OF INFORMATION COMMUNICATION TECHNOLOGY

The discussion about female migrant labour and their activities in their host country cannot be separated from issues of gender. As discussed previously, migration is heavily gendered and most Indonesian domestic workers in Hong Kong are women who need to work hard in the host country without the legal assurance of their citizenship, while maintaining connections and responsibilities to their families back home. Female migrants working as contract domestic workers with a maximum two-year contract are excluded from legal settlement in Hong Kong. They are not allowed to gain Hong Kong citizenship (Constable, 2007, p. 144; Constable, 2014, p. 10; Lai, 2010, p. 503). However, they are able to continually renew their contracts.

Many female migrant domestic workers have renewed their working contracts several times and many have lived in Hong Kong for more than 10 years and spent their time as provisional workers. They have remained in Hong Kong to fulfil both their own financial needs and those of their families back home. During this time, female migrant domestic workers with children must mother from a distance. The practice of 'transnational mothering' (a notion discussed by Bryceson & Vuerola, 2002; Madianou & Miller, 2012; and Yeoh & Huang, 2000) maintains their intimate connections with their family in the homeland. This transnational connection through the spirit of mothering is explained further as a clear concept of the gendered identity of being a submissive daughter, caring mother and obedient wife, often sacrificing oneself for the sake of the family (Yeoh & Huang, 2000, pp. 418–422). However, for some Indonesian female migrants, working abroad could also be perceived as a way to reduce social, economic, cultural and political pressure from their family back home.

However, the sense of self-sacrifice and mothering felt by female migrant workers is central; for migrant women, "continued links with the homeland are vital and necessary—to varying degrees for different groups of women—to a sense of self and negotiation of identities" (Huang & Yeoh, 2000, p. 394). Female migrant workers, in the deepest sense, still need to retain bonds with their families back home.

The recent method employed by migrant workers to retain a connection with their families is the use of communication technologies. Scholars agree that utilisation of ICT by migrant workers has significantly increased their capability to access and share information with fellow migrants and their relatives back home (Bunmak, 2012; Madianou & Miller, 2012; Thomas & Lim, 2011). Migrant workers have a range of options mediating their communication access, including the usage of transnational communication tools such as mobile phones (Thomas & Lim, 2011). These tools have succeeded in mediating migrant workers' interactions across national boundaries.

In the context of transnational family relationships, Mirca Madianou and Daniel Miller emphasise that the use of communication technologies is constantly changing (2012, p. 7). Before the common use of ICTs in the 21st Century, migrant workers sent letters and cassettes with recorded personal messages to communicate with their family. Nowadays this is done via phone calls through mobile phones; text messages, emails, web chats, and social media are also common channels for transnational communications. Progressively, there is a shift towards the use of communication technology as the medium for transnational family relationships.

A deeper analysis of communication technologies used by migrants is provided by Raelene Wilding (2006). Her study on migrants' activities in seven countries found that different types of media had various effects on the relational styles developed by migrants (2006). For instance, email is used to improve the quality of communication with the family, whereas cheap calls through mobile phones are used to increase the quantity of family connections. Mobile phones have also helped with the activities of transnational mothering, by allowing mothers to remotely assert their role in the distant family. Paradoxically, ICTs are beneficial not only in managing family harmony, but also increase the feeling of isolation, as "the intimacy of long-distance contact made the lack of face-to-face contact even more palpable" (Thomas & Lim, 2011).

Female migrant workers are in a difficult position in terms of connections with their homeland. Furthermore, the concept of 'family' can always be redeveloped in the context of migration. Female domestic workers, who have left their families to work abroad, struggle to develop a sense of 'family' in the host country. The distance that separates female domestic workers from their children, husband, parents and other family members initially forces them to regain a sense of family warmth, trust and dependency by building intense friendships with other domestic workers. Female domestic workers in the host country negotiate their identities as individuals, as workers, or even as members of the Indonesian community abroad, by developing connections with fellow migrant workers and joining migrant organisations.

Female domestic workers are routinely secluded in their employers' households, and it can be a challenge for them to negotiate their collective identity. Female domestic workers manage their limited days off to join organised activities, channelling their sense of belonging and solidarity, applied in the forms of migrant activism agendas such as mobilisation in street protests, or the weekly migrant domestic worker meetings in Victoria Park. The activist discourse engaged in by Indonesian female migrant workers in Hong Kong may be regarded, according to Lai (2011), as a way of "forgetting, resisting, and breaking from sanctioned feminine sacrifices to the well-being of the family back in the homeland" (p. 575). The existence of several organisations in Hong Kong may be a conduit for migrant workers' self-expression of diasporic identity.

4 MEDIA ACTIVISM AND MIGRANT WORKERS

The use of communication technology and the media by IMDWs to communicate and discuss campaigning issues on migrant rights and employment has provided important insights in the study of migrant and media activism. Before continuing the discussion on migrant worker media activism, I will begin with a discussion on the concept of mass media and new media.

Mainstream conventional media such as newspapers, television and radio are centralised and require high maintenance in the form of resources and investments (Lister, Dovey, Giddings, Grant, & Kelly, 2003, p. 12), whereas new media such as the Internet can be used both as point-to-point and point-to-mass message distribution with low investment.

Lister et al. consider interactivity as one of the main functions of new media, whereas conventional media can only offer passive consumption of products (2003, p. 20). The audience of new media is able to directly alter the contents (images and texts), interact with the authors of the contents or simply make comments on the discussion board. This is what is practised on social media platforms on the Internet, where users are encouraged to share personal and public information; for example, through social media with interaction via posts and comments.

For Henry Jenkins (2006), the rise of the participatory culture cannot be separated from the role of social media, which empowers users to produce their own content. As suggested by Daniel Trottier and Christian Fuchs (2014), we should be careful about theorising social media because, as a term, social media is complex. They argue that "all computing systems and therefore all web applications, and also all forms of media can be considered as social because they store and transmit human knowledge that originates in social relations in society". However, not all computing systems and web applications encourage direct communication; for example, Amazon is a tool for information on books and goods (Trottier & Fuchs, 2014, p. 5). Therefore, it is not simple to define social media. The important thing is, as Trottier and Fuchs assert, that social media is used to "support cognition, communication/networking and cooperation (communities, collaborative work, sharing of user generated and other content)" (2014, p. 7).

Furthermore, Sam Hinton and Larissa Hjorth (2013) in their book *Understanding social media*, emphasise social media use is never only about online activities; it contains offline modes of engagement. They emphasise that "the relationships that people have online are always shaping, and shaped by, the offline" (2013, p. 3). How we make friends and behave online, as Hinton and Hjorth suggest, is motivated by our offline lives (2013, p. 3). Hinton and Hjorth also focus on the issue of control and freedom (empowerment) in social media, which is not always empowering and not always controlling (p. 30).

The popularity of Facebook and Twitter has been developed by its users as tools to publish personal and organisational profiles. Several IMW unions in Hong Kong, including the IMWU and FLP, have also used social media to communicate issues and activism agendas to their members. It is important to note that IMWs in Hong Kong use both conventional and new media as tools for media activism.

Media activism, according to William Caroll and Robert Hackett (2006), involves "organised 'grassroots' efforts directed to creating or influencing media practices and strategies, whether as a primary objective, or as a by-product of other campaigns" (p. 84). Furthermore, James Gillett (2003) points out that media can be used to create spaces for marginalised or stigmatised identities, and that media activism "can be understood as a response to the failure of social institutions, particularly the mass media, to provide a forum for citizens to address problems and issues of common concern" (p. 610). Media activism, in this case, is advocacy for the community in a public sphere.

Contemporary media activism involves ICT. Pickerill points out that ICT is important in the movement of media activism as it provides "new spaces for social interaction free from the hierarchical and bureaucratic pressures of existing society" (2004, p. 172). Jenny Pickerill promotes understanding of ICTs as 'democratic properties' that could facilitate public concern in the political process without concern for the traditional government hierarchy (p. 172). Pickerill also emphasises that the adoption and utilisation of ICT by a group or a network reflects their existing organisational forms and identities. She notes three important constraints to the adoption and utilisation process by media activists: 1) financial difficulties, 2) differing skill attainment and 3) hardware and software problems (p. 172). These constraints should be handled properly, through planning and detailed organisation to maximise the functioning of media activism.

Regarding the use of ICT in media activism, Pickerill (2004) says that "a well-managed ICT can improve response times, aid the gathering (and the flow) of information" (p. 176) and "aid networking with other activist groups … and generate campaigns" (p. 183).

5 CONCLUSION

I have described and explained several key theories on the recent conditions of IMDWs. Key points have arisen from the literature review and discussion. Women's migrant labour is the result of growing economic and male employment insecurities, forcing households to rely to women to survive. Migration is exceedingly gendered, creates the opportunity for gender inequality and is open to exploitation.

Regarding Hong Kong's condition as a receiving country for migrant workers, Ignacio and Mejia (2009) consider that the country provides equal statutory labour rights, where IMDWs are supported to assemble and create organisations. Nevertheless, IMDWs may face termination by employers if they join strikes (Smales, 2010). Migrant organisation development is encouraged in Hong Kong. As advised by Briones (2009), migrant worker organisations are positioned as protectors of migrant workers and aim to raise public awareness and influence government policy regarding labour rules.

On the matter of migrant identity, the above discussion has given special attention to the concept of transnational mothering. The sense of self-sacrifice with respect to mothering felt by female migrant domestic workers in the context of diasporic identities is central, because for migrant women, continued links with the homeland are vital and necessary for a sense of self and negotiation of identity. To maintain a connection with their homeland, migrant workers use ICTs.

In interpreting media activism, Cammaerts et al. (2013) emphasise the importance of direct action but do not underestimate the power of passive participation through media technologies. Gillett (2003) points out that mass media can be used to create spaces for marginalised or stigmatised identities. Meanwhile, contemporary media activism involves new media that utilise ICTs. Pickerill (2004) emphasises that ICTs are important in the movement of media activism because they provide spaces for social interaction that are free from the pressures of society.

REFERENCES

Constable, N. (2007). *Maid to order in Hong Kong: Stories of migrant workers.* Ithaca, NY: Cornell University Press.

Constable, N. (2014). *Born out of place: migrant mothers and the politics of international labor*. London, England: University of California Press, Ltd.

Ford, M., Lyons, L., & van Schendels (Eds.). (2012). *Labour migration and human trafficking in Southeast Asia: Critical perspectives*. London, England: Routledge.

Fuchs, C. (2014). *Social media: A critical introduction.* Thousand Oaks, CA: Sage.

Indonesian Law Number 39 Year 2004 Concerning Indonesian Migrant Worker Placement and Protection Abroad.

Indonesian President Regulation Number 81 Year 2006 on the National Body of Indonesian Migrant Worker Placement and Protection (BNP2TKI).

Information Services Department. (2012). Hong Kong: The facts. Hong Kong government. Retrieved from http://www.gov.hk/en/about/abouthk/factsheets/docs/media.pdf.

ILO. (2011). C189—Domestic Workers Convention, 2011 (No. 189) Convention concerning decent work for domestic workers. *International Labour Organization.* Retrieved from http://www.ilo.org/dyn/normlex/en/f?p=NORMLEXPUB:12100:0::NO::P12100_INSTRUMENT_ID:2551460.

ILO. (2012, September 5). Landmark treaty for domestic workers to come into force. *International Labour Organization.* Retrieved from http://www.ilo.org/global/about-the-ilo/newsroom/news/WCMS_189191/lang--en/index.htm?utm_source=feedburner&utm_medium=feed&utm_campaign=Feed%253 A+ungen+(UN+gender+equality+news+feed).

ILO. (n.d). Ratifications for Indonesia. *International Labour Organization.* Retrieved from http://www.ilo.org/dyn/normlex/en/f?p=1000:11200:0::NO:11200:P11200_COUNTRY_ID:102938.

Jenkins, H. (2006). *Convergence culture: Where old and new media collide*. New York, NY: New York University Press.

Lai, M. (2010). Dancing to different tunes: Performance and activism among migrant domestic workers in Hong Kong. *Woman Studies International Forum, 33*, 501–511. http://dx.doi.org/10.1016/j.wsif.2010.07.003.

Madianou, M. & Miller, D. (2012). Migration and new media: Transnational families and polymedia. London, England: Routledge.

Thomas, M. & Lim, S.S. (2011). Migrant workers' use of ICTs for interpersonal communication: The experience of female domestic workers in Singapore. *Media Anthropology*. Retrieved from http://media-anthropology.net/thomas_lim_migrant_workers_ICT.pdf.

Wilding, R. (2006). 'Virtual' intimacies? Families communicating across transnational contexts. *Global Networks, 6*(2), 125–142. http://dx.doi.org/10.1111/j.1471-0374.2006.00137.x.

Yeoh, B., & Huang, S. (2000). 'Home' and 'away': Foreign domestic workers and negotiations of diasporic identity in Singapore. *Women's Studies International Forum 23*(4): 413–429. http://dx.doi.org/10.1016/S0277-5395(00)00105-9.

Urban Studies: Border and Mobility – Kerr et al. (Eds)

Interagency collaborative team in broadcasting management at the border area of Sintang Regency, Indonesia

L.H. Kurnia, I.W. Midhio & T.B. Prasetyo
Indonesia Defense University, Bogor, West Java, Indonesia

ABSTRACT: Sintang is one of the five border regencies in West Kalimantan. Its remote location makes this area to have limited access to information and communication. The Ministry of Communication and Informatics of Indonesia as the leading sector in the development of access to information and communication has launched a program to solve this problem. However, the implementation of this program has not achieved the expected results due to weak coordination and program planning. Therefore, this study was conducted to formulate an integrated team and a special coordination channel in terms of broadcasting management on the border area. In this study, we use qualitative approach with field observation, in-depth interview, and literature study to collect the research data. The result of this study formulated an Interagency Collaborative Team led by the Directorate General of PPI Kemenkominfo, which consists of broadcasting stakeholders.

Keywords: information, communication, Sintang, broadcasting management

1 INTRODUCTION

Sintang is one of the five border districts in West Kalimantan. The geographical structure of Sintang Regency, especially in the border area, includes hills and is dominated by production and protected forests (BPP Sintang, 2015: 2). This geographical structure is one of the obstacles to develop Sintang's border area in the fields of access to transportation, health, education, and information communication. Therefore, National Agency of Border Management (BNPP) categorized Central Ketungau and Ketungau Hulu Sub-District in the development of priority locations for two periods, 2010–2014 and 2015–2019 (bnpp.go.id, 2016).

From the perspective of strengthening the values of Indonesia and improving the national identity of the community on the border of Sintang Regency, the presence of Public Service Broadcasting (LPP) embedded with vision and mission is crucial. LPP TVRI and RRI as Public Service Broadcasting centers have special principles and functions in disseminating Indonesian values, reflecting diversity and national identity, being a flag carrier, unifying the nation, and providing access to all citizens (rri.co.id, 2016). These functions are in line with the statement of McQuail et al. (1992: 9) about Public Service Broadcasting that was designed to "serve the audience and social institutions within the national territory, center-peripheral in form of organization, expected to protect national language and culture and (however implicitly) to represent the national interest. As an aspect of their national character, broadcasting institutions were also usually monopolistic or quasi-monopolistic in their form of control". In terms of establishment and maintenance of such a national identity, Public Service Broadcasting plays a vital role. Public Service Broadcasting Television has functions to contribute, formulate, and spread national culture in national integration, as well as to make the country a united nation, "making the nation as one man" (Reith, cited in Scannelll, 1990:23, cited in Bulck and Poecke, 1996: 224).

The presence of LPP in Sintang Regency is represented by RRI Sintang that has been built since 2002 and relay station of TVRI. Unfortunately, the transmitter signal of the TVRI or

RRI Sintang tower has difficulty reaching all Sintang border societies due to obstruction by hilly land contours. To overcome this, the Ministry of Communication and Informatics of Indonesia (Kemenkominfo) in 2011 through Minister Instruction No. 01/INST/M. KOMINFO/03/2011 initiated a community radio program to reach areas that have difficulties in accessing information and network of TVRI and RRI. In its implementation, Kemenkominfo through Universal Service Obligation (KPU/USO) program designed nine programs to improve access to information and communication in the border area.

However, these programs encountered several problems and constraints so that their main purposes have not been achieved. On the basis of previous research, Kurnia (2018) describes the complexity of broadcasting management problem in Sintang Regency, West Kalimantan. This study divides the factors hampering the development of information access through broadcast media into five key result areas that must be addressed for the achievement of information dissemination in the border region. These key result areas are planning, coordinating, implementation and monitoring programs, maximizing of broadcaster agencies, and regulation. On the basis of these key result areas, Kemenkominfo's program determines the success of the implementation of the program in border areas. This study recommends setting up a special team and coordination line consisting of broadcasting stakeholders. Therefore, this paper aims to further elaborate this recommendation by using the interagency collaborative team concept on broadcasting management in Sintang Regency.

2 METHODOLOGY AND ANALYSIS TOOL

In this study, we use qualitative research method by field observation, in-depth interview, and literature study to collect the research data. To analyze and process the results of field data, we use the concept of integrated collaborative team to analyze and provide input in the management of broadcasting on the border so that dissemination of information can be accessed by all border communities, especially in Sintang Regency, West Kalimantan.

The concept of integrated collaborative team

> "The ICT model is designed to enable organizations to work together in ways that generate the structural and process supports associated with successful implementation and sustainment of innovations" (Hurlburt, et.al, 2014: 161). Interagency collaborative team (ICT) is a model designed to formulate multiple organizations or institutions to work together in a structural system that can support successful implementation and sustainability of the program. On the basis of the results of field research, it is necessary to have a team that integrates and collaborates stakeholders in the management of broadcasting on the border area. The formation of this team is aimed to make stakeholders, who previously did not coordinate well and talk, work together in accordance with the authority and responsibility. The main barriers to successful interagency collaboration are cultural barriers, procedural/regulatory barriers, and personal barriers (Barnett, 1995: 74). Therefore, in order to achieve successful interagency collaboration, a special approach is needed, such as interagency collaborative team, as a means to communicate the strategic planning and collective visioning.

3 DISCUSSION

As will be discussed in this paper, broadcasters include television and radio, which directly touch the border point of Sintang Regency and have a goal to channel the information needed by local residents as well as to distribute the Indonesian cultural values and national identity of Indonesia. These broadcasters are LPP RRI Sintang, Community Radio, which operated in Senaning-Ketungau Hulu, and LPP TVRI Pontianak Station.

3.1 *RRI Sintang*

RRI Sintang, located in Oevang Oeray Road, Baning, Sintang, Indonesia, was established in 2002 with the aim to occupy the blank spot area that is not covered by RRI Central broadcast. RRI Sintang is the first RRI established on the border of West Kalimantan, even long before RRI Entikong. Sintang is chosen because it is considered as a strategic location in the middle of West Kalimantan and directly adjacent to Sarawak, Malaysia. RRI Sintang is expected to foster a sense of nationality of the people in the border area by broadcasting data about Indonesia's programs, information, and development (Ngatno, Head of RRI Sintang, 2016). RRI Sintang's coverage is highly dependent on the size and power of the transmitter. The constraint currently facing RRI Sintang is strengthening of the RRI transmitter, which takes two times greater electric power than the power of the transmitter. On the basis of interviews with Hcad of RRI Sintang in February 2016, there has been no coordination with Kemenkominfo for the development and strengthening of RRI Sintang. The distance from Sintang to the border of Ketungau Hulu and Ketungau Tengah is an obstacle for RRI Sintang reporters to be able to present at the border. In addition to long distances, the risks are also a constraint.

3.2 *TVRI station of Pontianak*

TVRI Station of Pontianak is a public broadcasting institution serving in the West Kalimantan area that covers the inland areas and border of Kalimantan. Currently, TVRI Station of Pontianak broadcasts terrestrially with relay transmitter station in more than 19 locations in West Kalimantan. Until 2014, TVRI Pontianak Station could reach the border area only through satellite broadcasting. However, due to corruption cases conducted by West Kalimantan provincial officials related to the leasing of satellite transponders to expand the coverage of TVRI (Antaranews, 2014), TVRI Pontianak Station no longer receives the help of the West Kalimantan Provincial Government and can only broadcast terrestrially in the city of Pontianak with the duration of broadcast of only 4 hours per day. The cessation of satellite TVRI broadcasts in West Kalimantan has led to many complaints by people across West Kalimantan. M. Arif (36), for example, from Singkawang City, who could not watch TVRI special broadcast about events in local Kalbar, said "Yes, it is also sad usually we could broadcast local news of West Kalimantan through TVRI. But we couldn't broadcast it anymore" (Pontianak Pos, 2014).

3.3 *Community radio*

Community radio empowered by Kemenkominfo is located in a remote village, especially on the border, with the aim of providing an informative and entertaining event to the local community. The development of community radio is based on the instruction of the Minister of Kominfo (No. 01/INST/M.KOMINFO/03/2011) on the implementation of information village in the border region of Indonesia. At the time of the study, the number of community radios that were running was 76, in every area of information village programs throughout Indonesia. Community radio is managed by the Community Information Group (KIM) established by the Kemenkominfo through the Directorate General of IKP (Information and Public Communication), which consists of local community. Community radio (Rakom) is a community broadcasting institution that became one of the Kemenkominfo programs to open access to information for people in the border area. In Law No. 32 of 2002 on Broadcasting, in section 6 of Article 21, it is stated that the community broadcasting institution is independent and non-commercial, with low transmit power and limited area coverage, in order to serve the interests of the community. The purpose of the establishment of this community radio is to educate and promote the community to achieve prosperity, by implementing an event program that includes culture, education, and information that describes the identity of the nation.

Kurnia (2018) stated the following main issues in managing Rakom and Border Community Information Group (Kimtas) in Sintang Regency: (1) Inadequate human resources and infrastructure (Kimtas manager is given equipment by the center but is not considered an operational support tool for the running of the program). (2) Managers receive tools for broadcasting at the sub-district office or certain locations according to the assignment, but after that, there is no further monitoring and sustainability by the central government. HR does not get further training and direction if the tools are damaged. (3) The handover of the tool is done directly by third party to the community without coordination with the provincial or local government. There is no grant process or delivery of aid assets to local governments, so there is no allocation of funds for the maintenance of these assets. (4) Managers refuse to continue, manage, and maintain assets, because there is no cost of sustainability support to pay for electricity and diesel. The district government cannot provide assistance for that matter, since the asset status is not the property of the region. (5) The difficulty of Rakom sustainability in the border area is also due to Rakom managers who are not getting compensation or salary. (6) An operator or Rakom manager is not a person who has experience in the field of broadcasting, so Rakom goes without the procedures and broadcasting agenda. The one-year-old Rakom program in Sintang, and in some other border areas, indicates the desire of the community to manage and develop community radio. However, the government that initiated the existence of Rakom did not maximally organize the program. In the second year, Rakom equipment has been damaged, and the sub-district lacks the funding and ability to repair the tools. In addition to the constraints of equipment resistance, human resources became an obstacle in managing Community Radio (Kurnia, 2018).

On the basis of the problems that have been submitted, it is observed that the main problem of sustainability of Rakom program is the lack careful planning by central government for program maintenance and sustainability; lack of coordination and synergy between central government, local government, and other stakeholders related to the management of Rakom; and the limited space for the management of Rakom in obtaining financial resources for the sustainability of Rakom. In addition to the lack of careful planning, these problems arise because of the absence of a clear SOP to the local government and related authorities.

3.4 *Interagency collaborative team*

Therefore, on the basis of the issues presented in the previous section, it is necessary to establish a special team in terms of broadcasting management on the border area, which aims to meet the needs of information access for border communities and the distribution of Indonesian values in order to strengthen the national identity of communities in the border area of Indonesia. The team is structured on the concept of interagency collaborative team. The ICT team formulated in this study was led by the Directorate General of PPI Kemenkominfo, who oversees the Heads of Provincial Communications and Informatics. Directorate General of PPI Kemenkominfo is responsible for ensuring assistance in the regions by monitoring and coordinating with third parties as providers of procurement services, provincial government, district government, broadcasting agencies, and sub-district heads whose territory is the locus of aid programs.

DG PPI Kemenkominfo's duty is to ensure that the program runs synergistically and communication and good cooperation between the provincial government and third-party service providers are established. The third party is responsible for the implementation of the program to the Directorate General of PPI and financing the Directorate General of BP3TI Kominfo. A third party or service provider is working with the provincial government to provide assistance and maintenance programs and improvements. The provincial government in this position has a supervisory function of the implementation of the program responsible for reporting it to the Directorate General of PPI Kemenkominfo, who should have direct responsibility for monitoring and implementing the program, not as long as it is immediately released to third parties and resulted in the number of programs that are not continuous and inappropriate target.

The provincial government facilitates the need and reviews the implementation of broadcasting management program from KPID, LPP TVRI, and local government. Local gov-

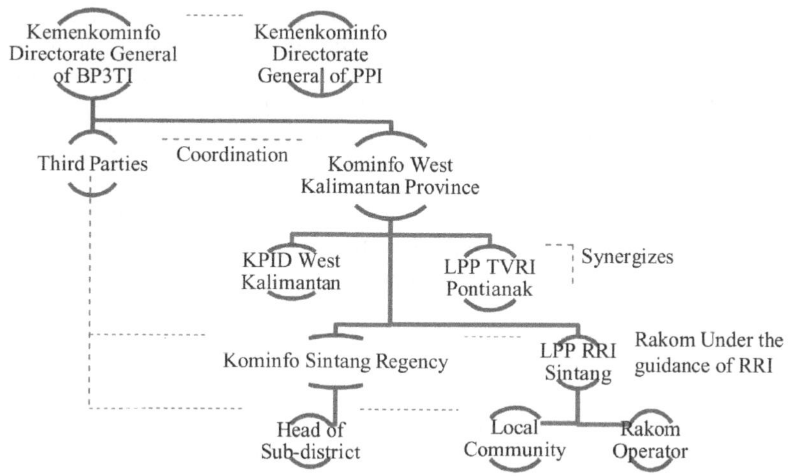

Figure 1. Interagency collaborative team chart of border broadcasting management in Sintang Regency. Source: Processed by researcher, 2018.

ernment through Head of Kominfo's Division in Sintang Regency facilitates the need and reviews the implementation of the central program for RRI Sintang and Community Radio. This team can coordinate remotely, as long as each institution performs its duties and responsibilities. Coordination meetings also need to be conducted periodically to review stakeholder needs, problems encountered, and comprehensive program evaluation. KPID Kalbar as an institution that oversees and monitors the development of broadcasting on the border of West Kalimantan can play a role in providing rewards to broadcasting institutions that have contributed to help disseminate and open access to information in the border region. On May 2015, Kemenkominfo, TVRI Kalbar, RRI Sintang, and West Kalimantan Governor signed an MoU for the development and strengthening of LPP TVRI and RRI in West Kalimantan. However, until this research was conducted, there has been no MoU sustainability or coordination on what steps or programs will be implemented. This is an opportunity that should be used by RRI and TVRI to strengthen its presence in the border region. Through this ICT border broadcaster, RRI and TVRI can question the forum during a coordination meeting on the sustainability of a memorandum of understanding signed in 2015.

RRI Sintang aimed to foster the implementation of Community Radio program in border sub-district. Coaching is within the scope of broadcasting training, journalistic training, tool maintenance, and other things that support the running of Community Radio broadcasting programs. The Community Radio operator also plays a citizen journalist for RRI Sintang and TVRI Pontianak, which reports on the conditions and information surrounding the border sub-district. With the concept of citizen journalist, Rakom managers can get more intensive input from RRI Sintang or TVRI Pontianak. In addition, the broadcasting location of Community Radio installed transmitter RRI Sintang relay. Thus, if Rakom operators cannot broadcast, the Rakom broadcast program will continue to be filled by RRI Sintang broadcast. Sintang RRI program can be enjoyed by the border community that has not been covered by broadcasting RRI Sintang. Rape managers can also form a community and invite local residents to alternately broadcast and can be a container of information and communication for local residents.

As for the strengthening of TVRI, it requires an increase in competitiveness or competitive advantage in the competition with private broadcasting institutions, which are more popular and more accessible to the border community. Through this team, TVRI Pontianak Station can convey its need to be able to broadcast back satellite in West Kalimantan region. TVRI Pontianak Station faces the same problem as RRI Sintang, that is, the lack of human resources as a reporter and delivery channel of news related to border area. Through this team, TVRI Pontianak can synergize with RRI Sintang in terms of information flow and actual data on the border. News from areas collected by RRI Sintang can also be distributed by TVRI Pontianak Station (or vice versa). This will further disseminate information about the border area. As a fellow public broadcasting institution in the border area, communication and good synergy between RRI Sintang and TVRI Pontianak should be established. News and information gathered by RRI Sintang and Community Radio can be news material to be informed and broadcast by TVRI Pontianak located at provincial level, and vice versa. Border broadcasting management strategy through the establishment of ICT is present to answer the problems of coordination and synergy between institutions, lack of human resources RRI Sintang, lack of capacity Community Radio managers in terms of broadcasting, weakness of transmit RRI and TVRI, and weak supervision in the implementation of the program.

4 CONCLUSION

In order to overcome the problems encountered in the implementation of managing broadcasting and to broaden access to information in border areas, it is necessary to set up a special team and coordination line consisting of broadcasting stakeholders. This team integrates the interests and responsibilities of stakeholders in meeting the need for information access through media broadcast. It is achieved by using the concept approach of interagency collaborative team, which is led by Directorate General of PPI Kemenkominfo with members of related institutions such as DG BP3TI, third party, KPID, Diskominfo West Kalimantan Province, TVRI Pontianak Station, Diskominfo Sintang Regency, RRI Sintang, Rakom management, and local communities. The Directorate General of PPI Kemenkominfo coordinating with the Directorate General of BP3TI Kemenkominfo is in procurement with the third party. Asset relief from Kemenkominfo was granted and became a regional asset. The local government plays an important role in coordination and serves as a bridge between the nationals and the community. Community Radio is under the guidance of RRI Sintang. Its management receives journalistic coaching and becomes citizen journalist for RRI Sintang & TVRI Pontianak, as well as invites local community to play an active role in dissemination of information in the border area of Sintang.

REFERENCES

Badan Pengelola Perbatasan Kabupaten Sintang. (2015). *Profil Kawasan Perbatasan Kabupaten Sintang 2015*. Sintang: Author.

Badan Pusat Statistik Kabupaten Sintang. (2015). *Kabupaten Sintang dalam Angka 2015*. Sintang: Author.

Bennet, B. G. (1995). Visioning for the Future: What Can Educational Leaders Do to Build a Shared Commitment to Interagency Collaboration? *Journal of School Leadership*, 69–86.

Bulck, H. d., & Poecke, L. V. (1996). National Language, Identity Formation and Broadcasting, Flanders, the Netherlands and German-speaking Switzerland. *Wuropean Journal of Communication*, 217–233.

Hurlburt, M., Aarons, G. A., Fetter, D. W., Gunderson, L., & Chaffin, M. J. (2014). Interagency Collaborative Team Model for Capacity Building to Scale-Up Evidence-Based Practice. *Children and Youth Services Review*, 160–168.

Kurnia, L. H. (2018). Building national identity in the border areas: The critical success factors analysis in the management of community radio in Sintang, West Kalimantan. *The Asia-Pacific Research in Social Sciences and Humanities* (pp. 239–246). Depok: Universitas Indonesia.

Law No. 32 of 2002 on Broadcasting. (2002).

McQuail, D. (1992). *Media Performance, Mass Communication and the Public Interest*. London: Sage.

Minister Instruction No. 01/INST/M.KOMINFO/03/2011. (2011).

Morley, D., & Robins, K. (1989). Spaces of Identity: Communications Technologies and the Reconfiguration of Europe. *Screen*, 10–34.

Pejabat Pemprov Kalbar Ditahan Karena Dugaan Korupsi. (2014, July 11). Retrieved from Antaranews: http://kalbar.antaranews.com/berita/324448/pejabat-pemprov-kalbar-ditahan-karena-dugaan-korupsi.

Pontianak Pos. (2014, April 12). Polisi-Jaksa Bidik Proyek Satelit TVRI.

Urban Studies: Border and Mobility – Kerr et al. (Eds)
© 2019 Taylor & Francis Group, London, ISBN 978-1-138-58034-3

Border broadcasts and national identity representation in Entikong, West Kalimantan

L. Tjahjandari & T.I. Setyani
University of Indonesia, Depok, Indonesia

L.H. Kurnia
Indonesia Defense University, Bogor, Indonesia

ABSTRACT: This study focuses on the border broadcasting and national identity in Entikong, West Kalimantan. This study aims to describe the presence of national broadcast in border area compared to Malaysia's broadcast and to analyze national identity construction related to the issues. This study uses qualitative research methods by using descriptive and social analysis. The results of this study are 1) the majority of border communities can access & fond of national broadcast, only small percentage can access and fond of Malaysia's broadcast, 2) the presence of national broadcasts brings positive impact and benefit, especially in terms of knowledge, development, and a sense of pride to Indonesia, 3) broadcasts play a role in improving the sense of nationality and the use of Indonesian language, 4) Sense of nationality appears in the border community's hope and efforts to be able to access national broadcasts amidst all the limitations they experience.

1 INTRODUCTION

Border area is a significant area in the geostrategic and geopolitical map of Indonesia. However, the people in border area do not fully experience the benefits of development, when compared to the other areas, even though border area is essentially the face of the country. Implication of this problem is shown by the limited access of information to the national broadcast. The limited access is in contrast with the role of national broadcast in spreading Indonesian value. These values are embodied in the identity and attitude of society as part of a great nation. In the nationality narrative, identity is an important element in the formation of sense of unity as a nation. In the border area, border people have unique characteristic when compared to other Indonesian people in general. Border people, such as in West Kalimantan, have the same cultural identity background with the people in neighbouring country. This affected their dynamic daily life, at one hand they have the same cultural identity, but on the other hand they have difference national identity separated by the political borders between two countries.

Identity is a study that will always be actual to be discussed. Identity examines to see who we are and our sense of belonging. Identity construction is essentially a matter of defining a boundary or a difference and a category system. Humans and things do not necessarily gain an identity, but they get that identity through its relationship with other humans and the environment (Morley and Robins, 1989: 12, in Bulck and Poecke, 1996: 218). 'Difference is constitutive of identity' as presented by Fishman (1972: 52–53), identity is the result of self-identification: us and them. Language plays an important role in forming national identity. Many experts convey the importance of language in the process of forming a sense of nationality. Language becomes a symbol, which partially reflects the fact that language is the carrier of ideas, identities or national symbols. In other words, language is often a manifestation of ethnicity and the means to distinguish 'us' and 'them' (Fishman, 1977). Bachofer (2014) gave an example, when Martin Luther translated the Bible into German, it spread to several countries in Europe.

The Bible is the media to spread German language, which represented the German identity. Through this spreading, the people as receiver received message that they also have the same language, that proved and showed they are one community. This language influence in mass media and receiver interpretation in forming identity process is exemplified in that case. Similarly, in the broadcast in border area, broadcast pattern being consumed by the people will also influence the forming of collective identity. This collective identity is the one that will affect in the forming of individual and community identity. Anderson (1983) argues that the distinction between nations lies in the difference of perspective in imagining oneself and nation. Imagined communities is an idea conveyed by Benedict Anderson about a community that consists of different kinds of ethnic groups, religions, cultures, who never even meet or know each other, but imagine themselves belonging into a community called nation.

Broadcasting also plays an important role in the construction of nationalism in various countries. For instance, after the independence of Ireland, broadcast also played an important role as a flag carrier and bearer of cultural nationalism, through a radio namely, Radio Éirann (Barbrook, 1992: 205). The state-owned radio broadcasts news about politics and independence of Ireland. These programs are intended to attract the listeners by using old language to preserve the traditional rural culture using the latest electronic mass media (Mulryan 1988: 8; Ó Tuathaigh 1984: 102–3). Althusser also stated that there is a correlation between mass media and ideology. Mass media are utilized to address the society by placing individuals in particular social role and relation. This is contained and integrated in the whole process of ideology infiltration. Moreover, mass media, especially radio and television, should become an effective-efficient instrument to distribute and penetrate the values and dominant discourse in the people's mind, so that it will become political consensus. It shows the essential role of mass media to greet, treat, influence, and form the consensus, especially for the people located outside of information centre, in this case the people in the border area.

Based on this matter, it can be seen how strong the relationship between the presence of mass media and national identity construction in a society. Broadcast as a mass media is a kind of media that is more easily absorbed by all circles of society, due to its audio and audio-visual nature. The presence of broadcasts in the border region plays a vital role as flag carrier, information media and strengthening the values of Indonesian-ness for border communities. Therefore, this study entitled "Border Broadcasts and National Identity Construction in Entikong, West Kalimantan" was conducted to describe the presence of national broadcast in border area compared to Malaysia's and to analyze national identity construction related to the issues. Entikong Sub-district was chosen as a place of study due to its unique location, because it is directly adjacent to Malaysia which has same cultural background and one of the most densely border crossing points in Indonesia. Entikong Sub-district has long been a center of traffic flow between the two countries. Beginning with the operation of the Cross Boundary Observation Post (PPLB) in 1989, making Entikong Sub-district a more advanced border area than other border areas (Siburian, 2002: 88).

2 RESEARCH METHOD

This study uses qualitative research methods by using descriptive & social analysis. The data were obtained from field observation, in-depth interview, literature review, Focus Group Discussion (FGD), and questionnaire as additional instrument. The informants for this study are head of RRI Entikong, broadcasters of RRI Entikong, Entikong's sub-district head, & people of Entikong Sub-district. As for FGD, the attendees came from various backgrounds & have the capacity to discuss broadcasting in Entikong. As for seeing the broadcast consumption patterns, this study used a research instrument in the form of questionnaires distributed to a number of 144 respondents at SMK Yayasan Lintas Batas & SMKN 1 Entikong. The composition of respondents is 99% of vocational students & 1% of the general population. Respondent selection is based on the assumption that high school students represent the perspectives of young generation of border communities in viewing the present broadcasts and represent the community from various of villages in Entikong.

3 DISCUSSION

3.1 *Broadcasting overview in Entikong, Sanggau Regency*

Based on data obtained from West Kalimantan KPID in 2014, it is generally stated that in the area of West Kalimantan there is a blankspot area whose people cannot access national, local radio and television broadcasts. Blankspots are for national television broadcast amounted to 66% and 58% for radio broadcast area in West Kalimantan. The high number of blankspot broadcast area in West Kalimantan is in line with the results of research and field survey conducted by Monitoring Center of Communication and Information Ministry in 2011, regarding the overflowing of Malaysian television and radio broadcasting into the three border areas, namely Sajingan (Sambas Regency), Sanggau Ledo and Entikong (Sanggau Regency) which can be seen in Table 1.

People in Entikong Sub-district generally already have a satellite dish, so they can access the national television broadcasts. Previously, people only used UHF antenna, which only received Malaysian television and TVRI broadcasts. The transition from UHF antennas to satellite dishes occurred since the early 2000s. In addition to national television broadcasts, the public also get access to local television broadcasts, such as TVRI Pontianak, Ruai TV, and Sanggau TV. As for self-paid cable television, there are already incoming service providers, namely Sonia Vision and Astro owned by Malaysia. TVRI Pontianak Station was also broadcasted through satellite and could be enjoyed by the people in border area. However, in the year 2014 TVRI Pontianak Station stopped broadcasting through satellite due to the findings by the BPK. Since then, TVRI Pontianak Station broadcast only in the area of Pontianak. On the contrary, people in Entikong can actually access TVRI Timor Leste and TVRI Bali, because it has been broadcast using satellite. People in Entikong Sub-district also expressed hope for transmitters tower in the border area, so that they could watch television without the use of satellite dish and could watch live events on television, especially national sports such as football, badminton and boxing. They generally use satellite dish to receive national television broadcasts. Related to broadcast content, the people hope that the problems and portraits of the border areas can be discussed in many national broadcasts, whether newscastings or other television programs, such as entertainment and film.

As for radio broadcast, Sanggau Regency has RRI Sanggau which broadcasts at the district/regency level. RRI Sanggau was inaugurated back in April 2017. However, long before RRI Sanggau was established, RRI Entikong was already officially established and aired. RRI Entikong is the only RRI station located at the border sub-district level. RRI Entikong Broadcasting has been running well and independently. The broadcast programs that are aired also vary, such as Sanggau Panorama, Morning Home, Public Opinion and Friend Greetings. In addition there is a program, Rumah Kita, which became a favorite program of the listeners. Rumah Kita is program that elevates the culture of the local ethnic group and is broadcasted using the local language, such as Dayak Iban, Dayak Bidayuh and Javanese Language (to accommodate Javanese transmigrant groups). The Entikong people are also active in conveying their

Table 1. Comparation of Indonesian and Malayasian Broadcast Frequency in The Border Area of Sajingan, Sanggau Ledo, and Entikong.

		Monitored frequency					
		Sajingan		Sanggau Ledo		Entikong	
No.	Type of broadcast	Indonesia	Malaysia	Indonesia	Malaysia	Indonesia	Malaysia
1	FM Radio Broadcast	2	18	3	18	2	34
2	TV Broadcast	1	4	1	5	1	11
Total		3	22	4	23	3	45

Source: Balmon Kelas II Pontianak, 2016 in Kurnia, 2016.

aspiration in the segment of Public Opinion. As for the frequency of the broadcast frequency, RRI Entikong airs on 106.6 FM. RRI Entikong broadcasts even reach out and have listeners all the way to Kuching and Serian, Malaysia. RRI Entikong broadcast can also be heard not only through the radio, but also from the smartphone by downloading the RRI Play application from the Google Store app or by using a built-in radio application on the phone.

3.2 *Consumption patterns of television broadcasts*

In the category of television broadcast, 97% of respondents claimed to have a television and only 3% who do not have television. While for accessing television broadcasting, 88% of respondents stated they access national television broadcast via satellite dish, 6% using cable TV, only 2% using UHF antenna. Based on the results of in-depth interviews, the people in border area long for and need national broadcasts. Therefore, since the popular use of satellite dish, the local community fund theirselves to purchase the satellite dish. There are still some 2% of respondents who use UHF antennas. This may be due to unfavorable economic factors to purchase satellite dish, or limited access to remote areas on the Entikong border. The existence of the desire and efforts of the people in border area to be able to access the national broadcast signifies the sense of belonging of the people in border area towards Indonesia and reflects the identity of the people in Entikong border as part of the Indonesian nation.

Regarding accessibility to Indonesia's national broadcasts, 80% claimed that they could access Indonesia's national broadcast, 15% respondents are neutral, and 3% respondents claimed they could not access Indonesia's national broadcast. The number of 3% of respondents is in line with the number of 3% who do not own television and 2% respondents still using UHF antenna. There are possibilities that these 3% respondents, who can not access the national broadcast of Indonesia, are respondents who do not own television or satellite dish antenna. Whereas, for the accessibility to Malaysia's broadcast, about 24% respondents claimed they could access Malaysia's broadcast, 27% respondents are netral, and 45% respondents stated that they could not access the Malaysia's broadcasst. Based on FGD's result, the greatest aspirations conveyed by the border community is the desire for access to live sports broadcasts, such as football and regional sporting events. The border community is fond of watching alongside sports games, especially when Indonesia matches against neighboring countries, such as Malaysia. They expressed a strong sense of nationalism and unity that emerged especially when the match with Indonesia against Malaysia was underway.

In everyday life, 5% of respondents spend time watching television over 9 hours, 7% for 6 to 9 hours, 36% for 3–5 hours, while at most 51% of respondents spend only 1 to 2 hours watching television. This can be because the respondents are final year students of SMK, who spend more time in school or preparing practice and learning for the next day. As for the choice of television viewing time, 60% of respondents said watching television at night, 21% in the afternoon, 16% during the day, and only 2% watching in the morning. The majority of respondents who watch television in the afternoon and evening are due to the morning and afternoon, the respondent are still in school. About the level of favourite in watching television, 61% of respondents said they liked watching television, 32% said they were neutral and 4% said they were not fond of watching television. From this fondness, the researchers continue the question about the passion of watching Indonesian and Malaysian television broadcasts. The result is 74% said they enjoyed watching Indonesian national broadcast, 16% said they were neutral, and 5% said they did not like to watch Indonesian television. In comparison, the next question asked about the level of favourite Malaysian television broadcasting, the data obtained as follows, 21% of respondents claimed to watch the broadcasts of Malaysia, 29% said neutral, and 46% said not to watch the broadcast Malaysia.

3.3 *Consumption patterns of radio broadcasts*

In the category of radio broadcast, 43% of respondents claimed to have a radio and 56% do not have radio. This can be due to the transformation of sophistication in access to information, so that the public no longer has radio in the conventional form. This is seen in the next question which is 61% of respondents stated accessing radio via mobile phone, only 5% have analog radio, and 34% stated neutral and abstain. In daily life, 2% of respondents spend time listening to the radio over 18 hours, 2% for 12 to 18 hours, 6% for 6–12 hours, while at most 51% of respondents spend only 1 to 6 hours and 40% of respondents do not listen to the radio. The low level of listening to the radio can be due to the respondents which are the young generation who more fond of popular broadcasts on television. Meanwhile, based on field studies, in-depth interviews and FGD showed that loyal radio listeners, especially RRI Entikong, were mostly people of over 30 years of age, especially those who are living in the rural areas and working in the fields. As for the choice of radio listening time, 43% of respondents said listening to the radio at night, 20% in the afternoon, 19% during the day, and only 12% listening in the morning.

About the level of favourite in listening to the radio, 38% of respondents said they liked listening the radio, 40% said they were neutral and 16% said they were not fond of listening radio. Regarding accessibility to Indonesia's radio broadcasts, 62% claimed that they could access Indonesia's radio broadcast, 23% respondents are neutral, and 10% respondents claimed they could not access Indonesia's radio broadcast. A total of 62% of respondents who said they can access the national radio broadcast are respondents who have access to RRI Entikong's broadcast. While the rests are the respondents who can not access RRI Entikong due to the hilly geographical contour of Entikong. RRI Entikong signal type is the FM signal, which cannot reach areas obstructed by hills. Thus, although RRI Entikong has a very large coverage area reach to Malaysia, there are still rural areas in Indonesia's border area that cannot access RRI Entikong broadcast. As for access to Malaysian radio broadcasts, 21% of respondents stated that they could access the Malaysian broadcast, 27% stated that they were neutral, 45% stated that they could not access the Malaysian broadcast. Regarding the level of listening to national radio broadcasts, 61% of respondents said they enjoyed the national radio broadcasts, 22% said they were neutral, and 10% said they were not interested in listening to Indonesian radio broadcasts. Meanwhile, in comparison to the level of listening to Malaysian radio broadcasts, 21% said they liked listening to Malaysian radio broadcasts, 31% said they were neutral, and 42% said they were not interested in listening to Malaysian radio broadcasts. Thir fondness for listening to Malaysian radio broadcasts could be due to the background of the common cultural identity and daily local language usage of the border community in Entikong and the entertainment songs are more varied.

3.4 *Broadcasting and national identity in Entikong, West Kalimantan*

Regarding the usefulness of broadcasts in the Entikong border area, 88% of respondents felt the benefits of the broadcasts consumed, 91% felt the existence of broadcasts adding to the latest knowledge and information, 80% said the broadcasts increased self-motivation, and 88% said the broadcasts increase their love to Indonesia. Overall, the presence of broadcasts consumed perceived benefit and positive impact for border community. As for the correlation between the presence of broadcasts with a sense of nationality, 57% of respondents said the emergence of a sense of optimism towards Indonesia. Based on the results of field studies and in-depth interviews, the low level of optimism is due to news that reflects the negative side of the current state of the nation, for example corruption and criminality. Furthermore, a total of 74% of respondents felt the increase in knowledge about developments that occurred in Indonesia and 75% felt that those broadcast increasing their pride to Indonesia.

In terms of language and culture, a number of 68% respondents stated that the broadcasts made border communities more frequent in using Indonesian language, 52% stated that makes them want to learn foreign languages, 69% said they knew international developments, 34% felt an increased interest in other countries' culture, 34% felt that they wanted to live in another country. The high number of Indonesian language usage due to the influence of national broadcasting consumption, when viewed from the perspective of Fishman (1977), Althusser, and Bachofer (2014) it will show the role of language as a symbol of the bearer of the idea of nationality and national identity that is broadcast through national broadcasts. Language became a symbol of unifying the Indonesian nation which consists of various tribes and local languages. Indonesian language becomes a representation of national identity which is broadcast through national broadcasts. With the Indonesian-language broadcasts, then a border community, as receiver, received message that they also have the same language, that proved and showed they are one community. This language influence in national broadcast and receiver interpretation in forming identity process is exemplified in that case. These broadcast patterns, that are being consumed by the border community, will also influence the forming of collective identity as one nation and one community. While the desire and knowledge of the global world indicates that the border community is not less informed by the people in big cities. The respondents, who in this case are the youth on the border of Entikong, have the same desire and knowledge as the youth in general to be able to explore and increase their knowledge to compete in this globalized world.

The correlation between the broadcast consumed and the national identity can be seen from the efforts of the border community, that initially have not access to the national television broadcast, but attempted independently by purchasing a satellite dish to watch Indonesian broadcasting. In fact, they can enjoy the Malaysian broadcast by using ordinary UHF antenna, without the need to spend more to buy a satellite dish. A strong desire and need for national broadcasting can also be seen from their hope to get access to live broadcasts of football matches, especially the matches where Indonesia participate and compete. These points out the existence of a strong sense of belonging of Entikong border community towards Indonesia, despite its remote isolation in the rural and directly neighboring Malaysia, any access limitations which they experience, as well as the distance between the border area and the urban center, does not diminish the spirit and the sense of nationalism of the border community. Judging from the concept of the 'imagined community' (Anderson, 1983), the border community who experiences delays in development, limited access, far from urban centers and government, and have not interact directly with other Indonesians across the archipelago, but are able to imagine themselves as part of the Indonesian nation. These things, further strengthen the sense of nationalism owned by border communities, especially in the border area of Entikong.

4 CONCLUSION

In the case of consumption patterns of television broadcasting, almost all the border communities already own the television, expressed fond of watching television and attempted to access national broadcasts through purchasing a satellite dish. In some areas, border communities can access Malaysian television broadcasts. As for the pattern of radio broadcast consumption, only half of respondents who have analog radio and the rests access the radio through the built-in application of the mobile phone. Most border communities can access the national radio broadcasts from RRI Entikong and there are still some respondents who can access the Malaysian radio broadcast. Most people are fond border radio broadcasts Indonesia (RRI Entikong) and the rest fond of radio broadcasts Malaysia. The presence of national broadcasters in the border region Entikong perceived benefits and its positive impact on border communities. The existence of these broadcasts is felt to increase knowledge, information about the latest developments in Indonesia, and a sense of pride towards Indonesia. Broadcasts play a role in improving the sense of nationality and the use of Indonesian language in the daily life of the Entikong border community. A strong sense of nationalism appears on their eagerness and efforts to access national broadcasts Indonesia amid all the limitations they have.

REFERENCES

Anderson, B. (1983). *Imagined Communities: Reflections on the Origin and Spread of Nationalism.* London: Verso.

Badan Pusat Statistik Kabupaten Sanggau. (2016). *Kabupaten Sanggau dalam Angka 2016.* Sanggau: Author.

Barbrook, R. (1992). Broadcasting and National Identity in Ireland. *Media, Culture and Society, 6*, 203–227.

Bulck, H. d. B., and Poecke, L.V. (1996). National Language, Identity Formation and Broadcasting, Flanders, the Netherlands and German-speaking Switzerland. *European Journal of Communication*, *11*(2): 217–233.

Fishman, J.A. (1972). *Language and Nationalism: Two Integrative Essays. Rowley.* MA: Newbury House.

Fishman, J.A. (1977). Language and Ethnicity. In Giles, H. (Ed.) *Language, Ethnicity and Intergroup Relations.* London: Academic Press.

Morley, D. And K. Robins. (1989). Spaces of Identity: Communications Technologies and the Reconfiguration of Europe. *Screen, 30*(1–2): 10–34.

Mulryan, P. (1988). *Radio, Radio: The Story of Independent, Community and Pirate Radio in Ireland. Dublin*: Borderline Publications.

Ó Tuathaigh. G. (1984). The Media and Irish Culture. In Farell, B. (Ed.), *Communication and Community: The Case of Irish Radio.* Dublin: RTE.

Siburian, R. (2002). Entikong: Daerah Tanpa Krisis Ekonomi di Perbatasan Kalimantan Barat-Sarawak. *Antropologi Indonesia*, *67*, 87–93.

Urban Studies: Border and Mobility – Kerr et al. (Eds)
© 2019 Taylor & Francis Group, London, ISBN 978-1-138-58034-3

The slut-shaming phenomenon in social media: A case study on female English literature students of Binus University

Paramita Ayuningtyas & Abdul Aziz Turhan Kariko
Binus University, Indonesia

ABSTRACT: This study is aimed to investigate the issue of slut-shaming in social media further. In the digital era, social media becomes the borderless space, where people can socialize, interact, and express themselves. On the contrary, social media has also become the forum where cyber-bullying, particularly slut-shaming, happens. The victims of social shaming are mostly young women, including college students. On the basis of that background, this study was conducted to find more about the slut-shaming phenomenon in social media. To achieve this goal, research data were collected through questionnaires with female students of English Department, Bina Nusantara University, as the respondents. The obtained data were analysed both quantitatively and qualitatively. The findings show that females tend to feel insecure and uncomfortable in social media, even though the latter is an effective tool for women to voice up their issues. In addition, victim-blaming still exists in the Indonesian context. Slut-shaming also happens to women who do not follow the Indonesian values and norms. This study will offer a new perspective about slut-shaming in the Indonesian context and will be a helpful source in developing curriculum policy, specifically course material, to handle slut-shaming in cyberspace.

Keywords: Social media, Cyber-bullying, Slut-shaming, Internet, feminism

1 INTRODUCTION

Slut-shaming, a social stigma labelled to women who are considered to have "unconventional" sexual behaviours, is not a new phenomenon in society. For example, Nathaniel Hawthorne wrote about this theme in his famous novel *The Scarlet Letter*, published in 1850. Webb (2015) even mentioned that slut-shaming occurred in the era of Roman Republic, where married women were ideologically contrasted to sluts. In America, this issue became a main concern for feminists, since a research conducted by the American Association of University Women in 1993stated that three-fourths of female students in the eighth grade have experienced sexual abuses, either physical or verbally (Tanenbaum, 2015).

Slut-shaming happens not only in the real world, but also in cyberspace, especially in social media such as Instagram, Path, and Twitter. Ideally, social media can serve as a facility to express self and to interact with others. Moreover, social media are often promoted as a democratic venue, where everyone can voice their opinions (Tate, 2016). However, in reality, women have possibly become the target of slut-shaming in social media. In many instances, women in social media are labelled as sluts because they have posted pictures of themselves wearing revealing clothes or because they have non-conservative thoughts. One significant example of slut-shaming in social media is when Leslie Jones (an American actress) became a target of slut-shaming because she played a role originated by men in the reboot of the movie *Ghostbuster* released in 2016.

In Indonesia, female celebrities like Ayu Ting-Ting and Mulan Jameela have frequently become victims of slut-shaming in their Instagram accounts due to their behaviours and personal lives. Not only celebrities, but also ordinary women can become the target of slut-shaming, as stated by Nurvitasari (2016), who wrote that one of her friends experienced slut-shaming because of her appearance and her dating style in social media. There are several stories about sexual assault and social stigma on women in the cyberspace. Unfortunately, the data and statistics about this issue are not available yet, since there is no systematic reporting method for this kind of case (Nurvitasari, 2016).

The number of social media users is increasing every year, and women are active social media users, particularly on Instagram and Twitter. According to Brenner (in Herring and Kapidjic, 2015), female teenagers spend more time in social media than their male counterparts do. They generally use social media to communicate with their peers and to strengthen the existing friendships. Related to the issue of slut-shaming previously explained, this study intends to investigate further about this issue by focusing on the experience of female students in Binus University.

The stigma of slut-shaming has severe psychological impacts on the victims. In 2016, Tiziana from France sent her sexy video to her ex-boyfriend, who later spread the video on the Internet. Being unable to cope with the comments from netizens, Tiziana committed suicide. Focussing on the major psychological effects on slut-shaming, this study is conducted to explore this issue by limiting the scope only to female English literature students in Binus University. This study will offer a new perspective about slut-shaming in social media and will be a constructive source in developing curriculum policy, specifically course material, or technology to handle slut-shaming in cyberspace.

In this study, we apply both quantitative and qualitative methods, with female English students of Binus University as the respondents. Data were taken from the results of a questionnaire distributed to 40 students in September and October 2017. The questionnaire contained nine questions: seven closed questions and two open questions. Since this study is focussed on a small group of respondents, it does not intend to generalize the opinion and experience of female students related to slut-shaming, but to interpret the answers from an individual's perspectives and thoughts. From here, a further research with a bigger aim can be elaborated.

2 DISCUSSION

In this section, we present the result of the questionnaire as well as the discussion and interpretation of the data. The questionnaire was distributed to 40 female students from English Department, Binus University, in the age range of 17–22 years. The respondents were not asked to write their name, because we expected that by being anonymous, they could be more comfortable in answering the questions since slut-shaming is a sensitive issue for women. All of the questions are composed in Indonesian for easy understanding. The translation and the result can be observed in Table 1.

Table 1 summarizes the result of the closed questions, and the discussion of the result is as follows. Questions 1 and 2 are placed in the beginning of the questionnaire to know what social media that respondents use the most often and what reasons they have for using social media. Each social media platform has its own characteristics, and it could be the reason why respondents chose that particular platform instead of the others. Meanwhile, questions 3 and 4 focus on their experience in using social media, but still in a general sense. Still in a form of closed questions, questions 5–7 try to narrow down the topic into slut-shaming in social media. In questions 5 and 6, we would like to know whether the respondents have any experience of being slut-shamed in social media or not. Finally, question 7 is about their friends or relatives' experience.

For question 1, the result indicates that 35 students chose Instagram, 4 students chose Twitter, and 2 students chose Facebook. The total number of answers is 41, since apparently one respondent chose two options. On the basis of the result for question 1, it can be seen how respondents use a photo-sharing platform the most. This implies how visual contents seem to be their most important consideration in choosing social media platform. Next, the result of question 2 shows that students use social media to spend their free time (22), to find

Table 1. Answers for the closed questions.

No.	Questions	Answers
1.	Which social media do you use the most often?	a. Twitter: 4 b. Instagram: 35 c. Path: 2 d. Facebook: 0
2.	Your main reason for using social media:	a. To find friends: 6 b. To express self: 5 c. To spend time: 22 d. Other reasons: 7 (find photos, get updated, etc.)
3.	Have you ever felt unsafe or uncomfortable in social media?	a. Yes: 30 b. No: 10
4.	Before posting photos or status in social media, what consideration do you have?	a. Not wanting to feel embarrassed: 16 b. Fearing that it will hurt other people's feelings: 17 c. Seeing the benefits of the post: 6
5.	Have you ever experienced slut-shaming (labelled as a bad woman because of the way you dress or behave) in social media? (if the answer is not, go straight to question number 7)	a. Yes: 36 b. No: 4
6.	What did you feel after experiencing slut-shaming?	a. Sad b. Embarrassed c. Angry: 2 d. Afraid: 1 e. Other: 1 (indifferent)
7.	Has your friend/relative ever experienced slut-shaming?	a. Yes: 9 b. No: 29

friends (6), to express themselves (5), and for other reasons (e.g., to get updated and to find photos (7). Again, for question 2, a number of students chose more than one option. Most of the respondents consider social media as a way to spend their leisure time, but it can be seen how a number of respondents also use social media to interact with other people and also to express themselves. For youth, the Internet is the potential space for them to interact and to express their opinions more freely since in geographical and social terms, cyberspace is not limited by any boundary.

In question 3, the students were asked whether they have ever felt unsafe or uncomfortable in social media. The result shows 75% "yes" and 25% "no". The majority of respondents who answered "yes" seem to agree with Tate's statement that despite its democratizing capacity, cyberspace could be a dangerous space for women. Related to question 3, in question 4, we asked what aspects they consider before posting in social media. Social media is a public space, and if it is not set as locked/private, the content can basically be viewed by everyone in the Internet. Thus, some people might have considerations about what to post and what not to post in social media. The answers reveal that 16 students chose not wanting to feel embarrassed, 17 chose fearing that it will hurt other people's feelings, and 6 chose seeing the benefits of the post. Most of the respondents think that when posting in social media, they have to think about other people's feelings. They do not want their post to hurt and annoy other people. On the contrary, 16 students think what they post should not embarrass themselves.

Moving on to the fifth question, which focuses on slut-shaming in social media, 90% answered that they have never been slut-shamed, while 10% answered that they have experienced slut-shaming. When asked what they felt after being slut-shamed in question 6, two respondents answered that they felt angry, one said that she felt afraid, and one felt indifferent about it. Then, for question 7, on whether their friends/relatives have experienced slut-shaming, 29 respondents answered no, 9 answered yes, and 2 did not answer. It is somehow a relief to know that most of the respondents have never experienced slut-shaming. However, this also leaves a question: what kind of definition of slut-shaming do they have in mind? Do the respondents really understand which actions can be categorized as slut-shaming, since the first- and third-semester students in Binus University have not taken any course that include feminism and sexism topics in the syllabus (e.g., literary criticism)?

In question 8, we would like to get their opinion about the slut-shaming cases experienced by public figures like Ayu Ting Ting and Mulan Jameela (singers famous for their so-called love scandals) and Awkarin (an Instagram celebrity who often poses in revealing clothes). Meanwhile, question 9 asks the readers to give suggestions about what should be done to stop slut-shaming in social media. These last two questions are open questions because we wanted the respondents to elaborate their answers and from here, we could analyse their opinion towards slut-shaming phenomenon in social media.

We categorized the answers based on the similarities of their points of view in approaching the topic. Then, we picked answers that best represent the topic of this research and discussed them using feminist and cultural studies theories. The answer for question 8 can be four: victim blaming, netizens' responsibility, neutral tone, or gimmick and indifference. We will explain what each category is and present the answer that most represents the idea.

First, victim blaming means that the respondents' answers indicate that they blame the victims for the slut-shaming they receive. A total of 19 respondents think that it is because of the public figures' clothes or behaviour and that they deserve to be slut-shamed. A 19-year-old respondent stated that, "If they (the female celebrities) decided to wear revealing clothes, they have to accept the consequences. Indonesian society tends to be conservative." Another respondent (also a 19-year-old girl) emphasized that what the celebrities posted is against Indonesian norms and values. One point to underline here is that respondents think that public figures still have to hold onto Eastern/Indonesian norms and values despite the fact that cyberspace is not limited by geographical borders.

The second category is netizens' fault, which signifies that slut-shaming happens because netizens cannot control their behaviour and words in the Internet. This is the answer from a 17-year old respondent: "It (slut-shaming) needs to be stopped. Not only because slut-shaming ruins someone's reputation, but also because human beings do not have the right to judge other people. They must respect other people's rights." Another respondent (a 19-year-old girl) even said that, instead of doing slut-shaming and publicly shaming the women, netizens could give constructive comments through private messages.

Furthermore, one respondent's answer (a 22-year-old student) showed her awareness about gender stigma in the society, "... in Indonesia, it doesn't have to be Awkarin or any other artist, girls who wear revealing clothes, even just for a bit, will always be judged and labelled as sluts." She continued, "But then when boys do the same thing, they are praised and congratulated for it. It's unfair, and it's what the society thinks. Boys will be boys and girls will be sluts." The respondent's answer shows her concern about gender labels and gender inequality in society. Girls who look sexy will be immediately labelled as sluts; in contrast, boys will be praised when they can "conquer" many girls.

Table 2. Answers for the open questions.

No.	Questions	Answers
8.	What do you think about the slut-shaming experienced by female public figures like Awkarin, Ayu Ting Ting, and Mulan Jameela?	Victim blaming: 19 Netizens' fault: 16 Neutral-tone: 3 Gimmick: 1 Indifference: 1
9.	In your opinion, what should netizens do to stop slut-shaming in social media?	Both sides' responsibility: 2 Netizens' responsibility: 28 Women's responsibility: 9 Indifference: 1

Next, we categorized the answers that show slut-shaming is both netizens' and public figures' responsibility as neutral tone. The respondents in this category think that slut-shaming can be avoided if netizens and public figures are more careful in expressing themselves in the Internet. They also believe that the Internet is a free space where people can give any comment. One answer from an 18-year-old respondent represents this: "In my opinion, people's perspectives are different. Some people might think it's common and not their business. Some also think that it's not right and should be corrected. But everyone is free to give their opinion". Besides trying to sound neutral, this respondent also highlighted the democratic element of cyberspace, where anyone (despite their social and cultural background) can express their opinions freely.

Meanwhile, one respondent has an opinion that what the public figures do and the controversy around them is a promotional gimmick. She thinks that, "It's a common thing in entertainment industry. It's set up by their bosses (managers) so that they can be famous." By doing controversial actions, more people will know them and their popularity will increase. The last category – indifference – means that one respondent does not follow the news of Indonesian celebrities and does not have any interest in it.

For question 9, there can be four answers: both sides' responsibility, netizens' responsibility, women's responsibility, and indifference. In contrast to the result of question 8, which seems to victim-blame the female celebrities, the result of question 9 indicates the importance for netizens to be well-mannered and positive in giving comments. An 18-year-old respondent said that netizens should think "what if they are in the position of being slut-shamed". Another student adds that, "If netizens see a post that they do not like in social media, they'd better ignore it." A number of respondents also highlighted that netizens need to learn about ethics and have knowledge of using social media. The following answer represents the idea: "Slut-shaming can be reduced by educating the netizens about the right way to use social media."

Some respondents (nine students), however, seem to put the blame on the women only, as shown in these answers:

> *If they live in Indonesia, not in Europe, they have to dress appropriately if they do not want to receive negative comments.*
>
> *Don't be too vulgar in social media.*

The respondents seem to think that in order to stop slut-shaming, women's bodies should be regulated. They have to dress in a proper way, following the norms in Indonesia. This kind of answer indicates that it is women who provoke slut-shaming. There is no implication in their answer that shows how slut-shaming is an impact of gender inequality.

3 CONCLUSION

In America, for decades, slut-shaming has been an important issue. In Indonesia, along with the growing of social media, slut-shaming has become common, especially towards female celebrities. Unfortunately, this issue has not been explored much, as indicated by the lack of academic articles and news articles about it. Therefore, this study can be a valuable source in dealing with the issue before it grows more severe.

From the result of the questionnaire, it can be seen how using social media has become one way for the respondents to spend their free time. They even use it to express themselves, such as posting photos or sharing opinions. However, the result also shows that the respondents (all are female) still feel unsafe and uncomfortable in social media, while actually social media can be an effective tool for women to bring their issues to the forefront. Another thing that can be noted from the discussion of the result is that the action of victim-blaming still exists. There is still an opinion that women deserve to be slut-shamed if they do not dress and act following Indonesian values and norms. The respondents (who are also female) seem to internalize the dichotomy between good girls and slutty girls. Ringrose and Renold (in Armstrong et al., 2014) see this as a form of internalized oppression, which could be threatening for the movement of gender equality.

On the basis of our study results, we have a suggestion to help in dealing with the situation. The Internet has become a necessity for this generation of "digital natives" (a term coined by Ethan Zuckerman in his book *Digital Cosmopolitans in the Age of Connection*), yet this technology progress has not been accompanied by sufficient education about how to use it appropriately. To tackle this problem, we believe Internet education that incorporates gender equality needs to be included as part of the college curriculum in Binus University. One of the subjects that should be considered in the course outline is gender education in social media. Students need to be introduced to the issues of slut-shaming, gender-based hatred, and sexual harassment in the Internet. By educating the students (both male and female), it is expected that the young generation can have healthier and safer interactions in social media. Thus, cyberspace can be a harmless place for everyone, in general, and young women, in particular.

REFERENCES

Almazan, V.A., & Bain, S.F. 2015. College students' perception of slut-shaming discourse on campus in *Research in Higher Education Journal* 28.

Amichai-Hamburger, Y. 2017. *Internet Psychology: The basics*. New York and London: Routledge.

Armstrong, E.A., Hamilton, L.T., Armstrong, E.M., & Seeley, J.L. 2014. "Good girls": Gender, social class, and slut discourse on campus. *Social Psychology Quarterly 77:* 100–122.

Boyd, D.M, & Ellison, N.B. Social Networking Sites: definition, history, and scholarship in *Journal of Computer Mediated Situation* 13.

Herring, S.C. & Kapidjic, S. 2015. Teens, Gender and Self-Presentation in Social Media in *International encyclopedia of social and behavioral sciences*. Oxford: Elsevier.

Karaian, L. 2014. Policing "sexting": Responsibilization, respectability and sexual subjectivity in child protection/crime prevention responses to teenagers' digital sexual expression. *Theoretical Criminology* 18: 282–299.

Nurvitasari, A. 2016. Cyberbullied in Indonesia. *Magdalene.* Retrieved March 22, 2017 from: http://magdalene.co/news-813-cyberbullied-in-indonesia.html.

Rahimi, R., & Liston, D.D. 2009. What does she expect when she dresses like that? Teacher interpretation of emerging adolescent female sexuality in *Education Studies: A Journal of the American Educational Studies Association* 45: 512–533.

Tanenbaum, L. 2015. *I Am Not a Slut: Slut Shaming in the Age of the Internet*. New York: HarperCollins.

Tate, E. 2016. Challenging Women's Digital Agency: The Frequency of Slut-Shaming in Social Media in *The Journal* 1 (1): 37–41. (4). Retrieved March 22, 2017 from: http://theijournal.ca/index.php/ijournal/article/view/26477/19661.

Webb, L. 2015. Shame Transfigured: Slut-shaming from Rome to Cyberspace. *First Monday* 20 (4). Retrieved March 22, 2017 from: http://firstmonday.org/ojs/index.php/fm/article/view/5464/4419.

Urban Studies: Border and Mobility – Kerr et al. (Eds)
© 2019 Taylor & Francis Group, London, ISBN 978-1-138-58034-3

Mobile life, communication technology, and disreputable literacy

S. Hermininrgum
Universitas Brawijaya, Malang, Indonesia

ABSTRACT: By earning the G-20 membership, the position of Indonesia as a country of emerging economy in global arena is unquestionable. However, domestically, the results of its National Development Program still create distinctive gaps between rural–urban regions, provinces, and islands. In the context of the Human Development Index, as issued by UNDP in 2016, Indonesia ranks 133 in literacy. The study reported in this paper, therefore, aims to reveal one of potential roots of Indonesian disreputable literacy, which is portrayed in an urban area: Landungsari Village, Malang Regency, and East Java Province. The reading habit of the elementary-school-age children is the focus of this study, because as the core element of literacy, the reading habit will increase human capital, particularly within the intellectual domain. To disclose how globalization affects literacy, the data are analyzed by reviewing the outcome of the 3-year community services and the recent field survey. The findings obviously illustrate that the low interest in reading is triggered by the influence of mobile life, unconducive surroundings, and the misuse of communication device of advanced technology: smartphones.

Keywords: globalization, human development, Landungsari Village, literacy, smartphones

1 INTRODUCTION

Appadurai's idea (2010) on global capitalism as the grounds of globalization and the utilization of technology advancement as the representation of modernity implies that being modern is the concept of the globalization itself. Accordingly, Pieterse (1996) asserted that "modernity is a keynote in reflections on globalization", whereas Giddens (1990) stated that "globalization is the corollary of modernity". Oxford English Dictionary (1989) defines the word modern as "characteristic of the present or recent times, as distinguished from the remote past". This signifies that globalization refers to spatial-geographical and temporal-historical dimensions. The present and the past become an inevitable part of understanding globalization. This is caused by the fact that, during its progress, globalization, which is always linked with modern life, cannot be separated from the development of technology; covering either information or communication. All these fields play a tremendous role in promoting mobile life, which dissolves the barriers of not only time and space but also gender and age.

Economy and technology are progressing in an interconnected manner, as a sign of modernity in the era of globalization. The betterment of a country is nevertheless judged internationally by the economic growth and the utilization of technology. This criterion makes Indonesia one of the world's largest advanced and emerging economies. The achievements gained from Indonesian's national development program promoted Indonesia to a position worthy of international recognition. However, it is unfortunate that these achievements are not balanced out well with all-embracing development of the human life aspects. IPM (Human Development Index) implicitly remarks that the result of human development does not correlate strongly with the growth of national economy, which is achieved by the Indonesian government.

The results of the National Development Program of Indonesia are always faced with development gaps. This is a logical consequence from being an archipelago with thousands of islands, which therefore increases the gaps between the ever-developing Java Island and other islands, metropolitan cities, and remote areas, and even a rural–urban gap within a city. The term rural–urban itself becomes a little vague when it is associated with the existence of border, because in this globalization era, the advancement of technology (be it transportation, information, or communication technologies) makes all human activities progress rapidly. The continuous development in economy, as well as industrial advancement in all sectors play a significant role in increasing the rate of migration of Indonesian between different areas, islands, cities, or even villages. This mobile life, surely, also creates problems relating to the quality of human life, particularly in terms of proper environment, healthcare, and education.

In the field of education, literacy becomes the base of the intellectual field to answer the challenge of globalization. The skill of writing and reading is believed to enable humankind to reach the worldwide civilization. Nevertheless, from the 188 countries studied by United Nations Development Program (UNDP), it was claimed that literacy in Indonesia in 2016 decreased from the previous year: from 100th place to 133rd place. This is very different from the rank of a neighboring country, Malaysia, which is in the 59th place. It may suggest that the national development creates gaps not only related to geographical conditions, but also to the values of humanity, which are most likely being built through education. In summary, the result of infrastructure building in Indonesia does not correlate strongly with the result of human development. The low awareness level of the importance of reading and writing becomes a serious problem when, in reality, from the total of 250 million people of Indonesia, on average, only 0.001% practice the literacy culture (Sagitarius, 2017).

In Indonesia, the example of this kind of socio-cultural phenomena can be found everywhere, including Java. If there is a notion that Java is an island where the population is thoroughly contented with the results of development programs compared to other islands, it is actually caused by the Java-centrist stereotyping. Holistically, however, this opinion is not necessarily true. The main attractions of Java as a business and education center, political rivalry, and other crucial functions, especially economic growth, lead to the dense population of Java. This populous society then gives rise to numerous urban areas, which automatically triggers the problems that have to be faced by its dwellers. Take, for example, the impact of mobile life as the character of urban society on the socio-cultural behavior, which consists of the entangled problems of economy, health, environment, and education.

As the requirement to form a qualified human being, education relies on the literacy culture as its core. However, the rapid advancement of technology, synergized with the current of globalization, which is known to have no barrier, apparently brings a detrimental impact on human quality development. On the one hand, technology enables faster process in transportation, transfer of information, and smooth communication. However, this sort of instant lifestyle often neglects the values that shape human character. In response to this, this study was carried out to uncover the reality of the gaps of other aspect of development, more importantly the role of the bombardment of information technology in affecting human development, especially the construction of human capital through literacy.

2 METHOD

The social change linked to the mobile life and its effect on the cultural change of the urban people of Landungsari Village essentially associated with the premise proposed by Goldthrope (1992: 126) "that economic development and social mobility are positively associated; the more economically developed a society, the higher the rates of social mobility that will display". Although Goldthrope's argument correlates industrial societies and mobile societies, the general assumption considerably underlines the prominent role of economic development. It means that if this phenomenon is placed within the context of globalization, the rapid industrial progress will not be the sole determining factor of people's mobility, but it also involves other factors that become the part of characteristics of globalization, which principally relies on economy.

To explore the socio-cultural transformation, which is pulled apart into two opposite poles, namely the high usage of technology as a symbol of modernity and the low literacy level as the impact of economic growth of the urban society of Landungsari Village, the search of relational data is performed in two ways.

On the basis of the local government program, Rencana Strategis 2013–2019 Desa Landungsari, the main goal is to optimize the potency and resources of the village by exploring the existing internal and external factors. This indicates the importance of accurate plans to face the changes. Optimization is meant to increase the welfare of villagers by raising their incomes. The betterment of financial status is very likely to be the priority to clarify the position of Landungsari Village, which is categorized as "self-supporting" village within both regency and provincial levels. However, in reality, the human resources quality of the people in Landungsari Village cannot adequately support the necessary changes because Landungsari Village is not an industrial site that could absorb many workers. The total population number of 1,518 people (767 males and 751 females) in the age range of 7–18 years do not have formal education. Data from the Village Headman Office is used as the basis to observe the socio-economic and socio-cultural behaviors of the people of Landungsari Village. The expected changes of these behaviors will be observed by conducting fieldwork to elucidate the connection of its impact toward the reading habit of the children.

Before the field survey was done, the result of the community service, which was performed consecutively from 2014 to 2016, was reviewed. Basically, the community service activity also aimed to map the reading habits of the students at the state Elementary School (SDN) Landungsari II located in Dusun Klandungan, a densely populated sub-district but located quite far from the center of the village. The low reading habit is then compared with the recent survey conducted in the last two weeks of July 2017 in the two elementary schools. Both are state schools and the only elementary schools in Landungsari Village. One of the schools is the same school that the community service has helped 3 years ago. Meanwhile, the other school is located in Dusun Ndungan, which is closer to the center of the village and the main road connecting two cities. The comparison between the places and the communities in which students live in is done to trace whether the use of smartphones influences their reading habits differently. Exploring the reading habit of children between 7 and 12 years of age, who, theoretically, enter the second phase in the classification of communication stages after the non-verbal communication (Saville-Troike, 2006: 230–231) is used to trace the link of literacy culture, a supporting environment, and other factors that become the characteristics of globalization.

3 DISCUSSION

Before 20–30 years, Landungsari Village was an agricultural village, with most of the villagers doing farming and animal husbandry. However, the drastic changes happened in the 1990s with the establishment of some universities and institutes in its surrounding area. The unavoidable sharp increase of newcomers, especially students from neighboring cities or even islands, shaped the village, which was initially categorized as a rural–sub-urban area, and transformed it into urban area with crowded inhabitants. Because of its location adjacent to the main road of the City of Malang, which connects the Surabaya and Batu cities, Landungsari Village today is really citified. This becomes an interesting example because the economic growth is not always correlated with industrial advancement. In fact, there are other factors of globalization that support the dynamics of national economy in Indonesia, e.g., the mobile life, which is not even industrialization.

The backlash of the arrival of new migrants who continuously take up the space of Landungsari Village is the rise of income for local people, which, in fact, results in human progress. Lindsay's argument (in Harrison and Huntington, 2000: 283) that "human progress is not possible without economic growth" has gained importance in support of the changing orientation patterns of the people in this village. Previously, their way to generate income relied on farming and animal husbandry activities. However, for the sake of a prosperous future, today, they leave their agriculture to adapt with modernity. Almost the entire productive land in the village was transformed into buildings functioned as micro-economy entrepreneurship, which can quickly generate income, by means of food stalls, computer rental, groceries, stationeries, and, more importantly, rental rooms. Consequently, this shortcut economic achievement changed their culture—the oversimplified thoughts to take opportunity.

The subjective sense of culture makes the situation of the people of Landungsari Village critical. The poor performance of human resources in terms of education is shown by the fact that as high as 42.3% of the total local population do not have formal education. This phenomenon has indirectly become a portrait of a complicated urban society or even a torn-apart society: the right leg goes to the modern lifestyle, while the left one has difficulty in following that step, and even tends to remain unchanged. This reality clearly explains the discrepancy between the ideas of modern and non-modern. Referring to the socio-cultural factors that determine modernity, it is important to take into account King's assertion on group's exposures. He said, "to see societies as varying contributions of modern and non-modern elements, sometimes mutually indifferent, sometimes mutually supportive, and sometimes mutually hostile" (in Harrison and Hutingtton, 2000: 113). These three criteria are the unavoidable obstacles in constructing rural–urban society to be a good and stable society, as described in the case of Landungsari Village. Technology as the vehicle of globalization disrupts the social and cultural orders within the society.

Mobile life enjoys a high speed due to technology. Means of transportations, flow of information, and sophisticated communication tools blurred all borders of time and space. Smartphone is the product of technology that was created to accelerate and ease communication across geographical barriers, such as villages, urban-suburban areas, cities, islands, and even countries. In Indonesia, the possession of this technological product within the last decade has become the symbol of modern lifestyle, making the phrases must have, must buy, and must use daily mottos. In general, this lifestyle is, in fact, creating changes in the society's behavior, including that of elementary school students.

The study that was stressed on the usage of smartphone as a product of information technology by elementary school students shows a negative trend. Of the 250 students of SDN Landungsari I, 183 (73.2%) had smartphones, with 170 of them had their own, while the other 13 borrowed it from their parents or family members. However, in SDN Landungsari II, of a total of 72 students, 67 (93%) had smartphones: 53 had their own, and 14 borrowed from their family members. The ownership of smartphones, however, does not signify the positive connectedness with the students' reading habit. This can be traced from the students' favorite choice of the applications served in smartphones, namely Internet games. Undoubtedly, the function of smartphone such as for sending message, attending calls, taking pictures, or listening to music, embraces modern lifestyle: practical, economical, and fashionable. However, prioritizing playing games seems to be a large part of explanation why children literacy culture is decreasing from time to time.

Dealing with the reading habits of the students of SDN Landungsari I and II, the result of the survey underscores that, besides the influence of Internet games, the low literacy level happens because the bigger portion of their reading activity relies on reading for school instead of reading for pleasure (Figure 1). Different from reading for pleasure, which can strongly evoke their new ideas and help discover facts and experiences (Mikulecky and Jeffries, 1996), reading for school is done only to meet the targeted purposes or to keep up with school compulsory. Reading for pleasure is fundamental for shaping the intellectual quality of the students. Children's love for reading will stimulate their critical thinking, which can henceforth be used to manifest their insights into writing competency. In brief, when reading for pleasure does not become the part of children's daily routine, their literacy culture can be classified as deficient.

Unfortunately, students' engagement with smartphones in both of the elementary schools in Landungsari has totally taken over their interest in reading. This condition is aggravated by the absence of paragons, unconducive environment, and unavailability of proper facilities to motivate children to read. Even to the students of the Elementary School of SDN Landungsari I who do not have smartphones (about 17%) and those of SDN Landungsari II (7%), the dream of buying, having, and using smartphones is stronger than the spirit of reading for pleasure. Thus, it can be said that smartphones are also playing a prominent role for children in celebrating globalization.

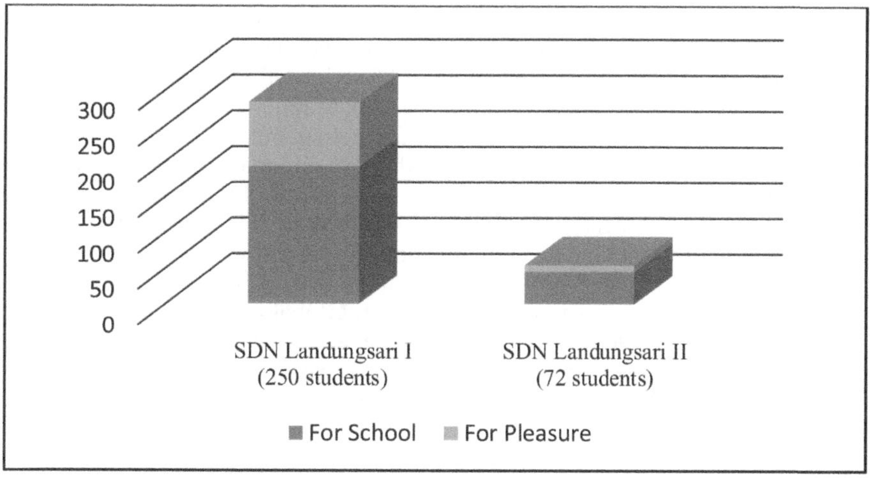

Figure 1. Students' reading habits.

It is true that the island of Java to this point is still securing its position as the center of national development. However, the tempting infrastructures as the result of development would be paradoxical when it is contrasted with the results of human development. As a magnetic attraction for the migrants from all over Indonesia, Java contributes to the complicated socio-cultural phenomena, especially because of the presence of myriads of urban areas. As a citified village, Landungsari is one of the prototypes of the crowded urban in East Java, Indonesia, which represents the impact of imbalance results of Indonesian development program.

The results of the study carried out in Landungsari Village reveal one of the salient examples where smartphones have enabled the change of the socio-cultural behavior of Indonesian children since the young age. The time when they start to get engaged in educational institutions, which forge their way to the future, is interrupted by the globalized lifestyle. The negative impact of uncontrollable smartphone usage by elementary-school-age children within the rural–urban society of Landungsari Village contributes to the decrease of the human capital. The fundamental level of literacy culture and the surroundings, which has the view that the achievement of national development is material prosperity, contributes to the misleading concept that weakens human values. In a certain extent, this indicates that Indonesians are in a so-called trap of the paradoxical National Development Program. The result of the program, whose core is in economic growth in accordance with the global standard, seems to occupy its own pole, which is the opposite of the pole of the achievement of the less taken-care-of human resources development. There is neglect toward the human values because the Indonesian children, as represented by the children of Landungsari Village, in the age group of 7–12 years lost their golden period of constructing human capital by literacy activities.

4 CONCLUSION

There is an entanglement between the hidden impact of globalization and the values of humanity. Improper response to globalization, the misuse of communication devices, particularly smartphone as a product of advanced technology by children of elementary schools is worrisome. The fulfillment of modern lifestyle in a wrong time and wrong place is a tremendous disaster for the future Indonesia. Because of the outpouring of technology, the current of social changes continues to move rapidly, and the children as the precious springs of the future are not being nurtured properly. The poor awareness of the rural–urban society, as exemplified by Landungsari people, toward the significance of literacy as the main supporting pillar of human capital casts a huge threat. The fact that children's reading habit is declining year by year, as a result of modern lifestyle, definitely cannot be trivialized if Indonesia truly wishes to be part of the global society. It requires awareness by all members of the Indonesian society that the reading habit is a golden milestone for literacy culture. In addition, there is a need for an understanding that the national development of Indonesia is not merely a rapid emerging economy, but also does attach to the improvement of human capital. Furthermore, one of its predominant aspects is reputable level of literacy.

In conclusion, the Indonesian government should be attentive toward the crucial role of literacy as the underpinning element of human capital because it will threaten the future of a holistic Indonesian human resources development. Globalization, which attributes to the ever-growing economy, modern and mobile lifestyle, and the unwise utilization of technology, has indirectly created polarization between economic values and humanity values. This confirms the dispute of the paradoxical foci of globalization effect: positive and negative.

REFERENCES

Appadurai, A. (2010) *Modernity at Large: Cultural Dimensions of Globalization.* Minneapolis: University of Minnesota.

Bahuet, C. (2016) Index Pembangunan Manusia (IPM): United Nation Development Program. In Sagitarius, A. M. 2017, May. Wong Indonesia, Mung 0,001 Persene Kang Seneng Maca. *Penyebar Semangat* 20: 9–10.

Book Reading 2016, Pew Research Center. Retrieved July 10 2017 from: http://www.pewresearch.org.

Central Connecticut State University. Retrieved July 8 2017 from: http://www.ccsu.edu/wmln/rank.html.

Giddens, A. (1990) *The Consequences of Modernity.* California: Stanford University Press.

Goldthorpe, H. (1992) Employment, Class, and Mobility: A Critique of Liberal and Marxist Theories of Long-Term Change. In Haferkamp, H. and Smelser, N.J. (Eds.). *Social Change and Modernity*. Los Angeles: University of California, 122–147.

Hidayat, B. (2017, May 8–14) Dukungan Badan Bahasa Dalam Gerakan Literasi Nasional (GLN). *Tempo,* p. 65.

King, A.D. (1997) The Times and Spaces of Modernity (or Who Needs Postmodernism?). In Featherstone, M. Lash, S. & Robertson R. (Eds.). *Global Modernities*. London: Sage, 45–68.

Laksana, A.S. (2017, May 7) Gerakan Literasi Tanpa Campur Tangan Negara. *Jawa Pos*, p. 6.

Landungsari, K.D. (2013) *Rencana Strategis Desa Landungsari 2013–2019*. Unpublished manuscript.

Lindsay, S. (2000) Culture, Mental Models, and National Prosperity. In Harrison L.E. and Huntington, S.P. (Eds.). *Culture Matters: How Values Shape Human Progress*. New York: Basic Books, 282–293.

McCraken, G. (1990) Consumption, Change, and Continuity. In McCraken, G. (Ed.). *Culture & Consumption: New Approaches to The Symbolic Character of Consumer Goods and Activities.* Bloomingtoon and Indianapolis: Indiana University Press, 93–103.

Mikulecky, B. S. and Jeffries, L. (1996) Reading for Pleasure. In Chamas, E. (Ed.). *More Reading Power.* USA: Addison-Wesley Publishing Company, Inc., 1–12.

Oxford English Dictionary, 6th ed. (1989). Oxford: Oxford University.

Pieterse, J.N. (1996) Globalization as Hybridization. In Featherstone, M. Lash, S. & Robertson R. (Eds.). *Global Modernities*. London: Sage, 45–68.

Saville–Troike, M. (2006) *The Etnography of Communication: An Introduction.* Oxford: Blackwell Publishing.

Urban Studies: Border and Mobility – Kerr et al. (Eds)
© 2019 Taylor & Francis Group, London, ISBN 978-1-138-58034-3

Author index

For Product Safety Concerns and Information please contact our
EU representative GPSR@taylorandfrancis.com Taylor & Francis
Verlag GmbH, Kaufingerstraße 24, 80331 München, Germany